FOREST HEALTH AND PROTECTION

McGraw-Hill Series in Forestry

Avery and Burkhart	*Forest Measurements*
Castillon	*Conservation Of Natural Resources: A Resource Management Approach*
Dana and Fairfax	*Forest Policy*
Daniel and Helms	*Principles Of Silviculture*
Davis and Johnson	*Forest Management*
de Steiguer	*The Age Of Environmentalism*
Duerr	*Introduction To Forest Resource Economics*
Ellefson	*Forest Resources Policy: Process, Participants And Programs*
Harlow, Harrar, Hardin, and White	*Textbook Of Dendrology*
Heathcote	*Environmental Problem Solving: A Case Study Approach*
Klemperer	*Forest Resource Economics And Finance*
Laarman and Sedjo	*Global Forests: Issues For Six Billion People*
Nyland	*Silviculture: Concepts And Applications*
Panshin and De Zeeuw	*Textbook Of Wood Technology*
Sharpe, Hendee, Sharpe, and Hendee	*Introduction To Forest And Renewable Resources*
Shaw	*Introduction To Wildlife Management*
Sigler	*Wildlife Law Enforcement*

FOREST HEALTH AND PROTECTION

Robert L. Edmonds
James K. Agee
Robert I. Gara
College of Forest Resources
University of Washington

Boston   Burr Ridge, IL   Dubuque, IA   Madison, WI
New York   San Francisco   St. Louis
Bangkok   Bogotá   Caracas   Lisbon   London   Madrid   Mexico City
Milan   New Delhi   Seoul   Singapore   Sydney   Taipei   Toronto

McGraw-Hill Higher Education

A Division of The **McGraw-Hill** Companies

FOREST HEALTH AND PROTECTION

This book is printed on acid-free paper.

1 2 3 4 5 6 7 8 9 0 DOC/DOC 0 9 8 7 6 5 4 3 2 1 0

ISBN 0–07–021338–0

Vice president and editorial director: *Kevin T. Kane*
Publisher: *Edward E. Bartell*
Sponsoring editor: *Robert Smith*
Marketing manager: *Debra A. Besler*
Project manager: *Jill R. Peter*
Production supervisor: *Laura Fuller*
Design manager: *Stuart D. Paterson*
Senior photo research coordinator: *Lori Hancock*
Supplement coordinator: *Sandra M. Schnee*
Compositor: *Electronic Publishing Services, Inc., TN*
Typeface: *10/12 Times Roman*
Printer: *R. R. Donnelley & Sons Company/Crawfordsville, IN*

Cover designer: *Nicole Dean*
Cover image: *Digital Stock*

Figure 18.3, page 419: From *An Introduction to the Study of Insects,* 5th Edition, by Donald J. Borror, Dwight M. DeLong, and Charles A. Triplechorn, copyright© 1981 by Saunders College Publishing, reproduced by permission of the publisher.

The credits section for this book begins on page 591 and is considered an extension of the copyright page.

Library of Congress Cataloging-in-Publication Data

Edmonds, Robert L.
    Forest health and protection / Robert L. Edmonds, James K. Agee,
  Robert I. Gara. — 1st ed.
  p. cm.
    Includes bibliographical references and index.
    ISBN 0–07–021338–0
    1. Forest health.   2. Forest protection.   3. Forest ecology.
  I. Agee, James K.   II. Gara, Robert I.   III. Title.
  SB761.E34   2000
  634.9'6—dc21   99–40276
                                                          CIP

www.mhhe.com

CONTENTS

Preface   *xv*

1 THE CONCEPT OF FOREST HEALTH   1

Introduction   1
Definitions of Forest Health   2
Characteristics of a Healthy Forest   3
Symptoms of Forest Health Problems   4
Possible Causes of Forest Health Problems   6
Current Health Status of the World's Forests   6
The Influence of Forest Management and Other   9
Human Activities on Forest Health: Case Studies   9
    The Inland Western United States   9
    The Influence of Introduced Pathogens and Insects
      on Forest Health   11
    The Influence of Air Pollution on Forest Health   12
Forest Health Monitoring Networks   15
    The U.S. National Forest Health Monitoring Network   15
    Forest Health Monitoring in Western Europe   17
    The Acid Rain National Early Warning System
      (ARNEWS) in Canada   18
Conclusions   19
References   20

2 ECOLOGICAL PRINCIPLES   23

Introduction   23
Forest Biomes   24
    Forest Zonation   25
    Forest Classification by Current and Potential Vegetation   27
Production Ecology   29
Succession   33
    The Complex Nature of Disturbance   34
    Multiple Pathway Succession   36
    Ecological Diversity and Sustainability   42
    Stability and Equilibrium   43

v

Ecosystem Management   44
    Coarse and Fine Filter Approaches   45
    Island Biogeography and Forest Health   46
References   47

3 FIRE AS A PHYSICAL PROCESS   49

Introduction   49
The Elements of Fire Behavior   49
    Weather   50
    Topography   54
    Fuels   54
Fire Behavior Prediction   58
    Surface Fire Behavior   58
    Crown Fire behavior   63
    Spotting   66
    Geographic Information Systems and Fire   67
    Fire Danger Rating   67
References   69

4 FIRE ECOLOGY AND FIRE REGIMES   71

Introduction   71
Fire Adaptations of Plants   71
Predicting Tree Mortality from Fire   74
Fire Regimes of North America   77
Ponderosa Pine Forests   80
    The Fire Regime   81
    Stand Development Patterns   81
Coastal Douglas-Fir Forests   83
    The Fire Regime   83
    Stand Development Patterns   84
Western Subalpine Forests   87
    The Fire Regime   88
    Stand Development Patterns   89
Longleaf Pine Forests   92
    The Fire Regime   93
    Stand Development Patterns   93
Jack Pine Forests   94
    The Fire Regime   95
    Stand Development Patterns   96
References   98

5 ORGANIZING FOR FIRE MANAGEMENT   102

Introduction   102
Evolution of Fire Policy   103
Responsibilities for Fire Protection   106

Fire Prevention and Detection   108
    Fire Prevention   108
    Fire Detection   110
Fire Suppression   112
    Fire Control Organization   112
    Fire Suppression Strategies   114
    Fire Equipment   117
    Safety in Suppression   121
Wildland Fire Rehabilitation   123
The Urban Wildland Interface   125
References   127

6 FIRE STRATEGIES FOR FOREST HEALTH   130

Introduction   130
Forest Structure and Fire Behavior   131
    Increasing Height to Crown Base   131
    Reducing Surface Fuels   132
    Managing Crown Structure to Prevent Fire Spread   133
    Fuel Treatments for Forest Health   135
Prescribed Fire   139
    Firing Techniques   139
    Methods of Ignition   142
    The Prescribed Fire Plan   143
    Prescribed Fire and Fuelbreaks   147
Managed Wildland Fire (pnf)   148
    Planning for Managed Wildland Fire (pnf)   150
    Operations   152
    The Future of Wilderness Fire   153
Constraints To Fire Use   155
    Smoke   155
    The Escaped Fire   157
    The Fine Filter for Special Resources   157
References   158

7 WIND AND FOREST HEALTH   160

Introduction   160
Wind as a Physical Process   160
Wind Effects on Trees   163
    Crown Morphology   163
    Stem and Root Shape   165
    Windsnap (Stem Breakage) and Windthrow (Uprooting)   165
    Wind Effects on Forest Succession   168
    Case Study: Pacific Northwest Spruce-Hemlock Forests   169
    Case Study: Northeast Hardwood–White Pine–Eastern Hemlock   170
    Case Study: Southern Pines   172

Forest Management Strategies 173
   Shelterbelts 173
   Windthrow Hazard Identification 174
   Site Preparation and Planting 176
   Thinning 176
   Pruning 176
   Clearcuts 177
   Stream Buffers, Interior Leave Patches,
     and Other Retention Strategies 180
Synergistic Effects 181
References 182

8 INTRODUCTION TO DISEASES 185

Introduction 185
Definition of Disease and Types of Diseases 187
A Brief History of Forest Pathology 188
Impact of Diseases 190
The Disease Triangle, Square, or Tetrahedron 191
The Concept of Pathogens, Parasites, Saprophytes,
   and Symbiotic Relationships 192
Signs and Symptoms of Disease 193
Linkages Between Diseases, Insects, Fire, and Wind 195
Diseases in Urban Forests and Street Trees 196
Forest Pathology Books 196
References 197

9 ABIOTIC (NONBIOLOGICAL) AND ANIMAL-CAUSED INJURIES 199

Introduction 199
The Pattern of Abiotic Injuries 199
Injuries Caused by Temperature Extremes 200
Low and High Moisture and Low Soil Oxygen 202
Global Change 203
Air Pollution 204
   Types of Air Pollutants 204
   Dispersion of Air Pollutants 207
   Damage Caused by Air Pollutants 209
   Effects of Acid Precipitation 212
   Management of Air Pollution Problems 212
Herbicides 214
Mechanical Injuries 215
Nutrient Deficiencies 217
Nutrient Imbalances and Elemental Toxicities 220
Fire 221
Salt Injury 221

Animal Damage   222
    Seed Eaters   223
    Voles   224
    Deer and Elk   224
    Tree Squirrels   224
    Porcupines   224
    Black Bears   225
    Pocket Gophers   225
    Mountain Beavers   225
    Snowshoe Hares   226
    Livestock   226
References   226

10 DISEASE-CAUSING ORGANISMS   228

Introduction   228
Features of Disease-Causing Organisms   228
    Fungi   228
    Parasitic Plants   231
    Bacteria   233
    Viruses and Viroids   233
    Phytoplasmas   234
    Nematodes   235
Classification, Life Cycles, and Diseases Caused by Fungi   237
    Zygomycota   240
    Ascomycota and Deuteromycetes (Imperfect Fungi)   241
    Basidiomycota   245
    Oomycota   247
Spread and Ecology of Fungi   251
References   253

11 NURSERY DISEASES AND MYCORRHIZAE   254

Introduction   254
Selection of Nursery Sites   255
Abiotic Injuries   255
    Low-Temperature Injuries   256
    High-Temperature Injuries   256
Biotic Diseases   257
    Diseases Affecting Seeds, Roots, or Root Collar Areas   257
    Diseases Affecting Shoots   261
Management of Seedling Diseases   264
Mycorrhizae   266
    Roles of Mycorrhizae   270
    Management of Ectomycorrhizae   271
References   273

12 ROOT DISEASES   275

Introduction   275
Types and Causes of Root Diseases   276
    Structural Root Diseases   276
    Feeder and Vascular Root Diseases   277
Symptoms, Signs, Hosts, Distribution, and Spread
  of Root Diseases   280
    Structural Root Diseases   280
    Feeder Root Diseases   294
    Vascular Root Diseases   295
Interactions Between Root Diseases, Insects, and Fire   296
Root Disease Management   297
    Inoculum Reduction   297
    Use of Resistant Species   298
    Genetic Resistance   300
    Chemical Treatment of Stumps   300
    Biological Control   301
    Thinning   301
    Fertilization   302
    Avoidance of High-Hazard Sites   302
    Natural Climatic Controls   302
    Prescribed Fire   303
    Quarantine   303
    Doing Nothing or Passive Management   304
Modeling of Root Diseases and Disease/Bark Beetle Interactions   304
Conclusions   306
References   307

13 FOLIAGE DISEASES AND RUSTS   309

Introduction   309
Foliage Diseases   309
    Typical Symptoms   310
    The Life Cycle of Foliage Pathogens   313
    Specific Hardwood Foliage Diseases   313
    Specific Conifer Foliage Diseases   318
Rusts   323
    Rust Life Cycles   327
    Management of Rust Diseases   329
References   331

14 STEM AND BRANCH DISEASES   333

Introduction   333
Mistletoes   333
    Dwarf Mistletoe Infection Rating Scheme   338
    Ecological Relationships   338
    Management   339

Cankers   341
   Cankers on Hardwoods   341
   Cankers on Conifers   346
Galls and Other Malformations   347
Decay   348
   Types of Decay   348
   Decay Terminology   349
   Important Decay Fungi   350
   Spread of Decay Fungi   350
   The Concept of Compartmentalization of Decay in Trees (CODIT)   354
   Ecological Role of Decays   356
   Management of Decay Fungi   357
Vascular Wilts   358
   Dutch Elm Disease   358
   Oak Wilt   362
Stains   362
References   362

15 FOREST DECLINES   364
Introduction   364
The Concept of Decline Diseases   368
   The Environmental Stress and Secondary
     Organisms Theory   368
   The Concept of Predisposing, Inciting, and
     Contributing Factors   369
   The Climate Change or Climate Perturbation Theory 369
   The Air Pollution Theory   375
   The Ecological Theory   376
References   381

**16 MANAGEMENT OF FOREST DISEASES AND THE
DETERIORATION OF WOOD PRODUCTS**   383
Introduction   383
Forest Disease Management   383
   Disease Detection, Recognition, and Appraisal   383
   Major Management Strategies   384
   Silvicultural Management   384
   Integration of Disease, Insect, and Fire Management 390
Modeling Forest Diseases   390
Wood Products Deterioration   393
References   396

17 INTRODUCTION TO FOREST ENTOMOLOGY   398
Introduction   398
Historical Perspective   401
   Europe   403

North America   405
The Future of Forest Entomology   409
References   410

18   BASIC ENTOMOLOGY   413

Introduction   413
The Arthropoda   414
   The Chelicerata   414
   The Mandibulata   417
The Insects   418
   External Morphology   418
   Internal Morphology   421
   Insect Development   424
Insect Orders   426
   The Collembola (Springtails)   426
   The Ephemeroptera (Mayflies)   430
   The Odonata (Dragonflies and Damselflies)   430
   The Plecoptera (Stone flies)   430
   The Dictyoptera (Suborder Isoptera) (Termites)   431
   The Orthoptera and Other Dictyoptera (Crickets,
     Grasshoppers, Walking Sticks, Mantids, Roaches, etc.)   431
   The Thysanoptera (Thrips)   432
   The Hemiptera (Suborder Heteroptera) (True Bugs) 432
   The Hemiptera (Suborder Homoptera) (Aphids, Cicadas,
     Leafhoppers and Planthoppers, and Scales)   432
   The Coleoptera (Beetles)   433
   The Trichoptera (Caddis Flies)   433
   The Lepidoptera (Moths and Butterflies)   433
   The Hymenoptera (Ants, Parasitic Wasps, Bees,
     Hornets, Sawflies, and Others)   434
   The Neuroptera (Lacewings, Antlions, Alderflies,
     and Dobsonflies)   435
   The Diptera (Flies)   435
References   435

19   PRINCIPLES OF FOREST INSECT MANAGEMENT   436

Introduction   436
Public Forestry and Forest Health   436
Timber Management   439
Insect Pest Management   440
   Mechanical and Physical Control   442
   Chemical Control   442
   Semiochemical Control   447
   Biological Control   449

Integrated Pest Management (IPM)   457
References   459

20 INSECT DEFOLIATORS   462

Introduction   462
Detection, Evaluation, and Control   463
  Detection   465
  Applied Control   467
The Defoliators   469
  Eastern and Southern North American Forests   471
  Western North American Forests   481
References   488

21 BARK BEETLES AND THEIR MANAGEMENT   491

Introduction   491
  Associations with Fungi   491
  Biology of Bark Beetles   493
Management of Bark Beetles   495
  Damage Thresholds   496
  Detection   497
  Pest Control Strategies   497
The *Dendroctonus*   498
  The Southern Pine Beetle (*D. frontalis*)   498
  The Mountain Pine Beetle (*D. ponderosae*)   502
  The Douglas-Fir beetle (*D. pseudotsugae*)   504
  The Spruce Beetle (*D. rufipennis*)   505
  The Western Pine Beetle (*D. brevicomis*)   506
The Engraver Beetles (*Ips* and *Scolytus*)   508
  The Engraver Beetles (*Ips spp.*)   508
  The *Scolytus* Engraver Beetles   512
References   514

22 AMBROSIA BEETLES AND THEIR MANAGEMENT   516

Introduction   516
Ambrosia Beetles and Softwoods   517
Ambrosia Beetles and Hardwoods   522
References   524

23 WOOD PRODUCTS INSECTS   526

Introduction   526
  Borers of Living Trees   526
  Borers of Weakened Trees, Windthrows, and Logs   530
  Borers of Dry and Seasoned Wood   536
References   542

24 INSECTS OF SEED ORCHARDS, NURSERIES, AND YOUNG PLANTATIONS 543

Introduction 543
Seed Orchards 544
Nurseries 548
Plantations 553
References 560

25 FOREST INSECT QUARANTINE 561

Introduction 561
Regulations on Importation of Logs and Wood Products 566
References 569

Glossary 571
Credits 591
Index 593

PREFACE

Health is a good thing. When people are ill, they are usually diagnosed with symptoms such as a fever or unusual blood pressure or physical injury. The treatment in most cases, based on the diagnosis, is successful. People appreciate good health and advanced societies have provided a clear mandate for maintaining it.

Forest health also is a good thing. The diagnosis of the status of forest health, and its improvement if needed, is clearly a technical issue, but forest health is not as easily defined as issues of human health. Dead trees, for example, play an important role for wildlife in forest ecosystems, so some decadence in a forest is desirable. Even if people agree on a definition of forest health, and a diagnosis of poor forest health, there is no clear social mandate for maintaining forest health. We leave this latter issue for forest policy makers, and in this text focus on the technical issues concerning forest health and forest protection.

The idea for this book came out of our teaching efforts in the College of Forest Resources at the University of Washington. Over the years we have taught individual courses in our respective disciplines of fire (Agee), insects (Gara), and diseases (Edmonds), and we also offer a forest protection course integrating fire, insects, and diseases. Although there are excellent texts available for each discipline, no single integrative text is available. We finally decided to write a text to fill this need and to do it within the context of applied forest protection, which involves ecology, forest health, and ecosystem management. Just as modern forest management has taken an integrative approach embracing the concept of ecosystem management, there also is a need to take an integrated approach to forest health issues where fire, wind, insects, and diseases and their interactions are treated holistically.

This volume, to our knowledge, is the first integrative text in this field. The first two introductory chapters cover concepts of forest health and principles of forest ecology. Chapters 3 through 7 cover fire as a physical process, fire ecology and fire regimes, fire management, fire strategies for forest health, and wind and forest health. Chapters 8 through 16 discuss forest diseases and cover an introduction to diseases, abiotic and animal-caused diseases, disease-causing organisms, nursery diseases and mycorrhizae, root diseases, foliage diseases and rusts, stem and branch diseases, forest declines, and management of forest diseases and deterioration of forest products. Finally, Chapters 17 through 25 cover entomological aspects of forest health: an introduction to forest entomology; basic entomology; principles of forest insect management; insect defoliators; bark beetles and their management; ambrosia beetles and their management; wood products insects;

insects of seed orchards, nurseries, and young plantations; and forest insect quarantine. Where appropriate the interrelationships between insects, diseases, fire, and wind are stressed. A glossary of terms also is included to assist the reader.

Fire, wind, insects, and diseases are natural disturbance agents in forests. An understanding of natural disturbances in forests is essential to any forest management plan, whether the objective is timber production, wildlife conservation, or wilderness management. However, human activities have strongly influenced forests worldwide, and will continue to do so. Fire regimes have been altered by cutting, forest management, and fire suppression policies of the 20th century. When fires do occur, they are more commonly catastrophic and stand replacing. Native insects and diseases now often play increased roles in forest ecosystems because of the absence of natural fires. Alien insects and diseases, many introduced inadvertently, have reduced the health of North American forests. This is a concern for present as well as future generations; more of these introductions are inevitable.

In North America forest succession has been dramatically changed by fire suppression and introduction of pests, and the resulting forests now contain species mixtures that are different from those at the turn of the century. A similar situation has occurred in Australia and other countries. In addition, air pollution, including gases such as sulfur dioxide and ozone, acid rain, and excess nitrogen inputs have negatively impacted forests in many regions including the eastern United States and Canada, Eastern and Western Europe, Latin America, and Asia. In some areas trees have been killed directly by air pollutants or placed under so much stress that they are more susceptible to insects and disease organisms. The management approach to these forest health issues will vary by country, and by land management classifications, but the principles of the problem will be similar regardless of geography.

This book is intended for students of forestry, natural resources, and conservation in the United States and Canada. But we hope it will be appealing and appropriate for students in other countries. We have used international examples in the book where appropriate. The book is aimed at advanced undergraduate students who have some knowledge of forest ecology, although we have provided a brief introduction to ecological principles in introductory chapters. Graduate students also will find this book useful, but it is not intended to be a research review. Adequate documentation is provided throughout, but the book is not designed to be an exhaustive review of the primary literature. The course for this book is intended to be a one-term offering, but the material can easily be spread across two terms, giving more time to discuss chapters and reference readings in depth. We hope we will stimulate students to look in greater detail into the fascinating world of fire, wind, insects, and diseases and their interactions—that is, forest health and protection.

Finally, we would like to thank those people who contributed to the production of this book. They include Greg Filip, Oregon State University, Mark Petruncio, Heritage College, and R. Jay Stipes, Virginia Tech, along with several other anonymous reviewers whose comments were much appreciated, and the students in our forest protection class who commented on the manuscript. In addition, we would like to thank Rene Alfaro, Eric Allen, W. H. Bennett, R. F. Billings, J. H. Borden, Scott Cameron, Gary Daterman, Arnie Drooz, Andy Eglitis, Ed Holsten, Jack Nord, Imre Otvor, Timothy D. Paine, Roger Ryan, Wayne Sinclair, Bob Thatcher, T. R. Torgesen, and Boyd Wickman for providing photographs.

R. L. Edmonds
J. K. Agee
R. I. Gara

C H A P T E R

1

THE CONCEPT OF FOREST HEALTH

INTRODUCTION

Diseases, insects, and **abiotic agents** such as fire, wind, and drought are the major natural disturbance agents that change forest ecosystems; anthropogenic air pollution also strongly influences forests. Typically these factors have been treated separately in textbooks, but as we manage forests more as ecosystems, it is becoming apparent that we should not only examine their separate effects, but also their interactions. In this book we attempt to do this. However, for ease of organization, Chapters 2 to 7 are devoted to ecological principles, fire, and wind; Chapters 8 to 16 cover forest diseases, and Chapters 17 to 25 involve forest entomology.

Disturbance agents not only cause economic impacts through timber losses, but also influence species succession, species composition, distribution and abundance, forest structure and function, site conditions, and wildlife habitats (Sprugel 1991, Castello et al. 1995). Fire, wind, insects, and diseases strongly interact. For example, disease or insect killed trees are subject to fire, diseased trees may be windthrown or attacked by insects, and blown down or wind damaged trees may be susceptible to insects, diseases, and fire. These agents have always influenced natural forests, but in the past century their patterns and influences have been changed by forest management practices, including forest cutting and fire suppression. Future global warming might not only affect species composition, but also fire, wind, and ice storm frequencies, and disease and insect levels.

Forests not only provide a source of fiber for human needs, but they also provide water, fish and wildlife habitats, a source of biodiversity, and recreational opportunities. To do this in a sustainable manner, the role of disturbance agents must be understood. In this chapter we consider definitions of forest health, characteristics of a healthy forest, signs and possible causes of forest health problems, the current state of forest health in the world, the influence of forest management, air pollution and introduced pathogens and insects on forests, and the establishment of forest health monitoring networks.

1

DEFINITIONS OF FOREST HEALTH

"Forest health" or "forest ecosystem health" are terms that are increasingly being used in relation to the management of forest ecosystems (Kolb et al. 1994, Oliver et al. 1994, Sampson and Adams 1994, Vogt et al. 1997, Rapport et al. 1998). Ecosystem management implies that land is not managed for a single species and that it involves (1) ecosystem complexity, (2) biological legacies (including structures such as green trees, logs, and snags) that are important in reestablishing ecosystems after major disturbances, and (3) a landscape perspective (i.e., large space and time scales) (Christensen et al. 1996, Sampson and Knopf 1996, Kohm and Franklin 1997). We discuss this in more detail in Chapter 2.

There are many definitions of forest health, ranging from utilitarian to ecosystem perspectives as shown in Box 1.1 (Sampson 1996, Rapport et al. 1998). From a utilitarian perspective a forest is healthy if management objectives are satisfied and unhealthy if not (Kolb et al. 1994). Under this traditional line of thinking, insects and diseases are generally considered to interfere with intended human uses of forests. This utilitarian definition may be appropriate where the production of wood fiber is the main objective—for example, in poplar plantations or in intensively managed southern pine forests—but when managing for multiple objectives, this definition is too narrow and an ecosystem perspective with a broader definition is preferable. It is our intent to provide the scientific basis of forest health, leaving the policy implications aside since they are so closely tied to land management objectives (Oliver et al. 1994).

With so many definitions of forest health, it is not surprising that questions have been raised about the concept. For example, the community structure of forests in the southeastern United States was radically altered by the chestnut **blight** (Callicot 1995), but were energy capture, primary production, and nutrient cycling changed? It is possible that if one tree species replaces another the ecosystem may still remain healthy. But this may not always be the case, and significant reductions of keystone species can have catastrophic consequences on ecosystem health. For example, white pine blister rust has devastated the population of whitebark pine at high elevations in the western United States (Figure 1.1). Whitebark pine seeds are a major source of food for grizzly bears, red squirrels, and birds (Clark's nutcracker), and the absence of seeds could drastically affect these wildlife populations if no alternative food source is found (Schmidt and McDonald 1990).

Maintenance of biodiversity is an important aspect of ecosystem management, and it is argued that biologically diverse ecosystems are healthy ecosystems (Vogt et al. 1997). What is the relationship between biodiversity and ecosystem health? The chestnut blight appears to have reduced biodiversity with the loss of the chestnut, but if you add the new fungus there was no net loss in species richness (Callicot 1995). Tree biodiversity may in fact have increased because the chestnut was such a dominant species. Thus biodiversity measured only by species richness should not be the sole criterion of ecosystem health.

Box **1.1**

Definitions of a Healthy Forest

- A condition where biotic and abiotic influences on forests (e.g., pests, pollution, thinning, fertilization, or harvesting) do not threaten management objectives now and in the future.
- A fully functioning community of plants and animals and their physical environment.
- An ecosystem in balance.
- A condition of forest ecosystems that sustains their complexity while providing for human needs.
- A healthy forest is one that is resilient to changes.
- The ability of forest ecosystems to bounce back after being stressed.
- The ability of a forest to recover from natural and human stressors.
- A healthy ecosystem should be free from "distress syndrome," where this syndrome is characterized by reduced primary productivity, loss of nutrient capital, loss of biodiversity, increased fluctuations in key populations, retrogressions in biotic structure, and widespread incidence and severity of diseases and insects.

CHARACTERISTICS OF A HEALTHY FOREST

What the are the characteristics of a healthy forest? A healthy forest involves more than the health of individual trees, and dead trees occur in natural ecosystems (Figures 1.2 and 1.3). A degree of large-scale disturbance (windstorms, fire, etc.) is indigenous in ecosystems and must be included in any realistic definition of "naturalness" (Sprugel 1991). Scientists have also recognized that aquatic and terrestrial components of ecosystems are linked, and that forest management practices can strongly influence forest streams and anadromous fish (Naiman et al. 1992). Thus the health of forested watersheds should also be considered under the topic of ecosystem health. Measurements of ecologically healthy watersheds include water yield and quality, community composition, forest structure and the health of the riparian zone, fish production, wildlife use, and genetic diversity (Karr 1995).

To evaluate forest health we need good indicators, and Sampson (1996) suggested that the following list may indicate healthy conditions. This list is far from complete and other indicators are continually being evaluated.

1. Trees and understory plants should be vigorous and healthy in appearance. Species, age class distributions, and stand densities should be within historical ranges for the site, and growth and mortality should be consistent with the ecosystem type and the age of dominant trees.
2. Vegetation diversity should be balanced between the supply and demand of light, water, nutrients, and growing space.

Figure 1.1
Dead whitebark pine (*Pinus albicaulis*) in the Rocky Mountains in Montana killed by white blister rust. Death of the whitebark pine has caused ecosystem disruption since whitebark pine seeds are a major source of food for grizzly bears, red squirrels, and birds (Clark's nutcracker) and could drastically affect their populations.

3. The forest should be capable of tolerating and recovering from known disturbances (such as fire and wind).
4. Soil erosion should be minimal. Clean water should flow from streams except during extraordinary runoff events, and stream banks need to stable and riparian vegetation ample.
5. Aquatic species should be diverse and aquatic indicator species should be present in expected numbers.
6. Wildlife diversity and presence need to be appropriate for the ecosystem, especially in riparian zones.
7. Insect, disease, and fire frequencies should be within the normal ranges for the ecosystem.

SYMPTOMS OF FOREST HEALTH PROBLEMS

Symptoms of forest health problems or potential problems usually can be recognized (Sampson 1996). If trees appear stressed or sickly under normal weather conditions (including periodic droughts) and have high mortality rates with a lack of regeneration, then a forest health problem is suspected. Other symptoms of problems include tree density higher or lower than the historical range for the ecosystem involved; significant species of trees, understory plants, fish, or wildlife

Figure 1.2
This healthy old-growth Pacific silver fir (*Abies amabilis*) ecosystem in Mt. Rainier National Park, Washington, contains dying, standing dead, and downed logs that provide habitat for wildlife.

Figure 1.3
Natural mortality in a wave-regenerated balsam fir (*Abies balsamae*) forest near Franconia Notch, New Hampshire. Natural "mortality waves" move through the forest once every 50 to 70 years (Sprugel 1991). The waves move at a uniform rate and there are always freshly killed areas. Since consecutive waves are only about 100 m apart, an area of a few dozen hectares can contain all phases of the disturbance cycle.

either missing or present in higher proportions than historical ranges suggest; low numbers of terrestrial or aquatic indicators species; one or more plant or animal species excessively impacting or dominating the stage of succession involved; invasion of exotic species that negatively affects the forest in the absence of natural controls; and excessive erosion (Figure 1.4). Streams may run muddy in normal events, although this depends on soil type since streams in areas with high clay or soil organic matter content may be naturally turbid especially in tropical areas. Other indicators of health problems are the drying up of normally perennial streams, poor riparian vegetation, and eroding and unstable stream banks (Naiman et al. 1992).

POSSIBLE CAUSES OF FOREST HEALTH PROBLEMS

The major forest health issues involve species loss, insect and disease epidemics, excessive wildfires, air pollution, water quality and quantity problems, impacted wildlife populations, nutrient imbalances, and soil and watershed damage. These are caused by (1) large-area disturbances such as logging or overgrazing where these areas have been converted to uniform, even-aged overly dense forests; (2) an aggressive species that is not native to the region, thus disrupting species balance or ecosystem processes (e.g., the fungi causing white pine blister rust, chestnut blight and Dutch elm disease, and insects such as the gypsy moth and the Asian long-horned beetle); (3) a change in the environment such as a major change in nutrient cycles through atmospheric fallout (e.g., excess nitrogen), an increase in carbon dioxide or ozone, or a long-term climatic change at rates significantly faster than those that occurred naturally; and (4) dead and live fuels existing in such quantity, or arrangement, that any wildfire ignition results in adverse effects to the vegetation, soils, and biota of the ecosystem. It should be recognized, however, that coarse woody debris is an important functional component of forest ecosystems, and some needs to be left on the site after timber harvesting. The amount, type, distribution, and orientation, however, will vary from ecosystem to ecosystem and should fall within natural ranges.

Some particularly sensitive forest health issues revolve around the role and use of fire in ecosystems and many questions are still unanswered—for example, What fires should we let burn? Which fires should be suppressed? How can we reintroduce fire to ecosystems? Should salvage logging and thinning be permitted to reduce fuel loads despite the ecological benefits of maintaining coarse woody debris in forests?

CURRENT HEALTH STATUS OF THE WORLD'S FORESTS

Forests in both the northern and southern hemispheres have been extensively cut and changed dramatically from their natural conditions, and this should be taken into account in assessing forest health. Until relatively recently cutting was largely restricted to temperate zone forests, but it now occurs extensively in tropical and boreal forests. In the eastern and southeastern United States, large areas of forest

Figure 1.4

A clearcut in coastal British Columbia showing severe erosion, which might affect streams. The erosion gully starts in the uncut forest, but is exacerbated by the road cutting across the clearcut. Is this a healthy forest ecosystem? Only part of the system is heavily impacted and probably will recover with time as the vegetation recovers.

were cleared for agriculture in the 18th and 19th century. Much of this has reverted back to relatively healthy regrowth forest, although forest health problems occur in certain areas, particularly at higher elevations (Ayers et al. 1998). Forest health problems also occur in the inland western United States and Canada, but coastal forests in western North America appear to be relatively healthy (Campbell and Liegel 1996). Continuous forest cutting and reforestation has taken place over centuries in Europe, yet many of the European forests appear to be healthy—for example, in Scandinavia (Figure 1.5). Other areas in Europe, such as Germany, the Mediterranean, and some Eastern European countries have serious health problems (Innes 1993). Forest health problems also have been noted in New Zealand, Australia, Chile, South Africa, and other countries, including those in tropical areas (Rapport et al. 1998).

The least healthy forests in the world appear to be located in areas under stress, resulting from (1) air pollution; (2) suppression of natural fires, which has resulted in changes in tree species composition and increases in insect and disease severity; (3) drought; and (4) overutilization by humans. Forests at high elevations in Western and Eastern Europe and in the eastern United States have experienced considerable mortality. Air pollution and acid rain are thought to be contributing to this decline (Innes 1993). Forests in southern California also are being impacted by air pollution (Miller and McBride 1999). The stressed dry forests in the inland western United States—that is, those on the eastern slopes of the Cascade Range, the Sierra Nevadas, and the Rocky Mountains—are generally considered to be unhealthy (Monnig and Byler 1992, Mutch et al. 1993, Campbell and Liegel 1996). Here, root diseases, defoliators, tip moths, and wood and bark beetles cause extensive mortality. This is discussed in more detail later. Overutilization of forests for fuelwood in Asia and Africa has led to unhealthy forests where the landscape is devastated by erosion, flooding has increased, and it is difficult to reestablish vegetation. In Australia and New Zealand, there are major attempts to restore native

Figure 1.5
An intensively managed thinned Scots pine (*Pinus sylvestris*) forest in Finland. This stand appears very healthy.

forests to healthy conditions, and the federal government in Australia has decreed that at least 30% of the native forest be restored to conditions that existed before 1750. Native forests in Australia can no longer be cleared to establish exotic plantation species like they could in the past.

Vast areas of plantation forests of exotic species have not only been established in Australia, but also in New Zealand, South Africa, Brazil, Chile, Ecuador, Great Britain, Germany, and other countries. Exotic plantations in temperate zones have tended to be relatively healthy, even though they consist mostly of monocultures of relatively few species, such as radiata pine (Figure 1.6), other pines such as loblolly, slash and lodgepole, Douglas-fir, Sitka spruce, eucalypts, and poplars. However, there is great concern that monospecific plantations are very susceptible to health problems. Outbreaks of insects (Ciesla 1993) and diseases, particularly foliage and root diseases, have occurred occasionally in these plantations, and there is some concern that they may not be sustainable. Single-species exotic plantations in the tropics have not been very successful largely due to outbreaks of tip and shoot insects. Intensive forest management also implies genetic selection for fast-growing, high-quality trees. In many cases disease and insect susceptibility is increased by this practice. It certainly has been in agriculture. Tree species susceptible to rust diseases, such as pines and poplars, are particularly vulnerable. Hansen and Lewis (1997) discuss important diseases of conifer forests in Europe, eastern Siberia and the Russian far east, Fennoscandia, India, Australia, New Zealand, and Africa.

Exotic plant introductions also can have a direct influence on ecosystem health, where species, particularly from another continent, become aggressive and replace native plants. Examples of this are kudzu (*Pueraria thunbergiana*) from Asia and *Melaleuca* spp. from Australia in the southeastern United States, *Lantana* spp. from the northern hemisphere in the eucalpyt forests of eastern Australia, and Scots broom in the western United States and other areas in the world. Exotic introductions have strongly influenced the island forests of Hawaii.

Figure 1.6

A uniform radiata pine (*Pinus radiata*) plantation near Batlow, New South Wales, Australia, with row thinning. There is very little understory and little animal activity, although some animals from nearby native forests are present. Is this a healthy ecosystem and can it be sustained?

THE INFLUENCE OF FOREST MANAGEMENT AND OTHER HUMAN ACTIVITIES ON FOREST HEALTH: CASE STUDIES

The Inland Western United States

In the inland western United States, trees across wide areas of the landscape are dying faster than they are growing or being replaced (Monnig and Byler 1992, Mutch et al. 1993, Sampson and Adams 1994, Campbell and Liegel 1996). Because of this, tree mortality conditions exist that almost guarantee large and severe wildfires. In these ecosystems not only are the trees at risk but aquatic resources, wildlife, and other values also are affected. Managers of public and private forests are being challenged to take rapid preventative action to restore these forests to conditions more similar to their historical range of variability or at least to a socially desired condition.

How did we arrive at this situation? Patterns of fire, insect, disease, and wind disturbance have all been altered by forest management practices, particularly fire suppression. Plant species distributions, forest structure and composition, succession, biodiversity, and landscape pattern have all been changed, and the incidences of diseases and insects and the risk of catastrophic fire have increased.

The forests at greatest risk are composed of unsustainable combinations of tree species, densities, and structures that are susceptible to the drought and fire regimes that occur in the region (Campbell and Liegel 1996). However, the health problems in the forests dominated by ponderosa pine are different from those in forests dominated by lodgepole and white pine.

Ponderosa pine forests, which occur in the driest areas, are particularly susceptible to forest health problems. Prior to European settlement many of these forests consisted of large, widely spaced trees visited by low-intensity wildfire every 5 to 15 years (see Chapter 4). Burning by Native Americans also may have contributed to the appearance of these stands. With wildfire control, vegetation developed vigorously under the mature trees (Figure 1.7). On the driest sites an understory of ponderosa pine developed. On slightly wetter and cooler sites, shade-tolerant Douglas-fir developed. The understory then provided a "ladder" for devastating crown fires (Agee 1993). The problem was worse where selective logging quickly shifted the species mix from ponderosa and other long-needled pines toward firs. For example, in Idaho ponderosa and western white pine components declined 40 and 60%, respectively from 1952 to 1987 while true firs, lodgepole pine, and Douglas-fir increased 60, 39 and 15%, respectively. Unless these forests are restored to a species mixture more resembling pre-European conditions, we will have forests that are increasingly subject to disease and insect attack and catastrophic fire. Strategies to reduce fire problems are summarized in Chapter 6.

There are vast areas of lodgepole forests in the inland West that provide a different scenario. Lodgepole pine has evolved with outbreaks of mountain pine beetle and large-scale, intense, stand-replacing fires (Agee 1993). When stands are young the beetle poses little threat, but as trees age the inner bark thickens and they become more suitable hosts. As tree growth rates slow, the beetle population explodes and extensive mortality follows. Dead trees become fuel and sooner or later lightning or humans ignite the forest, and it burns in spectacular fashion; witness the Yellowstone National Park fires of 1988. The period between such fires typically varies from under 100 years to 300 years, depending on the site. Lodgepole pine in the inland West has adapted to intense fire, and a high proportion of its cones are serotinous (i.e., its resin-closed cones are opened by the intense heat from fires) (Agee 1993). These forests recover quickly from fire, and within five years there is a thick carpet of lodgepole pine seedlings, and many of the standing dead trees have fallen down (Figure 1.8). This is a healthy forest given the park's land management direction to allow natural processes to operate, but managing lodgepole pine forests is difficult since the public appreciates neither large-scale bark beetle mortality nor large-scale intense fires, particularly if there is a risk of loss of life and property. Thus risk assessment and management are important components of forest health and ecosystem management.

Mixed conifer forests with a strong western white pine component were common in the inland West at midelevations. Western larch, grand fir, and Douglas-fir also were present. Periodic low-intensity fires also burned through these forests naturally, maintaining the white pine. However, as the white pine aged to old-growth it became more susceptible to insects, particularly to mountain pine beetle, which killed trees, leading eventually to occasional high-intensity stand-replacing fires (Agee 1993). Root diseases also tended to thin stands and maintain white pine by selectively killing the more susceptible Douglas-fir and true firs. With European settlement and selective cutting of white pine, fire suppression, and the introduction of white pine blister rust in the early 1900s, the health of these forests dropped

Figure 1.7

A ponderosa pine (*Pinus ponderosa*) forest in eastern Oregon with large dead and dying trees due to mountain pine beetle. Fire suppression has allowed the development of a dense understory of small trees, which provide a fuel ladder for a stand-destroying crown fire.

dramatically (Campbell and Liegel 1996). Douglas-fir now is the main species in this zone, while western white pine is poorly represented. Root diseases and the Douglas-fir beetle are the major agents reducing productivity in the Douglas-fir and true fir stands. Restoring these ecosystems also is going to be difficult, and perhaps it can never be accomplished because of the presence of the blister rust. Some blister rust–resistant seedlings are available, but this resistance has to last for a long time to protect trees until maturity. Resistance in agricultural crops to rust fungi generally does not last long.

The Influence of Introduced Pathogens and Insects on Forest Health

Introduced pathogens and insects have taken a large toll on the health of forests throughout the world. A handful of diseases and insects have caused tremendous damage in North America. White pine blister rust, chestnut blight, Dutch elm disease, gypsy moth, and recently dogwood anthracnose have caused a large amount of mortality and changed tree species composition. *Phytophthora cinnamomi*, which may have originated in Southeast Asia or Northeast Australia, has now spread throughout Australia and the world. It has influenced the regional distribution of *Eucalyptus* spp. in Australia such that susceptible species occur on drier ridges, while resistant species occur in moist depressions and swales where the pathogen is more active (Old 1979). Plantations of radiata pine in the southern

Figure 1.8
A lodgepole pine (*Pinus contorta*) forest in Yellowstone National Park, Wyoming, in 1994 that was burned in 1988. Note the carpet of lodgepole pine seedlings. Many of the killed trees have already fallen. This is a healthy forest since a stand replacement fire is natural in this ecosystem.

hemisphere also are very susceptible to introduced pathogens from their original native ranges. For example, needle pathogens from North America cause much damage in New Zealand, Chile, and Ecuador. *Sirex noctilio,* a wood wasp, is of considerable concern in radiata pine in Australia. The pine wood nematode, which has caused serious damage to native pines in Japan, was introduced from the United States and its introduction is of concern elesewhere (Fielding and Evans 1996).

Serious quarantine and inspection measures were implemented by the United States and many other countries after the impacts of these introduced pests were noted, and they have been relatively successful. The gypsy moth was introduced from Europe in 1869 in Massachusetts before inspection measures and has been slowly expanding its range ever since to include the entire northeastern United States and portions of the South, Midwest, and even the West (Liebhold et al. 1997). It feeds on more than 300 shrubs and trees, but prefers oaks (Figure 1.9), and has caused serious defoliation in oak forests in the northwestern states (Figure 1.10), although it has yet to invade the most susceptible forests (Liebhold et al. 1997). There is also great concern that the Asian gypsy moth, which was recently discovered in both eastern and western North America, could cause even greater problems because it attacks conifers as well as hardwoods (the European gypsy moth rarely feeds on conifers). Also the female Asian gypsy moth can fly, unlike the European gypsy moth, which greatly enhances its ability to spread. With increased world trade of lumber and logs, we see the possibility of increased pest and pathogen movement. There is a very large concern that pests and pathogens from the southern hemisphere or Siberia imported on logs could impact North American forests.

The Influence of Air Pollution on Forest Health

As well as changing the forest through management, humans have inadvertently changed it through air pollution. Gases, like sulfur dioxide and ozone, acid rain, and excess nitrogen are the major air pollutants of concern. In the early days of industrialization, large areas of forest were killed or greatly impacted in North America, Europe, and even in Australia by the burning of high sulfur coal and ore smelting activities that produced toxic sulfur dioxide gas (Smith 1990). Sulfur

Figure 1.9

Gypsy moth larvae rapidly consume leaves of preferred oaks during the growing season.

(Photo courtesy Michael McManus, USDA Forest Service.)

Figure 1.10

The gypsy moth has defoliated large areas of oak dominated forests in the northeastern United States since its introduction in Massachusetts in 1869. It has spread widely since then and continues to spread.

(Photo courtesy Michael McManus, USDA Forest Service.)

dioxide is still adversely impacting forests in eastern Europe. Ground-level ozone now is thought to be one of the major air pollutants stressing forests throughout the world. It is produced as a result of photochemical oxidation of nitrogen oxides from automobile exhausts in the presence of hydrocarbons. Ozone causes major forest health problems where there are large concentrations of automobiles and high radiation intensities in the growing season, such as in southern California (Miller and McBride 1999), but it is a worldwide problem.

Acid rain is formed by the reaction of sulfur dioxide and nitric oxide gases with water vapor in the atmosphere, forming sulfuric and nitric acids, and was noted to be a problem in the 1970s in Europe and the eastern United States. Concern about the effects of acid rain on aquatic and terrestrial ecosystems led to the establishment of the National Acid Precipitation Assessment Program (NAPAP) in 1980 to coordinate acid rain research and report the findings to the U.S. Congress. As a result of research in the 1980s, it was determined that streams and lakes were being adversely impacted by acid rain (NAPAP 1993). However, it was suggested that there was little evidence for large-scale forest declines in the forests of North America due to air pollution and acid rain. Rather, there was evidence of smaller-scale single-species problems in several areas, such as red spruce and sugar maple decline in the eastern United States and Canada, as well as decline of ponderosa pine in southern California. In high-elevation areas in the eastern United States—like Clingman's Dome in the Great Smoky Mountains National Park, North Carolina— there was considerable fir mortality (Figure 1.11). This was initially attributed to acid rain, but the introduced balsam woolly adelgid is known to have caused a considerable amount of this mortality (Ayers et al. 1998). Conifer mortality also has been observed at Mount Mitchell in North Carolina, Whiteface Mountain in New York, Camel's Hump in Vermont, and Mt. Moosilauke in New Hampshire. The cause of tree death in these high-elevation areas continues to be under investigation, and air pollution continues to be blamed (Ayers et al. 1998). NAPAP (1998) concluded that sulfur and nitrogen deposition are causing adverse impacts in certain highly sensitive ecosystems, especially spruce-fir ecosystems, and there is concern that if deposition levels are not reduced in the South and West adverse effects may be develop in these regions in the long run. Forest declines and air pollution are covered in more detail in Chapters 9 and 15.

As a result of NAPAP research, the 1990 Clean Air Act Amendment included the Acid Deposition Control Program, which provided market incentives for controlling electric utility emissions of sulfur dioxide starting in 1995 (NAPAP 1998); NO_x reductions were not started until 1996. Emission reductions have significantly reduced precipitation acidity in the Midwest, mid-Atlantic, and northeastern states, but as yet nitrate concentrations have not been reduced. Many impacted lakes and streams have responded positively, but not the Adirondack lakes. There also is concern that the gradual leaching of soil nutrients, especially Ca, Mg, and K, from sustained inputs of acid rain could eventually affect the nutrition and growth of trees. This depends on rate of cation depletion, soil cation reserves, forest age, weathering rates, species composition, and disturbance history. Liming of forests has been implemented in Europe, particularly in Germany, to improve the nutrient status of soils (Huettl and Zoettl 1993). The impact of excess atmospheric nitrogen inputs to forests now is recognized as a serious pollution problem in western Europe (Wright and Rasmussen 1998) and the eastern United States (Aber et al. 1998). Nitrogen inputs beyond normal background levels change ecosystem functions and make trees more susceptible to cold temperature injuries, diseases, and insects.

Figure 1.11
Tree mortality on Clingman's Dome in Great Smoky Mountains National Park, North Carolina, originally assumed to be due to acid rain. However, most of the mortality was due to the introduced balsam woolly adelgid, perhaps interacting with acid rain.

FOREST HEALTH MONITORING NETWORKS

In response to concerns about acid rain and air pollution effects on forest health, a number of national forest health monitoring networks have been established, such as the U.S. National Forest Health Monitoring Network (Burkman and Hertel 1992, Stolte 1997), the Acid Rain National Warning Early Warning System (ARNEWS) in Canada (Hall 1995), and programs in European countries (Commission of the European Communities 1991). Monitoring networks also have been set up in many U.S. states, Canadian provinces, and Australian states. These systems typically use crown condition and mortality along with other variables to monitor forest health as discussed in more detail below.

The U.S. National Forest Health Monitoring Network

In 1990 the U.S. Department of Agriculture (USDA) Forest Service and the United States Environmental Protection Agency (EPA) initiated the cooperative Forest Health Monitoring (FHM) program to monitor the status, changes, and trends in all the nation's forest ecosystems in the context of EPA's Environmental Monitoring and Assessment Program (EMAP) (Burkman and Hertel 1992, Stolte 1997). Partners include the USDA Forest Service (State and Private Forestry, Research, and National Forest System), the USDI Bureau of Land Management, state foresters, and several universities. Information is obtained from two major sources: permanent forest plots established throughout the nation, and recurring insect, pathogen, and fire incidence data. The EMAP sampling grid includes 12,600 points across the contiguous United States, 2,400 points in Alaska and 56 points in Hawaii. About 4,600 plots are forested. Each 1 ha circular (56.4 m radius) forest health monitoring plot is centered about 158,000 acres (63,942 ha), and plot centers are located about 27 km apart. Inside the 1 ha plots are four circular 17.95 m radius plots with 7.32 m radius subplots in each; 2.1 m radius microplots are inside the subplots (Campbell and Liegel 1996). Plots are measured on 4-year cycles. How well trees in these plots reflect the surrounding forest is yet to be seen. Because of the low number of plots in this network relative to forest area, they may reflect change in conditions only at specific locations or provide a very broad regional assessment, and may never realistically detect incidence and severity of diseases or insect outbreaks.

The FHM program has four major components: detection monitoring (national or regional monitoring), evaluation monitoring (intensified monitoring in problem areas), intensive site ecosystem monitoring (monitoring to understand processes to improve predictive capabilities), and Research on Monitoring techniques. Detection monitoring involves lichen communities, ozone bioindicator plants, tree damage, tree mortality, tree growth, tree crown condition, tree regeneration, vegetation structure, plant diversity, fuel loading and down woody debris. Defoliation, discoloration, dieback, branch breakage, main stem breakage or uprooting, and causal agents are noted. Additional indicators are soil nutrition and toxicity, foliar nutrition and toxicity, and landscape characterization. The evaluation monitoring approach begins when regional Forest Health Protection and

FHM managers agree that a particular forest health change or trend warrants additional attention. There are four FHM regions: North, South, Interior West, and West Coast. The first regional implementation was in New England in 1991; California and Colorado were added in 1992, and Idaho, Minnesota, Texas, West Virginia, and Wisconsin in 1993. Other states have been added since including those in the Pacific Northwest.

Reports of forest health condition are made annually, and the program and updates are available on the World Wide Web at www.fs.fed.us/foresthealth/. Initial results from 20 states from Minnesota to Maine including nine major forest types (white-red-jack pine, oak-pine, elm-ash-cottonwood, spruce-fir, oak-hickory, maple-beech-birch, hard pine, oak-gum-cypress, and aspen-birch) suggest that there are no large-scale forest health problems (Burkman 1993). However, there are concerns about several problems in both natural and urban forested areas. The most notable ones are shown below:

- Gypsy moth defoliation, causing oak decline and mortality in oak-hickory forests
- Beech bark disease, causing extensive mortality to America beech
- Butternut **canker,** causing widespread loss of the butternut in oak-hickory and maple-beech-birch forests
- Dutch elm disease, continuing to cause mortality of American elms
- Chestnut blight, continuing to affect the American chestnut throughout its range
- Dogwood anthracnose, causing serious losses in the understory and ornamental tree species
- Eastern hemlock looper and hemlock woolly adelgid, affecting forest and ornamental hemlock
- White pine blister rust and white pine weevil, limiting eastern white pine planting
- Abiotic factors, including ozone damage on sensitive species such as eastern white pine, effects of acidic deposition on the high-elevation red spruce, drought-caused tree mortality (particularly in the Lake States), and fire associated with drought and insect defoliation

In the southern states the most notable problems involve littleleaf disease on shortleaf and loblolly pines, Annosus root disease on pines, fusiform stem rust on loblolly and slash pines, and southern pine beetle on loblolly, slash, and shortleaf pines (Stolte 1997). In the inland West Armillaria root disease, laminated root rot and Annosus root disease on conifers, white pine blister rust on five-needle pines, **dwarf mistletoes** on conifers, mountain pine beetle on ponderosa and lodgepole pine, western spruce budworm on Douglas-fir, spruce, and true firs, and high fire danger are the most notable problems (Campbell and Liegel 1996, Stolte 1997). West Coast forests are being impacted by laminated root rot, Annosus root disease, Armillaria root disease, Port-Orford-cedar root disease, Swiss needle cast on coastal Douglas-fir in Oregon, and ozone air pollution, particularly in California (Campbell and Liegel 1996, Stolte 1997, Miller and McBride 1998).

Forest Health Monitoring in Western Europe

Following concerns about acid rain damage to forests in Germany and other countries in the early 1980s, the Council of Ministers of the European Community (EC) adopted a resolution on the protection of the EC's forests against atmospheric pollution which took effect on January 1, 1987 (Commission of the European Communities 1991). The EC countries are Austria, Belgium, Denmark, Finland, France, Germany, Greece, Ireland, Italy, Luxembourg, the Netherlands, Portugal, Spain, Sweden, and the United Kingdom. In 1990 the non-EC countries Switzerland, Hungary, Czechoslovakia, and Poland were added. A common inventory methodology requires that a network of points be established following a systematic grid covering the entire forest area. A 16 × 16 km grid is used and the latitude and longitude of each point are provided by the Commission of European Communities to each EC country and member state. Countries also are encouraged to collect additional information on denser networks. At each grid intersection point falling in a forest, 20 to 30 dominant and codominant trees are selected for assessment. Crown defoliation and discoloration are the major indicators. In 1987 and 1988 defoliation was estimated in five classes, shown in Box 1.2.

Since 1989 defoliation has been estimated in 5% increment classes with class 0 = 0% defoliation, class 1 = 1 to 5% defoliation, class 2 = 6 to 10% defoliation, and so on. Defoliation is estimated in comparison to a tree with full foliage, which is a healthy reference tree in the vicinity. An example of different degrees of defoliation in Norway spruce in Germany is shown in Figure 1.12. Discoloration is estimated in five classes, shown in Box 1.3.

There is a major problem separating changes in crown density or coloration attributed to air pollution from those of other factors, since only a small fraction of the sample trees showed direct air pollution damage. However, air pollution might be a predisposing or contributing factor to forest decline (see Chapter 15 for a discussion of these factors). Additional information collected includes altitude, aspect, soil water availability, humus type, age of dominant trees, date of observation, tree number, tree species, and observations on easily identifiable damages. The total number of plots now involved is 2,005 with about 50,000 trees.

Box 1.2

Class	Degree of defoliation	Percentage of needle/leaf loss
0	Not defoliated	0–10
1	Slightly defoliated	11–25
2	Moderately defoliated	26–60
3	Severely defoliated	>60
4	Dead	

Figure 1.12
A thinned Norway spruce (*Picea abies*) forest in southern Germany. Shown are trees with different amounts of defoliation and crown thinness. Most of the trees would be rated as moderately defoliated, 26 to 60% needle loss.

BOX 1.3

Class	Degree of discoloration	Percentage of discoloration
0	Not discolored	0–10
1	Slightly discolored	11–25
2	Moderately discolored	26–60
3	Severely discolored	>60
(4)	(Dead)	

Observations in 1990 for the entire data set (EC plus non-EC) showed that 20.8% of the trees were damaged (defoliation >25%) with 13.8% with discoloration of over 10%. Conifers were more defoliated than broadleaves, but broadleaves had more discoloration. Highest discoloration was found in the Mediterranean region and lowest in the non-EC mountainous region. *Quercus suber* appears to be the most damaged species with severe defoliation. Most species, however, showed no change over the period from 1987 to 1990. Interestingly, many trees switched back and forth from one defoliation and color class another over the years, indicating that foliar symptoms are very dynamic.

The Acid Rain National Early Warning System (ARNEWS) in Canada
ARNEWS was established in 1984 to monitor the health of Canada's forests (Hall 1995). It is operated by the Forest Insect and Disease Survey (FIDS), which has a 60-year history of monitoring the effects of damaging factors on Canadian forests.

The FIDS provides an annual national overview of the condition of the 220 million ha of productive forest in Canada. ARNEWS consists of 150 permanent sample plots located across Canada. Altogether the system contains 11,700 trees: 3,500 hardwoods and 8,200 conifers. In each plot one or more assessments per growing season are made on tree mortality and causes, disease and insect conditions, abiotic damage, and acid rain symptoms. Observations are made on tree growth, crown structure, crown density, and foliage and soil chemistry every 5 years.

Assessment of tree health is based on three factors: (1) mortality levels—normal or greater than normal, (2) defoliation—Is it enough to affect tree health?, and (3) presence of symptoms of pollution damage. A crown classification system similar to that used in the European and U.S. networks integrates measures of defoliation and dieback. So far data have been compiled from 1986 for 12 species: eastern white pine, jack pine, lodgepole pine, tamarack, black spruce, white spruce, coastal Douglas-fir, inland Douglas-fir, balsam fir, trembling aspen, white birch, mountain paper birch, red oak, and sugar maple. Results indicate that Canada's forests are in a generally healthy state (Hall 1995). Mortality rates were about 1% annually for both conifers and hardwoods. There was some variation among species. It was very low in white spruce, the species with the widest distribution in Canada, but occurring in areas with little pollution. Sugar maple is of interest because of its suspected damage by pollution. Although it averaged less than 1% mortality, it did have 2.4% mortality in 1989 perhaps due to drought. In 1988 a special North American Sugar Maple Project (NAMP) was established jointly with the Canadian Forestry Service, Natural Resources Canada, and the United States Department of Agriculture Forest Service. Throughout the range of sugar maple in North America, 233 monitoring sites were established, 62 in Canada. Results from NAMP suggest that the condition of sugar maple crowns improved between 1988 and 1996. Three other species had average mortality rates near or above "normal mortality": balsam fir (1.8%), trembling aspen (2.3%), and birches (3.0%). Mortality in balsam fir was caused by the eastern spruce budworm, whereas that in aspen was due to the large aspen tortrix and unseasonable frosts. White birch near the Bay of Fundy appeared to be severely damaged by acid fogs. Most damage to crowns was caused by insects and diseases, particularly rusts, Armillaria root disease, foliage diseases, adelgids, weevils, eastern spruce budworm, and eastern hemlock looper. ARNEWS appears to be an effective early warning system for Canada's forests.

CONCLUSIONS

To provide for fiber, water, fish, wildlife, biodiversity, and recreational opportunities for future generations forests must be healthy. Forest health, however, is a difficult concept to grasp. Most of us can determine if a forest is unhealthy, but it is more difficult to define what a healthy forest is. Forests are continually changing, and fires, wind, insects, and diseases are important agents of change. Humans have had a tremendous impact on the world's forests by cutting and managing them and altering the operation of natural processes like fire. Introduced diseases and insects also have greatly influenced forest health throughout the world.

Concern about the effects of air pollution and acid precipitation on forests in North America and Europe resulted in the establishment of forest health monitoring networks that monitor many aspects of forests including the trees and other vegetation, soil, water, insects, diseases, soil organisms, and so on. Rather than focusing just on the management of individual pests and fire, forest health now is considered in the ecosystem context. Ecosystem management, however, is a "fuzzy" concept and we are still a long way from understanding how ecosystems function, how insects, diseases, wind, fire, drought, air pollution, and global change affect ecosystems, and how these factors interact.

Many forest health questions remain to be answered. For example,

What are good indicators of a healthy forest?

What is the desired condition of the forest?

Is the landscape the appropriate scale for managing ecosystems?

How do we balance commodity and other uses of the forest relative to forest health?

How do we reintroduce fire into ecosystems?

What is the risk of management inaction relative to fire, insects, and diseases?

Can models be developed or improved to help managers in decision making?

In this book we explore many of these questions.

REFERENCES

Aber, J., W. McDowell, K. Nadelhoffer, A. Magill, G. Bernston, M. Kamakea, S. McNulty, W. Currie, L. Rustad, and I. Fernandez. 1998. Nitrogen saturation in temperate forest ecosystems—hypotheses revisited. *BioScience* 48:921–934.

Agee, J. K. 1993. *Fire ecology of Pacific Northwest forests.* Island Press, Washington, D.C.

Ayers, H., J. Hager, and C. E. Little (eds.). 1998. *An Appalachian tragedy: air pollution and tree death in the eastern forests of North America.* Sierra Club Books, San Francisco, Calif.

Burkman, W. 1993. *Northeastern areas forest health report.* USDA Forest Service, Northeastern Areas, NA-TP-03-93.

Burkman, W. G., and G. D. Hertel. 1992. Forest health monitoring. *Journal of Forestry* 90(9):26–27.

Callicot, J. B. 1995. A review of some problems with the concept of ecosystem health. *Ecosystem Health* 1:101–112.

Campbell, S., and L. Liegel (technical coordinators). 1996. *Disturbance and forest health in Oregon and Washington.* USDA Forest Service, Pacific Northwest Station, General Technical Report PNW-GTR-381, Portland, Oreg.

Castello, J. D., D. J. Leopold, and P. J. Smallidge. 1995. Pathogens, patterns, and processes in forest ecosystems. *BioScience* 45:16–24.

Christensen, N. L., A. M. Bartuska, S. Carpenter, C. D'Antonio, R. Francis, J. F. Franklin, J. A. MacMahon, R. F. Noss, D. J. Parsons, C. H. Peterson, M. G. Turner, and R. G. Woodmansee. 1996. The report of the Ecological Society of America committee on the scientific basis for ecosystem management. *Ecological Applications* 6:665–691.

Ciesla, W. M. 1993. Recent introductions of forest insects and their effects: a global overview. *FAO, Plant Protection Bulletin* 41(1)3–13.

Commission of the European Communities. 1991. Forest health report 1991. Technical report on the 1990 survey. Commission of the European Communities, Directorate–General for Agriculture. Office for Official Publications of the European Communities, Luxembourg.

Fielding, N. J., and H. J. Evans. 1996. The pine wood nematode *Bursaphelenchus xylophilus* (Steiner and Buhrer) (=*B. lignicolus* Mamiya and Kiyohara): an assessment of the current position. *Forestry* 69:35–46.

Hall, J. P. 1995. ARNEWS assesses the health of Canada's forests. *Forestry Chronicle* 71:607–613.

Hansen, E. M., and K. J. Lewis (eds). 1997. *Compendium of conifer diseases.* American Phytopathological Society Press, St. Paul, Minn.

Huettl, R. F., and H. W. Zoettl. 1993. Liming as a mitigation tool in Germany's declining forests—reviewing results from former and recent trials. *Forest Ecology and Management* 61:325–338.

Innes, J. L. 1993. *Forest health: its assessment and status.* CAB International, Wallingford, United Kingdom.

Karr, J. R. 1995. Using biological criteria to protect ecological health, pp. 137–152 In D. J. Rapport, C. Gaudet, and P. Calow (eds.). *Evaluating and monitoring the health of large scale ecosystem.* Springer-Verlag, Heidelberg.

Kohm, K. A., and J. F. Franklin (eds.). 1997. *Creating a forestry for the twenty-first century: the science of ecosystem management.* Island Press, Washington, D.C.

Kolb, T. E., M. R. Wagner, and W. W. Covington. 1994. Concepts of forest health. *Journal of Forestry* 92(7):10–15.

Liebhold, A. M., K. W. Gottschalk, D. A. Mason, and R. R. Bush. 1997. Forest susceptibility to the Gypsy moth. *Journal of Forestry* 95(5):20–24.

Miller, P., and J. R. McBride (eds.). 1999. *Oxidant air pollutant impacts in the Montane Forests of southern California: a case study of the San Bernadino Mountains.* Springer, New York.

Monnig, E., and J. Byler. 1992. *Forest health and ecological integrity in the Northern Rockies.* USDA Forest Service, Northern Region FPM Report 92-7.

Mutch, R. W., S. F. Arno, J. K. Brown, C. E. Carlson, R. D. Ottmar, and J. L. Peterson. 1993. *Forest health in the Blue Mountains: a management strategy for fire-adapted ecosystems.* USDA Forest Service, Pacific Northwest Research Station, General Technical Report PNW-GTR-310, Portland, Oreg.

Naiman, R. J., T. J. Beeche, L. E. Benda, D. R. Berg, P. A. Bisson, L. H. MacDonald, M. D. O'Connor, P. A. Olson, and E. A. Steel. 1992. Fundamental elements of ecologically healthy watersheds in the Pacific Northwest coastal ecoregion, pp. 127–188. In R. J. Naiman (ed.). *Watershed management: balancing sustainability and environmental change.* Springer-Verlag, New York.

NAPAP (National Atmospheric Assessment Program). 1993. 1992 report to Congress. NAPAP, Washington D.C.

NAPAP (National Atmospheric Assessment Program). 1998. Biennial report to Congress: an integrated assessment. NOAA, Silver Spring, Md.

Old, K. M. 1979. Phytophthora *and forest management in Australia.* CSIRO, Melbourne, Australia.

Oliver, C. D., D. E. Ferguson, A. E. Harvey, H. S. Malany, J. M. Mandzak, and R. W. Mutch. 1994. Managing ecosystems for forest health: an approach and the effects of uses and values. *Journal of Sustainable Forestry* 2:113–133.

Rapport, D. J., C. Gaudet, R. Constanza, P. R. Epstein, and R. Levins (eds.). 1998. *Ecosystem health: principles and practice.* Blackwell Science, Malden, Mass.

Sampson. F. B., and F. L. Knopf. 1996. *Ecosystem management: selected readings.* Springer-Verlag, New York.

Sampson, R. N. 1996. *Forest health issues in the United States.* Forest Policy Center, American Forests, Washington, D.C.

Sampson, R. N., and D. L. Adams (eds.). 1994. *Assessing forest ecosystem health in the inland West.* Parts 1 and II. Haworth Press, NY.

Schmidt, W. C., and K. J. McDonald (compilers). 1990. *Proceedings—symposium on whitebark pine ecosystems: ecology and management of high mountain resources.* USDA Forest Service, Intermountain Research Station, General Technical Report INT-270, Ogden, Utah.

Smith, W. H. 1990. *Air pollution and forests: the interaction between contaminants and forest ecosystems.* Springer-Verlag, New York.

Sprugel, D. G. 1991. Disturbance, equilibrium, and environmental variability: what is "natural" vegetation in a changing environment? *Biological Conservation* 58:1–18.

Stolte, K. W. 1997. *1996 national technical report on forest health.* USDA Forest Service, Admin. Report FS-605, Southern Research Station, Asheville, N.C.

Vogt, K. A., J. C. Gordon, J. P. Wargo, D. J. Vogt, H. Asbjornsen, P. A. Palmiotto, H. J. Clark, J. L. O'Hara, W. S. Keaton, T. Patel-Weynard, and E. Witten. 1997. *Ecosystems: balancing science with management.* Springer-Verlag, New York.

Wright, R. F., and L. Rasmussen. 1998. Introduction to the NITREX and EXMAN projects. *Forest Ecology and Management* 101:1–7.

2

ECOLOGICAL PRINCIPLES

INTRODUCTION

On a small island, an introduced European rabbit population closely grazed the grasslands, some of which had been forested before being logged and farmed earlier in the century. The rabbits were a rich source of food for wintering bald eagles, who would swoop down on unsuspecting rabbits before they could duck into their burrow systems. Trees could not become established in the grasslands because of the heavy grazing pressure. Then for some unknown reason, the rabbits stopped reproducing, and the population fell dramatically. Bald eagles stopped foraging in the area due to lack of hunting success. Adjacent forests provided seed that allowed small Douglas-fir and lodgepole pine to invade the old fields, which were once forest, for several years, until the lack of grazing pressure allowed the grass to increase its cover and mulch to the point that trees could not become established. With the increased grass cover, the vole population increased, and with it hawk and owl numbers. Voles began to girdle many of the trees that had invaded the grassland, particularly the Douglas-firs, reducing their numbers. After about 8 years the rabbit population began to increase again, removing the grass cover, reducing the vole, hawk, and owl populations, and increasing the eagle population again. A narrow age group, or cohort, of lodgepole pine and Douglas-fir, survives in the grassland area that was historically forest, and grows larger each year. Eventually, now that they have passed a critical establishment phase of their life history, these trees will probably convert the old fields back to forest.

This true story illustrates a few of the many ecological principles at work in our wildlands. Some areas will have a tendency to become forest, while others will not, a principle called *site potential.* The process of change from grassland to forest, or vice versa, is called *succession.* Succession can follow many pathways, and is affected by life history characters of the plant species, and how different species interact with one another (Kimmins 1997). The species composition and structure of plant communities influence their function as wildlife habitat. Wildlife are not

23

simply passive responders to plant succession, but may have significant influences on plant succession and other wildlife species. The concept of feedback between ecosystem components is oftentimes underestimated. The triggering mechanism for the rabbit decline, although several possibilities were studied, was never determined, adding an element of surprise to the managers of this ecosystem when the decline began. In most forested ecosystems, disturbance includes not only wildlife influences but also those of fire, wind, insects, and disease, to name at least a partial list of potential disturbances. Different forest types will be influenced uniquely by various combinations of disturbance factors.

FOREST BIOMES

A forest **ecosystem** includes not only the dominant tree species but also the associated understory vegetation, vertebrate and invertebrate animals, and microbes (Perry 1994). Forest ecosystems usually are identified by the major tree species because they are the most visible component. There are many levels of forest classification, the broadest being the concept of the forest biome. A **biome** is a set of plant communities that have generally similar life-forms. At a global scale these can be defined in broad categories associated with global climate patterns: tropical rainforest (warm and wet), tropical seasonal forest (warm and seasonally dry), temperate rainforest (cool and wet), temperate deciduous forest (cool and summer wet), temperate evergreen forest (cool and summer dry), taiga (cold), and a variety of woodlands where tree cover is limited and shrubs or herbs codominate (Kimmins 1997). In some areas the seasonality of precipitation is as important as the annual total in defining the forest/nonforest boundary. In temperate latitudes lack of precipitation tends to define where forest gives way to shrublands or grasslands; at higher latitudes and altitudes, temperature tends to be a limiting factor for forest establishment and growth. In North America generalized vegetation formations (Figure 2.1) include a wide variety of prairie, shrub, woodland, and forest biomes (Barbour and Billings 1988). Each of the biomes will contain several to many somewhat different forest types. For example, the Pacific coast/Cascadian forests will contain types dominated by Sitka spruce, western hemlock, Douglas-fir, Pacific silver fir, mountain hemlock, coast redwood, and other species too (Franklin and Dyrness 1973). Slightly different characters of vegetation may be used in different classifications, so that any two maps of continental vegetation usually will differ to some extent.

   At the regional scale finer breakdowns of the community composition can be made. Climate will remain an important predictor variable, but landscape position, soil type, disturbance, and other gradient patterns will become important in defining vegetation types. The eastern deciduous forest (Figure 2.2) is composed of nine different regional components, each named after the forest community type that is most common on well-drained uplands in the region (Greller 1988). In the Rocky Mountains (Figure 2.3), vegetation zones are named after the tree species that is most shade-tolerant, although many other species may actually be present or dominant on the site (Pfister and Arno 1980). Both systems are discussed below in more detail.

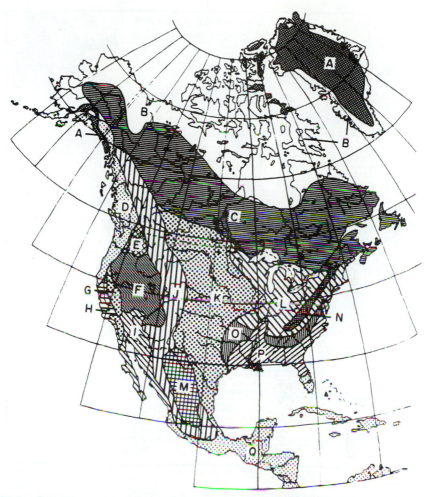

Figure 2.1

Generalized vegetation of North America. A = ice; B = Arctic tundra; C = taiga;
D = Pacific coastal/Cascadian forests; E = Palouse prairie; F = Intermountain deserts,
shrub steppes, woodlands, and forests; G = Californian forests and alpine vegetation;
H = California grasslands, chaparral, and woodlands, I = Mohave and Sonoran deserts;
J = Rocky Mountain forests and alpine vegetation; K = central prairies and plains;
L = mixed deciduous forests; M = Chihuahuan deserts and woodlands; N = Appalachian
forests; O = piedmont oak-pine forests; P = coastal plain forests, bogs, swamps,
marshes, and strand; Q = tropical forests.

Forest Zonation

If elevation were the only environmental variable on the landscape, forests might
actually be arranged like the central Idaho forest types shown in Figure 2.3. Many
other types of variation exist so that rarely are forest zones arranged like layers

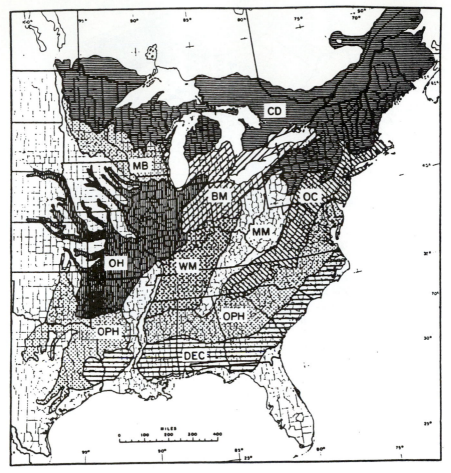

Figure 2.2
Map of the regional vegetation of the deciduous forest. CD = conifer with deciduous; MB = maple, basswood; OH = oak, hickory; BM = beech, maple; WM = western mesophytic; MM = mixed mesophytic; OC = oak-chestnut; OPH = oak, pine, hickory; DEC = deciduous and evergreen hardwoods, conifer.

in a cake (Figure 2.4*A*). Aspect, or the direction that a slope faces, influences solar radiation and moisture relations. The angle of the sun in the northern hemisphere makes south aspects (those facing to the south) drier and warmer. Lower-elevation forest zones representing warm environments are therefore found at higher elevations on south aspects than on north aspects (Figure 2.4*B*). Local topography also may influence temperature and moisture relations (Franklin and Dyrness 1973). Ridges represent relatively warm and dry environments, as they often have shallow soils and receive more solar radiation, so lower elevation zones interfinger up ridges. Conversely, valleys will receive less sunlight, may have deeper soils, receive

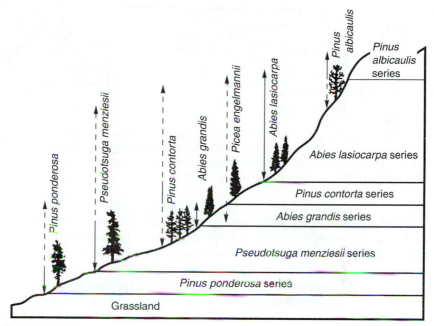

Figure 2.3

Forest types of central Idaho arranged on an elevational gradient. The forest series are defined by horizontal lettering, and species likely to be found in a series are identified by vertical arrows. The solid portion of each arrow denotes that portion of the range of each species where it is the most shade-tolerant and therefore has a forest series named for it. (Source: From USDA Forest Service.)

cold air drainage at night from higher elevation, and will usually have vegetation characteristic of higher elevations (Figure 2.4*C*). These patterns of environment will be reflected in patterns of vegetation, and may be important in predicting forest health issues. Foliar diseases such as that caused by *Hypodermella* spp. on western larch require high humidity, so will be more likely found on larch near canyon bottoms and valleys rather than ridgetops. Historic fire patterns usually are more frequent on ridges, but can be more intense in less frequently burned riparian streamside areas (Agee 1998).

Forest Classification by Current and Potential Vegetation

Three primary ways of classifying vegetation are used in North America. One is based on the current dominant vegetation and is typically called a **cover type classification** (SAF 1980). The second is based on that vegetation which would dominate eventually in the absence of disturbance, or **potential vegetation** (Daubenmire 1968). The third is an **ecosystem classification** (Kimmins 1997), relying on more than vegetation to define classes. There is no single best classification scheme, and for most purposes more than one is necessary to effectively manage forest ecosystems.

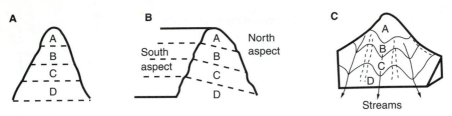

Figure 2.4

Forest zonation rarely occurs in uniform strata (**A**) because of effects of aspect (**B**), slope, soil variation, cold air drainage, and other factors. Interfingering of lower-elevation zones up ridges with higher-elevation zones down valleys resulting in a typically wavy pattern (**C**) of forests on a mountainside.

Cover type classifications indicate the major species currently present on a site. There is no implication as to whether that cover type is temporary or permanent. The most widely used cover type system is that of the Society of American Foresters (SAF 1980). Cover types are defined for eastern and western forests, and currently 145 types are recognized. A forest cover type must have over 25% tree cover, occur over significant land area, and will be named for up to three dominant tree species that comprise more than 20% of the relative basal area of the stand, total basal area being defined as the sum of the cross-sectional area at breast height of all the trees in the stand. Examples of SAF cover types would be silver maple–American elm (62), longleaf pine–scrub oak (71), or lodgepole pine (218). Cover types may change over time with or without disturbance, so a cover type map for an area will likely be a shifting mosaic over time, as timber harvest, insect outbreaks, fires, or other disturbances occur.

Potential vegetation classifications are named for those species likely to eventually dominate if the site is undisturbed. Such species may be minor dominants at present. Potential vegetation is the dominant vegetation that would exist today if human influence was removed and resulting change in plant communities was telescoped into a single moment. Essentially, this definition is based on those species that would dominate in the absence of disturbance. A national approach in the United States is the classification of potential vegetation by Kuchler (1964) or the ecoregion classification of Bailey (1978). In the northwestern United States, a more specific potential vegetation classification known as the habitat typing utilizes this approach. Potentially dominant overstory species, defined on the basis of shade tolerance, define a **forest series** (see Figure 2.3), and when overstory species are linked with **indicator species** in the understory, a **plant association** is defined. Indicator species are plants that are common and "indicate" site temperature and/or moisture. For example, in the western hemlock forest series, sword fern indicates a moist, productive site; the shrub ocean spray indicates a drier, less productive site. A plant association allows inferences about environment for tree establishment and growth, and it defines the trajectory of plant change over time. For example, the SAF cover type "lodgepole pine" carries no information about what changes may occur in the future, but if it is in the Douglas-fir series, perhaps a Douglas-fir/huckleberry plant association, Douglas-fir will eventually dominate this warm, productive site. If the lodgepole pine cover type were in the subalpine fir series, perhaps a subalpine fir/beargrass plant association, then subalpine fir will slowly increase over time on this cold, dry site. Cover types define "where we are"; potential vegetation defines "where we're going," although the journey may be long and usually will be disrupted by disturbance.

Ecosystem classifications are commonly used in Canada, and are similar to the potential vegetation approach except that environmental parameters are explicitly defined instead of being inferred from vegetation. In British Columbia a system known as biogeoclimatic classification focuses on vegetation-soil units using climatic relationships as a framework (Kimmins 1997). A biogeoclimatic zone might be, for example, the coastal western hemlock biogeoclimatic zone, with dry, mesic, and wet subzones. Below this level subzones are further classified based on gradients of soil moisture and fertility. To these subunits are then applied recommendations associated with resource management activities such as tree species to be planted, or types of slash burning activities.

Forest classifications, at a minimum, allow us to place similar forest types into some logical order. Correct forest classifications allow managers to accurately define what trees are on site today, what might be there some time in the future, and over that time how the forest may grow, provide wildlife habitat, or be susceptible to forest health declines.

Production Ecology

Forest production might include everything a forest produces: wood, water, wildlife, recreation, and other products. In an ecological context it can again be interpreted broadly to include the various functions that the forest serves. However, production ecology is normally constrained to evaluating the energetics of forest ecosystems, including the storage of energy in living and nonliving components of the system, as well as the energy flow through the system (Perry 1994). Energetics are important because they are the basis for the organization of ecosystems. Energy stored or flowing between various components of the ecosystem is the foundation of most forest management. However, knowledge of the energy inputs, storage, transfer, and outputs alone, although important, is not a sufficient basis for forest management.

There are two major types of organisms in forest ecosystems, organized by trophic categories, or how they utilize energy (Figure 2.5): **producers** (autotrophs), and **consumers** and **decomposers**. Producers, as the name implies, fix energy from independent sources, such as the sun, and provide the energy sources for consumers. Consumers rely on organic molecules like carbohydrates, fats, and proteins, and are of several types. **Herbivores** consume green plants and often are called primary consumers. These include animals such as deer, elk, bison, grasshoppers, and spruce budworms. **Carnivores** consume herbivores and utilize the energy stored within them. **Omnivores** utilize plants and animals as energy sources, and **detrivores** utilize the energy found in dead plants and animals (detritus). Such simple

ECOSYSTEM

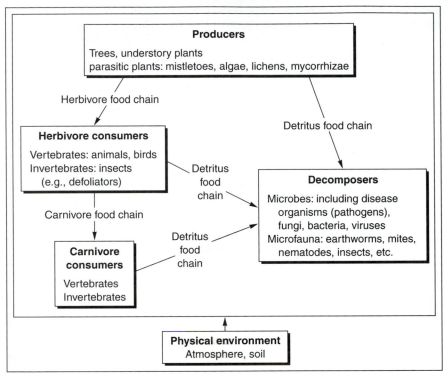

Figure 2.5
Producers in a forest ecosystem fix energy from the sun. Consumers include herbivores, carnivores, and omnivores. Decomposers (detritivores), producers and consumers, the energy transfers between them, and interactions with the physical environment define the ecosystem.

classifications can mask the gradient between categories, when some organisms shift food supply from one category to another. Carnivorous plants are a good example of organisms that are primary producers but also are carnivores.

These trophic levels can be arranged, one above the other, into what is known as an **ecological pyramid** (Kimmins 1997). For the biomass of forest ecosystems, the pyramid is upright, with producers, particularly the trees, forming the wide bottom of the pyramid, and primary and various secondary consumers (herbivores, carnivores) forming smaller biomass layers above that. This relationship may not hold for *numbers* of organisms, but trees are so large in comparison to most other components of the system that they are a dominant biomass category wherever forest ecosystems are found.

The function of forests for myriad values depends on the primary production by the plants. Although only about 2% of the sun's energy is captured by photosynthesis and stored as chemical energy, it is sufficient to run the grand chemical factory we call the forest ecosystem (Perry 1994). Forests are very efficient compared to other vegetation in fixing solar energy. Total photosynthesis is also known as **gross primary production**. Because a certain amount of that energy is used in respiration, the difference is known as **net primary production**:

Gross primary production (GPP) = Net primary production
(NPP) + Plant respiration

Net primary production can be divided into three further categories: the accumulation of dry matter into tree stems, branches, and other biomass, or the annual increment to the **standing crop;** herbivore respiration; and decomposer respiration. Up to 80% of GPP is utilized as respiration, so as little as 20% is left to accumulate as standing crop. The relation between NPP and standing crop is not constant. Young forests, with low standing crop, may have high NPP, whereas old-growth forest, with increasing biomass available for decomposers (logs, forest floor) will have high standing crop but low NPP. An old-growth Douglas-fir forest, for example, may lose over 90% of its GPP in respiration. Although forests cover less than 50% of the earth's land surface, they are responsible for about 80% of the terrestrial NPP (Perry 1994).

Disturbances by fire, insects, and disease can significantly change productivity. Fire is essentially instantaneous respiration of organic material, and periodic low intensity fires may limit available energy for combustion, creating a cyclic stability in ecosystem energetics (see Chapter 4, ponderosa and longleaf pine). If the fire is intense and scorches the foliage, it eliminates the ability of the stand to photosynthesize and can initiate major changes in the standing crop by consuming logs and forest floor.

The productivity of forest ecosystems can be compared by ranking the net primary production of each ecosystem type (Whittaker 1975). Generally, productivity increases with increasing radiation from Arctic to tropic areas (Table 2.1). Note that sufficient variation exists that some temperate forests have higher NPP than some tropical forests. GPP in tropical areas usually is much higher than in temperate or boreal regions, but because of high respiration rates in the warm tropics, NPP may not exceed that in temperate forests. For the most part, energetics are used not so much to compare ecosystems as to understand how they function, in terms of dry matter and nutrient cycling.

Productivity is difficult to measure. Most studies have concentrated on aboveground production only, which from the perspective of utilizable wood production may be appropriate, but this ignores important belowground processes (Figure 2.6). In comparing "poor" and "good" productivity Douglas-fir stands, for example, the total of aboveground and coarse root production was 8.4 t ha^{-1} (Mg ha^{-1}) versus 15.3 t ha^{-1}; but if fine root production was included, the total biomass for the poor site was 15.4 t ha^{-1}, 87% of the 17.8 t ha^{-1} on the good site (Keyes and Grier 1981). Fine root turnover is annually high, with up to 12 t ha^{-1} in coniferous forests.

Herbivores may under normal conditions consume small proportions of ecosystem production, but under epidemic levels can have major impacts on productivity. Under epidemic conditions, aphids tended by ants can consume 25% of the standing crop of foliage and even higher proportions of certain nutrients.

TABLE **2.1** | **Net Primary Productivity for Forests of the World**

Forest ecosystem	Net primary productivity	
	t ha⁻¹ yr⁻¹	
	Range	Mean
Tropical rain forest	10–35	22
Tropical seasonal forest	10–25	16
Temperate evergreen forest	6–25	13
Temperate deciduous forest	6–25	12
Boreal forest	4–20	8
Woodland and shrubland	2.5–12	7

Source: From Whittaker (1975).

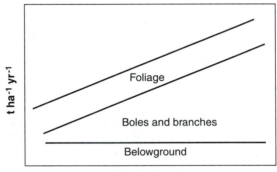

Total NPP (t ha⁻¹ yr⁻¹)

Figure 2.6
The proportion of net primary productivity (NPP) devoted to belowground production decreases on more productive sites (right side of diagram), although the absolute amount remains relatively stable over a wide range of NPP. Bole and branch biomass in both absolute and relative amounts increases on sites with higher NPP.

Insects normally consume 3 to 10% of leaf production in a stand, but outbreaks of defoliators can completely defoliate stands in some forests, limiting NPP and possibly killing some species.

Detrivores, or decomposers, vary by type and species between forest ecosystems. In terrestrial ecosystems they account for most of the energy flow once it has been fixed by the primary producers (Perry 1994; Figure 2.7). Both bacteria and fungi are primary decomposers in forests, but the rate of decomposition will vary across environments. The residence time for coniferous leaf litter in high-elevation forests may be 70 years or more, compared to less than a year in many tropical forests.

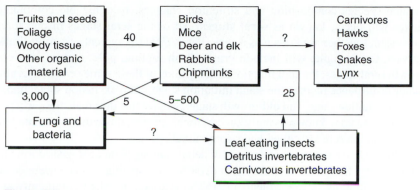

Figure 2.7
Energy flow through the food web of a northern hardwood ecosystem, expressed as kilocalories/m²/year. By far the most energy goes directly from producers to bacteria and fungi. The wide range (5–500) shown for insects is associated with occasional outbreaks of leaf-eating insects. Small but unknown amounts of energy go to higher trophic levels (hawks, foxes, etc.).
(Source: Broadly adapted from data in Gosz et al. 1978.)

Carnivores are for the most part minor energy shunts in forest ecosystems (Figure 2.7). As the trophic level increases, the energy dynamics are much smaller proportions of the total energy cycling of the ecosystem. Because herbivores are generally pretty digestible for carnivores, the assimilation efficiency (proportion of energy consumed that is digested and assimilated by the organism) is high for carnivores compared to herbivores trying to digest plant material. Compare a horse scat, for example, to that of a coyote or weasel; much of the grass consumed by the horse is still recognizable, whereas the carnivore scat is largely just bone and hair.

Although energetics are an important analysis tool for forest ecosystems, energy flows do not identify many important relationships. Grazing animals, for example, may selectively remove trees and shrubs when the plants are very small, energetically unimportant but ecologically significant. That removal will have major effects on subsequent change in forest ecosystems, affecting their species composition, structure, and function.

SUCCESSION

The process of change in forest ecosystems over time is known as **succession. Primary succession** occurs in ecosystems that have not been influenced previously by organisms, such as an outwash plain from a receding glacier, or a new volcanic surface. **Secondary succession** occurs on areas previously influenced by organisms; recovery after timber harvest, fire, insect attacks, windthrows, and other disturbances are examples of secondary succession (Kimmins 1997). Most forest succession issues, including those associated with forest health, are issues of secondary succession. During the process of forest ecosystem change at one location, which will include species composition and structural changes, recognizable community stages will occur, known as **seral stages**. All of the seral stages in one forest ecosystem together are known as a **sere** (Perry 1994). A typical sere after timber harvest, for example, will include grass, shrub-sapling, pole, mature, and old-growth seral stages. Parallel functional definitions of these same seral stages after major disturbance include the stand initiation stage, stem exclusion stage, understory reinitiation stage, and old-growth stage. Species that dominate the early portion of the sere are known as early successional species; those that dominate later portions of the sere are called late successional species.

Succession has been a major topic of debate among plant ecologists during the 20th century. Early in the century two persistent competing theories were developed. The first, by F. E. Clements (1916), envisioned the plant community as a superorganism, with many properties of the individual organism. Progression towards a regional stable vegetation was inevitable, and this regional vegetation was known as the **climax** and largely defined by climate. The theory was called the *monoclimax theory* because of the single successional endpoint. All plant communities in the region, under this concept, would converge toward this endpoint if given enough time. Each seral stage would alter the community and provide an environment suitable for the next stage to develop. The second theory was the individualistic theory proposed by H. A. Gleason (1926). Gleason maintained that plant communities were aggregations of individual species, each with unique character, and largely unpredictable. Under these assumptions no single climax stage was possible, and the concept of a plant community was deemed a "flight of imagination."

These early theories, as rigid as they appeared and often based on theory without empirical evidence, nevertheless form the backbone of today's models of succession. The concept of potential vegetation, introduced earlier in this chapter, is a descendant of the monoclimax theory and its intermediate, the polyclimax theory, which had less stringent convergence assumptions. The individualistic theory was further developed by Frank Egler (1954), who defined two major successional pathways (Figure 2.8) in abandoned agricultural fields, depending on the life histories of the species that could potentially grow on the site. Egler's relay floristics model fits primary succession well, where one plant community changes the environment and allows a quite different plant community to replace it. His initial floristics model fits many secondary successional sequences: many of the species are initially present, but different relative growth rates, plant longevity, and shade tolerances regulate the succession of plant species on a disturbed site. In forest ecosystems both early and late successional species may be present across much of the sere, but relative dominance will shift over time.

The Complex Nature of Disturbance

Recent attention to plant succession has focused on mechanisms of species change, and there has been a trend away from defending grand, unifying models of plant ecology (Christensen 1988). One of the major driving forces behind this change is the recognition of the complex interactions of disturbance in ecosystems. Disturbance is simply defined as a break in the settled order of things, but as most forest ecosystems are not "settled orders," it may be better defined as "any

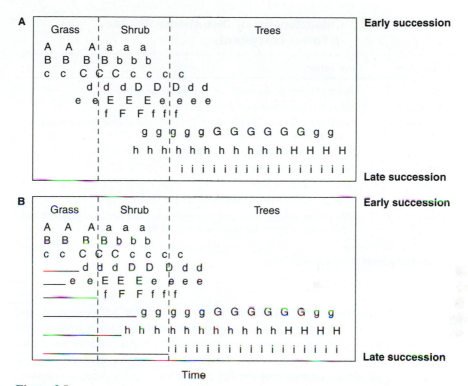

Figure 2.8
Relay (**A**) and initial (**B**) floristics occur in most successional sequences. Letters represent different species. Capital letters represent dominance of that species, lowercase represent the species as common, and a line represents minor presence only. Relay floristics are most commonly seen in primary succession, where one community alters the environment sufficiently, perhaps by fixing nitrogen or adding organic matter, allowing another community to succeed it. Organisms in the later communities are not found in the earlier successional communities. Initial floristics are most commonly observed in secondary succession after disturbance. Most species are found across the time sequence but relative dominance varies over time.

relatively discrete event in time that disrupts ecosystem, community, or population structure and changes resources, substrate availability, or the physical environment" (Pickett and White 1985). This definition, while an improvement, still leaves much to be desired. The definition of "relatively discrete" is not clear: Must it be a several-minute-long crown fire, or may it be an insect attack that continues for several years? Does it originate outside the ecosystem (exogeneous) or from inside the ecosystem (endogenous)? A physical disturbance like fire may be of exogeneous origin, but its effects may depend very much on the endogenous state of the ecosystem (fuel buildups, tree species and size). A biotic disturbance such as a bark beetle attack is endogenous in origin, but its effects may depend on presence or absence of exogenous factors. Any definition of disturbance has to remain generalized, but the effects of disturbance can be ordered by several characteristics (Table 2.2).

T A B L E **2.2** | **Characteristics of Disturbance Regimes in Forest Ecosystems**

Descriptor	Definition
Type	Agent of disturbance: wind, fire, insects, disease, etc.
Frequency	Mean number of events in a time period
Magnitude	Described as intensity of the event: streamflow in a flood, energy release rate of a fire, deposition from a volcano
Predictability	Variation, usually in frequency or intensity of an event
Extent	Area disturbed per time period
Timing	Season of disturbance
Synergism	Effect on occurrence of other disturbances
Severity	Effect on ecosystem, measured in various ways: number of organisms killed, proportion of basal area removed, erosion amounts, etc.

Source: Adapted from Pickett and White (1985).

Disturbance frequency may depend on a combination of exogenous and endogenous factors (Agee 1993). Frequent fires in ponderosa pines forests were often lightning-ignited, but were constrained by lack of fuel to a minimum return interval of 2 to 3 years. Southern pine beetles often become epidemic during climate-related stimuli such as drought, but require susceptible pines of a certain size. Disturbance magnitude may similarly be regulated by internal and external factors. The variability in both of these (predictability) will be important in predicting ecosystem effect. The extreme event (a short fire return interval or an intense defoliator attack) will be as important as the average frequency or magnitude. Disturbance extent is critical for those species killed by the disturbance if they must rely on seed. Although sprouting species can immediately regrow after fire or flood, heavy-seeded species with no alternative mechanism for dispersal (mechanisms such as seed banks or bird-dispersed seeds) may locally be extirpated for a considerable time after mature individuals are killed. Season of disturbance can be important. Some species are more susceptible to damage in certain seasons, such as foliage after bud flush, or fine root kill at a time when root reserves are low. Synergistic effects, or second-order effects, can multiply the effects of a first-order disturbance. Bark beetle attack after windthrow or fire can result in damaged trees being killed before they can recover. Insects that topkill and retard tree leader growth, such as white pine weevil, will have implications in later decades for windthrow as the recovered trees often have two dominant stems, both of which are less resistant to breakage (Chapter 7).

Multiple Pathway Succession

The multiple combinations of disturbance character create multiple pathways for successional change in forest ecosystems. **Multiple pathway models** can incorporate individualistic species character into potential vegetation frameworks, and by so doing meld together successors of the two early general theories of plant

T A B L E **2.3** | **Life History Characteristics of Species Used in Graphical Succession Simulations[a]**

| Species | Longevity | Life history character | | | |
		Early growth	Shade tolerance	Fire tolerance[b]	Susceptible to pine bark beetle
A	H	H	L	H/H	Yes
B	M	L	H	L/H	No
C	L	H	M	L	No
D	M	M	H	L	No
E	L	H	L	L	Yes
F	M	L	H	L	No

[a]See Figures 2.9, 2.10, and 2.11.
[b]Tolerance shown is for saplings and mature trees.
Abbreviations: H = high, M = moderate, L = low.

succession. Within a plant association, for example, three types of information are used to predict successional change: method of persistence, conditions for establishment, and critical life history stages (Noble and Slayter 1980). Method of persistence defines the recovery potential of the individual species (e.g., sprouting) or its reproductive strategy after disturbance. Conditions for establishment define the species capable of responding to various degrees of shading, relative early growth characteristics, and so on. Critical life history stages include reproductive age, longevity, developing bark for fire resistance, and so on. Successional trajectories, measured in terms of species composition and/or forest structure, will depend on the influence of various disturbances on tree species and sizes present at the time of disturbance. Rather complicated flow diagrams can be generated for multiple pathway successional sequences, but simpler systems can be designed for most forest ecosystems.

Using the species characteristics in Table 2.3, practical applications of succession with or without disturbance can be outlined (Figures 2.9, 2.10, and 2.11). Each set of diagrams shows overstory and understory relative density (0–1); in the absence of disturbance, the most shade-tolerant species will increase in the understory and then in the overstory. Disturbance alters the successional trajectories by affecting overstory or understory species. In the fire scenario, an example where frequent fire once occurred, species A is maintained as a dominant by the low thinning effect (removing understory trees) of the frequent fires, maintaining microsites where species A can become established. In the wind scenario, wind thins from above, removing both species C and D, but opening the understory for reestablishment of species C. In the insect scenario the demise of early successional species E is hastened by its removal by bark beetles. All of the "no disturbance" scenarios assume that no major disturbance occurs, and are included

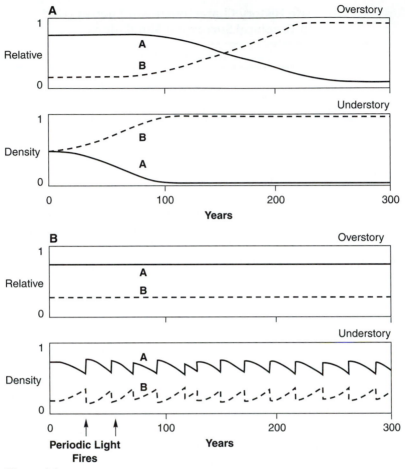

Figure 2.9
A, In the absence of disturbance, shade-tolerant species B increases first in the understory, and slowly replaces species A in the overstory as older trees die. **B,** In the presence of frequent, low-intensity fires, which thin the stands from below, both species A and B coexist on the site, as growing space is periodically opened for the less shade-tolerant species A, and smaller size classes of species B are selectively removed from the stand by fire.

to show successional trajectories rather than accurate projections. In fact, removal of one disturbance from a forest ecosystem may increase the frequency or magnitude of others. Removal of fire in western forest ecosystems, for example, has increased the scale and magnitude of insect and disease problems.

Because of its myriad characteristics, disturbance is not an easy process to characterize. The volcanic history of Mount St. Helens in Washington State suggests that depth of tephra deposit, for example, may not be an accurate predictor of event severity (Figure 2.12). The layer called Wn deposited 1 to 3 m of tephra around the bases of trees, yet many of the Douglas-firs survived and are rooted in

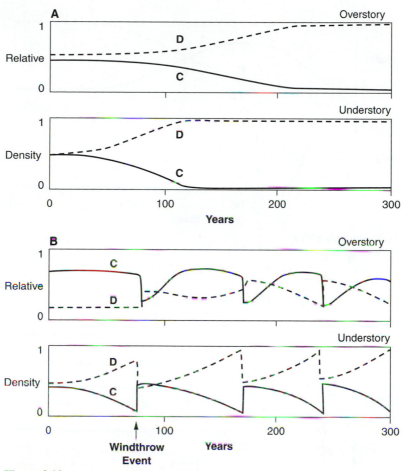

Figure 2.10
A, In the absence of disturbance, shade-tolerant species D increases first in the understory, and slowly replaces species C in the overstory as older trees die. **B,** In the presence of periodic windthrows, which thin the stands from above (e.g., take larger trees), enough understory openings exist that both species codominate the site.

pre-Wn deposits. Other trees were established after more recent events. The 1980 eruption deposited only minor amounts of tephra, but the associated heat killed all of the trees and blew over many of them (Yamaguchi 1993). Similarly, in coast redwood forests, the world's tallest trees are found on alluvial flats subject to periodic silt depositions. After each deposition event, new roots developed up from the old system and then laterally from the stem. Because coast redwood is more tolerant to such deposition than any of its associated species, it was able to exploit the nutrients and water on the site by essentially abandoning buried root systems. In heavily logged watersheds with unstable soils, however, the deposits began to include coarse gravel, into which the redwoods could not grow new roots, because

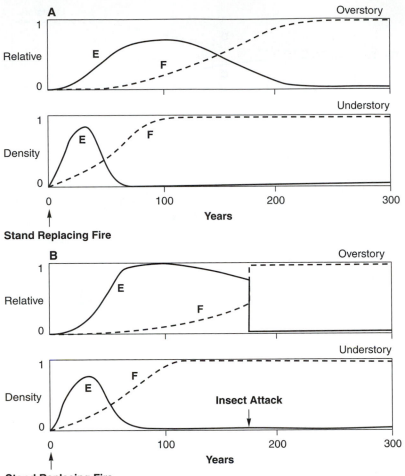

Figure 2.11

A, After a stand-replacing fire, but with no further disturbance, shade-tolerant species F slowly increases in the understory and replaces species E (a pine species) in the overstory as older trees individually die. **B,** A more realistic scenario is that rather than a slow replacement of species E, a species-specific bark beetle attack will remove the pine, and the transition to a pure stand of species F is accelerated. Another stand-replacement event might be necessary to open enough growing space for species E.

the gravels were too well drained. At the same time, the water table rose because of the gravel-choked streams, and many streamside redwoods began to die. The quality of the deposit, as well its quantity, were critical to understand the effects of periodic flooding (Agee 1980).

A common but incorrect perception is that disturbance always sets succession back. After severe events like a crown fire or a mudflow, succession does begin anew. But many disturbances are selective in nature, and in fact can remove early

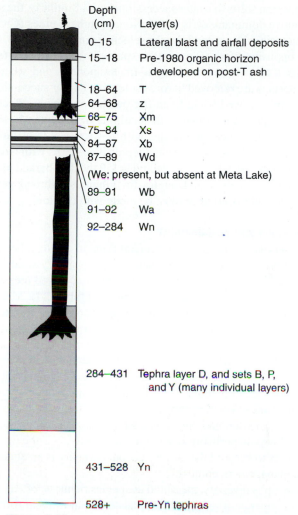

Depth (cm)	Layer(s)
0–15	Lateral blast and airfall deposits
15–18	Pre-1980 organic horizon developed on post-T ash
18–64	T
64–68	z
68–75	Xm
75–84	Xs
84–87	Xb
87–89	Wd
(We: present, but absent at Meta Lake)	
89–91	Wb
91–92	Wa
92–284	Wn
284–431	Tephra layer D, and sets B, P, and Y (many individual layers)
431–528	Yn
528+	Pre-Yn tephras

Figure 2.12

A 5 m deep soil profile showing volcanic (tephra) deposits near Mount St. Helens. Up until 1980, trees rooted in layers shown were alive. The larger tree at the bottom survived 2 to 3 m of deposition (the Wn layer) during its life, and many trees like this were alive at the time of the 1980 eruption. All trees at this site were killed by the 1980 eruption, although tephra deposition was much less in 1980.

successional species from the forest and leave later successional species (Foster 1988). In northeastern forests hurricanes often will topple the dominant trees (Chapter 7), such as earlier successional oaks and pines, and leave the shorter, more shade-tolerant species in place. Although forest structure is altered, the species composition shifts to a later successional mix. Similarly, the southern pine beetle can remove dominant, early successional southern pines (Chapter 21) and favor later successional hardwoods by the selective removal of pines. Other insects can have the opposite effect. The western spruce budworm (Chapter 20) occurs in forests with a true fir and/or Douglas-fir component, and where these later successional species are removed in forests, leaving earlier successional pine and larch, succession is moved back. Even fire can have variable effects on succession. Where lodgepole pine grows with subalpine fir (Chapter 4), intense crown fires are possible, and succession begins with few residual trees left, clearly setting succession back. However, where lodgepole pine grows with Douglas-fir, less severe fires can occur, and they will selectively kill the thin-barked lodgepole pine, favoring the later successional Douglas-fir. Obviously, broad generalizations about succession and disturbance must be made with caution.

Ecological Diversity and Sustainability

Ecological diversity has been a controversial topic in ecology, because it is hard to define, used in different ways, and usually counts a very small proportion of the true ecosystem diversity (Magurran 1988). Well-accepted measures of diversity include species, stand, and landscape indices (Table 2.4). Most are defined on the basis of vascular plants and animals, which constitute a minor proportion of the diversity of the world's ecosystems. Invertebrates and microbial species are almost always ignored, as are most functional categories: deciduous versus evergreen, carnivores versus herbivores, and so on. Biodiversity, however it is measured, is important not only for its potential uses to benefit humans directly but also because we know so little about how ecosystems function. The cancer-treating properties of the bark of Pacific yew (*Taxus brevifolia*) is a recent example of direct human benefit from what had been considered a pretty but nonessential conifer by many. Aldo Leopold probably said it most succinctly: ". . . the first rule of intelligent tinkering is to keep all the pieces . . ." but diversity is an elusive concept to tie down in a management context.

Maximum alpha diversity, measured as species richness, often occurs in early succession in temperate forests, but later successional tropical forests may have the highest species richness. Structural diversity usually is highest in late successional forest. Beta and gamma diversity usually are associated with the variation in environmental conditions in a region. A broad plain usually will have lower beta and gamma diversity than a diverse mountainous region because of the higher range of temperature, moisture, and soil conditions in the latter area.

Increasing attention to structural diversity has resulted in much more knowledge about the role of deadwood in forest ecosystems. In many areas it not only provides critical habitat for many invertebrates, small mammals, small and large carnivores and omnivores, and birds, but also may provide sites for nutrient retention, nitrogen fixation, and moisture availability during the dry season (Harmon et al. 1986). Most forest management plans now include schemes to retain at least some deadwood (snags and logs) for their important function in the forest environment. These "biological legacies" may be critical to maintain biodiversity in managed forests.

T A B L E **2.4**	**Types of Diversity Found in Forest Ecosystems**

Type of diversity	Definition
Genetic diversity	Distribution of genotypes within a population or species
Alpha diversity	Variously measured as *species richness*, or number of species in a community; *species evenness*, or distribution of relative frequencies of different species; or *structure*, the spatial arrangement of plants (clumped, uniform spacing; single- or multilayered)
Beta diversity	Local landscape diversity, measured by different communities using the same attributes as for alpha diversity
Gamma diversity	Regional diversity, measured by biome richness and evenness

Sustainability is a concept that implies maintaining something over time: timber production, ecosystem function, biodiversity, or other measures of forest ecosystem state or output (Aplet et al. 1993). The concept inherently means retaining functional services of the ecosystem, providing habitat for viable populations of the organisms native to the ecosystem, and a capacity to provide outputs for human consumption, be that recreation, water, wood, or other services. The primary focus of sustainability is toward the ecosystem; outputs are defined in the context of a sustainable set of states (over space and time) of the ecosystem: no net loss of productivity, and no loss of genetic potential (Franklin 1993). Obviously, our knowledge of ecosystems is rudimentary enough that the interpretation of sustainability is quite wide at present.

Stability and Equilibrium

Stability is the degree to which forest ecosystems can resist change or rebound from change (Perry 1994). In general, stability is thought to be a good thing, as it is subject to planning and more immune to surprise. The interaction between stability, diversity, and disturbance is complex. Higher diversity and stability are not closely linked. For example, late-succession mixed-conifer forests with ponderosa pine in the overstory and Douglas-fir and grand fir in the understory are structurally more diverse than single-storied ponderosa pine forests maintained by frequent fire. However, the single-storied stand is much more stable, as it is less likely to be catastrophically affected by wildfire or by defoliating insects such as spruce budworm. Generalizations about the relations between stability and diversity have to be tailored to site-specific conditions.

How stable can disturbed landscapes be? In part, this depends on the scale that we choose to evaluate. In boreal forests fires can be large and intense, and can burn stands of any age class. Some recent stand age models used an equilibrium negative exponential age class distribution to model the role of fire in regulating age classes in the boreal forest (Van Wagner 1978). The larger number of younger age classes is created from a range of older stands (not just the oldest) burning with stand-replacement intensity. At the stand level equilibrium does not exist; at the landscape level some boreal forests appear to have a stable age structure that fits the negative exponential distribution well. As the size of the largest expected disturbance event becomes large in proportion to the landscape, an equilibrium state of the landscape is difficult to achieve, even if climate is stable. In some drier forests fire created a cyclic stability in energy dynamics, and limited large-scale insect or disease problems. Bark beetles would kill small groups of pines (again, a small-scale event on a large landscape) and a new age class patch (sometimes only 10–20 m on a side) would appear. A quasi- (or some might say queasy!) equilibrium is implied by these dynamics, but scale is again important in the determination. In some ways finding an equilibrium condition is much like trying to focus a variable-power microscope on an unknown organism: One can find a focus but it may take some searching at variable magnification.

Each of the concepts above—diversity, sustainability, stability, and equilibrium—are fuzzy concepts, just like the concept of forest health. They nevertheless are commonly appearing as forest policy objectives on public lands and private lands. They need to be interpreted in the context of the local situation, and applied with objective criteria to be useful. Understanding the relationships between all four concepts for a particular ecosystem type requires abandoning preconceived notions about anecdotes or examples from other forest ecosystems.

ECOSYSTEM MANAGEMENT

Ecosystem management, by its very definition, is an arrogant phrase. How could anyone possibly begin to manage the complex webs that we call ecosystems? A quick visual image might resemble Charlie Chaplin's famous pie factory scene, where he is quickly overcome by having to do too much directing of the assembly of each pie. Pies end up being partially completed, then falling off the assembly line. Consider trying to "manage" the producers, consumers, and decomposers of a typical forest ecosystem, with many sizes and shapes of pies and other things moving along many conveyor belts!

What is ecosystem management? The Ecological Society of America (1996) defined it as follows:

> Ecosystem management is management driven by explicit goals, executed by policies, protocols, and practices, and made adaptable by monitoring and research based on our best understanding of the ecological interactions and processes necessary to sustain ecosystem composition, structure, and function.

The dominant themes of ecosystem management (Table 2.5) provide an excellent framework for forest health issues (Grumbine 1994). The first three are directed to understanding how ecosystems work, a critical first link in assessing forest health problems and solutions. The next three focus on the technical aspects of assessment and action. The national Forest Health Assessment Program, for example, is targeted to track forest health issues over space and time. Monitoring actions are designed to assess the effectiveness of management programs. The last four

T A B L E **2.5** | **Dominant Themes of Ecosystem Management**

Theme	Description
Hierarchical context	Use "systems" perspective, and avoid focus on only one level of organization (genes, species, landscapes, etc.). Seek connections between levels.
Ecological boundaries	Must work across administrative boundaries. Use ecological boundaries.
Ecological integrity	Protect native diversity: viable populations, species reintroductions, and ecosystem patterns/processes.
Data collection	Increase research and inventory databases. Better manage existing data.
Monitoring	Actions must be tracked for results; monitoring provides feedback loop for management.
Adaptive management	Assumes scientific knowledge is provisional and focuses on management as an experiment.
Interagency cooperation	Use of ecological boundaries implies cooperation between adjacent landowners/managers.
Organizational change	Implementation of ecosystem management may require minor to major organizational (institutional) change.
Humans embedded in nature	People are a part of the ecosystem, influencing ecological processes and affected by them.
Values	Values play an important role in defining ecosystem management goals.

Source: Adapted from Grumbine (1994).

themes of ecosystem management are institution- and people-oriented. Disturbances such as diseases, fires, and insects rarely recognize political boundaries, and traditional organizations may not be effectively organized to deal with forest health issues, either from the spatial nature of the problems or the typical fiscal-year funding approach for most management agencies. Finally, as much as we seem to ignore this, people are part of ecosystems, and influence, as well as define, the problems and solutions of forest health and all other issues.

Coarse and Fine Filter Approaches

The notion of coarse and fine filters was developed for biological diversity conservation (Hunter 1990). A **coarse filter** approach involves managing a variety of ecosystems, and maintaining natural functioning of those ecosystems, as a way of providing habitat for native species of all kinds. It assumes that by perpetuating representative arrays of ecosystems and their processes, that native biodiversity will be maintained. An example of a coarse filter approach might be a series of nature preserves, but obviously these systems cannot be representative of all lands and species in the United States. Coarse filter approaches are being adopted for industrial forests, cooperatives of smaller nonindustrial forest landowners, and others. By maintaining processes such as fire (see Chapter 6) and allowing them to play as natural a role as possible, or recognizing that fire may not be the most appropriate tool and addressing the problems caused by its absence (perhaps substituting thinning as a silvicultural tool), many biodiversity objectives will be met.

Other biodiversity objectives, however, cannot be met, and this is where the fine filter approach is needed. **Fine filter** approaches focus on individual species that are not "caught" by the coarse filter. A good forest health example might be inventorying the degree to which the exotic disease white pine blister rust has infested whitebark pine population in the northern Rocky Mountains and Cascades, and developing and possibly introducing rust-resistant strains of whitebark pine (Arno and Hoff 1989). Most forest ecosystems will be managed using a combination of coarse and fine filter approaches, and forest health strategies can be designed to complement both types of strategies.

Island Biogeography and Forest Health

In the 1960s a theory of diversity was developed for islands by MacArthur and Wilson (1967). They postulated relationships between island size and distance from the mainland to describe species richness. Large islands had more niches and could maintain larger populations of each species than small islands, where extinctions had higher probabilities. Islands near a source of immigrants (such as a continent) would have higher immigration rates of species than more remote islands. This balance between extinction and immigration was dependent on the size and spacing of islands, and was supported by a variety of empirical data. This theory has been broadly applied to forest fragments as a means of conserving biological diversity. Of course, the matrix for real oceanic islands is seawater, a rather hostile environment for terrestrial organisms.

The "islands" in terrestrial landscapes are parks, wilderness areas, old-growth, or other similarly defined areas. The matrix is a managed landscape that is almost always a much softer edge than the matrix of water at the edge of a true island. Elements of species or structural diversity can be retained on matrix lands to help various populations disperse or migrate between "islands." Planning using the island biogeography concept usually has assumed that once lines are drawn around the core areas, and if they are adequately spaced and large enough, that the major species depending on these strategies will be able to maintain viable populations.

Most planning efforts have ignored forest health issues. What happens in the South when a southern pine beetle epidemic expands through the area? In drier inland Northwest forests, what happens when large wildfires burn through these areas? In the northeastern United States, what impact will hurricanes have on old-growth structures? Three competing approaches are currently vying for adoption: (1) do not actively manage and live with the catastrophic event as a reasonable alternative to active management, which itself may not be successful; (2) actively manage core areas to prevent catastrophic loss; or (3) develop a shifting mosaic of reserves, so if one is lost, another can be "grown" in its place. Each of these strategies probably has a proper role in these conservation efforts, but none is applicable to every forest ecosystem. The remaining chapters in this book are intended to give the reader a better understanding of major forest protection and health issues so that more site-specific and intelligent decisions can be made.

REFERENCES

Agee, J. K. 1980. Issues and impacts of Redwood National Park expansion. *Environmental Management* 4:407–423.

Agee, J. K. 1993. *Fire ecology of Pacific Northwest forests*. Island Press, Washington, D.C.

Agee, J. K. 1998. The landscape ecology of western forest fire regimes. *Northwest Science* 72 (Special Issue): 24–34.

Aplet, G. H., N. Johnson, J. T. Olson, and V. A. Sample. 1993. *Defining sustainable forestry*. Island Press, Washington, D.C.

Arno, S. F., and R. Hoff. 1989. *Silvics of whitebark pine (Pinus albicaulis)*. USDA General Technical Report INT-253.

Bailey, R. G. 1978. *Descriptions of the ecoregions of the United States*. USDA Forest Service, Intermountain Region, Ogden, Utah.

Barbour, M., and D. Billings. 1988. *North American terrestrial vegetation*. Cambridge University Press, Cambridge, England.

Christensen, N. L. 1988. Succession and natural disturbance: paradigms, problems, and preservation of natural ecosystems, pp. 62–86 In J. K. Agee and D. R. Johnson (eds.). *Ecosystem management for parks and wilderness*. University of Washington Press, Seattle.

Clements, F. E. 1916. *Plant succession. An analysis of the development of vegetation*. Carnegie Institute Publication 242, Washington, D.C.

Daubenmire, R. 1968. *Plant communities*. Harper and Row, New York.

Ecological Society of America. 1996. The report of the Ecological Society of America committee on the scientific basis for ecosystem management. *Ecological Applications* 6:665–691.

Egler, F. E. 1954. Vegetation science concepts. I. Initial floristic composition—a factor in old field vegetation development. *Vegetatio* 4:412–418.

Foster, D. R. 1988. Species and stand response to catastrophic wind in central New England. *Journal of Ecology* 76:135–151.

Franklin, J. F. 1993. The fundamentals of ecosystem management with applications in the Pacific Northwest, pp. 127–144 In G. H. Aplet, N. Johnson, J. T. Olson, and V. A. Sample (eds.). *Defining sustainable forestry*. Island Press, Washington, D.C.

Franklin, J. F., and C. T. Dyrness. 1973. *Natural vegetation of Oregon and Washington*. USDA Forest Service General Technical Report PNW-8.

Gleason, H. A. 1926. The individualistic concept of the plant association. *Bulletin of the Torrey Botanical Club* 53:7–26.

Gosz, J. R., R. T. Holmes, G. E. Likens, and F. H. Bormann. 1978. The flow of energy in a forest ecosystem. *Scientific American* 238:93–101.

Greller, A. M. 1988. Deciduous forest, pp. 287–316. In M. Barbour and D. Billings (eds.). *North American terrestrial vegetation*. Cambridge University Press, Cambridge, England.

Grumbine, E. 1994. What is ecosystem management? *Conservation Biology* 8:27–38.

Harmon, M. E., J. F. Franklin, F. J. Swanson, P. Sollins, S. V. Gregory, J. D. Lattin, et al. 1986. Ecology of coarse woody debris in temperate ecosystems. *Advances in Ecological Research* 15:133–302.

Hunter, M. L., Jr. 1990. *Wildlife, forests, and forestry: principles of managing forests for biological diversity*. Regents/Prentice-Hall, Engelwood Cliffs, N.J.

Keyes, M. R., and C. C. Grier. 1981. Above- and below-ground production in 40-year-old Douglas-fir stands on low and high productivity sites. *Canadian Journal of Forest Research* 11:599–605.

Kimmins, J. P. 1997. *Forest ecology*. Macmillan, New York.

Kuchler, A. W. 1964. Potential natural vegetation of the coterminous United States. *American Geographical Society Special Publication 36*.

MacArthur, R. H., and E. O. Wilson. 1967. *The theory of island biogeography.* Princeton University Press, Princeton, N.J.

Magurran, A. E. 1988. *Ecological diversity and its measurement.* Princeton University Press, Princeton, N.J.

Noble, I. R., and R. O. Slayter. 1980. The use of vital attributes to predict successional changes in plant communities subject to recurrent disturbance. *Vegetatio* 43:5–21.

Perry, D. H. 1994. *Forest ecosystems.* Johns Hopkins Press Baltimore.

Pfister, R. D., and S. F. Arno. 1980. Classifying forest habitat types based on potential climax vegetation. *Forest Science* 26:52–70.

Pickett, S. T. A., and P. S. White. 1985. *The ecology of natural disturbance and patch dynamics.* Academic Press, New York.

Society of American Foresters. 1980. *Forest cover types of the United States.* Society of American Foresters, Bethesda, Md.

VanWagner, C. E. 1978. Age-class distribution and the forest fire cycle. *Canadian Journal of Forest Research* 8:220–227.

Whittaker, R. H. 1975. *Communities and ecosystems.* Macmillan, New York.

Yamaguchi, D. K. 1993. Forest history, Mount St. Helens. *National Geographic Research and Exploration* 9:294–325.

3

FIRE AS A PHYSICAL PROCESS

INTRODUCTION

Wildland fire is a natural process that has many forest health implications. Fires in some ecosystems were essential regulating processes that maintained forest health, whereas in other ecosystems fires created forest health problems. Later chapters will deal in more detail with the ecological context of fire and forest health. This chapter focuses on the principles of fire as a physical process. Unlike many natural disturbance processes, such as volcanoes and hurricanes, the elements of the process of fire can be managed to mitigate ecological impact. We can ignite fires under specified conditions of fuels, weather, and topography to control the energy release rate of the fire, known as fireline intensity. We can alter fuel complexes so that the probability of a surface fire moving into the crowns of the trees is significantly reduced (Chapter 6). Fire management, in this broad context, will have major effects on tree survival, growth, and vigor, all of which are important elements of forest health.

THE ELEMENTS OF FIRE BEHAVIOR

Fires need fuel, heat, and oxygen to sustain the process of **combustion** (Figure 3.1). The fire triangle includes these three "legs" of combustion. The combustion process includes four major phases: preheating, flaming combustion, charcoal combustion, and cooling (Pyne et al. 1996). In the preheating phase, heat is applied to fuel particles and raises their temperature, volatilizing gases which escape. In the presence of oxygen, these burning gases will produce flaming combustion. As most of the gases escape, charcoal combustion at the surface of the fuel particle will create glowing combustion, or smoldering. The cooling phase completes the combustion process. Most fire suppression strategies are designed to remove at least one of the three "legs" of the fire triangle. Firelines remove fuel; water and aerial fire retardants remove heat and oxygen.

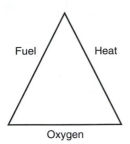

Figure 3.1
The fire triangle. The combustion process requires each leg of the triangle, and all fire suppression efforts are an attempt to remove one of the legs of the triangle.

Heat transfer occurs through four processes: **radiation, convection, conduction**, and **mass transfer** (Agee 1993). Radiation is heat transmitted through electromagnetic wave motion, and is the typical warming heat received from a fireplace. Convective heat is that transferred through currents in liquids or gases; the movement of hot air up into tree canopies is largely a convective process. Conduction is the movement of heat from one molecule to another; soil and cambial heating are modeled as conductive processes. Mass transfer is the movement of heat by active firebrands, called *spotting*. If heat transfer by one of these processes is sufficient, and adequate amounts of fuel and oxygen are present, the combustion process will continue.

The combustion process varies tremendously, both in terms of its rate of spread across the landscape, and rates at which energy is released. These characters of fire are known as fire behavior, and they are a function of the sides of the fire behavior triangle: weather, fuels, and topography. If sufficient information is available for each of the sides, the behavior of the fire can be predicted. In this section the elements of fire behavior will be summarized, and later integrated into predictive models. This summary, from a forest health viewpoint, is very brief, and the reader should refer to the reading list to become more familiar with these principles.

Weather

Fire weather is that portion of the annual climate of a region or locality that is subject to the ignition and spread of fire (Schroeder and Buck 1970). Fire weather can occur in almost any season, but severe fire weather is typically confined to one or two seasons of the year. These seasons are relatively predictable by region, due to the general circulation of the earth's atmosphere on a seasonal and annual basis. The climate of the earth is the result of complex, three-dimensional mixing of the earth's atmosphere as the earth rotates, such that weather is ever-changing in time and space. Unequal heating of the earth's surface by the sun sets into motion a net transport of heat energy from the equatorial regions of the earth to its polar regions. If the earth did not rotate, this heat movement could be represented by a simple convective model showing heat rising at the equatorial region, moving northward and southward towards the poles, cooling on the way, and returning as surface flow.

Because the earth does rotate, with uneven partitioning of heat due to unequal amounts of land area and sea, more complex patterns of airflow around the earth result (Schroeder and Buck 1970). As heated air moves from the equator north in the northern hemisphere, it descends at about 30° latitude, creating a belt of high pressure as it descends, and moves both north and south at the surface. Because the earth is rotating, this flow appears to occur at an angle, with an apparent deflection to the east as the flow moves north, and an apparent deflection to the west as flows move to the south. This deflective force is called the **Coriolis force**. These general wind patterns are known, respectively, as prevailing westerlies and trade winds. Similar winds from the polar regions south to about 60° latitude are known as polar easterlies.

This simple pattern of circulation would result in permanent high pressure at 30° latitude and at the poles, with low-pressure bands at the equator and at 60° latitude. However, some areas have concentrations of higher or lower pressure than the surrounding region, and are known as pressure cells. Much of this variation is due to relative amounts of sea and land surface; for example, during summer months, colder sea surfaces have well-developed high-pressure centers at the surface; warmer land surfaces have lower pressure. The summer-dominant Pacific High, Bermuda High, Icelandic Low, and California Low are examples of semipermanent pressure cells that migrate seasonally and have substantial effect on regional fire weather (Schroeder and Buck 1970; Figure 3.2). These cells are associated with relatively predictable seasonal weather patterns that comprise the regional climates of the northern hemisphere. Much of the regional wind flow and precipitation patterns are the result of the relative positions of high- and low-pressure systems.

Low-pressure systems are characterized by inward and rising air motion in a counterclockwise direction (in the northern hemisphere). As the air rises, it cools because of a decrease in pressure at higher altitudes. This process of temperature change with no real gain or loss of heat from or to external sources is called **adiabatic** warming or cooling, and the temperature profile with elevation is known as the temperature lapse rate. As it cools the amount of water vapor it can hold (saturation vapor pressure) decreases, while the absolute humidity stays the same. As the ratio of actual to saturation vapor pressure rises, relative humidity increases. If the cooling continues, relative humidity may reach 100%, and further cooling will cause precipitation (Figure 3.3). Low-pressure systems are, therefore, associated with high probabilities of precipitation. High-pressure systems are associated with divergent, descending air, that warms as it descends so that few clouds and little precipitation is likely. In the Pacific Northwest, onshore gradient winds in summer are the result of the Pacific High offshore, and the Bermuda High is associated with onshore winds in the southeastern United States.

Frontal winds, associated with the succession of different air masses across the land, can affect wind direction and wind speed. Warm fronts, where a warm air mass is replacing a cold one, usually have steady winds and gradual wind shifts from southeasterly to southwesterly (clockwise). Cold fronts also have clockwise shifts of wind direction, typically from southwest to northwest, and with increases in wind speed and gustiness. Cold front passages were associated with many of the large fire runs in the 1988 Yellowstone fires (Romme and Despain 1989), and were a contributing factor in the deaths of 14 firefighters in the 1994 South Canyon fire in Colorado (see Box 5.2).

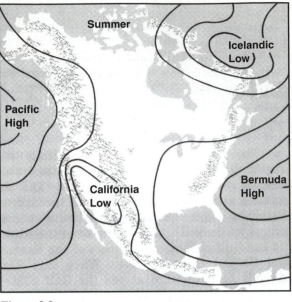

Figure 3.2

Semipermanent pressure cells may direct regional airflow seasonally and have significant influence on seasonal moisture, temperature, and windflow.

(Source: From Schroeder and Buck 1970.)

The winds from these large-scale weather systems, known generally as **gradient winds**, interact with landscape characteristics such as valley and mountain systems to create the actual wind profiles at the surface. For example, a gradient wind from the northwest may be deflected to parallel a west-facing valley by the mountains framing the watershed. In addition, local winds, caused by differential heating of land and water surfaces, will interact with the gradient winds to increase or decrease their effect at the surface. Interior mountain systems commonly have diurnal cycles of these convective local winds. After sunrise, early morning heating of high-elevation areas begins an upslope air movement as the heated air rises. During the day the winds shift to upvalley. Early evening cooling of the higher-elevation areas causes downslope winds that later become downvalley winds late at night. Standing at midslope one would experience a counterclockwise rotation of winds: upslope, upvalley, downslope, downvalley.

If the gradient wind is blowing in the same direction as the local slope and valley winds, a stronger wind will result. If the local wind is tending the opposite direction, the stronger of the two may predominate, but the wind will typically be of lower velocity than when gradient and local winds are blowing in the same direction.

One wind of particular importance to fire weather is the **foehn wind**, a gradient wind moving from high to low pressure at the same time it is decreasing in altitude (Schroeder and Buck 1970). These winds can be strong, and due to adiabatic

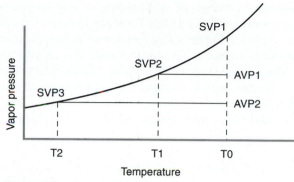

Figure 3.3
Cooling and warming can change the moisture status of an
air parcel. The ratio of the actual vapor pressure (e.g., AVP1)
to the saturation vapor pressure at that temperature (SVP1)
is known as relative humidity. An air parcel that starts with
an actual vapor pressure at AVP1 will cool along the line
AVP1-SVP2, reaching saturation at SVP2 and temperature
T1. Any further cooling, caused perhaps by the parcel being
lifted over a mountain, will move the parcel along the curve
SVP2-SVP3, and condensation will occur as the saturated
parcel continues to cool. If the parcel then warms after
reaching T2, perhaps descending the mountain, it will track
along the line SVP3-AVP2. If the parcel descends enough to
warm to the same temperature (T0) it began with, its actual
vapor pressure, and relative humidity, will be lower than it
started with.

warming usually are warm and of low relative humidity, sometimes lower than 5%.
In southern California these winds are known as Santa Anas; in the Sierra Nevada,
Mono winds; and in the Rocky Mountains, Chinooks. Foehn winds are associated
with almost all the large fires in the Pacific coastal states.

Precipitation occurs when air masses become saturated with moisture. Moisture can be added to the air mass to achieve saturation, but the more important
process is the cooling of the air mass. This usually is accomplished by lifting of
an air mass to higher altitude, by thermal lift, frontal action, or orographic (topographic) lifting. In each case decrease in pressure with altitude causes the air mass
to cool at 10°C per 1,000 m rise until saturated. This temperature lapse rate is
known as the dry adiabatic lapse rate. After saturation the air mass cools at the
wet adiabatic lapse rate of 5°C per 1,000 m, lower because of the release of energy
by the condensation of water. In western North America orographic lifting of maritime polar air produces precipitation across the Coast Range, the Cascades and
Sierra Nevada, and the Rocky Mountains. Each mountain range receives more precipitation on its western slope, and each range to the east receives progressively
less precipitation, unless the air mass is lifted to higher altitudes. The regional
occurrences of high- and low-pressure systems across the seasons create different regional precipitation patterns across the North American continent.

Fire climate may be thought of as the synthesis, over long periods, of daily fire weather (Schroeder and Buck 1970). There are 15 major fire climates across North America (Table 3.1; Figure 3.4), a function of geographical features of the continent, nearness to oceans, and continental scale atmospheric circulation discussed earlier. Mean annual temperature declines with increasing latitude. Annual precipitation is variable, with maxima in the coastal mountains of the Pacific Northwest and minima in the Great Basin, the Southwest semidesert, and the Arctic. The eastern seaboard has intermediate amounts of precipitation.

Topography

Topography has both direct and indirect effects on fire behavior. Its direct effects are primarily caused by slope gradients and the channeling of winds. Indirect effects include the influence of aspect, elevation, and orographic effects on precipitation that result in different vegetation types and fuel conditions. Slope is a direct input into fire behavior models. As slope gradient increases, upslope fuel particles are closer to the flames, the source of radiant heat, and the fuel particles preheat more rapidly. As slope steepens, fires spread more rapidly upslope because of the increased radiation effect (Rothermel 1983).

Physiography can alter the direction of wind and change its velocity. Wind may increase in narrow valleys as the wind stream is funneled through them. Along major streams incised through mountain ranges, winds will typically be strong. The Columbia River separating Washington and Oregon has a gorge that connects the coastal area with the interior Columbia Basin, and is a favorite place for windsurfers because of normally strong upvalley winds during summer caused by interior thermal low pressures. The surfers are blown upstream and can float back to their starting location by removing their sails and drifting on the river current. During foehn winds, however, the gorge acts as a funnel for downstream windflow toward the coast. These winds are associated with hot, dry air that blows wildfires out of control and blows windsurfers downstream so they must paddle back upstream at the end of their run. Mountain saddles are places of wind concentration as the path of least resistance, and first-order stream channels also may funnel winds, such that riparian areas may sometimes burn hotter than the adjacent slopes.

Fuels

All fuel is a form of stored energy from the sun. This chemical energy may be released in several ways. It may be decomposed by microorganisms, generally so slowly that the oxidation process does not create significant heat energy. Occasionally, if the decomposing biomass is sufficiently insulated, such as a tightly packed pile of fresh hay, enough heat will be retained that spontaneous combustion results. Fire is simply a rapid form of decomposition, in some cases releasing centuries of chemical energy storage in a matter of minutes.

Fuels vary widely in their ability to burn. They may have different amounts of chemical energy; for example, pitch has about 50% more energy per gram than foliage or wood, and all fuels have noncombustible mineral content that leaves the ash seen after "complete" combustion. However, for most applications, heat energy of fuels is treated as a constant, at about 18.5 kJ g^{-1} (Burgan and Rothermel 1984).

T A B L E **3.1** | **Fire Climates of North America**

Region	Vegetation	Annual ppt range (cm)	Fire season	Critical fire weather
Interior Alaska Yukon	Spruce-aspen tundra	25–40	May–Sept	Very long days in summer, maxima ppt in summer
North Pacific coast	Dense coniferous forest	50–500	June–Sept	Foehn winds in late summer, early fall
South Pacific coast	Grass in lowlands Brush at midelevation Conifers at high elevation	25–150	June–Sept to year-round	Summer drought, foehn winds in late summer to November
Great Basin	Sagebrush-grass at low elevation Conifers at higher elevation	25–100	June–Oct	Summer ppt light; hot dry days with unstable air; cold front winds
Northern Rocky Mountains	Heavy pine, fir, spruce	25–150	June–Sept	High lightning, Low humidities, foehn winds, cold front winds
Southern Rocky Mountains	Brush, pine in lowlands Fir-spruce in uplands	25–100	June–Sept	Foehn winds in spring–fall, cold front winds
Southwest	Grass, sage, chaparral pine in uplands	15–70	May–June Sept–Oct	High lightning, southwesterly airflow,
Great Plains	Grass, cultivated fields, isolated timber	25–100	Apr–Oct	Foehns from Rockies, Bermuda High push west
Central and northwest Canada	Prairies in south Conifers, hardwoods intermixed elsewhere	20–80	Spring to fall	Large geographic area— potential fire anytime
Subarctic and tundra	Scrub spruce—south Open tundra—north	25–60	Midsummer	Strong winds, low humidity
Great Lakes	Aspen, fir, spruce in south	>80	Apr–Oct	Peaks spring and fall—highs cause low humidity, cold front winds
Central states	Mixed pine-hardwoods Agricultural lands	50–120	Apr–Oct	Spring and fall, similar to Great Lakes
North Atlantic	Spruce to north Hardwoods to south	100–120	Apr–Oct	Spring and fall, similar to Great Lakes
Southern states	Pines in coastal plain Hardwoods in bottomlands Mixed conifer-hardwood in uplands	100–180	Spring–fall	Dry cold fronts, Bermuda High subsidence
Mexican central plateau	Grass, brush in lowlands Pines in uplands	25–50	Summer	Moist air from Gulf of Mexico and Pacific Cool, high elevation

Source: Summarized from Schroeder and Buck (1970).

In some ecosystems almost all the fuel is available for burning, and nothing may be left but ash. More typically, fires leave a substantial biological legacy behind: residual amounts of forest floor or logs or snags, and perhaps live trees, too (Table 3.2). A "blackened" landscape is itself evidence of incomplete combustion.

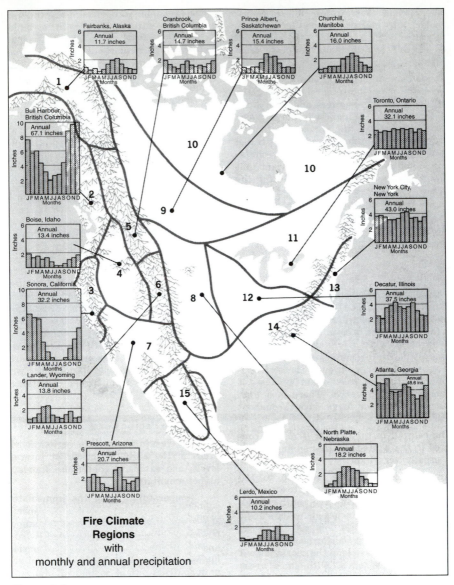

Figure 3.4

Fire climates of North America. Refer to Table 3.1 for a description of the types.
(Source: From Schroeder and Buck 1970).

The amount of fuel consumption depends on the rate that a fuel particle can gain or lose heat and moisture, a function of the surface area to volume ratio of the particle, and the moisture content of the particle as the fire arrives. Fire behavior models use inputs of **fuel load** (the total amount of fuel) by size class (an index to surface area/volume ratio), and **fuel moisture** to determine available fuel for the fire (Albini 1976). Live and dead fuels are considered separately by the models.

T A B L E **3.2** | **Fuel Consumption by Fire in Forest and Shrub Ecosystems[a]**

Fuel category	Before fire kg ha[-1]	Percent consumed
Old-Growth Douglas-Fir Forest, Pacific Northwest[b]		
Stems	645	5
Branch	53	10
Foliage	12	100
Understory plants	7	75
Snags and logs	215	30
Forest floor	51	80
TOTAL	983	16
Chaparral, Southern California[c]		
Live shrub <0.64 cm	9.9	95
Live shrub >0.64 cm	10.2	10
Dead	10.3	95
Litter	9.5	45
TOTAL	39.9	39

[a]Total fuel loading may be a poor index to available fuel. About 15% of the total above-ground biomass is consumed in the Douglas-fir ecosystem, whereas at least three times that proportion is consumed in the chaparral ecosystem.
[b]Source: From Fahnestock and Agee (1983)
[c]Source: From DeBano and Conrad (1978). Other reports from chaparral show more complete combustion ranging up to 80–90% consumption.

Small fuel particles have high **surface area/volume ratios**, and therefore have higher rates of heat and moisture exchange with the atmosphere than large fuel particles. This is why campfires are started with small twigs, and sustained by larger pieces that burn for longer periods of time. All dead wood, regardless of its size, is hygroscopic, gaining and losing moisture to the atmosphere as the relative humidity of the atmosphere changes. Moisture content is defined on the basis of dry weight, so that a fuel particle having 50 g of water for each 100 g dry weight is said to have a moisture content of 50%. For a given temperature and relative humidity, there is a specific **equilibrium moisture content** toward which all fuel particles will move, gaining or losing moisture along the way. The rate of change, however, will differ substantially, depending on the surface area/volume ratio, or size class, of the fuel. For fire behavior purposes this rate of change is modeled as an exponential function, so that in a given time period the moisture content moves 63% of the way toward the new equilibrium moisture content (Schroeder and Buck 1970). Different-size classes of fuel will have different time periods over which this change occurs (Table 3.3). The smallest fuel particles experience the 63% change in 1 hour; logs will have timelags exceeding 1,000 hours. If there were fuel pieces of each timelag class in a controlled environment chamber, and all had stabilized at an equilibrium moisture content of 20% (about 26°C at 90% relative humidity), and then equilibrium moisture content was changed to 10% by reducing relative humidity to 55%, the 1-hour timelag fuels would be at 14% in 1 hour, 11% in the second hour, 10.3% in the third hour, and so on. The larger fuels would experience the same moisture content changes in 10-, 100-, and 1000-hour increments of time. In the first hour only the 1-hour timelag fuels would experience a significant shift in moisture content, because they have the highest surface area/volume ratios.

Over a typical day temperature and relative humidity will shift significantly, and 1-hour timelag fuel moistures will closely track these changes. From about 2 to 6 P.M., these fine fuels will be at the lowest fuel moisture of the day, and fire behavior usually is most severe during this time. Usually, humidity recovery at night will increase fine fuel moisture and dampen fire behavior. The large fuels will change their moisture status more slowly, and 1,000-hour timelag moistures often are an index to seasonal trends rather than daily weather.

Dead fuel moisture is a function of atmospheric potential; live fuel moisture is a function of soil potential, too, as recharge from soil moisture and water stored in the stem of the tree is possible. Live fuel is less flammable than dead fuel, as the moisture content live fuel rarely dips below 70%, whereas dead fuel reaches saturation at about 30% and usually is much lower. Much more energy needs to be expended to vaporize and preheat fuel particles with high fuel moisture. New foliage in conifers often begins at 200 to 300% moisture content, dropping to 100 to 120% by the end of the dry season. Older evergreen coniferous foliage and small twigs tend to have smaller rates of changes, fluctuating around 100% (Agee 1993). Deciduous trees, with all new foliage each year, maintain higher crown fuel moisture than evergreen trees, so they are less flammable. As live fuel moisture decreases during the dry season, the probability of crown fire increases. If lichens are important canopy dwellers, they provide a very flashy crown fuel, as they often have very high surface area/volume ratios and may have moisture contents below 10%.

Fuel arrangement (porosity and depth) is another important variable in predicting fire behavior. It is associated with how aerated the fuelbed is, and is measured by packing ratio, the proportion of the fuelbed volume occupied by fuel particles. Packing ratio can be either too high (not well aerated) or too low (not enough fuel continuity) (Burgan and Rothermel 1984). For example, a rolled-up newspaper has a very high packing ratio, and does not burn easily. The same newspaper, with pages separated and spread over the volume of a room, also may not burn. But if the pages of the paper are crumbled and set in a pile, they will burn rapidly, representing an optimum packing ratio.

Fire Behavior Prediction

Surface Fire Behavior

Models of fire behavior attempt to integrate the effects of fuels, weather, and topography on the spread and intensity of open-burning fires (Rothermel 1972, Albini 1976). Fire spread can be measured in several types of units: m sec^{-1}, m min^{-1}, ft min^{-1}, m hr^{-1}, and chains hr^{-1}. It usually refers to the rate of spread in the

TABLE **3.3** | Interpretations of Fireline Intensity

Type of Fire	Flame length (m)	Fireline intensity		Interpretation
		SI units kW m⁻¹	English units BTU sec⁻¹ ft⁻¹	

Type of Fire	Flame length (m)	SI units kW m⁻¹	English units BTU sec⁻¹ ft⁻¹	Interpretation
Ground fire	0	0	0	Smoldering fire, may be in peat or forest floor, no definable fireline intensity.
Surface fire	<1	<258	<75	Most prescribed fires in this range. Direct control by hand possible. Monitoring at head of fire possible.
Understory fire	1–3	258–2,800	75–812	Control with bulldozers may be successful. Fires too intense for direct attack. Monitoring usually remote.
Crown fire	>3	>2,800	>812	Erratic fire behavior expected. Fire may torch into crown.

downwind direction, but fires also are spreading in flanking directions (at right angles to the wind) and as backing fires (against the wind). **Rate of spread** needs to be scaled down to estimate spread in these directions. Spread over short periods of time usually is modeled as an ellipse (Figure 3.5), although fires that spread over long periods of time rarely have this shape because of changes in windspeed and direction (Pyne et al. 1996).

Intensity of fires usually is expressed in the unusual units of heat release per unit time per unit length of fireline. Fires are modeled as line phenomena, so that intensity represents the rate of heat release along the leading edge of the fire. Common units in the management literature are English: BTU sec⁻¹ ft⁻¹; the ecological literature uses SI units: kW m⁻¹. In the United States both systems are likely to remain in use for the foreseeable future. **Fire intensity** can be expressed as a function of flame length, and a shortcut estimate in SI units (Chandler et al. 1983) is

$$\text{Fireline intensity (kW m}^{-1}) = 3 \,(10 \times \text{Flame length [m]})^2$$

As fireline intensity increases, fire behavior becomes more erratic and unpredictable (Table 3.3). These ranges of fireline intensity are defined in very general terms, and there are broad transitions between each class. **Ground fires** may be among the most difficult to control, as they often burn in well-insulated materials, and water applied may evaporate before extinguishing combustion. Peat bogs, ancient sloth deposits in caves, and thick forest floors are typical ground fire fuels. Occasionally, fires may spread slowly from log-to-log combustion along the ground over periods of weeks, and these also are considered ground fires.

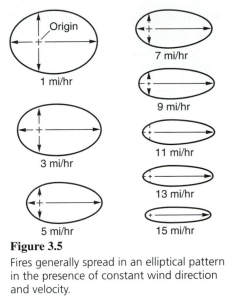

Figure 3.5

Fires generally spread in an elliptical pattern in the presence of constant wind direction and velocity.

(Source: From Rothermel 1983.)

Surface fires and **understory fires** (Figure 3.6) are those for which our modeling capability is most robust. Most prescribed fires used for forest health purposes will be in this range. As fireline intensity increases to flame lengths over 3 m, unusual fire behavior may result, and spotting (the transfer of firebrands ahead of the fire front) and crown fire behavior may result. There is relatively limited capability at present to model the behavior of these fires.

Surface fire behavior is modeled using an empirical fire spread model developed in the 1970s and available now in PC versions (Rothermel 1972, Burgan and Rothermel 1984). The **BEHAVE** model requires five types of inputs (Figure 3.7). The **fuel model** may be one of 13 stylized fuel models developed for North American vegetation (Table 3.4), or one developed from site-specific fuels information. In most cases the stylized models do a reasonable job of simulating typical fuel complexes to be expected. These models include information on fuel loading by size class, surface area/volume ratio by size class, energy content, packing ratio, and fuel depth. Fuel sizes above 100-hour timelags (>7.6 cm) are not considered by these models, as these fuels do not usually contribute to the leading edge of surface fires, although the consumption of the larger fuels may be important in estimating ecological effects. The user must specify the moisture content of dead fuels, and if important for the fuel model, moisture content of live fuel. Slope and windspeed also are required. Windspeed is estimated at midflame height, not the usual measured 20 ft (~6 m) windspeed, so it is scaled downward 40 to 90% to account for effects of surface roughness on windspeed, and also may be scaled down due to canopy cover. These five types of inputs contain enough information on fuels, weather, and topography that their integrated effect on fire behavior can be predicted.

Figure 3.6
Low surface fireline intensity in this forest is maintained by prescribing the fire with high fuel moisture, so that available fuel is limited.

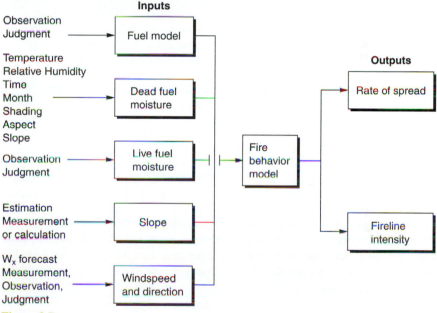

Figure 3.7
Inputs to the fire behavior model are integrated to produce fire behavior characteristics. Outputs other than fireline intensity and rate of spread are available but not shown here.

The model outputs will vary whenever any of the input values change. Fireline intensity and rate of spread for three sets of fuels and weather, with topography held constant (Table 3.5), exhibit a tremendous range. For these fuel types the outputs would change considerably if the inputs for weather (wind and fuel moisture) or slope were changed. However, they show some general trends. In light fuels, such as the Model 1 annual grass example, rate of spread can be high because of fine fuels, but fireline intensity is limited by light total fuel loads. In

TABLE **3.4** | **Fuel Models for Fire Behavior**

Model number	Description
Grass and Grass Dominated Complexes (Cured Fuels Only)	
1	Short grass (.3 m). Western grassland, not grazed.
2	Timber (grass and understory). Open pine grassy understory, wiregrass/scrub oak.
3	Tall grass (.6 m). Bluebunch wheatgrass, bluestems, broomsedge, panicgrass.
Chaparral and Shrub Fields	
4	Chaparral (2 m). Mature (10–15 yr) chaparral, manzanita, chamise.
5	Brush (.6 m). Salal, laurel, vine maple, alder, mountain mahogany, young chaparral.
6	Dormant brush, hardwood slash. Alaska spruce taiga, shrub tundra, low pocosins.
7	Southern rough. Southern rough (2 yr), palmetto-gallberry.
Timber Litter	
8	Closed timber litter. Tightly compacted short-needled conifer litter. Not much branch/log fuel.
9	Hardwood litter. Loosely compacted long-needled pine or hardwood litter.
10	Timber (litter and understory). Larger dead fuel as well as litter. Some green understory. Old-growth hemlock-fir.
Logging Slash (Western Mixed Conifers)	
11	Light logging slash (< 90 t/ha [Mg/ha]). Most needles have fallen, compact. Partial cuts and clearcuts.
12	Medium logging slash (90–270 t/ha). Most needles have fallen, slash somewhat compact.
13	Heavy logging slash (>270 t/ha). Most needles have fallen, slash somewhat compact.

Source: From Albini (1976).

brush fuels, such as the Model 4 chaparral example, rate of spread will generally be less than for grass fuels, but fireline intensity can be much higher; most of these fuels are small diameter, with higher total loading than the annual grass model. In the Model 10 old-growth Douglas-fir example, rate of spread is lower than the more flashy fuel types, but surface fireline intensity can be high, and may transition into crown fire behavior if threshold conditions are met or exceeded.

If a fuel treatment is proposed for an area, the fire behavior resulting from pre- and posttreatment conditions can be simulated. Fuel buildup over time can be monitored so that subsequent treatments can be scheduled on the basis of reaching threshold levels of fire behavior under simulated severe fire weather conditions. Although the outputs from BEHAVE need careful interpretation, they provide considerable insight for evaluating alternative treatments of a particular fuel complex.

These general characteristics of fuel types can be used to understand the fire behavior differences between the fuel types in a region (Williams and Rothermel 1992). The northern Rocky Mountains forest types, for example, have different seasonal fire behavior characteristics (Figure 3.8). These differences will influence wildland fire control efforts, and provide general guidance for the application of prescribed fire programs. The ponderosa pine/grass types will carry wildfires earlier in the season than the other types, because they are drier, lower-elevation

| T A B L E **3.5** | **Fire Behavior Outputs Using BEHAVE for Three Fuel Types and Specified Weather and Topography**[a] |

Fuel type and output variable	Midflame windspeed (m hr⁻¹ [km hr⁻¹])		
	2 (3.2)	6 (9.6)	10 (16.1)
Fuel Model 1 (Annual Grass)			
Flame length (ft [m])	3 (0.9)	6 (1.8)	8 (2.4)
Fireline Intensity (BTU sec⁻¹ ft⁻¹ [kW m⁻¹])	59 (17)	270 (78)	504 (146)
Rate of Spread (ft min⁻¹ [m min⁻¹])	39 (11.8)	175 (53)	327 (99)
Fuel Model 4 (Chaparral)			
Flame length (ft [m])	15 (4.5)	25 (7.6)	33 (10)
Fireline Intensity (BTU sec⁻¹ ft⁻¹ [kW m⁻¹])	1907 (553)	6009 (1742)	11542 (3347)
Rate of Spread (ft min⁻¹ [m min⁻¹])	42 (12.7)	132 (40)	254 (77)
Fuel Model 10 (Timber with Understory)			
Flame length (ft [m])	4 (1.2)	6 (1.8)	9 (2.7)
Fireline Intensity (BTU sec⁻¹ ft⁻¹ [kW m⁻¹])	113 (33)	330 (96)	626 (181)
Rate of Spread (ft min⁻¹ [m min⁻¹])	5 (1.5)	15 (4.5)	28 (8.5)

[a]Slope is held constant at 30%, and fuel moisture is specified at 1-hr timelag, 5%; 10-hr timelag, 7%; 100-hr timelag, 10%, live woody, 90%. As BEHAVE produces English units, these are shown first, with SI units in parentheses.

types with well-aerated fuelbeds (Figure 3.8A). Large "windows" for application of controllable prescribed fire in this type exist in both spring (Figure 3.8B) and also in fall (not shown). The high-elevation subalpine fir type, with short growing seasons, has compact fuels and is moist well into the summer in most years, so that wildfire behavior is not significant until late season, when it can become extreme (Figure 3.8C). If prescribed fire under controlled behavior is desired, the window of opportunity is very narrow, because these types move from nearly incombustible to crown fire behavior over a short period of time. The Douglas-fir types exhibit fire behavior characteristics intermediate between the other two forest types.

Crown fire behavior

Crown fire potential is dependent on conditions that initiate crown fire and those that sustain it. If either of these conditions is present in a stand, crown fires may occur (Figure 3.9). A fire moving through a forest stand may spread as a surface fire, a crown fire, or an intermediate mix of both (Van Wagner 1977). Crown fire initiation is a function of crown base height and heat of ignition (largely dependent on crown fuel moisture content):

$$I_0 = (Czh)^{3/2}$$

where

I_0 = critical surface fire intensity
C = empirical constant of 0.010
z = crown base height (m)
h = ignition energy (kJ kg⁻¹), varies primarily by foliar moisture

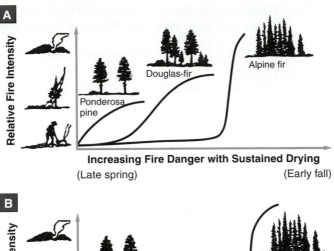

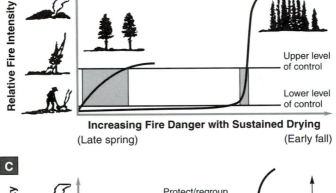

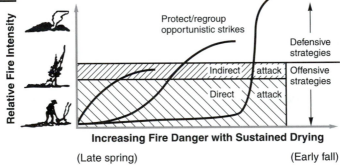

Figure 3.8

In the Rocky Mountains, different fuel types occur from low to high
elevation, presenting unique patterns of moisture and temperature
(Williams and Rothermel 1992). **A,** Fire behavior will vary considerably
over a period of increasing fire danger for grass, brush, and timber-
dominated fuel types. **B,** Prescribed fire opportunities will be extended
in the grass types compared to timber-dominated types. Another firing
window for the grass types occurs in fall (not shown). **C,** Tactical opportunities for fire suppression differ by type. Direct attack with hand crews
will be more difficult early in the season in the grass types, but the need
for purely defensive strategies is generally less when compared to the
timber-dominated types, where crown fires may occur late in the season.

Figure 3.9
Crown fires release large amounts of energy and may develop their own weather patterns when the power of the fire exceeds the power of the wind.

Various combinations of crown moisture content and crown base height (Figure 3.10) produce a range of fireline intensities over which crown fire initiation is possible. It is important to note that levels of fireline intensity below these critical levels may still result in stand-replacement fires, where the entire stand is killed. Within the ranges of crown fuel moisture often encountered, crown fire potential appears more sensitive to height to crown base. When actual fireline intensity (I) exceeds I_0, crown fire behavior is likely.

The crown of a forest is similar to any other porous medium in its ability to carry fire and the conditions under which fire will or will not spread (Van Wagner 1977). Crown fuels that will interact in the crown fire process include needles, lichens, and fine branches, and are dependent on sufficient heat energy to maintain the crown fire process. Limiting factors to crown fire spread can be developed by considering the net horizontal heat flux (E) into the fire:

$$E = Rdh$$

where

E = net horizontal heat flux, kW m^{-2}
R = rate of spread m sec^{-1}
d = crown bulk density kg m^{-3}
h = heat of ignition kJ kg^{-1}

Crown fire spread will continue above a certain minimum mass flow rate, defined as the product of rate of spread (R) and crown bulk density (d):

$$S = Rd = E/h$$

If either the mass of fuel entering the flame or the rate of spread is sufficiently high, then the mass flow rate (S) may be above the minimum (S_0) mass flow rate necessary to sustain crown fire, estimated at $S_0 = 0.05$. Therefore, even if weather is favorable for crown fire spread, crown fire potential can be limited by keeping $S < 0.05$. Crown fuel density is the variable most easily controlled by the manager, and we will return to criteria for such management later (Chapter 6).

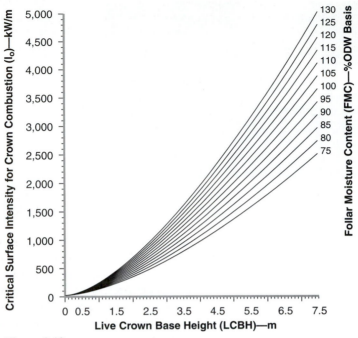

Figure 3.10
The relation between critical surface fireline intensity, crown foliar moisture, and height to the base of the live crown. For a live crown height of 5 m and crown moisture of 100%, a flame length of 2.5 m (roughly 1,900 kW m⁻¹) would be necessary to begin the crown combustion process.

(Source: From Rothermel 1983.)

Crown fires have been defined in three classes: *passive*, where $I > I_0$ but $S <$ 0.05; *active*, where $I > I_0$ and $S > S_0$; and *independent*, where active conditions occur but the crown produces enough heat to sustain crown fire activity without energy input from the surface fire. Passive crown fires will exhibit torching but not sustained runs of fire through the crown. Active crown fires are typical of running crown fires, and may have fireline intensities exceeding 50,000 kW m⁻¹. Independent crown fires are uncommon.

Spotting
Spotting is a form of mass transfer, when firebrands are lifted by convective updraft and carried over the main perimeter of the fire (Pyne et al. 1996). Spotting is most commonly associated with pulses of convective heat from individual tree torches, debris piles, or crown fires. Spotting distances can be quite long. In the 1991 Oakland fire, spots carried easily over eight-lane freeways. In very intense fires with tremendous convective uplift in Washington and Australia, spot fires have been located up to 25 km from the main fire front. Small cones or cone

scales, pieces of bark, grass clumps, and other such fuels are common firebrands carried over the fire perimeter. Spotfires can significantly increase the apparent rate of spread of a fire by igniting fuels ahead of the main fire front. Many firefighter injuries and deaths have been attributed to spot fires igniting fuels that trap the firefighters in unexpected situations.

Prediction of spotting behavior is still very imprecise (Albini 1979). Models have been produced to predict the maximum distance burning embers would travel over flat and undulating terrain, depending on tree height, particle size, burnout rate, and time or distance traveled. Smaller particles may be lofted higher but will burn out more quickly.

Geographic Information Systems and Fire

The spatial and temporal simulation of the spread and behavior of forest fires under varying conditions of fuels, weather, and topography is available through a PC-based computer model called **FARSITE** (Finney 1998). The PC version requires the support of a geographic information system (GIS) to generate, manage, and provide spatial data layers containing information on fuels, weather, and topography (Figure 3.11). Eight cell-based layers are required: elevation, slope, aspect, fuel model, canopy cover, height of base to live crown, crown bulk density, and tree height. These are combined into a "landscape" file. A weather "stream" also is required: inputs of fuel moisture, hourly windspeed and direction, and daily precipitation and temperature. Ignition location(s) must be specified, and the program will then spread surface fire, simulate spotting, and spread fires as crown fires if the minimal conditions for crown fire are met. Graphical and tabular output and cell-based (raster) output in GIS ASCII format, is possible for fire arrival times, fireline intensity, flame length, rate of spread, and heat/unit are in metric or English units.

The FARSITE model was initially developed to support decision making for park and wilderness managers. Based on expected or historic weather patterns, fire spread and intensity can be simulated over time, providing information for monitoring strategies, fire suppression strategies, assigning priorities to multiple fires, and other uses. It can be used to simulate either prescribed fires or wildfires, and has important forest health management implications for planning. FARSITE appears to hold much promise for the future. In 1998 version 3.1 was produced and the output appears to simulate actual fire perimeters quite well.

Fire Danger Rating

Fire danger rating is a broad-scale method of predicting fire potential. It is a system that allows relative rating of ignition, spread, and intensity of initiating fires through and between seasons. Early systems used simple indicators such as relative humidity, but these became more complex and parochial over time, until more standardized, national systems were developed in the 1970s (Deeming et al. 1977). The U.S. National Fire-Danger Rating System (NFDRS) has three major indexes related to fire potential: the Spread Component (SC), the Energy Release Component (ERC), and the Burning Index (BI) (Figure 3.12). In the U.S. system,

Elevation
Slope
Aspect
Fuel model
Canopy cover
Canopy height
Crown base height
Crown bulk density

Figure 3.11
The geographic information system layers required for FARSITE
(Source: From Finney 1998).

predictions are worst-case, because measurements are taken at midday at midslope positions on southerly or westerly aspects where possible. The weather observations and fuel model information are integrated into the three major indexes. The values may suggest increased alert status for fire suppression crews, or may be used as guides for when prescription "windows" for prescribed fires are approaching. Burning Index is scaled directly to predicted flame length (feet) times 10, so a BI of 30 is equivalent to a 3 ft flame length.

The Canadian Fire Weather Index System (FWI) is similar in concept to the NFDRS (Both use daily weather observations such as temperature, wind, relative humidity, and precipitation, but these are combined somewhat differently into the indexes [Pyne et al. 1996]). The Initial Spread Index (ISI) is similar to SC, the Buildup Index is similar to the ERC, and the Fire Weather Index (FWI) is similar to the BI. However, the FWI predicts fire behavior under closed stands rather than open, "worst-case conditions," and is formulated for a generalized pine forest rather than the 20 fuel models available for NFDRS. As jack pine and lodgepole pine are commonly found across Canada, this single fuel model approach probably works well for those conditions.

Fire behavior depends on the interaction of fuels, weather, and topography. In a management context, topography is relatively fixed. Weather is fickle. Managers can prescribe conditions under which prescribed fires burn, but are powerless to control conditions under which wildfires burn. Fuel is the only leg of the fire behavior triangle that is controllable, and both potential surface and crown fire behavior can be influenced in positive or negative ways by fuel treatments. The

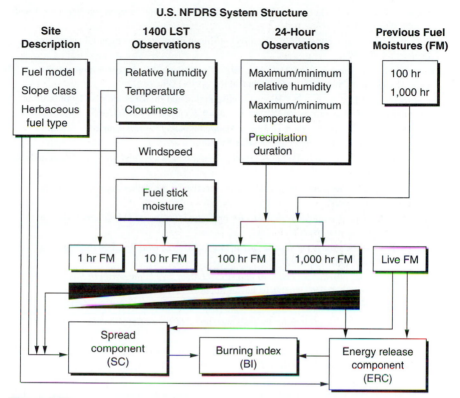

U.S. NFDRS System Structure

Figure 3.12
The National Fire-Danger Rating System shows site conditions, weather observations, their incorporation into intermediate indexes, and final indexes. The black wedges show the relative weighting of the different timelag class fuel moistures into the indexes.
(Source: From *Introduction to Wildland Fire*, by Pyne, Andrews and Laven, copyright 1996, John Wiley & Sons. Reprinted with permission of John Wiley & Sons, Inc.)

forest health issues that are fire-related largely revolve around fire behavior, either because of the direct effects of the fire on the ecosystem, or indirectly through fire's synergistic relation with insects and disease.

REFERENCES

Agee, J. K. 1993. *Fire ecology of Pacific Northwest forests*. Island Press, Washington, D.C.

Albini, F. A. 1976. *Estimating wildfire behavior and effects*. USDA Forest Service General Technical Report INT-30.

Albini, F. A. 1979. *Spot fire distance from burning trees—a predictive model*. USDA Forest Service General Technical Report INT -56.

Burgan, R. E., and R. C. Rothermel. 1984. *BEHAVE: fire behavior prediction and fuel modeling subsystem: FUEL subsystem*. USDA Forest Service General Technical Report INT-167.

Chandler, C., P. Cheney, P. Thomas, L. Trabaud, and D. Williams. 1983. *Fire in forestry*, vols. 1 and 2. John Wiley and Sons, New York.

DeBano, L. F., and C. E. Conrad. 1978. The effect of fire on nutrients in a chaparral ecosystem. *Ecology* 59:489–497.

Deeming, J. E., R. E. Burgan, and J. D. Cohen. 1977. *The national fire-danger rating system—1978*. USDA Forest Service General Technical Report INT-39.

Fahnestock, G., and J. K. Agee. 1983. Biomass consumption and smoke production by prehistoric and modern forest fires in western Washington. *Journal of Forestry* 81:653–657.

Finney, M. A. 1998. *FARSITE: Fire Area Spread Simulator, model development and evaluation*. USDA Forest Service Research Paper RMRS-RP-4.

Pyne, S. J., P. L. Andrews, and R. D. Laven. 1996. *Introduction to wildland fire*. John Wiley and Sons, New York.

Romme, W. R., and D. Despain. 1989. Historical perspective on the Yellowstone fires of 1988. *Bioscience* 39:695–699.

Rothermel, R. C. 1972. *A mathematical model for predicting fire spread in wildland fuels*. USDA Forest Service Research Paper INT-115.

Rothermel, R. C. 1983. *How to predict the spread and intensity of wildfires*. USDA Forest Service General Technical Report INT-93.

Rothermel, R. C. 1991. *Predicting behavior and size of crown fires in the northern Rocky Mountains*. USDA Forest Service General Technical Report INT-438.

Schroeder, M., and C. C. Buck. 1970. *Fire weather*. USDA Agriculture Handbook 360.

Van Wagner, C. E. 1977. Conditions for the start and spread of crown fire. *Canadian Journal of Forest Research* 7:23–34.

Williams, J. T., and R. C. Rothermel. 1992. *Fire dynamics in northern Rocky Mountain stand types*. USDA Forest Service Research Note INT-405.

4

FIRE ECOLOGY AND FIRE REGIMES

INTRODUCTION

Ecological effects of forest fires are a second-order effect of the presence of disturbance: First, the behavior of the fire must be predicted (Chapter 3), and from that behavior certain types of fire effects can be predicted (Figure 4.1). From the behavior of multiple fires, which may be quite variable over time, ecological effects on a community or ecosystem will result but be predictable at a much lower level of resolution. Even though the level of resolution decreases, the effects of multiple fires at the ecosystem level, collectively described as a fire regime, have been an important coarse filter (see Chapter 2) in the evolution of those systems.

FIRE ADAPTATIONS OF PLANTS

Many vegetation types in North America and the world have experienced fire on a regular basis: The earth is a fire environment (Pyne 1995). The massive fires of 1988 in and around Yellowstone National Park were simply another event in the region that had been equaled or eclipsed in this century and in previous centuries. The "worst" fire year in memory in Oregon (1996), including range and forest fires, was less than half of the historic average annual burned area of previous centuries of the forested part of the state alone (Agee 1990). This story is repeated almost everywhere, and the relatively smoke-free lens through which we view the present is indeed myopic.

Plant species have evolved several **adaptations** to the presence of fire (Figures 4.2, 4.3, 4.4, and 4.5). Some are able to survive the typical fire that burned; others are killed but have unique reproductive strategies stimulated by fire (Table 4.1). These adaptations are commonly to a particular type of fire, not just

71

State of the Art

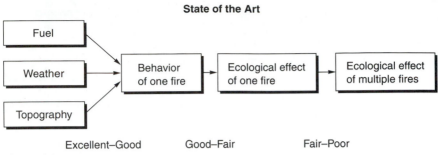

Fuel

| Weather |

| Topography |

→ | Behavior of one fire | → | Ecological effect of one fire | → | Ecological effect of multiple fires |

Excellent–Good   Good–Fair   Fair–Poor

Figure 4.1
Prediction of fire effects on an ecosystem may be difficult. Variability in fire behavior prediction, synergistic effects such as insect attack, and variability in fire characteristics each introduce complex effects not fully capable of prediction.

Figure 4.2
Fire adaptations: Thick bark protected the large ponderosa pines in the background as the smaller, thinner-barked firs in the foreground were killed by a prescribed fire.

Figure 4.3
Fire adaptations: Some plants killed by fire have seeds with hard coats that sit in the soil waiting to be cracked by the heat of a passing fire. This ceanothus plant germinated at the edge of a pile burn; when mature, its roots will harbor symbiotic bacteria that will fix atmospheric nitrogen into the ecosystem.

one line long

Figure 4.4
Fire adaptations: Serotinous, or late-opening, cones are a seed-bank strategy in the crown of certain tree species. These lodgepole pine cones have opened after passage of a crown fire through the stand.

Figure 4.5
Fire adaptations: Sprouting is a common adaptation to fire. Root suckering by aspen in Alaska is expanding the former clone (the dead, white stems in background).
(Photo courtesy of L. B. Brubaker.)

its presence (Whelan 1995). For example, the **serotinous cone** trait usually is found in ecosystems where fire is infrequent but of high intensity, while the thick bark trait is most common where fires are frequent but of low intensity.

Another way to classify fire adaptations is by life-history character of the species (Rowe 1983). Five categories are recognized. **Avoiders** are those species with no adaptation to fire, and are typically late-successional, shade-tolerant species. **Endurers** may be top-killed by fire but vegetatively reproduce, so the plant survives the fire but must regrow its crown (sprouters of Table 4.1). **Evaders** are killed by fire but have seed bank traits that allow the species to persist on the site (seed bank strategies of Table 4.1). **Invaders** are highly dispersive pioneer species that often have rapid growth (other traits, Table 4.1). **Resisters** are species that can survive fire in relatively intact form (thick bark of Table 4.1). Intense fires tend to favor *endurers, evaders,* and *invaders;* low-intensity fires favor *resisters,* and fire exclusion favors *avoiders.*

T A B L E **4.1** | **Fire Adaptations of Vegetation**

Trait	Function	Examples of species
Seed Bank Strategies		
Soil	Hard-coated seeds remain dormant in soil for many decades until fire cracks seed coat.	Ceanothus, manzanitas
Tree crown	Late-opening, or serotinous, cones remain closed in crowns. Fires that kill the tree melt the resin covering the cone scales and allow seed to fall on fresh ash.	Pitch pine, sand pine, jack pine, lodgepole pine, black spruce
Thick Bark	Protects cambium against heat from surface fires.	Ponderosa pine, coast redwood, giant sequoia, longleaf pine
Sprouting		
Epicormic	Regrowth from dormant buds protected by bark on branches and bole.	Coast redwood, pitch pine, eucalyptus, oaks
Basal sprouting	Resprouting from roots, basal buds, rhizomes, or lignotubers.	Oaks, aspen, many shrubs
Other Traits		
Fire-stimulated flowering	Increased reproductive effort in immediate post-fire years.	Many grasses and forbs
Rapid growth	Raises apical meristem above typical fire scorch height.	Longleaf pine
Wind-borne seeds	Allows colonization of interiors of large burns.	Fireweed, aspen

Source: From Kauffman (1990).

PREDICTING TREE MORTALITY FROM FIRE

When a crown fire burns across a stand, most conifers are killed and many hardwood species will either be top-killed or have to resprout a new crown. Fires of lower intensity will have selective effects, different by tree species and size classes. A conifer has three primary ways it may be injured by fire: **crown scorch, cambial heating** through bark, and root heating (Figure 4.6). Each type of damage is predicated on the general rule that tissue death occurs at 60°C for 1-minute duration, although higher temperatures require shorter times, and lower temperatures also will cause tissue death if extended longer. Predictive models of conifer mortality have been developed that use these principles to estimate tree mortality from known fire behavior (Peterson and Ryan 1986, Ryan and Reinhardt 1988).

Crown damage is caused by hot gases rising from the flaming zone and encountering leaves before the gases have mixed sufficiently with ambient air to cool (Van Wagner 1973). There is a close relationship between fireline intensity (I, kw m^{-1}) and scorch height (h_s, m):

$$h_s = 0.148\, I^{2/3}$$

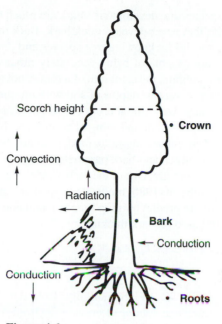

Scorch height

Convection

Radiation

Conduction

• **Crown**

• **Bark**

← Conduction

• **Roots**

Figure 4.6
Models for tree mortality assume a
temperature of 600°C that is applied to
the crown primarily through convection,
to the stem primarily through conduction,
and to roots primarily through conduction.

This relationship may be adjusted for air temperatures other than the assumed
25°C, no-wind condition. High temperatures at the time of the fire require an
upward adjustment, and higher windspeeds (for the same *I*) require a downward
adjustment because of higher mixing rates with cooler incoming air. For a given
scorch height, trees will be affected differently depending on their size and phenology. Large trees with high crowns will experience much less scorch than small
trees. Trees with large buds, or scorched after bud hardening in the fall, will experience less damage than trees with small buds or those burned right after bud burst.
The fractional volume of crown kill is used in models that predict tree mortality
(Peterson and Ryan 1986).

Stem damage is primarily a function of bark thickness. Other properties of
bark, such as density or moisture content, are much less important than bark thickness (Agee 1993). Based on typical flame temperatures at the base of trees and
the thermal diffusivity of bark (Peterson and Ryan 1986), the critical time for cambial kill (t_c, min) can be modeled as a function of bark thickness (x, cm):

$$t_c = 2.9\,x^2$$

If a small tree has bark 1 cm thick, mortality will occur in 2.9 minutes. If a large
tree has bark 10 cm thick at its base, it will take almost 5 hours for cambial

mortality to occur. Species that develop thick bark are much more tolerant of surface fires than species that never develop thick bark. Bark thickness can be estimated for many tree species by knowing tree species and diameter.

Root damage is not capable of being accurately modeled at present (Agee 1993). Unlike a stem cambium, the cambium of a root is not a fixed distance from a flame but is insulated by various depths of soil with unique thermal properties. Roots also come in many sizes. Smoldering combustion of the forest floor is a function of duff moisture content and mineral content; as both increase, duff is less likely to smolder. Current tree mortality models do not directly consider roots.

Models utilize tree crown and bark parameters, together with fire behavior, to predict tree mortality from fire. One of the early models developed for the inland Northwest, the Peterson/Ryan (1986) model developed an equation based on fractional fuel consumption to predict burning time (t_1), and compared this to critical times for cambial kill based on bark thickness:

$$P_m = c_k^{\,(t_c/t_1 - 0.5)}$$

where

P_m = probability of tree mortality (from 0 to 1)
c_k = fractional volume of crown kill (from 0 to 1)
t_c = critical time for cambial kill (min)
t_1 = duration of lethal heat (min)

Mortality from a moderate-intensity fire using this equation is shown in Table 4.2. A second model of tree mortality, the Ryan/Reinhardt (1988) model, uses similar parameters but a different equation form. This model forms the basis of FOFEM, an acronym for First Order Fire Effects Model (Reinhardt et al. 1997), adapted for personal computer use. The output displays the probability of mortality by species and size class (Table 4.3) for a range of flame lengths, and will show pre- and post-fire density by species.

The model of fire mortality used in FOFEM has been incorporated in successional models that evaluate change over time in forest stands. Most of the models are "gap-phase" models that simulate forest dynamics at the scale of a gap in the forest (~400 m^2). Developed initially in the early 1970s for eastern forests where gaps are a common scale of reproduction, they have been adapted in several western forest models: SILVA for California mixed-conifer forests (Kercher and Axelrod 1984), and FIRE-BGC for Columbia Basin forests (Keane et al. 1996). Their incorporation of disturbance is an essential part of accurate prediction of ecosystem dynamics, particularly in western forests. However, all of the current models have significant limitations for simulating fire effects. None of them account for patchiness of fire: Each fire burns every square meter of the simulation area. The effects of forest floor consumption and root mortality are not considered, and synergistic effects, such as post-fire insect attack, cannot at present be modeled with any consistency. Nevertheless, the emerging models continue to improve, and each sets the stage for newer generations of more powerful models.

Figure 4.7

A ponderosa pine stand, once open, is now crowded with small trees that would have been removed by previous low-intensity fires. The small trees lower the effective crown height and increase the chances of a crown fire under dry summer weather conditions.

FIRE REGIMES OF NORTH AMERICA

Wildland fire is a classic disturbance agent, and encompasses myriad combinations of fire frequency, magnitude, extent, and timing in the ecosystems of North America. It also has significant interactions with other disturbance agents, such as insects, disease, and wind, either as a precursor to those other disturbances or as a result of them. Yet fire is not independent of the ecosystems in which it occurs. It is influenced by the amount, arrangement, and structure of variable fuel complexes, and its vegetation effects are dependent on the adaptations to fire covered earlier in this chapter. A simpler paradigm is needed to meld these multivariate characters of fire disturbances into a more comprehensible framework, and that paradigm is the concept of the fire regime.

A **fire regime** combines the effects of fire disturbance into several broad categories. Several systems may be employed, but a usable system capable of application over broad areas is the one based on historical fire severity. Three broad categories of **fire severity** may be defined, based on the physical characters of fire and the fire adaptations of vegetation: low, mixed or moderate, and high (Agee 1993). A low-severity fire regime is one where the effect of the typical historical fire is benign. Fires are frequent (often <20 years), of low intensity, and the ecosystems have dominant vegetation well adapted to survive fire (*resisters*). At the other

TABLE **4.2** | **Probability of Mortality for Each Species in a Douglas-Fir Stand[a]**

Species	Diameter (cm)	Basal area (fraction of stand)	t_c/t_l	c_k	Mortality P_m	Fraction of stand
Douglas-fir	23.9	0.80	0.78	0.38	0.75	0.60
Subalpine fir	8.6	0.02	0.02	1.00	1.00	0.02
Grand fir	8.6	0.01	0.06	1.00	1.00	0.01
Lodgepole pine	18.2	0.10	0.05	0.32	1.00	0.10
Ponderosa pine	54.1	0.04	4.38	0	0	0
Engelmann spruce	13.2	0.01	0.20	0.94	1.00	0.01
Western larch	34.0	0.01	1.90	0	0	0
Western redcedar	32.0	0.01	0.56	0.41	0.95	0.01
Western hemlock	23.0	<0.01	0.51	0.52	0.99	<0.01
					0.75	

Source: From Peterson and Ryan (1986).
[a]An "average stand" was developed from inventory data from the Northern Rocky Mountain Region of the USDA Forest Service. Fire characteristics include scorch height of 10 m, fuel model 10, and moderate fuel moisture.
Abbreviations: t_c/t_l = ratio of the critical time for cambial kill to the actual time that lethal heat is present, c_k = proportional volume of crown kill, P_m = probability of tree mortality

end of the spectrum is the high severity fire regime, where fires are usually infrequent (often >100 years) but may be of high intensity. Most of the vegetation is at least top-killed, and vegetation may be composed of *evaders*, *endurers*, or *avoiders*. In the middle is the mixed- or moderate-severity fire regime, where fires are of intermediate frequency (25–100 years), range from low to high intensity, and have vegetation with a wide range of adaptations reflecting the variability in this fire regime.

Historic landscape character was in part due to fire patterns imposed on forest type mosaics defined by climate, soils, aspect, slope, and time. Typical landscape ecology characters such as patch size, corridors, and edge within a single potential vegetation type were largely a product of disturbance (Agee 1998). Fire, as one important type of disturbance, had effects on all of these landscape characters, depending on the fire regime (Table 4.4).

The five examples of fires and ecosystems in this chapter illustrate each of the fire regimes and its landscape effects. Ponderosa pine (*Pinus ponderosa*) and longleaf pine (*Pinus palustris*) are classic examples of the low severity fire regime. Jack pine (*Pinus banksiana*) and subalpine forests are widely distributed examples of the high-severity fire regime. Coastal Douglas-fir (*Pseudotsuga menziesii*) forests are illustrative of a high-severity fire regime that grades into a moderate-severity regime in its southern reaches. Our management of these forests in the

TABLE **4.3**	**FOFEM Output[a]**
	Fire Effects Calculator

Tree mortality module: Stand tree mortality
Region: Pacific West
Cover type: Sierra Nevada mixed conifer (SAF 243)
Flame length (ft): 4.0

Probability of mortality for each species/diameter entry

Species code	Diameter (in.)	Number of trees	Probability of mortality	Mortality of equality number
PINPON	30	30	.04	1
PINPON	20	20	.09	1
PINPON	10	10	.33	1
PINPON	5	10	.82	1
PINPON	2	10	1.00	1
ABICON	20	20	.15	1
ABICON	10	50	.44	1
ABICON	5	100	.95	1

Average mortality probabilities for 2 ft flames by species/diameter entry

Species code	Tree dbh	Flame lengths (ft)									
		2	4	6	8	10	12	14	16	18	20
PINPON	30	.0	.0	.0	.0	.4	.9	.9	.9	.9	.9
PINPON	20	.1	.1	.1	.3	1.0	1.0	1.0	1.0	1.0	1.0
PINPON	10	.3	.3	.7	1.0	1.0	1.0	1.0	1.0	1.0	1.0
PINPON	5	.6	.8	1.0	1.0	1.0	1.0	1.0	1.0	1.0	1.0
PINPON	2	1.0	1.0	1.0	1.0	1.0	1.0	1.0	1.0	1.0	1.0
ABICON	20	.2	.2	.2	.5	.9	1.0	1.0	1.0	1.0	1.0
ABICON	10	.4	.4	.9	1.0	1.0	1.0	1.0	1.0	1.0	1.0
ABICON	5	.7	1.0	1.0	1.0	1.0	1.0	1.0	1.0	1.0	1.0
AVERAGES	12	.4	.5	.6	.7	.9	1.0	1.0	1.0	1.0	1.0

Stand Tree Mortality
Tree mortality indexes:
Average probability of mortality: .58
Number of trees killed by the fire: 144
Average tree diameter (dbh [diameter at breast height]) killed by fire (in.): 6.4
Average probability of mortality for trees 4+ in dbh: .54
Total pre-fire number of trees: 250

Abbreviations: PINPON = ponderosa pine and ABICON = white fir.
[a]The output is specified both for the flame length of 4 ft and for 2 ft flame lengths between 2 and 20 feet. Because outputs are exclusively in English units, SI equivalents are not provided.

TABLE **4.4** | **Landscape Ecology Indices and Fire**

Landscape ecology measure	Type of fire regime		
	Low severity	Moderate severity	High severity
Patch size[a]	Small	Medium	Large
Ecological Edge[b]	Small	Large	Medium
Pre/post-fire patch similarity[c]	High	Moderate	Low
Islands/residuals	Regeneration occurs in small patches in a matrix of mature patches mostly unaffected by the fire. →		Residuals occur as small islands or corridors in a matrix of more open landscape where all vegetation has been top-killed.

[a]Patch size. Most ponderosa pine and mixed-conifer ecosystems in low-severity fire regimes have clumped distributions and these patches can be as small as 15–20 m on a side. High-severity fire regimes can have patches many thousands of hectares in size.

[b]Ecological edge. Edge usually is described as some ratio of patch perimeter to area. In low-severity fire regimes, fires create edge only in the small regeneration patches. In moderate-severity fire regimes, larger patches of low, moderate, and high severity, all intermixed, create maximum edge; high-severity fire regimes have much of their edge at the perimeter and are therefore intermediate in rank between the low- and moderate-severity fire regimes.

[c]Patch similarity. The pre- and post-fire similarity of a patch is highest where fire effects are most benign, in the low-severity fire regimes. Some of the patches in the moderate-severity fire regime are very similar to the pre-burn condition; others are quite different. In the high-severity fire regimes, pre- and post-fire patch character are most dissimilar.

Source: From Agee (1998).

20th century has transformed many historical low-severity fire regimes to high-severity fire regimes by removing fire from the landscape, reducing forest health, and creating significant social costs. Approaches to these problems will be discussed in Chapter 6.

PONDEROSA PINE FORESTS

Ponderosa pine occupies the drier portions of the western United States capable of supporting trees. Ponderosa pine is found as an early seral dominant in many plant associations, as well as a climax dominant in several plant associations. The focus of this discussion of ponderosa pine is on those communities where ponderosa pine is the potential (climax) vegetation, which cover only a small part of the wide geographical distribution of the species.

Ponderosa pine forests are found as a vegetation zone separating closed forest communities, often Douglas-fir or white fir (*Abies concolor*) forests at higher elevation, from woodland or grassland communities in drier or lower-elevation locales. In most presettlement ponderosa pine forests, there was a substantial perennial herbaceous component, together with pine needles created a flashy fuel that encouraged frequent widespread burning.

The Fire Regime

Substantial fire history information exists for ponderosa pine communities in the southwestern United States. The early work in the Southwest was based on individual point samples and found fire return intervals of 4 to 12 years. Because individual fires do not often scar every tree with an existing **fire scar**, this technique is relatively conservative as a point frequency estimate of fire. In Arizona a mean fire return interval of 1.8 years was calculated on ponderosa pine sites by compositing the cross-dated records of fire-scarred trees over a 150 ha area; the lowest mean fire return interval for a single tree was 4.1 years (Dieterich 1980). Individual trees in ponderosa pine forest on the Warm Springs Indian Reservation in Oregon were found to have mean fire return intervals of 11 to 16 years (Weaver 1959).

The intensity of these fires is inferred by the pattern of scarring but not killing residual trees, and from early accounts of free-ranging fire in this forest type. Fires tended to be of low intensity, rarely scorching the crowns of the older, mature trees. However, even in low-severity fire regimes, intense fires may sometimes occur, possibly due to longer than normal fire return intervals allowing litter and understory fuels to build up, or due to very unusual fire weather, such as strong dry winds.

Fire is linked with other disturbance factors in ponderosa pine forests. The most clearly linked is pre-fire or post-fire insect attack. Pre-fire insect attacks may have much to do with the clumped pattern of tree groups in these forests. Post-fire attacks on scorched and weakened trees are by western pine beetle (*Dendroctonus brevicomis*), mountain pine beetle (*D. ponderosae*), red turpentine beetle (*D. valens*), or pine engraver beetles (*Ips* spp.) (Furniss and Carolin 1977). Crown scorch at levels above 50% is associated with 20% or more mortality by western pine beetle in mature trees. Younger trees can survive more than 75% scorch with about 25% mortality (Miller and Keen 1960, Dieterich 1979).

There is little direct evidence of interactions between various pathogens of ponderosa pine and fire. Fire may help to control dwarf mistletoe (*Arceuthobium* spp.) infestation by pruning dead branches and consuming tree crowns that have low-hanging witches' brooms. Little decay may be present around fire scars, which may exist for the life of the tree. Fire might scar roots and encourage root rot formation, but this has not been documented in detail. Conversely, fire often burns out old stumps, which harbor decay organisms, and in presettlement ponderosa pine forests, likely had a net beneficial effect for live trees by locally sterilizing the soil around stumps and logs that burned.

Stand Development Patterns

In ponderosa pine forests, soil moisture usually is more important than light for the establishment of ponderosa pine seedlings (Barrett 1979). Available soil moisture, particularly in spring and early summer, is critical to the survival of ponderosa pine seedlings. Presence of other understory vegetation, such as perennial grasses or mature trees, can reduce height growth of established ponderosa pine compared to open-grown trees, making the small pines more susceptible to thinning by periodic fire.

Spatial pattern of age classes suggests periodic tree establishment at a small scale (Cooper 1960, White 1985). Stands of mature trees have variable clump sizes, with the clumping usually at 0.06 to 0.13 ha, equivalent to clumps 25 to 35 m on a side (Figure 4.7). The process of stand development is a result of the shade intolerance of ponderosa pine, periodic good years for seedling establishment associated with years of above-normal precipitation, and frequent fire. Gaps in the forest, created by mortality of an existing small even-aged group by bark beetles, allow the shade-intolerant pine to become established when a good seed year and appropriate climate coincide. In this opening the stand of young trees will be protected from fire because of lack of fuel on the forest floor, while other frequent fires will burn under mature stands and eliminate any reproduction there. As the trees in the opening continue to grow, they provide enough fuel to carry the fire and thin the stand. Within a group relatively uniform spacing is the result of moisture competition and a tendency for closely spaced, small trees to be selectively killed by fire.

In presettlement forests, as a young group and adjacent clumps continue to grow and underburn at frequent intervals, fire maintains the clumped pattern over time. When a clump becomes old, insect attack, possibly associated with one or more drought years, may attack one tree and then another in the clump. Disease also may attack members of a clump, especially if their roots are grafted to one another. A relatively rapid breakup of an old clump may occur, implying that this pattern is likely to remain spatially stable over time. Current knowledge suggests that as older clumps break up, they may do so gradually, so that the new clump may not closely resemble the older clump in age distribution or spatial extent (White 1985). Rather than the whole group dying simultaneously, one or two trees within a group might die and contribute additional fuel to the forest floor. When they burn, limited sites for regeneration are created, and this pattern, repeated over many decades for an old-growth group, will result in large within-group age ranges (30–250 years), implying that the spatial nature of clumps may vary over time on the landscape (Agee 1993).

Early ponderosa pine forests had substantial understory vegetation. In the Southwest and central Rocky Mountains, mountain muhly (*Muhlenbergia montana*) were common. In the Pacific Northwest, Idaho fescue (*Festuca idahoensis*) and bluebunch wheatgrass (*Agropyron spicatum*) were common (Wright and Bailey 1982). Several factors have been responsible for decline of perennial grasses in ponderosa pine forests over time. Livestock grazing since the late 1800s has reduced perennial grasses while encouraging annual grasses such as cheatgrass (*Bromus tectorum*), and allowing shrubs to invade. Even before a fire exclusion policy was enacted, some areas apparently experienced declines in fire activity, thought to be due absence of sufficient grass fuels. Due to fire exclusion and logging, many stands now have higher tree density and generally smaller, more uniformly dispersed stems, creating dense overstory canopies. Fire exclusion has not allowed thinning or pruning of the trees or topkill of shrubs by frequent fires (Agee 1993, Covington and Moore 1994).

The typical fire in ponderosa pine forests had little effect on the herbaceous component besides removing the cured material above the ground. Removing the accumulated needles and top-killing shrubs generally aided the grasses and forbs, and stimulated flowering. Bunchgrasses normally recover in 1 to 3 years, and most forbs recover quickly. Bulb and tuberous species are rarely harmed by fire. Competition from grasses and forbs in areas protected from domestic livestock limits tree regeneration to areas scarified by fire.

Ponderosa pine forests have experienced significant ecological change since fire suppression began. In most forests selective removal of the large "yellowbelly" pines has eliminated the visually obvious clumping pattern of the older forests, and replaced it with more uniformly spaced, smaller "black bark" forests. In unharvested areas establishment and growth of ponderosa pine in the understory of the larger pines also has reduced spatial diversity, and created fuel ladders to the larger trees. Where these larger trees are old and of low vigor, they may have been removed in past decades by insect attack. They are vulnerable to significant increases in insect attack either from prescribed fire or wildfire.

COASTAL DOUGLAS-FIR FORESTS

Much of the coastal area where Douglas-fir grows is called the "Douglas-fir region" because of the dominance of Douglas-fir over the area, but it is a late-successional or climax species over only a small portion of this region. To the north, western hemlock (*Tsuga heterophylla*) and western redcedar (*Thuja plicata*) are common late-successional species (Franklin and Dyrness 1973). To the south the Douglas-fir forests have high tree species richness; commonly, five to seven conifers will be found in association with three to five hardwood species, rather unimpressive by eastern United States standards but very impressive for the West. The Douglas-fir/hardwood type includes tanoak (*Lithocarpus densiflorus*), Pacific madrone (*Arbutus menziesii*), and canyon live oak (*Quercus chrysolepis*), with chinquapin (*Castanopsis chrysophylla*) on cooler sites (Sawyer et al. 1977). The dominance of Douglas-fir at the time of European settlement was largely due to the presence of disturbance, primarily by fire, for many centuries before such settlement. Pollen records establish that Douglas-fir dominance in past millennia often coincided with charcoal peaks in the pollen profile.

The Fire Regime

Native American ignitions are known to have been important in certain forest and grassland types of the Pacific Northwest. However, west of the Cascades in the coastal Douglas-fir zone, the role of aboriginal burning is much less clear. At present, the case for widespread aboriginal fires throughout coastal forests is not convincing (Agee 1993). In the southern portion of the region, however, a significant component of Native American burning probably added significantly to lightning ignitions. Burning of tanoak was done to "kill diseases and pests" and clean the ground under the trees so that the acorns could be picked up more easily. Douglas-fir/hardwood forests with less tanoak were probably also frequently burned along ridgetops to maintain travel corridors and openings for hazel (*Corylus cornuta*) and beargrass (*Xerophyllum tenax*) production, both of which were used for basket-weaving material 1 to 2 years after a site was burned (Lewis 1973).

There is a north-south regional gradient in lightning ignitions, largely due to increased lightning frequency and decreasing summer precipitation patterns from northern Washington to southern Oregon. A regional average fire return interval for coastal Douglas-fir forests has been estimated at 230 years with considerable spatial variability in this estimate due to higher ignition frequency to the south (Fahnestock and Agee 1983). Significant temporal variability is also characteristic of these fire return intervals, so that the 230-year average is of limited utility as a parameter of fire frequency across the region. The notion of a **"fire cycle,"** or a return interval of regular frequency, is not as meaningful as in drier forest types, where with some measure of variability a roughly cyclic occurrence of fire can be assumed. Rarely in a Douglas-fir stand is the fire record long enough or regular enough to infer a cyclic pattern of fire, particularly in the presence of climatic shifts that would alter any "cycle" in operation. Fire return intervals are probably much longer than 250 years in the moist Douglas-fir forests of the Coast Range of Oregon, the Washington Cascades, and the Olympics. In the central Oregon Cascades, a natural fire rotation of 95 years to 145 years occurred over the last five centuries, well below that of the moist Douglas-fir forests of Washington (Morrison and Swanson 1990). Other stands aged 500 years or older exist without much evidence of recurrent fire. In southern Oregon and northern California, presettlement mean fire return intervals ranged from 10 to 20 years, with longer intervals near the coast and at higher elevation (Taylor and Skinner 1998).

Fire intensity often is high under conditions that allow fire spread in coastal Douglas-fir forests. Extensive crown fires in old-growth forest may occur, as in the Tillamook fire (Heinrichs 1983), but extensive mortality can occur from fires of an intensity that scorches the crown but does not consume it. Independent crown fire behavior in thick-canopied young stands may be an additional significant type of fire. Toward the southern part of the region, fire severity becomes patchier (Figure 4.8), and low-intensity underburns were common in the southern portion of the region.

Stand Development Patterns

In moist Douglas-fir forests, long early seral tree recruitment (e.g., 75–100 years for Douglas-fir) has been documented after disturbance by fire (Hemstrom and Franklin 1982). On these sites the regeneration period is decades long and probably represents some regeneration from trees that initially colonized the burn and grew large enough to produce viable seed to help completely restock the stand. Lack of seed source, brush competition, and/or reburns have been identified as factors delaying regeneration on such sites. Early-season crown fires, or crown scorch fires in poor seed years, may be associated with a lack of early regeneration after a fire. Late-season fires, even if crown fires, occur when seed is mature, and Douglas-fir seed in cones can survive crown fire events. Patterns of reburns on the Tillamook fire of 1933 at 6-year intervals (1939, 1945, 1951), at Mount Rainier in the late 19th century, and at the southern Washington Yacolt burn of 1902 are evidence these sites will reburn (Agee 1993). High surface fire potential during early succession in Douglas-fir forest was identified as a "vicious

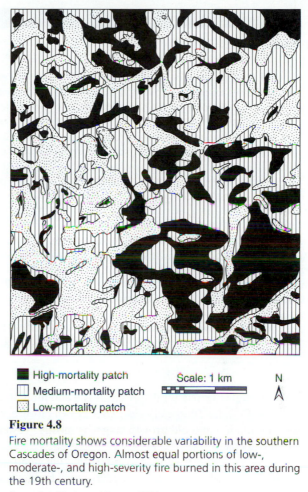

High-mortality patch
Medium-mortality patch
Low-mortality patch

Scale: 1 km

N

Figure 4.8

Fire mortality shows considerable variability in the southern Cascades of Oregon. Almost equal portions of low-, moderate-, and high-severity fire burned in this area during the 19th century.

(Source: From Morrison and Swanson 1990.)

cycle" of positive feedback, encouraging rhizomatous bracken fern (*Pteridium aquilinum*). Herbs and shrubs, once established, may extend the stand-initiation stage for decades. Given sufficient sources for reignition (e.g., the original Yacolt and Tillamook burns and all reburns are thought to have been human ignited), the reburn hypothesis is likely to be true in certain areas. However, it is not clear whether reburns were a common event prior to European settlement in the moist portion of the Douglas-fir region.

Douglas-fir regenerates well when partially shaded but without significant root competition, so that severely burned areas often are not ideal spots for Douglas-fir regeneration. Tree regeneration may be immediate if the fire is of crown scorch rather than crown consumption intensity, leaving current year canopy seed intact. Fires are an important process opening growing space for shade-intolerant

BOX 4.1

The Tillamook Burns

The moist forest of western hemlock and Douglas-fir along the coast of Oregon appear unlikely candidates for holocaustic wildfire, but most have a fire origin some-time in the last millennium. Surprisingly the highest flammability in these forests may be directly after a fire, as the initial fire may consume only 10 to 15% of the bio-mass and leave much of the rest in more flammable condition than it was before the fire. In 1933 logging in the wet Coast Range of Oregon was hard and dirty work. When a logging cable ran along a downed log in the summer, it sparked a holo-caust that was to repeat itself several times. The Tillamook fire of 1933 provided a spectacular firestorm: Almost 90,000 ha burned in one great 20-hour fire run. When the flames finally died, those involved could not have known that large fires would revisit the Tillamook on a 6-year cycle. Fueled by the debris created by the 1933 fire, another human-caused fire burned the Tillamook in 1939, spotting from snag to snag and burning fallen branches. This encouraged the growth of rhizomatous bracken fern, which grows to 2 m height there and is quite flammable when it cures in late summer. In 1945 the cycle continued, and it repeated again in 1951. A con-centrated snag-falling and tree planting finally reclaimed the Tillamook landscape, which now is a state forest covered by vigorous second-growth Douglas-fir. Most visitors passing through now never suspect that the occasional sentinel snag is evi-dence of the one of the most catastrophic fire episodes in Northwest history.

Douglas-fir. Where Douglas-fir is a seral species, it is a common practice to date disturbance events by Douglas-fir age classes. After crown closure, regeneration in the understory reinitiation and old-growth stages is mostly western hemlock, with some western redcedar.

In stand replacement fires it was once assumed that the stand would be ini-tially dominated by Douglas-fir, and after 100 years or so, hemlock would begin to enter the understory (Munger 1940). This may be a pattern on drier sites in coastal Douglas-fir forests, but on mesic to wet sites an initial floristics model (see Chapter 2) is more common: hemlock and Douglas-fir colonize together. Because hemlock grows more slowly, it may appear to be younger than the Douglas-fir, but small hemlocks or western redcedars may in fact be the same age as larger Douglas-fir.

Understory successional patterns in the decades after fire are a function of disturbance intensity and the life-history traits of the species involved, a combi-nation of deterministic and stochastic processes. Many of the species in these forests are persistent after disturbance. Understory response after fire may be partly deterministic, as most of the pre-burn shrubs and herbs will recolonize the site. However, patterns for some species will vary by site, disturbance intensity, predisturbance population conditions, postdisturbance propagule dispersal, and local weather. At the local scale, then, stochastic processes also may be operating in post-fire understory dynamics (Halpern 1989).

As stands enter the stem exclusion stage, overstory tree density begins to decline, with fewer but much larger Douglas-fir in the old-growth stage. Projecting successional patterns past the typical life span of Douglas-fir results in a stand dominated almost exclusively by western hemlock. This transition to climax western hemlock forest has rarely occurred due to the fire return interval in these forests.

In drier Douglas-fir/hardwood forests, Douglas-fir is likely to remain a dominant after fire in a mature stand where Pacific madrone and tanoak are understory species. These associated species in the presence of a moderate fire are top-killed (*endurers*) and have to resprout from the ground level. However, if a young stand with the same species mix is burned again in one or two decades, Douglas-fir as a small tree will not be fire-tolerant, and as an *avoider* will be killed, leaving the madrone and tanoak to resprout again and dominate the post-fire stand for many decades. Eventually a mixed forest of Douglas-fir overstory and tanoak understory will dominate these Douglas-fir/hardwood forests. All-sized forests with multiple-aged cohorts are common.

The patterns of stand development in Douglas-fir/hardwood forests represent variable fire severities and the ability of the different species to take advantage of post-fire conditions. Beginning after a stand replacement fire, the Douglas-fir regenerating on the site may survive several moderate severity fires that thin the Douglas-fir (*resisters*), remove the *avoiders*, and top-kill the associated hardwoods such as madrone and tanoak (*endurers*), unless the fire is of low intensity. Several reoccurrences of such fires will create a stand with several age classes of Douglas-fir (some of which are large), and an age class of Douglas-fir and hardwoods representing regeneration after the last disturbance. In stands with the oldest cohort above 250 years old, forest structure may superficially appear much like a more moist Douglas-fir forest that has western hemlock as a major codominant (Agee 1993).

WESTERN SUBALPINE FORESTS

The subalpine forests of the West often are dominated by mountain hemlock (*Tsuga mertensiana*) in the coastal regions, and subalpine fir (*Abies lasiocarpa*) forest series in more continental climates; both have high-severity fire regimes. Subalpine forests and associated parkland and alpine zones cap the crest of the major western mountain ranges. The elevation range of the subalpine forest zone increases from north to south and from west to east, and the dominant tree species differ along this gradient. The wetter, more coastal forests of the mountain hemlock zone are found up to 1,700 m in northern Washington to 2,000 m in the southern Oregon Cascades. Subalpine fir forests, the continental analog of the coastal forests, often begin their lower limit at those elevations. Prolonged winter snowpack is characteristic of these subalpine zones, with 7 to 8 m of snowpack in wetter locations (Franklin and Dyrness 1973).

Mountain hemlock is a major climax dominant in the high-elevation forests west of the Cascade and Sierra Nevada crest, with Pacific silver fir (*Abies amabilis*) and subalpine fir as codominants. At lower elevations Pacific silver fir (north) or red fir (*Abies magnifica*) (south) are the climax dominants. Subalpine fir forests are found throughout the Rocky Mountains and in the eastern Cascades of Washington and Oregon. Lodgepole pine (*Pinus contorta*) is the most ubiquitous early seral species in subalpine fir forests. Whitebark pine is characteristic of cold, dry environments. It often is found on southerly or westerly aspects in shallow, rocky soils.

The Fire Regime

None of the tree species of subalpine forests are well adapted to survive fire, although the patchy nature of fire in whitebark pine forests may allow some pines to survive (Arno 1980). Environments are marginal for tree establishment and growth, so a fire disturbance that kills most or all of a stand can create almost permanent meadows, or open parklands that persist for decades to centuries.

Fires are infrequent in subalpine forests. Because these forests grade into parkland and alpine areas, areal estimates of fire return intervals are often not possible. The high mortality associated with most fires leaves few fire-scarred trees so that point estimates of fire return intervals are seldom possible. Estimated fire return intervals range from over 1,500 years in coastal British Columbia (Lertzman and Krebs 1991) to 100 years in drier subalpine fir forests and less than 50 years in some whitebark pine forests (Arno 1980).

Fire behavior estimation in subalpine forests is complicated by the erratic, often weather-driven nature of these fires. The interaction of live crown fuels as a heat source is important in crown fire generation. Fire behavior estimates based on dead fuel loads on the forest floor surface will generally underestimate the fire potential of subalpine forests. In normal to somewhat unusual years, a fuel-limited situation for crown fire development exists, but in very unusual years (once in multiple decades?) large-scale crown fires will burn regardless of the fuel situation (Johnson 1992).

Most fires are stand replacement events because of the lack of fire resistance of the major tree species. Lower-intensity fires may occur in whitebark pine forests, as fire scars often are present in these stands (Arno 1980). The extent of subalpine fires is partly a function of fire weather, but more importantly depends on the distribution of the subalpine forest. Subalpine forest often is patchy and may grade into herbaceous parkland, or rock, or snow, which may not allow fire to carry across to the next forest patch along continuous burning fuels or by spotting. Lower-elevation forests may be less affected by strong winds and less capable of sustaining long fire runs, although fuels may be more continuous. In whitebark pine forests most fires appear to be small and many may burn only one clump. Crown fires can occur in subalpine forests anytime foliar moistures are low, and may be aided by lichen growth draped within the canopy.

As subalpine forests usually burn infrequently, it is not surprising that other disturbances may be very important in forest stand dynamics. In long-undisturbed mountain hemlock forests, individual standing tree mortality may be important; in other mountain hemlock forests, laminated root rot (caused by *Phellinus weirii*) may cause doughnut-shaped, multiple hectare rings of mortality (Dickman and Cook 1989). White pine blister rust (caused by *Cronartium ribicola*), an introduced canker disease, may have deleterious effects on western white pine and whitebark pine. Mountain pine beetles (*Dendroctonus ponderosae*) may be important in stand development patterns by selectively eliminating pines; balsam woolly adelgid (*Adelges piceae*) may kill true fir and spruce bark beetles (*Dendroctonus rufipennis*) may kill Engelmann spruce. Widespread tree death will increase dead fuel loads for decades.

Stand Development Patterns

Fire creates larger patches of mortality in mountain hemlock forests than other disturbances, which usually operate at the scale of a tree or small stand. None of the tree dominants are resistant to fire or adapted to grow well in open, recently burned environments. Post-fire communities may be dominated by shrubs or herbs for a century or more. South-facing slopes are much more likely to burn than north-facing slopes.

Subsequent tree establishment is a function of seed sources and climate. Most subalpine fir seed falls within 20 m of the seed tree, whereas mountain hemlock seed travels about 40 m. Early colonizing trees may play a crucial role in restocking more central areas of a burn as they mature and are able to produce seed (Agee and Smith 1984). Post-fire residuals are rare after fire in mountain hemlock forests, but when they do occur, they can have a significant impact on species composition of the new stand. Larger burns will remain treeless for a longer time because of the seed source effect (Figure 4.9). In older post-fire stands, initial stocking from surviving residuals has a significant effect on later recruitment, suggesting that early colonizers will have a large effect on later successional patterns.

Subalpine fir is a *fire avoider*, possessing thin bark, a low crown with persistent branches, and shallow rooting habit, making it very sensitive to fire. Where associated trees have some advantage in the presence of fire, such as the serotinous (or semiserotinous) cones of lodgepole pine, subalpine fir may be replaced on the site. It may persist as a suppressed understory species or recolonize the site so slowly that before subalpine fir can replace the pine another fire will allow another generation of lodgepole pine to dominate the site. At higher-elevation sites in the Rocky Mountains, trees may be eliminated for a century or more, depressing timberline below the physiological limits of tree growth (Stahelin 1943). Some sites may revert to ribbon forests, where forest regeneration is associated with windward areas of trees blown free of snow (Billings 1969). On wetter Cascade Mountain sites with little or no lodgepole pine, no tree species may be adapted to the post-fire conditions, and meadow creation is the likely result of a fire.

Seedling survival for subalpine fir, lodgepole pine, and Engelmann spruce is likely to be higher where seeds contact bare mineral soil. Decayed wood is sometimes a good substrate, but needle litter is not preferred. Subalpine fir will do better than Engelmann spruce on needle litter, because spruce is susceptible to fungal attack. On burned sites, protection from full sunlight, or browsing, or frost damage, may restrict successful seedling establishment to *safe sites* consisting of shrubs, existing small trees, or logs.

Patterns of stand development in subalpine fir forests are affected by two major factors: relatively slow tree recruitment after disturbance, and relatively

Figure 4.9
Large burns often remain treeless in subalpine areas for long times. This 60-year-old burn in the Olympic Mountains of Washington remains almost treeless.

short-lived dominant species. On young burn sites the rate of recruitment increases for almost a century, documenting a very long **stand initiation stage** (Oliver and Larson 1990). Mature sites (125–250 years old) have roughly bell-shaped age-class distributions with considerable variation. The bell shape suggests that there was a time in the past when tree recruitment peaked, usually 75 to 100 years after disturbance. Individual sites show sustained recruitment, a slow decline, or several peaks. These patterns are symptomatic of a very individualistic tree recruitment by site. It suggests that the **stem exclusion stage** (when regeneration ceases and some existing stems are naturally thinned from the stand) is not well defined in these forests. The shape of the age-class distributions on old sites appears to be closer to young than to the mature sites. Rather than bell-shaped distributions, old sites show substantial recent recruitment, indicative of the **understory reinitiation stage** of stand development. The **old-growth stage** does not really have time to develop, even in truly old forests, because the dominant old subalpine firs are reaching the average maximum life span (250-plus years) at about the same time that the understory reinitiation stage is under way. Stand development apparently truncates at the understory reinitiation structural stage.

In the eastern Cascades and Rocky Mountains, where lodgepole pine often is part of the stand at the time of the fire event, a tremendous amount of seed can be released by the opening of serotinous or semiserotinous cones, and first-year post-fire tree regeneration may be sufficient to restock the stand (Figure 4.10). Both subalpine fir and Engelmann spruce are common co-climax associates in such stands. Post-fire tree regeneration of spruce and fir may be slow and unpredictable, depending on the presence of a seed source, a good seed year, and above-normal growing season precipitation. Lodgepole pine reaches peak density and basal area within 50 years, and may dominate the site up to 150 years (Fahnestock 1976). The lodgepole pine forests of Yellowstone, burned by the great fires of 1988, are regenerating to vigorous lodgepole pine. Recurring fire has kept lodgepole pine in a dominant role across the Yellowstone landscape. Spruce and fir will

Figure 4.10
A recent crown fire in this lodgepole pine–subalpine fir forest will allow lodgepole pine to regenerate rapidly because of the serotinous cones of the pine, which open and drop copious viable seed.

eventually increase over time on mesic sites (Romme and Knight 1981), and subalpine fir is the major climax dominant over much of the area. Occasional lodgepole pines will persist up to about 400 years in these fir-spruce forests.

An early lodgepole pine stage may not occur if the fire return interval has been so long that lodgepole seed source has disappeared from the site. This requires fire-free intervals over 300 years. In such cases subalpine fir and Engelmann spruce will comprise the dominant tree component, but like *Tsuga mertensiana* forests without lodgepole pine, the tree recolonization time may be extended for decades to centuries. Conversely if fires occur at short intervals, lodgepole pine may be the sole conifer dominant, and the sites may be difficult to identify as subalpine fir forests.

Whitebark pine may grow in pure stands as the climax dominant or in association with subalpine fir and Engelmann spruce as an early seral species. Whitebark pine has wingless seeds, so dispersal by Clark's nutcracker (*Nucifraga columbiana*) is essential for recolonization of the interior portions of burns (Tomback 1986). In large severe burns seed dispersal by nutcrackers may give whitebark pine an advantage over other conifers that rely on wind dissemination of seed. Where whitebark pine is seral to subalpine fir, it may colonize the site first, along with lesser numbers of subalpine fir. Subsequent establishment is primarily by subalpine fir (Morgan and Bunting 1990). Historically the replacement took so long that fire was likely to occur before whitebark pine was eliminated from the site. With the introduction of white pine blister rust on whitebark pine caused by *Cronartium ribicola* (see Chapter 13), the dynamics of subalpine forests across the range of whitebark pine will be forever changed.

Subalpine fires tend to be erratic and unpredictable. Although they are infrequent, fires have been important in shaping the landscapes we see today. Many subalpine meadows bordering forest were created by fire, and provide diverse habitats for wildlife and recreationists. Fire has been important in the distribution and abundance of lodgepole pine, and in maintaining habitat for whitebark pine at treeline. The fire suppression period during the 20th century has not had overly

Box 4.2

Yellowstone Aflame!

The summer of 1988 was a summer of flame, one that consumed not only the forests of Yellowstone but also the natural fire policy of the federal land management agencies (see Chapter 6 for natural fire policy). The forests of Yellowstone sit on a broad volcanic plateau, and consist largely of lodgepole pine forests that appear to have regenerated after a series of severe forest fires in the early 1700s. In the early 1970s the National Park Service began to allow some natural fires to burn, under monitoring, within Yellowstone National Park, recognizing that fire was just as important an ecological process as was the famed Old Faithful geyser. For 15 years the plan worked well, but in 1988, an unusually dry year bereft of the usual summer rain, a combination of natural fires and human-caused wildfires raced across the park and adjacent lands. Stands that had been previously attacked by bark beetles, as well as healthy stands of all ages, burned with some of the most intense fire behavior ever recorded. About half of the park burned. The fires burned from June into September, when a light snowfall finally dampened the ardor of the flames. After 10 years the burned areas now are tourist attractions, bright with wildflowers and young, vigorous lodgepole pines, just as they were after the large fires of the 1700s. Yellowstone was, is, and always will be a high-severity fire regime, and the challenges of managing intense fire will remain with us.

significant impacts to date on landscape structure in subalpine zones because of the fairly long fire return intervals. A continued policy of fire exclusion should not have major impacts in this zone for continued decades. Subtle yet significant shifts may occur in species composition over time, and exclusion of a natural process such as fire may be at odds with park and wilderness mandates.

LONGLEAF PINE FORESTS[a]

Longleaf pine occurs across 25 million ha of the Coastal Plain region of southeastern North America. Its range includes long growing seasons (300–350 days) and 100 to 150 cm of evenly distributed annual precipitation, with 2- to 4-week droughts during which fires may occur (Christensen 1991). High lightning frequencies are typical of this region, but because the thunderstorms usually are associated with substantial precipitation, fires may not be ignited by most cloud-to-ground strikes. Burning by Native Americans was probably an important ignition source previous to the 1800s (Komarek 1974). Longleaf pine occurs in two major vegetation types: the xeric sandhill type, along with turkey oak (*Quercus laevis*), and the more widespread pine savanna type, where slash pine (*P. taeda*) and shortleaf pine (*P. elliottii*) also occur. Wiregrass (*Aristida stricta*) is a

[a]Portions of this chapter on longleaf and jack pine are reproduced in part from *Fire and pine ecosystems*, Chapter 11.

common grass associate in both types, and broomsedge (*Andropogon* spp.) in the pine savanna type. These grasses, together with fuel from other herbs and the long needles of longleaf pine, can provide ample fuel for fire spread within a year or two of a previous fire.

The Fire Regime

Longleaf pine has a low-severity fire regime. Fires are frequent but of low intensity, with fire frequency exhibiting a negative feedback effect on fire intensity. In the xeric sandhills, minimum historic fire return intervals are probably 3 to 5 years, because of limitations on continuous fuels. Historic fire return intervals may increase to 30 to 40 years where wiregrass is absent or where longleaf pine has been locally extirpated by logging. "Natural" fire frequencies have been described as from 2 to 10 years over most of the range. This has been supported by three lines of other evidence (Platt et al. 1991): the rapid invasion of pine savannas by hardwoods if fire is excluded for a few years, the high frequency of lightning, and a large number of herbaceous and woody species that resprout vigorously and are capable of surviving fire. Localized disturbance (lightning) can spread widely across the landscape through understory grass and combustible leaf litter of the pines.

At such high fire frequencies, limited biomass can accumulate between fires, so that fire intensity is limited by the available energy. In savanna landscapes fires must have spread widely across the landscape, but historical fire extent has not been studied in this type. With possible conditions for fire spread occurring year-round, fires in any season are possible. Prescribed fire studies have included comparisons of autumn, winter, spring, and summer fires.

Stand Development Patterns

Longleaf pine has a unique adaptation to a high-frequency, low-intensity fire regime: the **"grass stage."** Longleaf pine has large seeds that prefer mineral soil seedbeds, so an immediate post-fire environment is considered favorable. Early growth is concentrated in root development, particularly a long taproot. The apical bud is protected near ground level by a bunchgrasslike clump of needlelike leaves, and the bud has fire-resistant scales (Chapman 1932). The needles, while in the grass stage, are subject to infection by brown spot disease (caused by *Scirrhia acicola*), which is controlled by fires scorching the needles. Most pines in the grass stage will survive these low-intensity surface fires, while other tree species will be killed. After 3 to 5 years in the grass stage, the apical bud begins to grow rapidly, and in 3 to 4 years will have reached a height where it is protected against scorch from an average fire. The zone between 0.3 m and 1.5 m is the most critical for the bud; below that level the grass stage protects the bud and above that level lethal temperatures rarely occur.

Fire is an essential process in the growth and development of longleaf pine stands. The herbaceous layer comprising the understory of longleaf pine savannas, in the absence of fire, will be replaced quickly by hardwoods on productive sites (Christensen 1981). In the northern coastal plain, loblolly pine may replace longleaf pine, while further south shortleaf pine may become dominant with longer fire return intervals. Frequent fires control the hardwoods and other conifers and maintain the savanna. In its development from a seedling, through sapling and pole sizes over the first 25 years of its life, longleaf pine may survive as many as 10 fires.

Once in the pole stage, a frequent fire regime will have little effect on long-leaf pine density, as the thick bark developed by young trees will protect cambial tissues (Figure 4.11). Effects of fire on subsequent tree growth have produced both positive and negative results in comparison to control stands where fire has been excluded. Effects are less apparent in older stands. Maintenance-prescribed burns to reduce competition for pines and reduce fire hazard have been suggested at 3-year intervals (Sackett 1975), well within the natural range of disturbance variability in these ecosystems.

Some longleaf pine stands are "islands" or "hammocks" that are isolated from more continuous savanna landscapes. These areas are cut off from other savanna areas by broad-leaved forests that occupy moister areas. These hammocks probably burn less frequently, but may occupy sites dry enough that longleaf pine can still maintain landscape dominance. When these areas do burn, the accumulated fuel energy release kills competing vegetation and leaves longleaf pine as a dominant, preventing conversion to other vegetation types (Myers 1985).

A mature forest patch can be maintained for centuries without further regeneration, but there is little information available on historical patterns of stand development. Most of the natural stands were cleared as a result of logging and agriculture beginning in the early 1800s (Platt et al. 1988). Adult tree density is negatively associated with pregrass-stage longleaf pine, and juveniles in areas of low needle litter accumulations are larger and have higher post-fire survival than juveniles near mature trees. This negative feedback between mature trees and regeneration results in a strongly aggregated pattern of regeneration in openings. This pattern is less evident as stands develop over time. In a Georgia forest relatively undisturbed except by frequent fire, tree recruitment at the landscape level (60 ha) was almost continuous over a two-century period, suggesting a very stable landscape structure over time in the presence of fire (Platt et al. 1988).

JACK PINE FORESTS

Across the northern part of North America is a widespread forest region known as the boreal forest. Although there is considerable variability in the composition and structure of the boreal forest from east to west and north to south, it is characterized by long, cold winters, and short, warm summers. Closed boreal forest in more southerly areas, where jack pine is a dominant on coarse-textured and nutrient-poor soils, gradually change to the north into more open forest stands that are interspersed with bogs and some closed canopy forest, a mosaic called the *taiga*. The major tree species in the southerly boreal forest along with jack pine are black spruce (*Picea mariana*), white spruce (*Picea glauca*), paper birch (*Betula papyrifera*), and quaking aspen (*Populus tremuloides*).

Figure 4.11
This longleaf pine stand has been maintained through prescribed fire, and is disease- and fire-resistant as a result.
(Photo courtesy USDA Forest Service.)

The Fire Regime

Very long days during June and July and temperatures exceeding 33°C can create severe fire weather (Viereck 1983). Such weather encouraging intense fires generally begins with several weeks without rain, occasional days of low humidity and high day temperatures, and lightning. Lightning is the cause of about 90% of the area burned in the North American boreal forest (Stocks and Street 1983). Large crown fires may occur in spring months when conifer foliar moisture dips just before new buds flush, making the crown more flammable. This depression of foliar moisture may be due to an influx of carbohydrates in the leaves, increasing dry weight, rather than a loss of water.

An average fire return interval of between 50 and 150 years is typical of these forests, with more fire in the southern boreal forest and fire frequencies exceeding 300 years for white spruce forests. Fires in the boreal forest can be very large, ranging from thousands to hundreds of thousands of hectares; often large fires are clustered in particular years of severe fire weather (Johnson 1992).

The predominance of crown fire in the boreal forest is due to the continuous surface fuel layer, composed of dead foliage, moss and lichens, and fine shrubs with dead twigs, along with aerial foliage of adequate mass, bulk density, resin and wax content, and moisture content to sustain combustion (Van Wagner 1983).

Box 4.3

The Dilemma of the Natural Fire Regime as a Guide for Boreal Forest Management

Recent discussions on biodiversity have suggested that if managers were to better mimic natural disturbance regimes, a better "coarse filter" would exist and many biodiversity problems would be solved. A problem specific to high-severity fire regimes, and to jack pine forests, is that the patch size created by natural fires was quite large (hundreds to many thousands of hectares), with unburned islands usually comprising under 5 to 10% of the total area. The public reacts negatively to large clearcuts, and in many places silvicultural alternatives to avoid clearcutting have been proposed. Is it possible to use the natural fire regime as a guide to management?

Malcolm Hunter (1993) has proposed a "compromise strategy" where larger-cut blocks be aggregated, but separated by uncut corridors. This retains the spatial pattern caused by fire, and also provides stringers of uncut forest within the cut matrix. It could provide better screening than continuous large areas of harvested terrain. Patch size should vary, as it does in the natural system, and fine-scale features such as coarse woody debris also need to be addressed within cut blocks. Hunter concludes that whether or not to imitate natural disturbance regimes is as much a question for ethicists as for scientists.

Low-intensity fires occur, but are usually few and small, and have little effect compared to large, high-intensity crown fires. Deep organic layers may smolder after the passage of the fire front. This may have ecological significance, but is relatively unimportant to the rate of fire spread. Fires will burn through an ericaceous shrub layer (which often has high heats of combustion in its foliage) and ignite lichens on lower branches of trees, which will then ignite the tree crown (Figure 4.12).

Jack pine stands are the most likely to sustain surface or crown fires over the season. Black spruce is generally less flammable but under severe fire weather will carry fire similar to the pine stands. Sustained rates of spread of 100 m/minute and fireline intensities of 50,000 to 150,000 kW/m have been measured in northern Alberta (Kiil and Grigel 1969). The influence of stand age on flammability continues to be debated. There is little quantitative evidence to test this assumption, although there is variation in surface fire and crown fire potential over time. Recent research in jack pine stands suggests that fire behavior can be severe in both immature and mature pine. Heterogeneous terrain, caused by fragmentation of the forest by water bodies, may be associated with variation in fire extent.

Stand Development Patterns

After burning the serotinous cones of jack pine open and spread seed over the competition-free forest floor. Cones subject to 450°C for 30 seconds or 350°C for 3 minutes show no decrease in seed germination (Beaufait 1960). The rapid growth of jack pine seedlings allows them to overtop competing vegetation, and they can

Figure 4.12
Intense fire is common in jack pine stands, and regeneration will occur from seed stored in serotinous cones.
(Photo courtesy Canadian Forest Service.)

produce viable seed within 5 to 7 years. Younger trees may exhibit substantial polymorphism (both open and closed cones), which may aid recolonization; older trees may have old cones open, which may extend the species presence past one life cycle of jack pine if open ground is nearby.

Vegetation in the southern boreal forest is well adapted to burning. Jack pine has serotinous cones. Quaking aspen and balsam poplar (*Populus balsamifera*), as *fire endurers*, reproduce prolifically from root suckers; paper birch, even though it will stump sprout, generally uses the *fire invader* strategy of reproducing from seed. White spruce and balsam fir (*Abies balsamea*) are *fire avoiders*, with little adaptation to fire, and require refugia to contribute seed to the burned areas (Rowe and Scotter 1973). If fires are spaced any more than 10 years apart, jack pine usually will dominate the tree component of the post-fire stand. Two crown fires in quick succession may allow aspen and birch to dominate the post-fire stand; long fire-free intervals will encourage replacement of hardwoods by spruce. Of the major species in the boreal forests, the pines (jack and lodgepole pines) are rated as being most likely to successfully regenerate after fire, based on seed retention on tree, early season seed production, frost hardiness, unpalatability, seedling growth rate, and ability to tolerate full exposure. Black spruce has a *fire evader* semiserotinous cone habit. Fire-killed spruce can retain viable seed in its canopy for several years, and substantial seed may also lie on the forest floor (LeBarron 1939).

Deep forest floors are typical of boreal forests. Fires in duff with moderate fuel moisture may consume 5 cm of the forest floor; extended smoldering associated with low duff moisture may allow most of the forest floor to burn (Van Wagner 1983). Species relying on underground roots or rhizomes or buried seed will be detrimentally affected with increasing forest floor consumption (Flinn and Wein 1977). *Fire endurer* shrubs such as huckleberries (*Vaccinium* spp.) or *Ledum* spp., which rely on rhizomes, sprout best after moderate-intensity fires; deep burns will encourage *fire invader* species like fireweed (*Epilobium* spp.) to seed in.

Successional relationships after fire in the boreal forest are complex; a typical sere in the more southerly boreal forest where jack pine is dominant is described below to illustrate typical post-fire succession. Jack pine, at the southerly edge of the boreal forest, is found in a landscape mosaic of red pine (*Pinus resinosa*) and white pine (*P. strobus*), generally on the poorest sites. On the associated pine sites that burn with two closely spaced crown fires, red and white pine may be eliminated and jack pine and paper birch will capture the growing space. If the area is burned repeatedly at short intervals, the pine will disappear and sprouting hardwoods and brush, with the *endurer* strategy, will dominate the site (Heinselman 1973).

A unique attribute of young jack pine stands is their low branching habit above typically well-drained soil. Such sites are excellent habitat for the endangered Kirtland's warbler (*Dendroica kirtlandii*), which nest in the ground and use the low branches for perches. Patches of 30 ha or more are the minimum size for Kirtland's warbler breeding sites, well within the usual crown fire patch size for jack pine (Miller 1963).

At the landscape level it is tempting to think of jack pine ecosystems being in some sort of dynamic equilibrium, so although there is major instability and change at the individual stand level, there is a more stable set of stand ages at the landscape level. This is unlikely to occur for several reasons. First, jack pine stands are but one element of a southerly boreal patch mosaic that includes other stand types with different disturbance histories. Second, the patches that burn are not burned uniformly. Small fires tend to be uniform crown fires with no unburned islands; larger fires are characterized by more unburned islands, larger islands, and more edge. Stringers of unburned forest may be associated with slight depressions or higher soil moisture (Quirk and Sykes 1971). Finally, as is true for most fairly long-lived conifer species, changing climate in the southern boreal forest has changed disturbance regimes so that no equilibrium has existed in the last several centuries (Baker 1992).

REFERENCES

Agee, J. K. 1990. The historical role of fire in Pacific Northwest forests, pp. 25–38. In J. Walstad, S. R. Radosevich, and D. V. Sandberg (eds.). *Natural and prescribed fire in Pacific Northwest forests*. Oregon State University Press, Corvallis, Oreg.

Agee, J. K. 1993. *Fire ecology of Pacific Northwest forests*. Island Press, Washington, D.C.

Agee, J. K. 1998. The landscape ecology of western forest fire regimes. *Northwest Science* 72 (Special Issue):24–34.

Agee, J. K., and L. Smith. 1984. Subalpine tree establishment after fire in the Olympic Mountains, Washington. *Ecology* 65:810–819.

Arno, S. F. 1980. Forest fire history of the northern Rockies. *Journal of Forestry* 78:460–465.

Baker, W. L. 1992. Effects of settlement and fire suppression in landscape structure. *Ecology* 73:1879–1887.

Barrett, J. W. 1979. *Silviculture of ponderosa pine in the Pacific Northwest: the state of our knowledge*. USDA Forest Service General Technical Report PNW-97.

Beaufait, W. R. 1960. Some effects of high temperature on the cones and seeds of jack pine. *Forest Science* 6:194–199.

Billings, W. D. 1969. Vegetational pattern near alpine timberline as affected by fire-snowdrift interactions. *Vegetatio* 19:191–207.

Chapman, H. H. 1932. Is the longleaf pine a climax? *Ecology* 13:328–334.

Christensen, N. L. 1981. Fire regimes in southeastern ecosystems, pp. 112–136. In H. A. Mooney, T. M. Bonnicksen, N. L. Christensen, J. E. Lotan, and W. A. Reiners (eds.). *Fire regimes and ecosystem properties: proceedings of the conference.* USDA Forest Service General Technical Report WO-26.

Christensen, N. L. 1991. Vegetation of the southeastern coastal plain, pp. 317–364. In M. G. Barbour, and W. D. Billings (eds.). *North American terrestrial vegetation.* Cambridge University Press, Cambridge, England.

Cooper, C. F. 1960. Changes in vegetation, structure, and growth of southwestern pine forests since white settlement. *Ecological Monographs* 30:129–164.

Covington, W. W., and M. M. Moore 1994. Southwestern ponderosa pine forest structure: changes since Euro-American settlement. *Journal of Forestry* 92:39–47.

Dickman, A., and S. Cook. 1989. Fire and fungus in a mountain hemlock forest. *Canadian Journal of Botany* 67:2005–2016.

Dieterich, J. H. 1979. *Recovery potential of fire-damaged southwestern ponderosa pine.* USDA Forest Service Research Note RM-379.

Dieterich, J. H. 1980. *Chimney Springs forest fire history.* USDA Forest Service General Technical Report RM-220.

Fahnestock, G. R. 1976. Fires, fuel, and flora as factors in wilderness management: the Pasayten case. *Tall Timbers Fire Ecology Conference* 15:33–70.

Fahnestock, G. R., and J. K. Agee. 1983. Biomass consumption and smoke production by prehistoric and modern forest fires in western Washington. *Journal of Forestry* 81:653–657.

Flinn, M., and R. Wein. 1977. Depth of underground plant organs and theoretical survival during fire. *Canadian Journal of Botany* 55:2250–2254.

Franklin, J. F., and C. T. Dyrness. 1973. *Natural vegetation of Oregon and Washington.* USDA Forest Service General Technical Report PNW-8.

Furniss, R. L., and V. M. Carolin. 1977. *Western forest insects.* USDA Miscellaneous Publication 1339.

Halpern, C. B. 1989. Early successional patterns of forest species: interactions of life history traits and disturbance. *Ecology* 70:704–720.

Heinrichs, J. 1983. Tillamook. *Journal of Forestry* 81:442–444, 446.

Heinselman, M. L. 1973. Fire in the virgin forests of the Boundary Waters Canoe Area, Minnesota. *Quaternary Research* 3:329–382.

Hemstrom, M. A., and J. F. Franklin. 1982. Fire and other disturbances of the forest in Mount Rainier National Park. *Quaternary Research* 18:32–51.

Hunter, M. L. 1993. Natural fire regimes as spatial models for managing boreal forests. *Biological Conservation* 65:115–120.

Johnson, E. A. 1992. *Fire and vegetation dynamics: studies from the boreal forest.* Cambridge University Press, Cambridge, England.

Kauffman, J. B. 1990. Ecological relationships of vegetation and fire in Pacific Northwest forests, pp. 39–52. In J. Walstad, S. R. Radosevich, and D. V. Sandberg (eds.). *Natural and prescribed fire in Pacific Northwest forests.* Oregon State University Press, Corvallis, Oreg.

Keane, R. E., P. Morgan, and S. W. Running. 1996. FIRE-BGC—A mechanistic ecological process model for simulating fire succession on coniferous forest landscapes of the northern Rocky Mountains. USDA Forest Service Research Paper INT-RP-484.

Kercher, J. R., and M. C. Axelrod. 1984. A process model of fire ecology and succession in a mixed-conifer forest. *Ecology* 65:1725–1742.

Kiil, A. D., and J. E. Grigel. 1969. *The May 1968 forest conflagrations in central Alberta.* Canadian Forest Service Information Report A-X-24.

Komarek, E. V. 1974. Effects of fire on temperate forests and related ecosystems: southeastern United States, pp. 251–277. In T. T. Kozlowski, and C. E. Ahlgren (eds.). *Fire and Ecosystems*. Academic Press, New York.

LeBarron, R. K. 1939. The role of forest fires in the reproduction of black spruce. *Proceedings of the Minnesota Academy of Science* 7:10–14.

Lertzman, K. P., and C. J. Krebs. 1991. Gap-phase structure of a subalpine old-growth forest. *Canadian Journal of Forest Research* 21:1730–1741.

Lewis, H. T. 1973. *Patterns of Indian burning in California: ecology and ethnohistory*. Ballena Press Anthrological Papers 1. Ballena Press, Ramona, Calif.

Miller, H. A. 1963. Use of fire in wildlife management. *Tall Timbers Fire Ecology Conf.* 2:18–30.

Miller, J. M., and F. P. Keen. 1960. *Biology and control of the western pine beetle*. USDA Miscellaneous Publication 800.

Morgan, P., and S. C. Bunting. 1990. Fire effects in whitebark pine forests, pp. 116–170. In W. C. Schmidt, and K. J. McDonald (comps.). *Proceedings: Symposium on whitebark pine ecosystems: ecology and management of a high-mountain resource*. USDA Forest Service General Technical Report INT-270.

Morrison, P., and F. J. Swanson. 1990. *Fire history and pattern in a Cascade Range landscape*. USDA Forest Service General Technical Report PNW-GTR-254.

Munger, T. T. 1940. The cycle from Douglas fir to hemlock. *Ecology* 21:451–459.

Myers, R. L. 1985. Fire and the dynamic relationship between Florida sandhill and sand pine scrub vegetation. *Bulletin of the Torrey Botanical Club* 112:241–252.

Oliver, C. D., and B. C. Larson. 1990. *Forest stand dynamics*. McGraw-Hill, New York.

Peterson, D. L., and K. C. Ryan. 1986. Modeling post-fire mortality for long-range planning. *Environmental Management* 10:797–808.

Platt, W. J., G. W. Evans, and S. L. Rathbun. 1988. The population dynamics of a long-lived conifer (*Pinus palustris*). *American Naturalist* 131:491–525.

Platt, W. J., J. S. Glitzenstein, and D. R. Streng. 1991. Evaluating pyrogenicity and its effects on vegetation in longleaf pine savannas. *Tall Timbers Fire Ecology Conference* 17:143–161.

Pyne, S. J. 1995. *World fire—the culture of fire on Earth*. Henry Holt, New York.

Quirk, W. A., and D. J. Sykes. 1971. White spruce stringers in a fire-patterned landscape in interior Alaska, pp. 179–198. In *Fire in the northern environment*. USDA Forest Service, Pacific Northwest Region, Portland, Oreg.

Reinhardt, E. D., R. E. Keane, and J. K. Brown. 1997. *First Order Fire Effect Model: FOFEM 4.0, users guide*. USDA Forest Service General Technical Report INT-GTR-344.

Romme, W. H., and D. Knight. 1981. Fire frequency and subalpine forest succession along a topographic gradient in Wyoming. *Ecology* 62:319–326.

Rowe, J. S. 1983. Concepts of fire effects on plant individuals and species. In R. W. Wein, and D. A. Maclean (eds.). *The role of fire in northern circumpolar ecosystems*. John Wiley and Sons, New York.

Rowe, J. S, and G. W. Scotter. 1973. Fire in the boreal forest. *Quaternary Research* 3:444–464.

Ryan, K. C., and E. D. Reinhardt. 1988. Predicting post-fire mortality of seven western conifers. *Canadian Journal of Forest Research* 18:1291–1297.

Sackett, S. 1975. Scheduling prescribed burns for hazard reduction in the Southeast. *Journal of Forestry* 73:143–147.

Sawyer, J. O., D. A. Thornburgh, and J. R. Griffin. 1977. Mixed evergreen forest, pp. 359–381. In M. Barbour, and J. Major (eds.). *Terrestrial vegetation of California*. California Native Plant Society Special Publication 9, Sacramento, Calif.

Stahelin, R. 1943. Factors influencing the natural restocking of high altitude burns by coniferous trees in the central Rocky Mountains. *Ecology* 24:19–30.

Stocks, B. J. and R. B. Street. 1983. Forest fire weather and wildfire occurrence in the boreal forest of northwestern Ontario, pp. 249–265. In R. R. Riewe, and I. R. Methven (eds.). *Resources and dynamics of the boreal zone*. Association of Canadian Universities Northern Studies, Ottawa.

Taylor, A. R., and C. N. Skinner. 1998. Fire history and landscape dynamics in a late-successional reserve, Klamath Mountains, California, USA. *Forest Ecology and Management* 111:285–301.

Tomback, D. R. 1986. Post-fire regeneration of krummholz whitebark pine: a consequence of nutcracker seed caching. *Madrono* 33:110–110.

Van Wagner, C. E. 1973. Height of crown scorch in forest fires. *Canadian Journal of Forest Research* 3:373–378.

Van Wagner, C. E. 1983. Fire behavior in northern conifer forests and shrublands, pp. 65–80. In R. W. Wein, and D. A. Maclean (eds.). *The role of fire in northern circumpolar ecosystems*. John Wiley and Sons, New York.

Viereck, L. A. 1983. The effect of fire in black spruce ecosystems of Alaska and northern Canada, pp. 210–219. In R. W. Wein, and D.A. Maclean (eds.). *The role of fire in northern circumpolar ecosystems*. John Wiley and Sons, New York.

Weaver, H. 1959. Ecological changes in the ponderosa pine forest of the Warm Springs Indian Reservation in Oregon. *Journal of Forestry* 57:15–20.

Whelan, R. 1995. *The ecology of fire*. Cambridge University Press. Cambridge, England.

White, A. S. 1985. Presettlement regeneration patterns in a southwestern ponderosa pine stand. *Ecology* 66:589–594.

Wright, H. A. and A. W. Bailey. 1982. *Fire Ecology: United States and southern Canada*. John Wiley, New York.

5

ORGANIZING FOR
FIRE MANAGEMENT

INTRODUCTION

The fire started as a typical prescribed fire operation. Piles of logging debris were being ignited to reduce slash loads before replanting jack pine (Simard et al. 1983). It was early May in Michigan, rain had fallen within the previous week, and the piles were burning with flame lengths to 5 m. A few spot fires from burning embers had started outside the piles, and had been controlled. Suddenly a new spot fire began in standing timber and began to spread more rapidly. Other spot fires started across an adjacent highway, and the fire intensity began to increase with a roar. The fire began crowning and in about 3 hours traveled over 10 km, burning 10,000 ha, killing a tractor-plow operator, and destroying 44 residences. The Mack Lake fire of 1980 is a good example of an incomplete prescription, in this case a burning crew being unaware of an approaching cold front. It is but one of many examples of how fire management strategies must be closely meshed, in this case prescribed fire with fire suppression. The use of managed fire is integrally linked to the capability for fire suppression.

The experience of this past century suggests that fire suppression cannot stand on its own, either. The most expensive, high-technology fire-fighting forces in the world cannot begin to stop raging wildland fires in fuel buildups accumulated over multiple fire cycles erased from many forests. When those wildland fires encounter areas where fuels have been treated, either by manual removal or use of fire, less intense fire behavior is common and control, if desirable, often is possible. The reality of our experience, slow learners as we may be, is that all of these strategies are linked together (Pyne et al. 1996). The particular mix of strategies depends on the mix of land management objectives in the region, as fire management organizations now are typically interagency and interregional. Organizing for one purpose may help organizing for all.

The appropriate response to a wildland fire depends on an evaluation of risks, hazards, and values, the components of the fire management triangle (Figure 5.1). **Risk** is defined as the chance of a fire starting. With little or no risk, the fire management problem is minimal. Where risk increases, intensified fire prevention and

Figure 5.1

The fire management triangle. Most fire management programs are designed to balance the risks, hazards, and values within the planning area.

detection efforts are likely to be very cost-effective. **Hazard** can be generally defined as the amount, condition, and structure of fuels that will burn if a fire enters an area. Hazard reduction through management such as prescribed fire may be an effective way to decrease the fire management problem. **Values** are the net change in resource condition when a fire occurs. Historically values affected by fire were considered only to be potential losses; for example, a fire might destroy structures, or damage timber. Recently values such as wilderness character, where fire presence may be essential to maintenance of that character, or improvement of wildlife habitat or forest health, have been interpreted as potentially positive values that may result from properly managed fire. These changes are part of the transition from the era of fire control to the era of fire management, where more complex assessments of the costs and benefits of fire help define appropriate mixes of fire management strategies.

EVOLUTION OF FIRE POLICY

Humans are relative newcomers to the fire environment we call earth. Yet we have an exceptionally long cultural history with fire; it is both friend and foe, sometimes in the same event (Pyne 1995). Major policy shifts usually have been the result of large, damaging wildland fires; such fires do not always precede policy shifts, but such shifts do not occur without them.

The beginnings of institutionalized forest fire policy began around the turn of the 20th century, and were largely driven by problems occurring on forest reserves, those massive federal land withdrawals that were later to become the national forest system (Dana and Fairfax 1980). Late in the 19th century, several large fire disasters had occurred but were overlooked or overshadowed by other events (Brown and Davis 1973). The 1871 Peshtigo and Michigan fires covered over 1.5 million ha and killed at least 1,500 people, but somehow Mrs. O'Leary's cow kicking a lantern into the hay in Chicago ("The Great Chicago Fire") received most of the press coverage on fire that week. In 1884 the Hinckley fire devastated cutover Minnesota and killed 418 people, who were trapped within coalescing escaped slash burns. But the control of fire on a national scale seemed impossible in those times, both technically and economically, and would have been resisted by many residents who started many of the fires to scrape out a meager living on the land. In 1891 President Cleveland established the first portion of the system of forest reserves from the public domain that would later become the national forests, but fire control on these lands, too, would remain elusive into the 20th century.

In the first decade of the 20th century, the seeds of a national forest fire policy were sown. As professional forestry developed in the United States, the Forest Service was created to manage the federal forest reserves. Soon the forest reserves became "national forests," and the act of 1908 institutionalized deficit spending for forest fire control, but without any standards for an economically sound fire policy. This lack of standards has persisted to the present. Nonfederal landowners, meanwhile, were working on their own to control and use fire. By 1910 a number of organized forest landowner associations had sprung up in the West, primarily for the purpose of forest fire prevention and control. In the South largely unregulated but controlled underburning was practiced in the expansive pinelands of the region by both small and large landowners, and the timber industry in California was practicing underburning on its ponderosa pine lands.

The big western fires of 1910 were to change these practices (Pyne et al. 1996). Millions of acres of land, much of it national forest in Idaho and Montana, burned that summer, and 78 firefighters died. New political lobbyist Gifford Pinchot, recently fired by President Taft as chief of the Forest Service, helped the Weeks Act pass Congress in 1911, enabling the federal government to buy land for watershed protection and help states with fire protection. California was the scene of a large public debate in *Sunset Magazine* in 1920 on the practice of "light burning," but the California Forestry Commission, established to resolve the debate, concluded that fire exclusion appeared more practical than light burning (Agee 1993). A policy of systematic fire protection, or complete fire control, developed by the early 1920s and cooperative funding to the states for fire control was authorized by the 1924 Clarke-McNary Act. Some southern states had their funding contingent on eliminating traditional underburning of pine forests. Funding from the Clarke-McNary and associated acts became the cornerstone of cooperative fire programs at the state level.

In 1935, the "**10 A.M. policy**" was adopted as standard fire suppression policy on all Federal lands (Pyne et al. 1996). This policy stipulated that the control goal for every wildland fire would be to achieve control by 10 A.M. the next morning, and if unsuccessful, by 10 A.M. the following morning, until the fire was controlled. The midmorning period was chosen as the period after humidity recovery at night when control might be most successful. This policy was to remain federal policy, applied to low-value lands as much as to high-value lands, for almost 40 years. In the early 1940s, the wildland fire problem in the South became serious enough, together with manpower shortages associated with World War II, that the Forest Service formally recognized the use of prescribed fire. However, underburning did not escape from the South for use on federal lands in other regions until the late 1960s.

The end of World War II ushered in the "glory days" of fire control. Surplus military equipment and the specter of the Cold War kept resources devoted to fire control, and fire prevention became a high priority with the advent of Smokey Bear

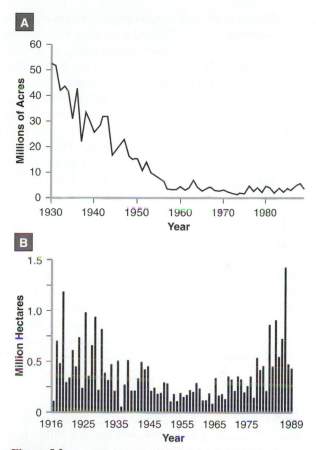

Figure 5.2

Wildland fire statistics can be deceiving. **A,** National trends from the 1930s onward show declining area burned with a few minor "bumps" in the 1980s **B,** Western trends have caused the "bumps" and appear to be accelerating the trend upward, with an additional 2+ million ha burned in 1996.

(Source: **A,** From MacCleery 1993; **B,** Agee 1993, derived from BLM sources.)

and a number of "Keep Green" organizations in western states. Fire statistics showed declines in burned area (Figure 5.2A), and the institutions were successfully achieving their goal of fire exclusion. Economics and environment eventually worked independently to reveal the shortsighted nature of the fire exclusion policy as unintended consequences (large, intense fires) began to increase.

In the late 1960s the National Park Service broadened the fire exclusion policy to recognize the natural role of fire in the evolution of the national park lands it administered (Kilgore and Briggs 1972). The Forest Service broadened its 10 A.M. policy in the early 1970s, and it evolved to a policy of "appropriate response" by the 1980s (Pyne et al. 1996). The use of fire, either prescribed or naturally ignited, and the ability to use flexible suppression responses (e.g., lower-priority response for backcountry fires where life and property threats are low), initiated a trend at the federal level of "integrated fire management," where the mix of control and use of fire could be applied in a site-specific manner. Prescribed fire flourished in the South but was not extensively used elsewhere.

The fires of 1988, including but not limited to Yellowstone, initiated a review of the natural fire policy. An interagency team responded that the policy was sound but its implementation had been flawed, and all natural fire plans were suspended pending incorporation of new management guidelines (Wakimoto 1990). Programs began slowly coming back on line in 1989 with more conservative criteria, discussed in Chapter 6.

Many of the western states began to experience the paradox of successful fire exclusion. The more successful that the fire suppression policy was in reducing wildland fire acreage, the more fuel accumulated on these lands that once burned frequently (see Chapter 4, Ponderosa Pine Forests), and subsequent wildland fires became more intense and uncontrollable. Fire statistics on western lands (Figure 5.2*B*) show a clear U-shaped trend, illustrating the initial effectiveness of fire exclusion through the 1950s and the difficulty of fire control over the last two decades. The fire management problem will remain high because the scale of the fuel buildup is so broad, and it will not be possible to treat the landscape in a very short period of time. Furthermore, many of these lands are becoming parceled into smaller and smaller fragments, and developed into homesites surrounded by hazardous fuels.

RESPONSIBILITIES FOR FIRE PROTECTION

Agencies at different levels of government have varying ability to develop integrated fire management approaches. Federal agencies, such as the Forest Service, Bureau of Land Management, National Park Service, and Fish and Wildlife Service, have many options available to them. They manage the land and can use or control fire to achieve wide ranges of land management objectives. The Forest Service, for example, may use prescribed fire to reduce fuels in frontcountry areas, allow prescribed natural fires in wilderness areas, and use fire suppression as a companion strategy in these areas and to protect life, property, and resources on all lands it manages.

State agencies have fire protection responsibility on nonfederal wildlands within each state. With the exception of Florida, states do not have the ability to implement prescribed fire on lands they do not own (Florida in 1977 passed the Hawkins Act, which allows the state to prescribe burn private lands when those lands present a hazard). Each state may contract its fire protection to others: the federal government for inholdings of state or private land within large federal reservations, or to the counties where the local fire protection capability is sufficient. Some states, such as California, have passed legislation to assist private landowners in prescribed fire operations, but this effort has generally been underfunded and not fully utilized by either landowners or the states. Between 1990 and 1995 six

Box 5.1

Tumbling Terminology

The National Wildfire Coordinating Group (NWCG), which coordinates interagency federal fire management, adopted new terminology for federal fire management in 1997. We have adopted much of the new terminology in this book, with cross-referencing as needed to the old terminology. The old terminology defined a *wildland fire* as any fire that occurred in a wildland. A **wildfire** was an unwanted wildland fire, a *prescribed fire* was the purposeful application of fire under predetermined conditions of fuels, weather, and topography to achieve management objectives, and a *prescribed natural fire* was a naturally ignited ignition that is allowed to burn under an approved management plan that includes monitoring. Under the new NWCG terminology, a **wildland fire** is any nonstructure fire, other than prescribed fire, that occurs in the wildland. It therefore includes what we used to call wildfires and prescribed natural fires in one term, but excludes prescribed fire and does not clearly define prescribed natural fire anymore. We redefine the old **prescribed natural fire** as a **managed wildland fire** that occurs in a designated zone within an approved fire management plan with appropriate prescriptions. **Prescribed fire** has essentially the same meaning under the NWCG terminology as it did under the old. Other fire management terms will be explained in the text.

This new terminology was adopted with little consultation with domestic or international colleagues. Confusion is sure to continue for years. As these changes by the federal agencies appear permanent, the NWCG must now change its name to the National *Wildland Fire* Coordinating Group, and the *International Journal of Wildland Fire*, that deals with all aspects of fire management, should perhaps change its name to the *International Journal of Wildland and Prescribed Fire*. We believe that some of the older terminology has a communication value that outweighs other considerations, and therefore we have chosen to refer to some terms by both the old and new names, and retain some terminology (such as confine, contain, and control) that have been eliminated in the new terminology.

Equivalent terms used in this text include:

This text		Old terminology
Wildfire	=	wildfire
Wildland fire	=	wildfire or prescribed natural fire
Prescribed fire	=	prescribed fire
Managed wildland fire (pnf)	=	prescribed natural fire

southern states (Alabama, Florida, Georgia, Louisiana, Mississippi, and South Carolina) passed legislation to authorize and promote prescribed burning by limiting nuisance actions and civil liability for damages or injuries from fire or smoke.

The definition of an appropriate mix of fire management strategies for forest health is most easily done on federal lands. Actions on state and private lands can be stimulated by federal programs, such as the incentive and cost-share program through the State and Private Forestry branch of the Forest Service, but these programs are woefully underfunded (National Research Council 1998). The forest health portions of State and Private Forestry are responsible for surveying for insect and disease problems, and cooperatively funding control programs for those forest health threats, but fire funding is still largely fire control–oriented at the state-private level. Federal assistance programs for fire control administered through the State and Private Forestry unit of the Forest Service provided almost $14 million to states in 1996.

FIRE PREVENTION AND DETECTION

Fire Prevention

The prevention of fire addresses the risk portion of the fire management triangle, with the intent of removing human-caused sources of wildland fire ignition. **Fire prevention** activities address the human-caused ignition problem, as lightning ignitions are essentially unavoidable. Across the United States about 90% of all fire starts are human-caused, although a smaller percentage of area burned is attributable to humans (Figure 5.3). In the Pacific States about 66% of all fire starts are human-caused; on Forest Service lands in the West, only about 33% of fire starts are human-caused (USDA Forest Service various dates). Generally, human-caused starts are higher closer to population centers and to roads or trails, so prevention strategies will differ by location. Agencies with state and private fire control responsibility usually have a higher proportion of human-caused fires than federal agencies managing less accessible and higher-elevation, lightning-prone lands.

Four conceptual approaches to fire prevention have proved useful (Table 5.1). Based on local fire cause determinations, strategies can be developed on a local basis to address chronic wildland fire causes. Campfire problems can be addressed through permits, where brief educational material or demonstrations can show campers safe ways to ignite, burn, and extinguish campfires, or by patrol, where uniformed personnel can visit campers and provide tips on safe fires. Debris fires can be reduced through similar permits and by suspending debris permits during severe fire weather. Machine fires can be addressed through spark arrestors, such as requiring them on chain saws. Arson or incendiary fires are very difficult to address through prevention techniques, but patterns of arson behavior often emerge and are solved by law enforcement.

One of the greatest public relations successes of any kind has been the Smokey Bear campaign, begun after World War II when a small bear cub was found in a New Mexico wildland fire with burned feet (Pyne et al. 1996). The bandaged bear soon healed and became a symbol of fire prevention: "Only YOU can prevent forest fires" was his deep-voiced plea. Smokey Bear has been marketed to schoolchildren and adults through posters, highway signs, pencils, key chains, T-shirts, and many other advertising "hooks" (Figure 5.4). The message has been simple and consistent over the years, although not entirely factual: Where lightning is a major cause of fires, "YOU" cannot do much about it, although

Area Burning in United States By Cause

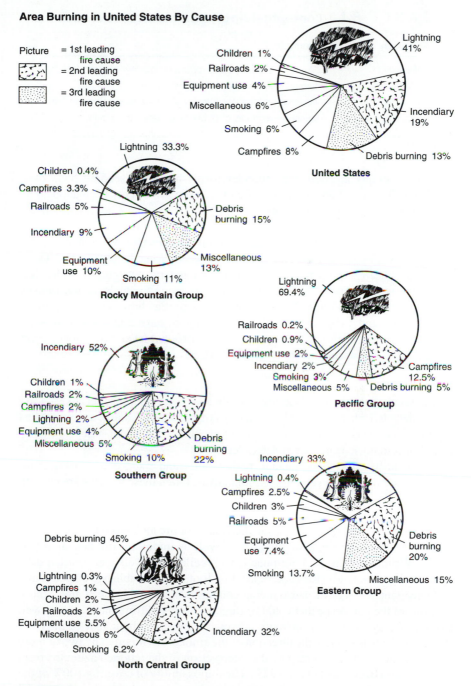

Picture = 1st leading fire cause

= 2nd leading fire cause

= 3rd leading fire cause

Lightning 41%

Children 1%
Railroads 2%
Equipment use 4%
Miscellaneous 6%
Smoking 6%
Campfires 8%

Incendiary 19%
Debris burning 13%

United States

Lightning 33.3%
Children 0.4%
Campfires 3.3%
Railroads 5%
Incendiary 9%
Equipment use 10%
Smoking 11%
Debris burning 15%
Miscellaneous 13%

Rocky Mountain Group

Lightning 69.4%
Railroads 0.2%
Children 0.9%
Equipment use 2%
Incendiary 2%
Smoking 3%
Miscellaneous 5%
Campfires 12.5%
Debris burning 5%

Pacific Group

Incendiary 52%
Children 1%
Railroads 2%
Campfires 2%
Lightning 2%
Equipment use 4%
Miscellaneous 5%
Smoking 10%
Debris burning 22%

Southern Group

Incendiary 33%
Lightning 0.4%
Campfires 2.5%
Children 3%
Railroads 5%
Equipment use 7.4%
Smoking 13.7%
Debris burning 20%
Miscellaneous 15%

Eastern Group

Debris burning 45%
Lightning 0.3%
Campfires 1%
Children 2%
Railroads 2%
Equipment use 5.5%
Miscellaneous 6%
Smoking 6.2%
Incendiary 32%

North Central Group

Figure 5.3

Fire origins across the United States. Nationally, human-caused fires are the most important source of area burned, but regionally the trends differ.

(Source: From USDA Forest Service.)

T A B L E **5.1** | **Conceptual Approaches to Fire Prevention**

Approach	Example
Self-interest	"Your tax dollars go to suppressing careless wildland fires."
Sentiment	Bambi, Smokey Bear: wildlife will be hurt.
Citizenship	Important during WWII, especially: be a patriot, prevent fires.
Compulsion	Temporary closures of public lands, law enforcement.

abortive attempts at "defusing" thunderstorms by cloud seeding have been attempted. The use of an animal as a fire prevention symbol has been adopted by many other countries and Canadian provinces: Alberta has a beaver, Quebec a chipmunk, and Mexico its own bear, Simone.

Another national media campaign is "Keep America Green," implemented at the state level by placing the state's name between "Keep" and "Green." These often have been a cooperative effort between federal agencies in individual states, the state forester, forest industry, and private individuals, using Smokey Bear literature and posters. Budgets for such activities have been declining in recent years, and some states, such as Washington, have abolished their Keep Green campaign, hoping that fire prevention at national level (the Smokey campaign) and the local fire district level will suffice.

Fire Detection

Emphasis on detecting wildland fires has always been important, because if a fire can be detected when very small, its chances of immediate control are high. Most fire management organizations use a mix of ground-based, aerial-based, and automated or remote-sensing–based detection.

Ground-based systems include both mobile and fixed detection systems. Mobile systems include vehicle and foot patrols. Often these systems are put into place when fire danger indices (see Chapter 3) reach certain levels. Campers often are credited with early fire reports. Permanent fixed detection systems were installed widely during the 1930s, and the system of fire lookouts on isolated peaks (Figure 5.5) became romantic outposts as well as important fire detection networks. Writers such as Jack Kerouac have recalled in novels their spartan days serving as fire lookouts (Kerouac 1958). Lookouts were established on peaks that had excellent views of the surrounding countryside, and their "seen area," those portions of the landscape that could be directly observed from the lookout tower, together with that of other lookouts, adequately covered about 75% of the land. When smoke was sighted, the direction of the smoke was reported by two or more lookouts and located on a map by the intersection of the reported azimuths from the lookouts (Brown and Davis 1973). The impossibility of covering 100% of the landscape, the declining usefulness of fixed lookouts under smoky conditions, and the advent of the airplane for detection caused many lookouts to be abandoned, but some are still used regularly.

Figure 5.4
Posters of Smokey Bear, the United States forest fire prevention symbol.
(Source: From USDA Forest Service.)

Figure 5.5
This lookout at the Oregon-California border is typical of many lookouts across the West still in use. Volunteers live in the structures during fire season and report fire azimuths by radio to local ranger stations.

Aerial detection has largely replaced lookouts in many areas of the country. Where lightning is an important cause of fires, aerial reconnaissance after thunderstorms is very effective at spotting fires. Although intermittent, aerial detection also is useful during periods of high fire danger. Once a fire is spotted, fairly exact location, burning conditions, and possible access routes can be relayed quickly to fire suppression forces. Data on remotely sensed lightning strikes (cloud-to-ground) now are routinely available in both Canada and the western United States (Pyne et al. 1996). Such information, while not specifically identifying fires, shows concentrations of lightning activity and where fires may be most likely to have started. Together with parallel information on storm tracks, fuels, and potential fire behavior, these data help to guide the fire control strategy of initial attack.

FIRE SUPPRESSION

Fire Control Organization
There are five functions necessary on any wildland fire: command, planning, operations, logistics, and finance. On very small fires, these functions may be filled by one or two individuals, but on large fires each function may be staffed by many people. The common nature of these functions on wildland fires and other emergency situations encouraged the development of a national system to deal with such incidents: the **National Interagency Incident Management System** (NIIMS).

The Incident Command System (ICS) was developed by NIIMS to manage a wide variety of emergency situations: earthquakes, floods, and fires, among others (Figure 5.6). It defines a modular system of organization that becomes more completely staffed as the size of the incident increases (Pyne et al. 1996). The Command function of ICS coordinates the other four groups within the ICS structure, making sure that resources are being directed and managed efficiently and effectively toward the objective. The task of the Planning function is to develop information to be used in a detailed fire suppression plan that will be updated as conditions change. The Finance function keeps the financial records for the fire, making sure that equipment, food, and firefighters are appropriately paid. The

Incident Command System Organization

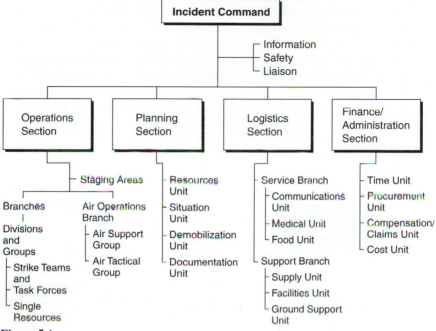

Figure 5.6

The Incident Command System has five major functions and is used in concept, if not direct application, by all fire control organizations.

(Source: From National Advanced Training Center, Marana, Arizona.)

Logistics function provides all needed support for the operation, including equipment, tools, food, camps, and medical services. Logistics includes the right stuff in the right amount in the right place at the right time. Given the fast mobilization on wildland fires, a good Logistics function is essential and very challenging, although not quite as glamorous as Operations. The Operations section directs the actions of ground and aerial resources actually used to suppress the fire, and focuses on effective tactics to suppress the fire. Units within Operations may be strike teams of engine crews, hand crews, bulldozers, and air attack units. ICS teams train as a cadre, and may come from multiple agencies, but can be mobilized on very short notice.

When wildland fires are first detected and attacked, the entire ICS organization may consist of a lead firefighter, or incident commander, and a brush truck or a few firefighters. When such fires escape initial attack and become larger, the ICS structure grows with it, with more individuals filling the functional roles within the ICS. Hundreds to thousands of people can be contained within a single ICS organization on large wildland fires. On very large events, there may be need to subdivide some of the functions beyond that shown in Figure 5.6, or to coordinate one incident with another (e.g., two large wildland fires in the same vicinity) in order to allocate limited resources appropriately between incidents.

NIIMS is a federal approach to emergency situation management, and although there are alternative approaches to emergencies by state and local fire organizations, all essentially define similar functions. ICS is the standard method of organization now in the United States due to its flexibility and use in wide variety of emergencies. It also is used in forest health applications other than fire. Spraying of urban areas for Asian gypsy moths has been done in the Pacific Northwest using ICS structure, with aerial support for application of bacterial pesticides (see Chapter 20) organized as strike teams under the Operations function.

Fire Suppression Strategies

Fire control strategies are planning and coordination activities designed to achieve control objectives consistent with agency policy and land management objectives. There are three basic suppression strategies used in wildland fire control: **confine, contain,** and **control** (Pyne et al. 1996). These terms are no longer used in the federal terminology (see Box 5.1), but will be used here as they clearly help to define the ways that firefighters approach the suppression task. A confine strategy defines certain geographical or fire prescription limits as the acceptable control strategy. A contain strategy involves construction of a fireline completely around a fire, although not always right at the fire edge, and a control strategy is the most aggressive, with a line around the fire and no chance that it will flare up again and cross the fireline. In practice, portions of large wildland fires will be suppressed using some of each of these strategies, and some portions of the line will progress through confine, contain, and control strategies through the life of the fire. Definition of suppression strategies is necessary to define appropriate tactical approaches for the fire. Tactical decisions are more specific in nature, and might include the exact location and width of the fireline, burnout operations, locations of helispots, and retardant drop heights.

Confine strategies are by their nature extensive fire control strategies, used primarily on public lands and on backcountry wildland fires. They may be the only suppression means available in a time of extensive regional fire activity, and are applied on the lower-priority fires. Confine strategies often are used in remote, inaccessible areas where more aggressive strategies will be expensive and of dubious value in control, either from cost-benefit or safety perspectives. Contain strategies are intermediate between confine and control in terms of aggressiveness, as the fireline is constructed at a distance from the actual flaming front. Safe and effective fireline construction is possible, from which firing out unburned terrain within the established fireline (called **backfiring**), or extinguishment if the flaming front is approachable, can occur. Control strategies are the most aggressive fire suppression strategies. Most initial attack tactics are attempts to construct the fireline directly adjacent to the fire to keep it small and from spreading. On high-value lands a higher proportion of control strategies will be applied than in backcountry areas.

Just as the strategies for fire suppression may be a mix of confine, contain, and control, fire tactics will include a mix of direct and indirect attack. During the

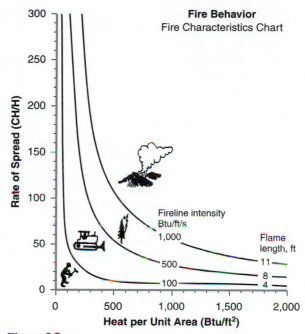

Figure 5.7

The fire characteristics chart (Andrews and Rothermel 1982). Several characteristics can be shown as a single point on the chart, and help to understand fire suppression interpretations. Hand crews can work in the area shown by the firefighter, while equipment can work under more severe fire conditions. Torching and crown fire are expected when surface fires become very intense. Tactics may depend on the evaluation of potential rate of spread and energy release effects on fireline intensity. This chart can also be related to the National Fire Danger Rating System (see Figure 3.12): Spread Component would replace Rate of Spread, Energy Release Component would replace Heat per Unit Area, and Burning Index (scaled to flame length) would replace Fireline Intensity

(Source: From USDA Forest Service.)

initial phases of the fire, the first firefighters on the scene will **size-up** the fire, or evaluate the character of the fire with the available suppression forces. The types of fuel burning, the direction of the fire, safety considerations, possible placement locations for firelines, and an assessment of additional forces needed are all a part of the size-up process. At that time a decision to use direct or indirect attack will be made. One useful chart (Andrews and Rothermel 1982) to assess different strategies is the fire characteristics chart (Figure 5.7). From fire behavior models, an estimate of rate of spread and heat per unit area can be obtained, and that point can be plotted on the chart. Short grass fuels would have high rates of spread but low heat per unit area; brush fuels would have intermediate spread and heat per unit area; slash fuels would have lowest rates of spread but high heat per unit area. Other aids to determining mode of attack are seasonal trends by different vegetation groups (Chapter 3, Figure 3.8*C*).

Direct attack is commonly used as part of the initial attack process. It is an aggressive tactic designed to put most of the effort right at the edge of the wildland fire and hopefully control it while very small. Three basic direct attack tactics are commonly used (Figure 5.8). The first is frontal attack, also known as "hitting the head" of the fire. The intent is to slow the spread of the fire along the perimeter where it is likely to be moving the fastest. This can be dangerous and generally is attempted by experienced crews only. Successful pinching off the rapidly moving head of the fire allows construction of a fireline along the more slowly spreading flanks and back of the fire. Flanking attack is a second direct control tactic, and proceeds by starting at the backside of the fire and building a fireline along the flanks toward the head of the fire. This usually is a safer environment for crews, particularly if they are inexperienced. The third direct control tactic is hotspotting, or hitting the vital points of fire spread first, and tying these portions of a fireline together later. On larger wildland fires the various direct attack tactics will be used simultaneously on different portions of the fire perimeter, and have both advantages and disadvantages (Figure 5.9).

Indirect attack is a tactic that attempts to achieve control by backing off the edge of the fire and establishing a fireline in places most likely to successfully hold as the fire approaches. From a safety perspective, or when fires are crowning, direct control is simply not possible or effective. Trails, roads, streams, meadows, or other breaks in fuel or terrain are likely places for control to be most effective. Generally there will be unburned area between the established line and the flaming front, so that this area must be burned out (backfiring) unless the fire is anticipated to approach the line with fire behavior incapable of jumping the line. If fire behavior decreases before the fire reaches the indirect control line, more direct control may be applied along the flaming front to achieve control more quickly and avoid extended periods of burnout within the line. As with direct attack, there are pros and cons of using indirect attack (Figure 5.10). Backfiring in particular is technical and sometimes risky, working best in uniform, light fuels. Ignition techniques are typically those used in prescribed burning (see Chapter 6).

The objective of all control tactics is to place a fireline adjacent to a cold, blackened area so there is no chance that the fire will jump the line (Chandler et al. 1983). Natural barriers such as streams, cliffs, and talus slopes may be used in place of a fireline if they are wide enough and fuels are discontinuous. Typically firelines are placed on ridgetops or along valley bottoms, and tied together with ridge-to-valley lines constructed when it is safe to do so. Often the downvalley edge of the fire will be safest during the time of day when upvalley winds are pushing the fire upvalley, and the upvalley portion safest in the evening when upvalley drafts cease. Regional physiography may, however, reverse this situation—for example, afternoon downvalley winds in east-facing canyons along the Pacific Coast due to sea breezes. A fireline is typically built uphill to keep the fire above equipment and crews.

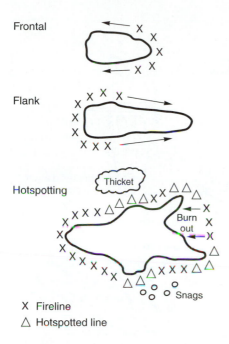

Frontal

Flank

Hotspotting

X Fireline

△ Hotspotted line

Figure 5.8

Three tactics of direct control: hitting the head or frontal attack, flanking, and hotspotting.

(Source: From Introduction to *Wildland Fire,* by Pyne, Andrews and Laven, ©1996, John Wiley & Sons. Reprinted with permission of John Wiley & Sons, Inc.

At the time a small wildland fire escapes initial attack, or a prescribed fire exceeds prescription, a formal evaluation of the alternative control options, called a **Wildland Fire Situation Analysis** (WFSA), is completed. The WFSA is the new (1997) term for the Escaped Fire Situation Analysis (EFSA), and more accurately represents the analysis; many fires that underwent an EFSA never were controlled to the point that they "escaped!" In essence, it is simply a formalized way of expressing a logical thought process about the variables involved in selecting a control alternative. In the WFSA the current fire conditions, both local and regional, are described, along with major economic, environmental, and social criteria to be met. Alternative strategic and tactical plans are outlined along with estimated probabilities of success and control time. Effects of alternatives and an evaluation of alternatives in relation to management criteria are summarized and the selected alternative justified. The WFSA is intended to be altered or updated as conditions change.

Fire Equipment

Equipment used to control and use fire has a long and inventive history. Forest fire equipment is intended to remove one of the three legs of the fire triangle: fuel, heat, or oxygen, or in the case of ignition devices, to add heat. There are many different tools that have been used in different regions, and most are flexible enough to combine several functions (Chandler et al. 1983).

The basic hand tools include shovels, rakes, Pulaskis (a combination ax and grub hoe), rake/hoe combinations such as the McLeod tool, and swatters or beaters. In heavier fuels tools like the Pulaski are most flexible; in light fuels swatters

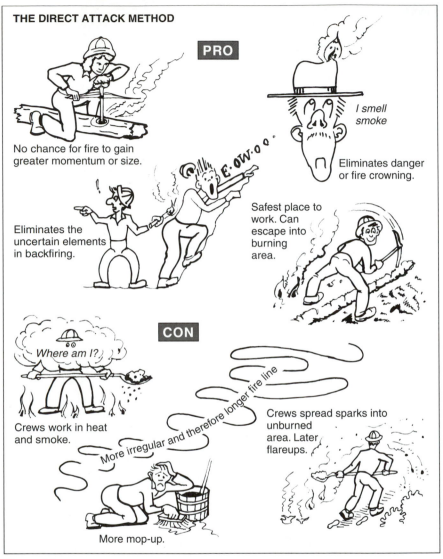

Figure 5.9
Pros and cons of direct attack.
(Source: From USDA Forest Service and California Division of Forestry, *Fundamentals of Forest Fire Fighting*.)

and beaters are very effective in smothering flames. Portable water containers fitted with hand-operated pumps and nozzles are commonly used on the fireline, and metal water containers have largely been replaced with rubber bags that carry about 20 liters of water. The nozzle has several settings that range from close spray to streams that carry some distance. Mechanized hand tools like the power saw have significantly increased production rates for fireline construction by hand-crews in heavy fuels.

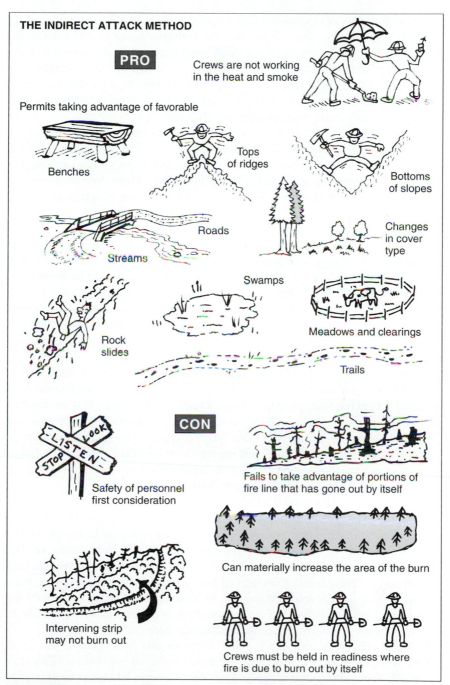

Figure 5.10

Pros and cons of indirect attack.

(Source: From USDA Forest Service and California Division of Forestry, *Fundamentals of Forest Fire Fighting*.)

A wide variety of mechanized equipment is available for fire control purposes (Pyne et al. 1996). For line construction the bulldozer is commonly used in the West, as it is better adapted to working in steep terrain than equipment like the tractor-plow, commonly used in the South and north-central states. Its effectiveness at linebuilding and clearing often requires substantial land rehabilitation after the fire, so bulldozers are used less in park and wilderness areas than in past decades, and judiciously used in frontcountry areas.

Explosives have been used to construct fireline in areas without a lot of heavy slash or extremely rocky terrain. A thick rope of explosive cord is placed along the ground and detonated from a distance. The need for specialized personnel to handle the explosive has restricted its use to larger fires where crew resources are limited, such as the Yellowstone fires of 1988, but it has seen widely scattered use across the West.

Fire engines adapted for wildland fire are used to spray water, foam, and retardant. The equipment used is generally of a type that is capable of negotiating the terrain in the local area, with smaller units being more maneuverable in rougher terrain. "Slip-on" units can be fit into the bed of trucks usually used for other functions. Other trucks are designed specifically for fire control purposes. Engine teams from various fire districts often train as teams and are used in ICS situations as "strike teams" that effectively work as a unit. The use of engines from different districts allows a quick response of engine teams without decimating the ability of any single local district for continued response at the local level.

The development of **fire retardants** is one of the great success stories of fire control. Early fire retardants had a clay base that contained boron and had the deleterious effect of retarding plant growth after fire. The development of ammonium phosphate and ammonium sulfate–based retardants, containing nitrogen and either sulfur or phosphorus, which are limiting elements for plant growth, have turned a source of environmental damage into a source of environmental rehabilitation through the fertilization effect of the retardant slurries. Nevertheless, these retardants must not be dumped into watercourses as they have been associated with fish kills. The use of foam for wildland fires is more recent. Foams are essentially water with added detergent, and have a stable bubble structure that wets fuels and acts as a heat sink. They have commonly been used for structural protection as they are less damaging to structures than retardant, which tends to be corrosive. The environmental effect of foam on fish populations has not been widely studied. Foams now are being used for firelines on both wildland fires and prescribed fires. Generally they are not intended to serve as a permanent fireline as they do not remove fuel but simply wet it. Line construction or burnout from a foam line to create blackline is preferred once the wetline has been established.

Aerial support is provided by both airplanes and helicopters (Pyne et al. 1996). A wide variety of aircraft have been used for aerial application of water or retardant. Some airplanes can scoop water directly from lakes, but most require loading from a land base. Accurate delivery of water or retardant is aided by a smaller lead plane that directs the larger craft to the desired location. Aircraft are strategically placed in regional centers so they can be directed to fires and arrive while the fire is small, to supplement initial attack efforts. They also are useful in extended attack situations. Helicopters equipped with buckets can drop water or retardants very precisely but usually are limited by the size of the bucket that can be carried underneath the craft. Generally aerial support is the most expensive fire control activity, with each drop of fire retardant from a fixed-wing plane costing multiple thousands of dollars.

Irrigation systems have been effectively used on wildland fires, particularly where a gravity feed will be possible. A large portable tank is assembled on safe, high ground, and filled by water trucks carrying water from caches or nearby water sources. Pipe is then assembled downslope and nozzles attached at close spacing along the fire perimeter. The pressure of gravity creates a strong, continuous water spray along the fireline.

Fire ignition devices are used in backfiring operations as well as for prescribed burning. The simplest tool is the match, but it suffers from a limited flaming life. Fuses similar to those carried in automobiles are the next step up in complexity and have a longer flaming life. The drip torch is perhaps the most common ignition device: a canister filled with kerosene or a gas/oil mixture and containing an extended metal tube with a wick on the end past which the fuel moves when the canister is tipped. It spreads a flaming line of fuel on the ground that remains ignited for a number of seconds. Larger canisters with pumps attached can be pressurized, creating flamethrowers that are essentially drip torches with pressure. They can be used to shoot flame at distance and concentrate heat to ignite heavier fuels or live fuels.

Aerial ignition can be done from airplanes or helicopters. A common aerial ignition device is the flying drip torch, a giant version of the hand-held drip torch but slung under a helicopter and operated from the cockpit. It drops copious burning fuel along the flight line when activated, and can be very useful in achieving rapid ignition in inaccessible terrain for slash burning or backfiring. A technique developed in Australia and now used in North America is the "ping-pong ball," a plastic ball filled with potassium permanganate that is injected with ethylene glycol, dropped from the plane, and then explodes from the chemical reaction in a short time. These work best in fine fuels that can be ignited with limited energy release from the ping-pong ball. They have been used to ignite large areas in a short period of time for prescribed burning, or to ignite backfires within contained wildland fires.

Safety in Suppression

Safety in the presence of fire, whether it is a wildland or prescribed fire, is the top priority in all fire management operations. Fatalities directly in the line of fire account for about two-thirds of all fatalities on fires, with aircraft, vehicle, medical, and miscellaneous categories making up the other third. Identifying the source of most of these situations resulted in the Ten Standard Firefighting Orders (Table 5.2) being issued in the 1950s, that were later supplemented with a series of "Watch Out!" situations that began in the 1960s with a list of 13, grew to 18, and now number over 20 (Pyne et al. 1996). A shorter list that combines many of

T A B L E **5.2** | **The 10 Standard Firefighting Orders**

F	Fight fire aggressively but provide for safety first.
I	Initiate all action on current and expected behavior of fire.
R	Recognize current weather conditions and obtain forecasts.
E	Ensure instructions are given and understood.
O	Obtain current information on fire status.
R	Remain in communication with crew members, your supervisor, and adjoining forces.
D	Determine safety zones and escape routes.
E	Establish lookouts in potentially hazardous situations.
R	Retain control at all times.
S	Stay alert, keep calm, think clearly, act decisively.

the Standard Orders is LCES: Lookouts, Communications, Escape Routes, and Safety Zones (Chamberlain 1998). When these directives are used as guidelines rather than orders, fatalities are more likely to occur (see Box 5.2 and Figure 5.11).

There are four common denominators of fire behavior on fatal fires (Wilson 1977). These denominators also are common on fires where nonfatal injuries occur. Crew members may be ignorant of the directives for safety, may choose to ignore them, or are caught off-guard by situations that appear more benign than they really are. They typically have occurred:

1. On relatively small fires or deceptively quiet sectors of large fires
2. In relatively light fuels, such as grass, herbs, and light brush
3. When there is an unexpected shift in wind direction or windspeed
4. When fire responds to topographic conditions and runs uphill

Adequate equipment for crew members is essential. Flame-resistant clothing, such as Nomex, offers protection against radiant heat and if subject to embers will char rather than flame. Sturdy boots, gloves, and hard hats complete the basic fire uniform. Fire shelters carried on the belts of firefighters reflect radiated heat when deployed and wrapped around a firefighter's prone body. Although firefighters joke about the deployed shelter looking like a baked potato, dozens of firefighters have been saved from serious injury when shelters reflected radiant heat and shielded the face and airways of trapped firefighters from flames and gases. Shelters are best utilized when firefighters have small open defensible spaces where they can cover themselves with the shelter. Vehicles often have been used successfully to shield firefighters as the fire passes. If the gas tank is not ruptured, the tank will not explode, and the cab, even if smoky, may be the best available place to ride out the pass of a rapidly moving fire.

Physical conditioning of fire personnel is essential for their safety. Working in sometimes difficult terrain, under hot and smoky conditions, can produce substantial stress on the body, even for individuals in good shape. Three tests have been used to measure aerobic capacity: a fixed time or distance run; a fixed distance hike with heavy pack; or the step test, a 5-minute exercise of stepping up

Box 5.2

The 1994 South Canyon Fire Tragedy

In early July 1994 a lightning-ignited wildland fire occurred just off Interstate 70 in Colorado. Known as the South Canyon Fire, it "blew up" four days later and killed 14 firefighters (Figure 5.11) Initially it was classed as a lower-priority fire, as at least 40 fires were ignited by the lightning storm that passed through the area. After several days initial attack was begun, and a helispot was constructed and several retardant drops were made. On July 6 the seven-person BLM/Forest Service crew and eight initial smokejumpers were joined by eight more smokejumpers and 10 Prineville, Oregon, "hotshots." In midafternoon on July 6, a dry cold front moved into the area, and the fire made several rapid runs with 100-ft flame lengths. At 4 P.M. the fire crossed the bottom of the drainage (see black "1" on Figure 5.11), pushed by 50 kph winds, moving onto steep slopes and burning through dense, highly flammable Gambel oak that had earlier underburned, desiccating the foliage. Within seconds the fire raced up the ridge with 60 m flames, catching the firefighters who had been building line downhill from the ridgetop. Eight of the ten Standard Firefighting Orders were determined by the investigative team to have been compromised by the crews. For a similar account of a fatal fire, the 1949 Mann Gulch fire in Montana, read Norman Maclean's 1992 book *Young Men and Fire*. See also Butler et al. (1998).

and down a 40 cm step 90 times a minute, with pulse rate measured 15 seconds after the 5 minutes has elapsed. Each of these tests can identify individuals who may have physical problems on active fires.

WILDLAND FIRE REHABILITATION

The damaging effects of wildland fires are to some extent unavoidable. Damage can arise from the direct effects of the fire, and direct effects of the firefighting effort. Direct effects of the fire include loss of plant and litter cover, eventual loss of root strength, and removal of physical barriers to soil movement such as logs. Direct effects of the suppression effort include bulldozer trails or hand line on steep slopes or through areas with sensitive plant species, retardant dumps into streams, and poor sanitation. There is substantial effort during the suppression phase of the fire to avoid these suppression impacts to the extent possible. A resources advisor often will be part of the ICS structure in the Planning section. As part of the suppression effort, a rehabilitation team will be formed to identify appropriate techniques, some of which are applied while the fire is still burning. In order for federal agencies to use emergency funds for rehabilitation, a plan must be complete within a short time after the fire is controlled and must be implemented immediately. Federal land management agencies have developed Burned Area Emergency Rehabilitation (BAER) teams to manage the rehabilitation planning process.

Fire Blowup

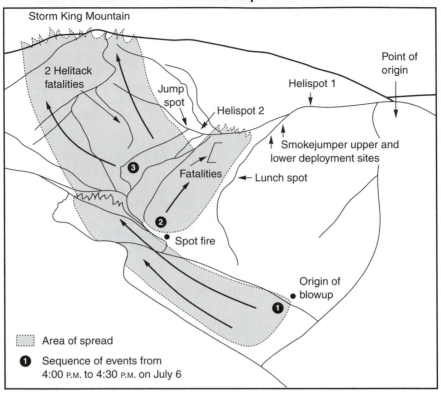

Figure 5.11

The blowup of the 1994 South Canyon fire. North is to the left. Circle 1 is the origin of the blowup. Circle 2 is the spot fire that ignited below 12 of the firefighters and overtook them. Circle 3 indicates a change in direction of the fire that overcame two Helitack crew members who initially chose to move northwesterly rather than over the ridgeline to the east and were cut off from that escape route when the fire at Circle 2 reached the ridgetop.

(Source: From USDA-USDI 1994.)

The usual objectives in rehabilitation are to prevent downstream damage and maintain site productivity. The treatments used in wildland fire rehabilitation include seeding grasses and forbs, mulching, contour felling, and checking dams in streams (Agee 1993). Probably the most controversial treatment is grass seeding. The intent of grass seeding is to provide immediate cover over the burned landscape, but grass cover doesn't appear until after the first few rains. It can provide a dense cover later, which may help stem erosion but also provides substantial competition to native plants, particularly those that recover by seed. Nitrogen-fixing plants like *Ceanothus* in the West can be essentially eliminated if grass seeding is heavy, and this can have damaging effects on long-term site productivity (Figure 5.12). Nitrogen, which is closely tied to site productivity, is volatilized by

Figure 5.12
Excessive grass seeding can have positive short-term effects but negative long-term effects by outcompeting native vegetation, including nitrogen-fixing plants.

fire, and obligate seeding species that fix nitrogen back into the ecosystem are choked out by grass competition and cannot help restore the nitrogen capital of the site. Grass seeding is generally not included in park and wilderness rehabilitation plans, but is usually considered on other lands. It is cheap to apply and the land looks green quickly, providing at least the appearance of recovery. Areas without pre-burn grass and shrub cover are the most effective areas to seed if grass seeding is selected as an appropriate rehabilitation technique.

Mulching is another way to provide cover, and can occur immediately after the fire. Mulching can be effective along roads, where machines can spread the straw, but is difficult to apply outside of developed areas. Contour felling of logs is an attempt to lay medium-sized logs across the slope to trap sediment moving downs-lope. In most applications their efficacy is questionable and the cost of felling these trees and filling small holes underneath them is very high. Check dams are also used to capture sediment, at the early stages of its entry into stream channels. Small dams of straw, or longer-lasting dams of logs or sandbags, can be constructed, but these must be maintained by periodic cleaning, and energy-dissipating devices must be placed at the base of the dam, so the dam is not undermined by erosion. Obtaining the expertise of a hydrologist and/or geomorphologist is essential if stream reha-bilitation is to be undertaken.

In recent years more attention has been given to providing structure for cavity-nesting animals. Many logs or snags with cavities are consumed in wildland fires, and although many snags often are created, they tend to be "hard" snags that will not develop cavities for many years. Construction of artificial cavities can provide needed habitat for birds and small mammals such as bats, rodents, and small carnivores.

THE URBAN WILDLAND INTERFACE

The spread of residential development into wildland areas is probably the most pressing fire management problem, as it is serious now and will become worse into the 21st century. Rural development in Virginia has increased 400% in the 1980 to 1990 period, and 8 of the 10 most destructive residential wildland fires in California history have occurred since 1980. The urban-wildland interface (or intermix [Pyne et al. 1996]) fire problem has a direct relationship to forest health because residential areas compete effectively for resources during fire suppression efforts, so that surrounding forest and shrubland fire tactics are not only compromised but also sometimes so low priority that much more area burns. Furthermore, presuppression forest health treatments intended to reduce fire hazard are more difficult in the urban-wildland interface: Prescribed fire may be more risky with high-value structures mixed among the wildland fuels, and residents may complain about smoke. The urban-wildland interface problem is not going to go away, but there are ways that the problems can be approached to protect the values in residences and surrounding wildlands (Fischer and Arno 1988). Standards for the protection of homes are well known (Figure 5.13), but few communities have incorporated them into zoning statutes, preferring to let the buyer beware.

Subdivisions should be planned with multiple outlets for escape routes, and the wider the road and lower the gradient, the higher the chance fire trucks can access the driveway. Driveways with turnarounds of 30 m are preferred so that equipment can enter and exit the driveway. Dead-end roads with no turnarounds place those homes at the bottom of the triage list. Homes with independent water sources (swimming pools, ponds, cisterns) can provide sprinkler-delivered water on gravity feeds or gas-powered pumps to the roof of the home. Hydrants close by are a plus, but may be overtaxed and lose pressure during a large fire event.

Most wildland fires ignite homes by igniting the roof, so having a non-combustible roof, such as tile, composition shingle, or metal, is preferred to wood shakes or shingles. A wood roof can be treated with a flame-resistant chemical, but this has to be repeated annually. Keeping the roof free of litter also is important. Noncombustible siding also is important, and enclosed eaves, decks, and balconies prevent heat from building up underneath combustible surfaces. Storing woodpiles away from the home is essential.

Landscaping needs to accommodate fire-safe principles. Although scalping all vegetation for several hundred feet is probably most efficient from a fire standpoint, it is not needed to provide reasonable safety during a fire event. Irrigated grass surrounding the house is preferred if water is available, but low density/volume native shrub cover also will suffice. Wildland grass should be mowed or "weed-whacked" when cured. Trees should be pruned up to 3 to 4 m so that it will be difficult for fire to jump into the tree crowns. Tree crowns should be spaced and not overlapping the roof, and mixtures of deciduous hardwoods with conifers is preferred to all conifers because the hardwoods have higher foliar moisture. If the house is situated in the middle of flat ground, roughly 20 m of treated landscape each direction from the house will provide adequate buffering; extended zones (up to 100 m) may be needed in hilly or mountainous terrain.

The larger landscape within which the homes are sited also may be treated to reduce surface and crown fire potential. Removal of surface fuels by pile burning or prescribed fire, and crown thinning by mechanical or manual techniques, or prescribed fire, may help fire control personnel stop wildland fires before they reach residences. Techniques to do this are discussed in Chapter 6. Even if such

PROTECTING RURAL HOMES FROM WILDFIRE

Roof

Fiberglass, metal or concrete tiles. Do not use shakes. If you do, treat them with fire retardant.

Keep roof and gutters clean of leaves and pine needles. Avoid flat roofs where they can collect.

Trees

No branches hanging over house.

Prune trees nearest to your house of limbs to 10 ft high.

Use fire resistant plants and shrubbery.

Debris

Keep brush, weeds, and debris at least 30 ft away from your house.

Do not pile firewood against your house or deck.

Water

Create a small pond, cistern, well, or hydrant for firefighting. Have a hose or pump if need be.

You

You and your family should practice fire drills and escape routes.

Figure 5.13

Positive measures can be applied to protect property and people. The concept of "defensible space" can be designed into new homes or retrofitted to older ones.

(Source: From *Chelan County Homeowners Guide*, Chelan County, WA.)

treatments are applied to surrounding landscapes, fires will remain a threat, and individual owners should not rely on firefighting forces to protect every residence. Homeowners can provide 90% of their own protection by applying these simple guidelines to their property (Box 5.3).

REFERENCES

Agee, J. K. 1993. *Fire ecology of Pacific Northwest forests*. Island Press, Washington, D.C.

Andrews, P. L., and R. C. Rothermel. 1982. *Charts for interpreting wildland fire behavior characteristics*. USDA Forest Service General Technical Report INT-131.

Brown, A. A., and K. P. Davis. 1973. *Forest fire: control and use*. McGraw-Hill, New York.

Butler, B. W., R. A. Bartlette, L. S. Bradshaw, J. D. Cohen, P. L. Andrews, T. Putnam, and R. J. Mangan. 1998. *Fire behavior associated with the 1994 South Canyon fire on Storm King Mountain, Colorado*. USDA Forest Service Research Paper RMRS-RP-9.

Chamberlain, P. 1998. LCES workshop. *Wildfire* 7(5–6):17–18.

Chandler, C. C., P. Cheney, P. Thomas, L. Trabaud, and D. Williams. 1983. *Fire in forestry*. John Wiley and Sons, New York.

Box 5.3

The Top 10 Techniques

The Firewise House
1. Use noncombustible roofing material.
2. If you have a combustible roof, treat it annually with a flame-resistant chemical.
3. Maintain access for fire trucks to enter and turn around.
4. Remove pine needles from roofs and gutters.
5. Enclose underfloor spaces and eaves.
6. Have a water source with independent gravity feed or gas-powered pump.
7. Store all flammables well away from the house.
8. Use noncombustible siding or treat annually with flame-resistant chemical.
9. Screen all vents with wire mesh no greater than 1/4 in. mesh.
10. Have your address clearly displayed from the road.

The Firewise Landscape
1. Reduce piles of debris near the house.
2. Don't stack firewood near the house.
3. Mow grass to keep fuel volume low to the ground.
4. Prune shrubs and/or trees to separate ground from aerial fuels.
5. Space trees so that crowns do not touch one another.
6. Plant low-volume shrubs.
7. Create a 10 m wide zone of very sparse fuel around the house.
8. Irrigate the landscaping in the vicinity of the house.
9. Favor less flammable vegetation (local lists are available).
10. Create a 30 m defensible space surrounding the 10 m primary zone, with increasing distances on steeper slopes.

Dana, S. T., and S. Fairfax. 1980. *Forest and range policy, its development in the United States.* McGraw-Hill, New York.

Fischer, W. C., and S. F. Arno (comps.) 1988. *Protecting people and homes from wildfire in the Interior West: proceedings of the symposium and workshop.* USDA Forest Service General Technical Report INT-251.

Kerouac, J. 1958. *The Dharma bums.* Viking Press, New York.

Kilgore, B. M., and G. Briggs. 1972. Restoring fire to high elevation forests in California. *Journal of Forestry* 70:266–271.

MacCleery, D. 1993. *American forests, a history of resiliency and recovery.* Forest History Society, Durham, N.C.

National Research Council. 1998. *Forested lands in perspective: prospects and opportunities for sustainable management of America's nonfederal forests.* National Academy of Sciences, Washington, D.C.

Pyne, S. J. 1995. *World fire: the culture of fire on earth.* Henry Holt, New York.

Pyne, S. J., P. L Andrews, and R. D. Laven. 1996. *Introduction to wildland fire.* John Wiley and Sons, New York.

Simard, A. J., D. A. Haines, R. W. Blank, and J. S. Frost. 1983. *The Mack Lake fire*. USDA Forest Service General Technical Report NC-83.

USDA Forest Service (various dates, annual). *Wildfire statistics*. USDA Forest Service, Division of Cooperative Fire Control. Washington, D.C.

USDA-USDI 1994. *Report of the South Canyon fire accident investigation team*. Joint Report, U.S. Forest Service and Bureau of Land Management, Boise, Idaho.

Wakimoto, R. H. 1990. The Yellowstone fires of 1988: natural process and natural policy. *Northwest Science* 64:239–242.

Wilson, C. C. 1977. Fatal and near-fatal forest fires: the common denominators. *International Fire Chief* 43(9):9–15.

6

FIRE STRATEGIES FOR FOREST HEALTH

INTRODUCTION

The wildland fire was crowning in 60-year-old ponderosa pine and Douglas-fir that naturally regenerated after railroad-logging of the old-growth pine in the 1920s. Under the severe fire weather conditions of 1994, the fire was killing all the trees in its path and threatening homes set among the trees. As the crown fire approached one residence, its behavior suddenly changed. The fire dropped from the tree crowns to the ground, became a low-intensity surface fire, and was extinguished at the edge of the house. Several years later the underburned trees were still green, a verdant dot amid a blackened landscape. The survival of this forest and its residence was not just luck or fate. The house had been constructed with a metal roof, and had a water source to help fire suppression. Just as importantly the forest surrounding the house had been treated with a forest health objective of decreasing fire potential, and the objective was achieved. This real example of integrated fire management illustrates the ways in which fuel management can ease the burden of fire control efforts and the effects of wildland fires. In this chapter we will discuss fuel treatment strategies for reducing wildland fire behavior, techniques of prescribed burning, and the use of managed wildland fire (pnf [prescribed natural fire]) in parks and wilderness.

The fire behavior principles in Chapter 3 clearly link the behavior of fire to the type, amount, size class, and arrangement of fuels on the landscape. The horizontal and vertical distribution of live and dead fuel is one way to describe forest structure. Changes in forest structure will affect the behavior of fire, but these changes often have complex influences on fire behavior and effects. Some treatments may increase fire behavior but reduce tree mortality, whereas others can have the opposite effect. A fire control officer may be concerned more with effects on fire behavior; a forester or wildlife biologist may be more concerned with fire mortality.

FOREST STRUCTURE AND FIRE BEHAVIOR

Wildland fire problems are increasing both in area burned and the average intensity of the fires, particularly in the western states (Brown and Arno 1991). Our fire behavior models deal primarily with surface fire behavior (Burgan and Rothermel 1984), but crown fire behavior is increasingly a major problem in wildland fire control and in ecological effects. Forest structure can be manipulated to reduce fireline intensity, probability of crown fire initiation, and chances of crown fire spread. Manageable fire behavior and effects can be produced by (1) managing surface fuels so that fireline intensity is limited, (2) reducing the probability that crown fires will occur, and (3) using silvicultural prescriptions that favor fire-tolerant trees (both species and size classes). These characters are not independent of one another, and will not be applicable to every type of forest. Some high-elevation forests, for example, have always burned with crown fire and are comprised of fire-sensitive trees with thin bark, so that attempts to make these forests resistant to fire are futile. In such forests, forest structure might be more successfully managed to reduce fire intensity than fire effects, because even surface fires will result in a stand replacement event (Agee 1993).

Extreme fire behavior can be reduced by implementing one or more of the strategies described below. Usually more than one will be necessary to achieve a reasonable standard of hazard reduction, and they are best accomplished by first reducing surface fuels and increasing base to live crown, with altering crown structure less of a stand-alone treatment. For example, opening the forest canopy by thinning may decrease crown fire potential movement but may increase surface drying of fuels and surface windspeeds, as well as possibly adding to surface fuel loads if fuel treatment does not accompany the thinning (van Wagtendonk 1996).

Increasing Height to Crown Base

Crown fires usually are initiated by flames reaching into the base of the live crown and then torching an individual tree or clump of trees. In Chapter 3 the equation necessary to initiate the crowning process was introduced (Van Wagner 1977):

$$I_0 = (Czh)^{3/2}$$

where

I_0 = critical fireline intensity, kW m^{-1}
C = a constant, 0.01
z = height to the live crown, m
h = heat of ignition, kJ kg^{-1}

In this equation the heat of ignition is largely a function of foliar moisture content, which will be maximized during bud flush and then decline over the season. In most forests late-season foliar moisture is 90% or more, expressed on a dry weight basis. Critical fireline intensities, expressed as flame lengths, necessary to initiate crown fire activity are shown in Table 6.1. Although foliar moisture is a variable, it is not manageable, whereas height to crown base can be manipulated.

TABLE **6.1** | **Flame Lengths Associated with Crown Fire Initiation Can Be Expressed Through Combinations of Foliar Moisture and Height to Crown Base**

Foliar moisture content (%)	Height of crown base m (nearest ft)						
	2 (6)	4 (13)	6 (20)	8 (26)	12 (40)	16 (53)	20 (66)
70	1.1 (4)	1.8 (6)	2.3 (8)	2.8 (9)	3.7 (12)	4.6 (15)	5.3 (17)
80	1.2 (4)	1.9 (6)	2.5 (8)	3.0 (10)	4.0 (13)	4.9 (16)	5.7 (19)
90	1.3 (4)	2.0 (7)	2.7 (9)	3.3 (10)	4.3 (14)	5.3 (17)	6.1 (20)
100	1.3 (4)	2.1 (7)	2.8 (9)	3.4 (11)	4.6 (15)	5.6 (18)	6.5 (21)
120	1.5 (5)	2.4 (8)	3.2 (10)	3.9 (13)	5.1 (17)	6.2 (21)	7.3 (24)

Prescribed fire, for example, will typically raise the height to the live crown by scorching understory foliage. The foliage falls off, usually will not resprout, and permanently increases the height to the base of the live crown. A low thinning (also known as thinning from below, taking the smallest trees and leaving the largest) using mechanical or manual equipment also will raise the height to crown base and thereby reduce the probability of crown fire initiation. Manual or mechanical pruning will serve the same function. Keeping I_0 large is an important way to reduce crown fire potential.

There is no truly objective way to define height to crown base. In an even-aged stand, it may be reasonably clear, but many stands will have multiple layers and the lowest layer may be quite sparse. A rule of thumb is to delineate the lowest continuous layer of foliage above the forest floor and use that estimate, to the nearest meter (foot), as the best estimate of base to crown height.

Reducing Surface Fuels

Surface fuel reduction and an increase in height to crown base are complementary techniques to reduce potential for severe fire behavior. The former will keep fireline intensity (I) low; the latter will keep critical fireline intensity (I_0) high. Prescribed fire is a common tool used in surface fuel reduction, and can consume 60 to 80% of the dead surface fuel load in a stand. However, it usually kills substantial live vegetation that adds to the dead fuel load. Up to two-thirds of the pre-fire dead load may be replaced over time by fire-killed shrubs and small trees in an initial prescribed fire in mixed-conifer forest (e.g., Thomas and Agee 1986). The height to crown base is increased at the same time, though, so that on balance fire hazard usually is decreased substantially by a single prescribed fire. Several anecdotal instances of wildland fires being controlled in areas that had been previously prescribed burned suggest that burning is a useful hazard reduction tool (Helms 1979). When applied periodically, it can maintain low fire hazard conditions in many forest types.

Managing Crown Structure to Prevent Fire Spread

The third way to ameliorate extreme fire behavior is to thin the crowns of the stand so that crown fire spread is less likely. In Chapter 3 (Crown Fire Behavior), the crown of a forest was described as similar to any other porous fuel medium in its ability to carry fire and the conditions under which fire will or will not spread. The net horizontal heat flux, E, into the unburned fuel ahead of the fire is a function of the crown fire rate of spread (R), crown bulk density (d), and ignition energy (h). The crown bulk density is described as those fine fuels (leaves, lichens, etc.) between the crown base and tree tops expressed on an area basis, and is the same concept as soil bulk density (mass per unit volume). If either R or d falls below some minimum level, crown fire propagation will cease.

One way to visualize the crown fire process is to picture the forest on a conveyor belt (Figure 6.1), moving the mass of crown fuel through a stationary flaming front at a rate equal to the fire rate of spread. The mass flow rate S, expressed in kg m^{-2} sec^{-1}, is the product of R and d:

$$S = Rd$$

In boreal forests a critical level of S, called S_o and equal to 0.05, has been found necessary to maintain crown fire spread (Van Wagner 1977). If either R or d is low enough that $S_0 < 0.05$, then crown fire propagation is unlikely to continue. If ranges of rates of spread (R) can be defined under critical fire weather, and assumed to be unmanageable, then ranges of crown bulk density (d), which can be managed, also can be defined. Measured crown fire rates of spread (R, m sec^{-1}) range from about 0.3 to 1.35 (Rothermel 1991), which are associated with critical levels of crown bulk densities (d, kg m^{-3}) from 0.17 to 0.04. Typical wind-driven crown fires have an R of about 0.67, corresponding to a critical d of about 0.075.

Crown mass and volume data are needed to calculate crown bulk density. The mass of a stand is calculated from the total foliar mass plus other fine fuels (fine twigs and lichens) likely to burn in a crown fire. Estimates of foliar mass often are available from allometric equations based on stem diameter, plus 10 to 20% extra for additional fine fuels. The volume is calculated from the base of the live crown to the tree tops, and can be estimated from tree height and live crown ratio. Defining the base to live crown can be difficult in patchy forests with variable crown structure, as both the base to live crown and crown bulk density may be difficult to objectively define. Some forest growth models may include calculations of foliar mass by 3 m (10 ft) "slices" through the canopy, from which calculation of crown bulk density is easy (e.g., Moeur 1985).

Limited empirical information exists to evaluate crown bulk density thresholds, and "thresholds" may not actually exist. Potential rate of spread and crown bulk density are probably positively correlated with one another, so that decreases in crown bulk density limit potential rates of spread. In the 1994 wildland fires around Wenatchee, Washington, changes in crown fire activity appear to have occurred around an estimated crown bulk density of 0.10 kg m^{-3} (Table 6.2). At levels of d near 0.10, understory fuel treatment appeared to limit crown fire potential. Stand 3 was untreated with $d = 0.10$ and sustained crown fire. Stand 4 with $d = 0.09$ had

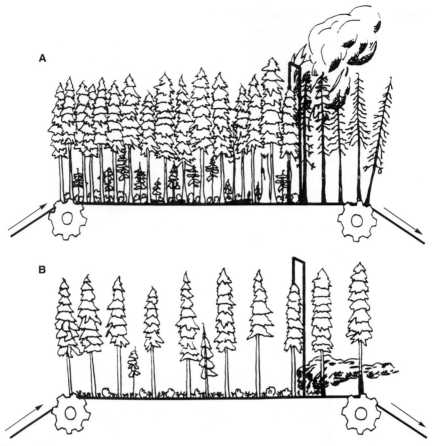

Figure 6.1
Critical conditions for mass flow rate (S_0) sustaining crown fire activity can be visualized by passing a forest along a "conveyor belt" through a stationary flaming front. **A,** Under severe fire weather and a high rate of spread, crown mass passes through the flaming front rapidly and achieves $S_0 > 0.05$ and crown fire occurs. **B,** Where crown bulk density is lower under the same rate of spread, critical levels of S_0 cannot be obtained and the fire remains a surface fire.

thinning of trees below 15 cm (6 in. diameter), pile burning of the debris, and pruning of residual trees to 3 m: A crown fire dropped to the ground as it entered this stand (Figure 6.2). The overstory was not influenced much by the treatment, but the surface fuel treatment decreased fireline intensity (I) and increased critical fireline intensity (I_0) needed to initiate crown fire activity. The treatment reduced energy flow into the crown sufficiently to terminate crown fire behavior. As windspeeds that drive the rate of spread may vary, there is a range, rather than a threshold, of crown bulk density below which crown fire activity is unlikely.

T A B L E **6.2** | **Relation of Estimated Crown Bulk Density to Crown Fire Behavior in Mixed-Conifer Stands of Ponderosa Pine and Douglas-Fir**

Location	Stand and treatment	Average stand diameter (cm [in.])	Average stand density (t ha⁻¹ [ac])	Estimated crown bulk density, d (kg m⁻³)	Comments on fire behavior
Stand 1	Unthinned	17 [7]	1700 [689]	≅0.15	Crown fire activity
Stand 2	Unthinned	17 [7]	1700 [689]	≅0.15	Crown fire activity
Stand 3	Unthinned	17 [7]	1000 [400]	≅0.10	Crown fire activity
Stand 4	Thinned	30 [12]	500 [200]	≅0.09	Crown fire stops
Stand 5	Thinned	30 [12]	250 [100]	≅0.05	Crown fire stops
Stand 6	Thinned	30 [12]	250 [100]	≅0.05	Crown fire stops
Stand 7	Thinned	24 [9.5]	225 [91]	≅0.035	Crown fire stops

[a]Stands are ranked in descending order of crown bulk density. Crown bulk density was estimated by assuming a single-storied stand with crown parameters calculated by using average stand diameter, base to live crown, tree height, and stand density.

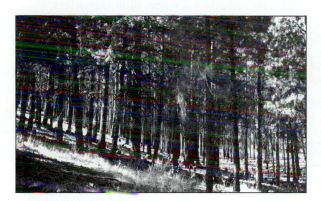

Figure 6.2
As a crown fire entered this relatively dense stand, it was unable to sustain its spread and dropped to the ground. The stand (Stand 4, Table 6.2) had been thinned from below and pruned, with the debris pile burned. Photograph is taken 2 years after the fire.

Fuel Treatments for Forest Health

There are myriad fuel treatments available for landscape applications. A recent model application of FARSITE (see closing portion of Chapter 3) simulated the effect of different fuel treatments on rate of spread and flame length (van Wagtendonk 1996) across a model landscape (Tables 6.3 and 6.4, Figure 6.3). **Prescribed fire** was the most effective treatment, with lowest spread and lowest flame lengths. The model, however, assumes that small live fuels killed by the treatment are removed from the site, so it overestimates the effect of the treatment. **Pile burning** shows little effect as surface fuels were not removed; however, base to live crown

T A B L E **6.3** | **Custom Fuel Models Built in BEHAVE (see Chapter 3) for Mixed-Conifer Fuel Treatments**

	Custom fuel model[a]			
Fuel variable	Model 14	Model 15	Model 16	Model 17
1-hr load (tons/ac)	3.0	1.5	4.5	1.0
10-hr load (tons/ac)	2.0	1.0	3.0	0.5
100-hr load (tons/ac)	2.0	1.0	3.0	0.5
Live load (tons/ac)	2.0	0.0	2.0	0.0
1-hr surface/volume ratio	2,000	2,000	2,000	3,000
Live surface/volume ratio	1,500	1,500	1,500	1,500
Depth	1.0	0.5	1.5	0.5
Moisture of extinction	35	35	35	12

[a]See Table 6.4 for application to fuel treatment scenarios.
Dead heat content is constant at 9,000 BTU/lb, live heat content is constant at 8,000 BTU/lb, adjustment factor is 0.5
Source: From van Wagtendonk (1996).

T A B L E **6.4** | **Fuel Treatment, Models, Crown Bulk Densities, Crown Base Heights, and Wind Reduction Factors for FARSITE Simulations**

Scenario	Custom fuel model (no.)	Canopy cover (%)	Crown density (kg m⁻³)	Crown base height (m)	Wind reduction factor
Control	14	81–100	0.3	1	0.22
Prescribed burn	15	81–100	0.3	2	0.22
Pile and burn	14	81–100	0.3	2	0.22
Cut and scatter	16	81–100	0.3	2	0.22
Biomass	14	50–80	0.15	1	0.32
Biomass/burn	15	50–80	0.15	2	0.32
Biomass/pile	14	50–80	0.15	2	0.32
Biomass/scatter	16	50–80	0.15	2	0.32

[a]Simulations were run with 95-percentile fire weather (exceeded only 5% of the time during the fire season).

was increased by treatment so crown fire potential is decreased. **Cut and scatter** increases fireline intensity across much of the area, even though it raises the height of base to live crown. Some torching, spotting, and crowning occurs under both pile/burn and cut/scatter treatments, but not under the prescribed burning scenario. All of the biomass (thinning) options had no crowning because of reduction in crown bulk density (under these weather conditions, crown bulk density in thinned stands was assumed to be a relatively high 0.15). Opening of the crown does allow

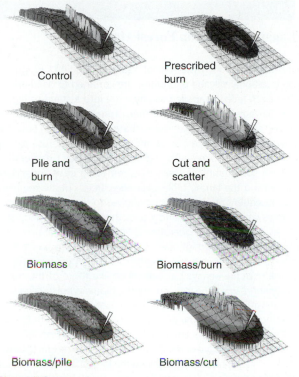

Control

Prescribed burn

Pile and burn

Cut and scatter

Biomass

Biomass/burn

Biomass/pile

Biomass/cut

Figure 6.3

Fire spread and flame lengths across a simulated 3,000 X 6,000 m landscape, with slope of 20% on near half of landscape and flat slope beyond. Ignition occurred at the point shown and the simulation was run for 24 hours using local summer weather. The control (Tables 6.3 and 6.4) is similar to NFFL Model 10 (see Chapter 3). *Prescribed burn* reduces surface fuels but does not open canopy. *Pile and burn* reduces understory live fuels by hand thinning but not surface dead fuels because of small burn piles. *Cut and scatter* is similar to pile burn but fuels are scattered rather than burned. *Biomass* options involve thinning the overstory and removing all trees cut (whole tree harvest) for all four options. It is shown as thinning the overstory only, and combining that with options already described for understory live vegetation and surface dead fuels. Each figure shows final fire spread (size of ellipse) and flame length (review copy: darker color is 0 to 1 m flame lengths, lightest colors are 4 to 5 m flame lengths).

(Source: From van Wagtendonk 1996.)

more drying of surface fuels and better air movement through the stand, so that surface fire behavior parameters increase, with resultant increases in either fire spread (see biomass/burn vs. prescribed burn) or flame length (biomass/cut vs. cut and scatter). Moderate overstory thinning combined with understory fuel removal appears to be an effective technique to manage wildland fire behavior.

Box 6.1

Salvage Logging Effects on Forest Health

Much debate has focused on the role of "salvage logging" in improving forest health. While traditional salvage operations are confined to dead and dying trees, the phrase has been recently applied to any entry into a stand of high fire, insect, or disease risk. Salvage logging can take many forms, some of which will improve forest health and some of which can exacerbate the situation. Let us look at three options in a typical old-growth mixed-conifer stand of ponderosa pine and true fir, formerly open, and now choked with understory trees (see Chapter 4). Option 1 is to do nothing. Option 2 employs a low thin, removing trees under 40 cm (15 in.) diameter and leaving the larger, older overstory. Trees cut are removed from the woods and a subsequent prescribed fire is applied to the stand. Option 3 takes the older, larger trees of highest value and leaves the middle and understory, scattering the slash resulting from the operation among the residual trees. Several years later a wildland fire burns through each of the stands. Which of these "salvage" options increases or decreases forest health from a fire perspective?

Option 1 will have fire behavior effects much like those of the control scenario of Figure 6.3; option 2 will appear much like the biomass/burn scenario, and option 3 will appear similar to the biomass/cut scenario. However, from a forest health perspective, the effect of the fire is the interaction of vegetation with fire behavior. The "no action" Option 1 will have significant understory mortality and under severe weather, potential for crown fire behavior. Subsequent tree mortality from insect population buildups is likely. Option 2 has retained the largest and most fire-tolerant trees and will be most fire-resistant of the three options. Forest health has been improved. Option 3 has the highest fire behavior in the stand with the smallest-diameter distribution of residual trees. This fire is most likely to be a stand replacement event; forest health has declined.

The phrase "salvage logging" as applied in a forest health context is difficult to understand for two reasons. First, the effects of salvage logging on forest health can vary considerably—both options 2 and 3 are commonly defined as salvage logging but have quite different effects on forest health. Second, salvage is a word in typical English usage that is most commonly applied to places like auto junkyards, where parts are removed from cars before the remaining hulks are crushed. Because we are interested in improving the health of the forest that remains after treatment, a better metaphor would be the auto body shop where cars are repaired, and as applied to forests, logging for restoration rather than salvage.

One option not considered in Figure 6.3 is the crushing of fuels created by thinning operations. Where harvester/forwarder machines are used, trees are cut and delimbed on site, with the cut material being placed in front of the machinery and acting as a cushion for the machinery as it moves slowly through the stand. This compacts the fuels and incorporates them partially into the soil, and from a fire behavior perspective is much better than a lop and scatter treatment if fuels are to be left on-site.

When understory or overstory trees are thinned, a subsequent effect that is commonly observed is "greenup," the recovery of herbaceous vegetation, shrubs, and small trees. This recovery adds fuels to the site, but that fuel may remain green and moist well into the fire season, particularly if the grasses are perennial. Such fuels may further dampen potential fire behavior while green (Figure 6.4), but will increase fire behavior if they cure. If the "greenup" lasts well into the summer, it will shorten the period of critical fire behavior for 2 months or more during the dry season.

PRESCRIBED FIRE

Prescribed fire is the skillful application of fire under predetermined conditions of fuels, weather, and topography to achieve management objectives (Pyne et al. 1996). These objectives may include (1) hazard reduction from natural fuel buildup or activity (thinning, pruning) fuels; (2) silvicultural objectives, either to prepare sites, thin or prune forest stands, or to manipulate species composition; (3) disease and insect control, either directly or by favoring tree species less susceptible to attack; (4) wildlife habitat management; (5) range and prairie management, particularly to control tree or shrub invasions; (6) mimicking the role of natural fire, or helping to restore that role, in natural area management; and (7) a variety of other objectives, including visual resources management (Hardy and Arno 1996, Tall Timbers Research Station, various dates).

Firing Techniques

The techniques used to ignite a prescribed fire have a large influence on the spread and intensity of fire, and also on its risk of escape and ecological effects. There are three primary ways the spread of fire can be encouraged: upwind/upslope, downwind/downslope, or sidewind/sideslope. These directions are called **heading, backing,** and **flanking** (Martin and Dell 1978). Depending upon the objectives of the burn, site conditions, and weather the day of the burn, any or all of these techniques may be appropriate. All of them require anchor points within which the fire can be started and stopped. The primary techniques incorporating these techniques follow (Table 6.5, Figure 6.5).

Heading Fire   A heading fire is one burning upslope or with the wind, and is the most intense and fastest-spreading type of fire. It usually is used in grass or brush and typically requires a wide fireline on the downwind side of the unit. This can be accomplished by plowing, as is done in the Southwest, or by blacklining. Blacklining is the burning of an area between two lines that have been wetted with water or foam. The wetlines will be spaced at a width exceeding the distance that spotting will occur, and must be very wide for heading fires. There is little control on a heading fire once it is ignited, and for that reason it is less commonly used than other techniques.

Backing Fire   The backing fire is not the same as a backfire used in fire control operations to secure an indirect attack line of defense. It is a fire ignited against

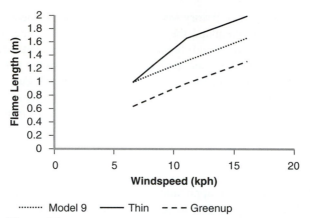

Figure 6.4

Dampening effect on flame length due to addition of live woody and herbaceous fuels after thinning. Thinning with partial slash cleanup increases potential fire behavior under a closed pine stand (NFFL Model 9—see Table 3.4), with the addition of 1 ton ac[-1] of 1-hr and 3 tons ac[-1] of 10- and 100-hr timelag fuels and an increase in fuel depth to 0.5 ft. But the addition of 1 ton ac[-1] of live herb and 2 tons ac[-1] of live woody fuels to the thinning slash ("greenup" denoted by triangles) decreases flame lengths under the moderate fire weather conditions simulated. If these green fuels cure during the season and become 1-hr timelag dead fuels, fire behavior will increase.

the wind or downslope, starting from a secure fireline such as a road, unvegetated ridgetop, or constructed fireline. Of all the techniques, it has the slowest rate of spread and lowest fireline intensity, and is therefore the most controllable. The line must be kept as much as possible 90° to the wind or slope. If one side moves much faster than the other, it may curl around and create a heading fire on the other side of the unit. On steep slopes burning debris such as pine cones may roll downhill and create heading fires. Because of the slow rate of spread of backing fires under most conditions, patience is essential in employing this technique.

Strip Head Fire   The strip head fire is one of the most versatile firing techniques. It begins as a backing fire from a secure fireline, and then a short strip of fire is ignited that is allowed to burn as a heading fire into the initial backing fire. Successive strips are then ignited in an upwind/downslope direction, with the width of the strips depending on the observed fire behavior of the previous strip. Fireline intensity will be lowest adjacent to the ignition line and highest as the short heading fire meets the backing fire moving from the previously ignited strip. When weather parameters are marginal but within prescription, the strip head fire technique can be adjusted to produce acceptable fire behavior by either narrowing or widening the strips.

T A B L E   **6.5** | **Firing Techniques for Prescribed Burning**

Technique	Where used	How done	Advantages	Disadvantages
Heading fire	Large areas, brush fields, clearcuts, under stands with light fuels	1. Backfire downwind line until safe line is created 2. Light head fire	Rapid, inexpensive, good smoke dispersal	High intensity, high spotting potential
Backing fire	Under tree canopy, in heavy fuels near firelines	1. Backfire from downwind line; may build additional lines and backfire from each line	Slow, low intensity, low scorch, low spotting potential	Expensive, smoke stays near ground, slow spread allows wind shift
Strip head fire	Large areas, brush fields, clearcuts, partial cuts with light slash under tree canopies	1. Backfire from downwind line until safe line created 2. Start head fire at given distance upwind 3. Continue with successive strips of width to give desired flame length	Relatively rapid, intensity adjusted by strip width, flexible, moderate cost	Need access in unit; 3 or more strips burning simultaneously may result in high-intensity fire interaction
Spot head fire	Large areas, brush fields, clearcuts, partial cuts with light slash, under tree canopies; aircraft or helicopters may be used	1. Backfire from downwind line until safe line created 2. Start spots at given distance upwind 3. Adjust spot distance to give desired flame length	Relatively rapid, intensity adjusted by spot spacing, can obtain variable effects from head and flank fires, moderate cost	Need access within area if not ignited aerially
Flanking fire	Clearcuts, brush fields, light fuels under canopies	1. Backfire downwind until safe line created 2. Several burners progress into wind and adjust their speed to give desired flame length	Flame length between backing and head fire, moderate cost, can modify from near backing fire to flank fire	Susceptible to wind veering; need good coordination among crew
Center or ring firing	Clearcuts, brush fields	1. For center firing, center is lighted first 2. Ring is then lighted to draw to center; often done electrically or aerially	Very rapid, best smoke dispersal, very high intensity, fire drawn to center and away from surrounding vegetation and fuels	May develop dangerous convection currents or long-distance spotting

Source: From Martin and Dell (1978).

Spot Head Fire   This technique is similar to the strip head fire but the ignition pattern is a series of spots rather than a line of fire. It is commonly used in aerial ignition of eucalyptus forests in Australia, and is likely to be utilized more if the scale of prescribed burning, particularly early spring burning, expands in the western United States. It creates a mosaic of intensities as each spot creates an elliptical spread pattern, reaching highest intensity where it merges with an adjacent spot head fire. The density of spot ignitions serves the same purpose as adjusting the width of strip head fires.

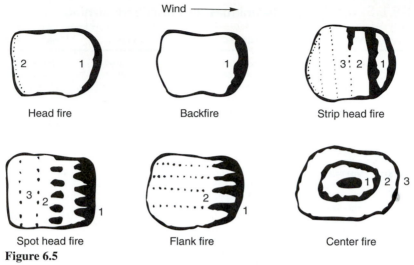

Wind ⟶

Head fire

Backfire

Strip head fire

Spot head fire

Flank fire

Center fire

Figure 6.5

Common firing techniques used in prescribed burning. Each ties into one or more anchor points from which the initial firing is safely conducted.

Source: From Martin and Dell 1978.

Flanking Fire   A flanking fire is ignited by several people simultaneously igniting parallel strips downslope or into the wind. The resulting fires have lowest intensities along the ignited lines and highest intensities where they merge with an adjacent strip. This technique is limited in use to light fuels, and is commonly used in grasslands and southern pine stands where regular prescribed burning has been practiced.

Center Firing   Center firing is designed to produce a hot fire that is drawn to the center of the unit. It has been used in slash burning and brushfields, where mortality of vegetation in the unit is not a concern, and there is not adjacent tree canopy to be scorched. A center ring of fire is ignited such that it draws toward the middle, and subsequently one or more concentric rings of fire are ignited in progression, each being drawn toward the convective column rising from the center of the unit. Where slopes are steep or wind is strong enough to affect the convection column, this technique is not preferred.

Mass Ignition   Mass ignition is designed to ignite the entire unit in a very short period of time, with the points of ignition merging quickly. It has been used in slash burning in an attempt to consume fine fuels that create fire hazard, and lift smoke in a defined convection column, without consuming significant amounts of large fuels that smolder and create substantial residual smoke.

Methods of Ignition

A variety of ignition devices are available for ground or aerial ignition (Pyne et al. 1996). The drip torch is the most common ground ignition device for light fuels.

The lighter walks along and dips the torch, spilling a mixture of gas and diesel through a lighted wick, after which the flaming mix falls to the ground. Fusees also are used, although they require crouching at each ignition spot and the fusee has to be held on the fuel until it ignites. Flamethrowers are very useful in heavy fuels like brush or slash, where a significant amount of heat is needed to begin the combustion process. The flying drip torch operates under the same principles as the hand-held torch, but it is slung under a helicopter and is much larger, being operated from the cockpit. As the helicopter moves across a unit, the trigger allows burning fuel to be ejected. This method is commonly used in slash burning, where internal access by ground crews is both inefficient and dangerous. Aerial ignition can also be accomplished by "ping-pong balls"—small plastic balls of potassium permanganate that are injected with ethylene glycol and immediately ejected from the aircraft, igniting from chemical reaction shortly after they reach the ground surface. This creates a pattern of spot head fires, and up to 4,000 ha/day can be covered by a single aircraft.

The Prescribed Fire Plan

Federal, state, and private land managers each have specific guidelines for preparing prescribed fire plans. They may be very detailed written plans, in the case of some federal agencies, to more informal written documents by some private landowners (Biswell 1989). The thought process often is quite similar in terms of the issues considered in the planning process. There usually is an overall set of land management objectives that guide all use of fire on the land: suppression, prescribed fire, and managed wildland fire (pnf). References will be made to the management objectives for the land, and to what end the use of prescribed fire is intended. The specifics of any single prescribed fire will not be relevant at this scale of planning, although the areas where prescribed fire might be applicable could be identified.

The unit burn plan will contain the specific goals of the prescribed fire, map information of topography, fuel types, control lines, and other criteria at a scale large enough to be meaningful; prescription criteria, including alternative firing plans, logistical information, and monitoring commitments (Martin and Dell 1978). On public lands an environmental assessment usually will accompany the plan.

Objectives   The objectives of the prescribed burn will guide the remainder of the plan, so the more specific they are, the easier the rest of the plan, including monitoring, will be. They may be stated in the form of what fire will remove, or what should remain after the fire.

Prescription Criteria   The specific prescription variables often are less important individually than how they react in an integrated sense to create a specific fire behavior or effect. Although temperature and relative humidity, as first-order variables, often are included in prescriptions, the second-order effects on fuel moisture and third-order effects on rate of spread and flame length are probably more important criteria to emphasize. Usually wide ranges of temperature and relative humidity are defined, with tighter ranges on the second- and third-order variables.

BOX 6.2

Prescribed Fire Objectives

Well-Stated Objectives

Favor Idaho fescue by burning after its phenology is complete but before its associated shrubs have gone dormant (this implies a specific season but does not say much about fire behavior characteristics).

Reduce 1- and 10-hour timelag fuels by at least 60% and 100-hour fuels by at least 40%.

Kill 90% of true firs below 2 in. dbh (diameter at breast height) but less than 15% of any pine size class above 6 in. dbh.

Poorly Stated Objectives

Reintroduce fire to the environment (says nothing about achieving anything).

Reduce fine fuels but maintain 100% of large coarse woody debris (may be operationally impossible, and not very specific for fine fuels)

Windspeed is a first-order variable that usually is tightly constrained, because of its effect on fire behavior, including spotting and escape, as well as its influence on fire effects such as scorch height. Flame length is the most used fire variable included in prescriptions, particularly under tree canopies. Duff moisture will be important if forest floor consumption or retention is a primary objective. Season will be critical if the objective is to excessively or minimally damage vegetation. For example, right after bud flush root reserves are low and new foliage is especially sensitive to heat damage (Agee 1993). Certain species will be dormant while others may still be physiologically active and more sensitive to fire. A single firing plan often is defined, such as a strip head fire or backing fire, but usually there will be some flexibility within the technique chosen in case of wind directions slightly different than those expected.

A variety of planning aids are available for prescribed fire. Historical weather and fire danger indices are available for most areas, so that historical potential fire characteristics such as flame length (related to Burning Index in NFDRS) can be displayed by various time periods during the fire season. One set of computer programs called RXWINDOW (Andrews and Bradshaw 1990) is tied into the BEHAVE package and essentially works backward from criteria such as flame length to define acceptable sets of weather variables that will produce the desired behavior. Other computer packages are available that help predict fuel consumption and smoke production (Ottmar et al. 1993). Most of the packages have been developed by USDA Forest Service researchers or practitioners, and are likely to see periodic improvements; most are adaptable to or were developed on PC platforms.

Logistical Information   Included in logistics are the size of the crew needed: who will be igniting, how many people will be on the line, the need for drip

Box 6.3

Developing a Prescription

A common objective for prescribed fire is to thin the stand from below, to reduce stand density and increase vigor of residual trees. In many mixed-conifer stands, pines will be the dominants and true firs (white or grand fir) will be the primary understory species. Assume that the following objective is the reason for the burn: Kill at least 70% of true firs below 12 cm (5 in.) dbh but less than 30% of any pine size class above 25 cm (10 in.) dbh (diameter at breast height). Using FOFEM (see Chapter 4, Table 4.3, all output in English units), the appropriate size classes of each species present in the stand is entered. A flame length of 2 ft will kill the true firs below 5 cm dbh; a flame length of greater than 4 ft will be required to kill more than 30% of the pines that are above 25 cm (10 in.) dbh. Because the fire behavior predictions are not exact, a prescribed flame length of 2 ft may actually turn out to be between 1 and 4 ft. In this case a prescribed flame length of 2 ft, given the likely "window" of variability around that average, appears to meet the objective.

Some trees are less susceptible to synergistic effects such as insect attack if burned in the autumn, so that the FOFEM results are more likely to be accurate in the autumn than after a spring burn. The intent is then to have an autumn burn, either a backing fire or a strip head fire, with a 2 ft flame length. Assuming the area has no slope and NFFL Fuel Model 10 (see Table 3.4), this can be achieved with 1-, 10-, and 100-hour timelag fuel moistures of 10, 12, and 12%, woody fuel moisture of 100%, and midflame windspeed of 1 mph. If wind increases to 2 mph, the flame length increases to 3 ft, which is becoming marginal for these objectives.

From historic archived weather records available through the Forest Service Weather Information Management System (WIMS) at Fort Collins, Colorado, a "window" of autumn dates can be defined when these conditions are most likely to be met, perhaps between September 25 and October 10.

torches, amount of fuel, air resources if applicable, fire trucks and/or hose, how much line construction is needed and how and when it will be constructed, and options for control if the fire moves out of prescription, either by escaping the line or exceeding one or more prescription criteria.

The Burn   If the steps leading up to the burn have been properly completed, the burn itself usually will be a success. The plan should be available to brief crews on burn objectives: Educated crews are much more likely to successfully execute the plan. Maps and photos should be distributed, and a current weather forecast and the ability to receive updated information should be available. Belt weather kits to monitor temperature, relative humidity, and windspeed should be assigned to the crew member responsible for periodic local weather monitoring. Firelines should be checked and reinforced as necessary. Notification of interested parties precedes ignition. Headquarters, local fire control agencies that are likely to receive calls, local residents who will see smoke, traffic controllers (if necessary) need to be contacted, and road signs need to be in place.

Ignition of a small test fire is important to test burning conditions, even if the conditions are predicted to result in acceptable fire behavior. Make sure to pick an area typical of the burn unit. If the fire behavior is acceptable, then an ignition strategy is implemented according to the burn plan. If for any reason the conditions change, the decision to ignite should be reevaluated; otherwise the fire plan should proceed as planned. Once firing is complete, mop-up and perimeter patrol will complete the burn phase of the plan.

Monitoring   Monitoring the burn is essential in order to evaluate the success of the burn in achieving the objectives for which it was set. Monitoring often is overlooked but is a critical part of the burn plan. Without feedback regarding how well the burn achieved its objectives, continued success in burning is unlikely. There are usually a variety of fine-tuning needed after a burn, either in the operation of the burn or its ecological effects.

Current resource management philosophies include adaptive management as a keystone element (Walters 1986). Adaptive management is simply the incorporation of learning by doing. What is learned by doing is fed back into the process to improve the next iteration, whether it is timber harvest operation, fisheries management, or forest health treatments. Monitoring is the step of the process that defines what was learned (Figure 6.6), but it usually is shortchanged in the budgetary process. Monitoring must be keyed to the objectives of the treatment. A monitoring plan usually will be specific to the ecosystem and agency involved, so that no single template will suffice for all burn plans. Monitoring variables will include those related to fire prescriptions, fire behavior, fire effects on resources, and economics.

The National Park Service (NPS), Western Region (1990), has a detailed set of monitoring requirements for its fire management programs. All prescribed fire programs must monitor Level 1, *Reconnaissance* conditions, including fire cause, location and size, vegetation type, fire behavior and weather, smoke conditions, and resources or safety constraints. Level 2, *Fire Conditions*, adds an array of new variables, including topography and weather, fire characteristics, and smoke characteristics. Monitoring frequency depends on fuel type and phase of fire spread. Level 3, *Immediate Post-fire Effects*, requires vegetation and fuel sampling at a minimum, and is carried out pre-burn, during the burn, and post-burn. Other sampling, such as soils and wildlife, will be important depending on objectives of the burning program. Level 4, *Long-term Change*, requires remeasurement of key variables at 1, 2, 5, and 10 years after the burn.

An excellent example of the value of monitoring is provided by the history of the fire management program at Crater Lake National Park in Oregon. In the late 1970s and early 1980s, prescribed fires were introduced to "get fire back into the ecosystem," and few records were kept of where the fires burned, under what conditions they burned, or what the effects were. In the mid-1980s, a pattern of dying ponderosa pines began to emerge, but records of season of burn, fireline intensity, and other variables were not available. Interviews with current and departed managers pieced a fairly complete record together, and research subsequently determined that late spring burning, even at low to moderate fireline intensities, was

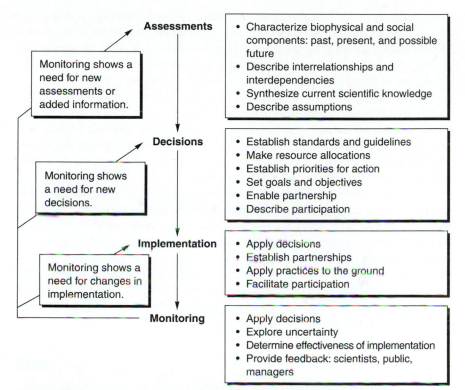

Figure 6.6

Adaptive management is a process of feedback to management of monitoring results, so that objectives and operations can be changed over time as appropriate. The "surprise" or unpredictable element of natural resources management, whether fire or other treatment, means that cookbook approaches are unlikely to be successful, but managers can learn by doing by allowing new "recipes" into the cookbook as more is learned about the ecosystem being managed.

Source: From USDA Forest Service.

associated with much of the tree death. Scientists hypothesized that fine root mortality just before the trees entered the dry season was likely to blame for the subsequent susceptibility of ponderosa pines to bark beetle attack (Swezy and Agee 1991). More complete monitoring records of the fires (perhaps through NPS Level 2, at a minimum) would have made the discovery process much easier, and helped to untangle the unexpected results of the prescribed fire program. Today very complete records on current prescribed fires at Crater Lake are available, using the NPS Western Region guidelines.

Prescribed Fire and Fuelbreaks

Landscape-level strategies for fuel management may involve a continuum of degrees of application using several techniques. One technique receiving renewed interest is the fuelbreak, a zone of reduced fuel using the principles summarized at the beginning of the chapter and often created with limited use of fire. At one time fuelbreaks were considered an alternative to prescribed fire, perhaps best exemplified by the 1930s Ponderosa Way, a fuelbreak the length of the Sierra Nevada intended to separate the chaparral and forest zones of the California mountains (Green 1977). Fuelbreaks in chaparral were initially designed as linear corridors, and in forest zones had quite "unnatural" appearances (Figure 6.7A), but do not need to be so obvious (Figure 6.7B). The fuelbreak in Figure 6.7A will require maintenance sooner than in Figure 6.7B because it is more open so that shrubs can regrow quicker, and has many small trees that, although pruned, will probably be girdled and killed if prescribed fire is used to maintain the break. Mostly large trees were left in Figure 6.7B, so that maintenance of the fuelbreak with prescribed fire will cause little tree mortality.

Recent evaluations of landscape-level fuel strategies, as in the Sierra Nevada Ecosystem Project, have advocated a new fuelbreak strategy known as the **defensible fuel profile zone** (DFPZ). Using the same principles of fuel reduction (thinning of the crown, increasing the height to the live crown base, reducing surface fuels), very wide (up to 400 m) DFPZs have been proposed, but the strategy then includes using the DFPZs as anchor points for area-wide treatment, including the use of prescribed fire (Weatherspoon and Skinner 1996). The concept, as applied on a large area basis with substantial use of prescribed fire, may well produce more insect- and disease-resistant forests, too.

MANAGED WILDLAND FIRE (PNF)

Fire is a natural process around the world, and is as natural in parks and wilderness as windstorms, avalanches, floods, and other disturbances. We do not need "let it blow" policies for hurricanes, or "let it flow" policies for floods. Why, then, do we need "let it burn" policies for parks and wilderness? Society has a complex cultural interaction with fire, and our ability to at least partially control it has led to its management in a quite different context than other natural disturbances. Many parks and wilderness areas were influenced by Native American fires, so that the "natural" structure and function of these ecosystems were in part of cultural origin. Is this natural? Should the objectives be process-oriented ("let fire do its thing") or object-oriented (defining specific desirable stand or landscape structures), or a mix of both? The philosophical debates continue about management objectives for wilderness (Lotan et al. 1985, Brown et al. 1995).

The formerly named "let-burn" fires were rarely that; most were carefully monitored, and often subject to "confine" or "contain" strategies (see Chapter 5) for various reasons. Such fires were renamed "prescribed natural fire," or pnf, and they were precisely that: conservative prescriptions for when natural-ignited fires could burn in parks and wilderness areas. A new term for pnf might be "managed wildland fire" and here it also will be tagged with "pnf" for clarity. This policy is specific to U.S. park and wilderness areas. Canadian resource managers have been more conservative, and use prescribed fire, together with confine and contain strategies on naturally caused wildland fires, to manage fires in their backcountry areas.

The management objectives for U.S. parks and wildernesses are not exactly the same, but the management direction is very similar: to allow natural processes

Figure 6.7

A, A traditional fuelbreak with canopy opening, surface fuel treatment, and pruning of small trees. The effect is very "unnatural" looking, and the removal of substantial large tree cover encourages rapid regrowth.

B, A fuelbreak designed to leave much of the large tree cover and remove most of the smaller trees (the fore-ground fir is not typical of the area). Surface fuels are also treated but almost all large trees are left.

to occur "untrammeled," or unaffected by human influence. Many natural processes proceed at their own pace: Glaciers, for example, grind down the hill whether managers care or not. Windstorms and volcanoes have incredible energy outputs and also are unstoppable. Large forest fires also can have incredible energy outputs, but they usually start as rather benign ignitions that are capable of control. The long history of fire control in North America constrained the natural role of fire in parks and wildernesses for many decades, and the policy changes in the late 1960s and early 1970s that allowed natural fire to burn in parks and wilderness were revisited after the large fire events of 1988 (Wakimoto 1990). Policy and operation for managed wildland fire (pnf) programs will continue to undergo rapid change in the future.

A healthy forest in the context of wilderness is one that is functioning "naturally": low-severity fire regimes burning with low severity, high-severity fire regimes burning with high severity, at "natural" frequencies, extents, seasons, and other characteristics of the natural fire regime (see Chapter 4). The interpretation of adequate forest health in the context of parks and wilderness inherently incorporates senescent forest structures and stand replacement fires. However, in many areas, we do not have a very precise understanding of the natural fire regime, so our judgments of changes due to fire exclusion in the 20th century, and the edge effects of different land practices around the borders of parks and wilderness (particularly important in small areas) are subjective. A case in point are Canadian boreal forests in parks, where prescribed fires are used to simulate the natural fire regime. Some have argued that the vast bulk of the area burned occurs under unusual weather conditions when control is almost impossible (Bessie and Johnson 1995), with an associated implication that a program to use regular prescribed burning is neither needed nor very "natural." Others have argued that suppression of fires that would later develop into large fires removes fire from the ecosystem, and that prescribed fires are therefore necessary to replace suppressed natural fires (Woodley 1995). In U.S. parks and wildernesses, managed wildland fires (pnf) are considered primarily in larger backcountry areas, and areas with fire return intervals long enough that the fire exclusion era of many decades had little impact on fuel accumulations and fire behavior.

There are over 40 million ha of wilderness in the United States in over 600 wilderness areas (Table 6.6). About 70% of the wildernesses are less than 40,000 ha each, but often they will abut a national park or another wilderness area, so that opportunities for interagency cooperation may ease the management constraints imposed by small areas (Parsons and Landres 1996). Prior to 1988 programs were expanding, with increasing numbers of fires and area burned (Figure 6.8). The suspension of all programs after the 1988 fires, and reauthorization of programs consistent with new management guidelines, have resulted in a slow recovery. In 1995 slightly over 10% of wilderness areas have approved plans to allow managed wildland fires (pnf) to burn: about half the National Park Service wildernesses, over 10% of Forest Service areas, few Bureau of Land Management areas, and no Fish and Wildlife Service areas (Parsons and Landres 1996). In Canada the Canadian Parks Service uses prescribed fire in lieu of managed wildland fire (pnf), because of their determination that unregulated wildland fire in these largely high-severity fire regimes would put values such as safety, property, and special habitats at risk.

Planning for Managed Wildland Fire (pnf)

Landscapes where managed wildland fire (pnf) is being considered are almost always areas designated for nature reserve status such as parks and wilderness areas (Fischer 1984). Reserve status alone is not an adequate justification for the use of managed wildland fire (pnf). Many parks and wilderness areas are small enough that two major constraints to the free play of natural forces such as fire are present. Both are edge effects. The first is that a natural fire within a small natural area has a much greater chance of escaping the area and causing problems on adjacent lands. The second is that historical fire extent in many areas probably included fire starts well outside the natural area that later burned into the area. In these adjacent areas such fires now are unlikely to be allowed to burn at all. A second set of constraints relate to how much ecological change occurred during the fire exclusion period. A naturally ignited fire now may not have "natural" ecological effects if a once wide-spaced pine stand now is choked with fire exclusion–era true firs. Rather than a low-intensity, low-severity fire, a naturally ignited fire now may well be a crown fire that kills all the trees in the stand.

T A B L E 6.6	Number of Wilderness Areas and Land Area Managed by Federal Agencies in 1995		

| | | Land Area | |
Agency	Number of Areas	Acres	Hectares
Bureau of Land Management	136	5,227,063	2,116,220
Forest Service	399	34,676,493	14,039,066
Fish and Wildlife Service	75	20,685,372	8,374,645
National Park Service[a]	44	43,007,316	17,411,869
TOTAL	654	103,596,244	41,941,798

[a]Legal wilderness only within parks, not total area in NPS System.

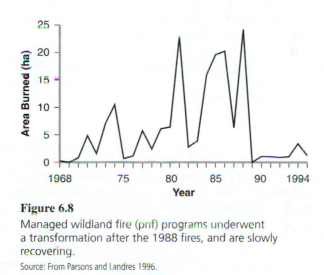

Figure 6.8

Managed wildland fire (pnf) programs underwent a transformation after the 1988 fires, and are slowly recovering.

Source: From Parsons and Landres 1996.

There also are concerns related to regional and national fire situations. Depending upon wildland fire danger, number of ongoing project fires, and competition for firefighting resources, suppression of all newly ignited natural ignitions (as well as all human-caused ignitions) may be mandated, and reclassification of existing managed wildland fires (pnf) to unmanaged wildland fires may occur if the area manager cannot certify that the fire is within prescription and is expected to remain so through the next 24-hour period with the resources and funds available. Because many of these fires burn for months, and become larger over time, the chances of reclassification increase over time, when the fire is likely to be larger and the availability of forces for suppression are either limited or needed elsewhere. Such fires reclassified to wildland fires may then be managed under a "contain" strategy, using natural barriers when possible. Emergency firefighting funds appropriated by Congress can be used for all wildland fires.

The result of planning is a general prescription that takes into account the acceptable areas within the park or wilderness, the acceptable number of fires in the area at one time, fire behavior criteria, individual fire size, and the potential need and availability of suppression personnel. A typical zoning strategy (Figure 6.9) might have a suppression zone around the edge of the area, a conservative prescription strategy (pnf Zone 2) adjacent to that, and the most liberal prescriptions (pnf Zone 1) in areas where fire escape potential is low and/or the fire regime is low to moderate severity.

A controversial aspect of planning is the extent to what fuel modification might be used in the boundary areas, whether inside or adjacent to the park or wilderness, in order to increase the probability of controlling natural fires as they reach the edge of the managed wildland fire (pnf) zone. Fuel modification might be accomplished by prescribed burning or by manual fuelbreak construction, using the principles outlined at the beginning of the chapter. The presence of defensible zones should result in more liberal prescriptions for managed wildland fire (pnf) (e.g., the zones in Figure 6.9 might be "pushed" closer to the boundary), but fuel modification may be considered "unnatural" by some. In Yosemite National Park (Figure 6.10), the presence of managed wildland fires (pnf) in the red fir (*Abies magnifica*) zone has itself resulted in natural fuelbreaks that have constrained the spread of subsequent fires.

Operations

When a wildland fire is detected and reported, a fire management team meets and goes through a step-by-step evaluation (Figure 6.11) to classify it. To qualify as a managed wildland fire (pnf), the fire must be of natural origin and be within preapproved and mapped zones where such fires are allowable. Then a plethora of other constraints are evaluated:

- Are there threats to life or property?
- Is air quality acceptable?
- Are there firefighting or monitoring resources now and in the foreseeable future to manage the fire?
- Are there special resources (cultural resources, endangered fish or wildlife, etc.) that may be at risk if the fire occurs or exceeds a given size?

If the answers to these and other questions do not direct the team to a suppression response, then the fire is declared a managed wildland fire (pnf), and monitoring crews are assigned to the fire. The decision chart is reevaluated on a daily basis to see if the fire is remaining within behavioral or size criteria earlier established, and whether or not the regional or national fire situation is benign enough to enable allocation of firefighting forces to the fire if it exceeds prescription.

The need for tighter prescriptions on managed wildland fire (pnf) is well illustrated by the situation at Yellowstone in 1988 (see Chapter 4). It should be noted that about half of the area burned in the Greater Yellowstone that year was from direct or indirect human-caused wildland fire (hunter, firewood, downed power lines) rather than of natural origin. Even so, the natural fires that were allowed to burn exceeded the ability of the park staff to adequately monitor and control

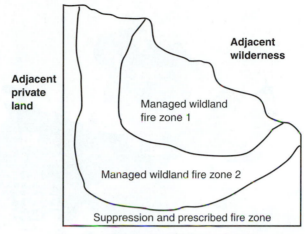

Figure 6.9

A hypothetical zoning scheme for a managed wildland fire (pnf) zone. Prescription criteria are identical in the adjacent wilderness. The "conditional" zone would have additional conservative fire behavior criteria because of the adjacent lands, and the "suppression zone" might have prescribed fire applied to buffer the managed wildland fire (pnf) zone and simulate the natural role of fire in the boundary zone.

(Figure 6.12). In mid-July the situation appeared manageable, with four fires burning. However, with the relatively level terrain of the Yellowstone Plateau, there were few natural firebreaks, and these fires, along with human-caused fires, continued to burn. By July 21 the fires were too large to control, and all existing and new starts were declared wildland fires, but ended up burning 500,000 ha by the end of the summer (Romme and Despain 1989). Over $140 million was expended on suppression efforts, even though much of the area burned was burning as it always had, with high-intensity fires, and was managed to maintain natural processes such as fire. Being natural was not enough to be socially acceptable, and the slow recovery of managed wildland fire (pnf) programs is directly linked to the 1988 fires.

The Future of Wilderness Fire

Park and wilderness areas are not closed systems: They affect surrounding areas and surrounding areas; even fire situations in other regions affect them. Preservation goals for these areas are elusive, and as we move forward in time, separating human-induced from natural change will become ever more difficult. In the short term, prescriptions for lightning fires will continue to be precise, including where they are allowed to burn, during what seasons, and how many at one time. Around the edges of these areas, fuel manipulation of some sort will become more common: prescribed fire, defensible fuel profile zones, and other strategies intended to widen prescription criteria for managed wildland fire (pnf). Prescribed fire may be employed more in wilderness, not only for fuel reduction and restoration of

Figure 6.10

Fifteen years of managed wildland fires (pnf) in the Illilouette Creek basin of Yosemite National Park have created a jigsaw-like mosaic of fires that have been largely constrained at the boundaries of previous fires. This is probably typical in low- and moderate-severity fire regimes.

(Source: From van Wagtendonk 1995.)

natural conditions, but as a primary tool of landscape manipulation in place of managed wildland fire (pnf). All of these strategies are employed to a limited extent now in parks and wilderness areas.

After the 1988 Yellowstone fires, a group of scientists evaluated the fires' ecological consequences (Christensen et al. 1989). They summarized our current dilemma:

> Increasingly, wilderness management has emphasized the preservation of natural processes rather than simply the preservation of natural features, objects, or scenes. . . . A great variety of future "natural" landscapes is possible. . . . We cannot escape the need to articulate clearly the range of landscape configurations that is acceptable within the constraints of the design and intent of particular wilderness preserves. . . . The questions that the next generation of wilderness managers must address are not whether manipulation is desirable, but what kind of manipulation is acceptable? By what means? For what purposes? On what scale? According to what social and political processes?

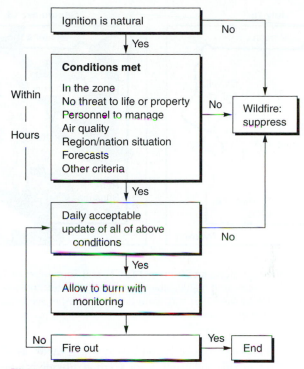

Figure 6.11

A flow chart of decisions is required for each managed wildland fire (pnf) by the fire management team. If any query is answered "no," the fire is declared a wildland fire and an appropriate suppression response is determined. The same (or slightly condensed) chart is then reevaluated each day the fire is burning to assure that it is still in prescription.

CONSTRAINTS TO FIRE USE

Appropriate resources, training, and expertise are required to effectively implement fire management plans. For both prescribed and managed wildland (pnf) fires, however, major additional challenges surface during the planning process, some of which have been briefly discussed earlier in this chapter. Three of the most common are discussed below.

Smoke

Smoke and fire have been common in past centuries and millennia across the North American continent. Totally "clean" air has not existed as long as physical and biological processes have existed on earth. Nevertheless, these arguments carry little weight when fire is used as a management tool. Management of smoke is an essential part of successful fire management (Sandberg and Dost 1990).

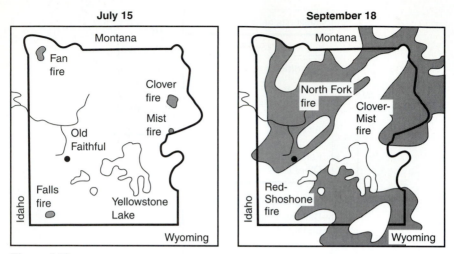

Figure 6.12
Growth of the 1988 Yellowstone fires from mid-July to mid-September of 1988.
The problem may have appeared manageable in early July. New technology, such
as FARSITE, would have allowed "gaming" with different weather scenarios to track
potential progress of the fires in advance.

Most of what is seen in smoke is condensed water vapor, as about 90% of emissions are carbon dioxide and water vapor. Carbon dioxide is a major contributor to global warming. The portion of the carbon not converted to carbon dioxide becomes particulate matter, carbon monoxide, and volatile organic matter. So most of the constituents of smoke are of concern to society: smoke affects visibility, it has adverse health effects, and may be contributing to long-term global problems.

Three primary strategies are used to minimize smoke impacts: avoidance, dilution, and emission reduction (Sandberg and Dost 1990). Avoidance is a strategy to simply direct smoke away from people. Burning so that smoke heads away from town rather than toward it is a good example of avoidance. Avoiding smoke is not always easily accomplished. A burn unit may be between two towns, or surrounded by rural homes. Sometimes smoke can be pushed above a low inversion layer and be carried off above any populated area. Dilution is a second strategy, accomplished by burning at sufficient distance from residences, or within a sufficient volume of air, such that the impact on populated areas is diluted. Emission reduction can be accomplished by not burning, or by consuming less biomass when a burn is conducted. Burning under higher fuel moisture can accomplish the latter, as can rapid ignition so that most of the consumption occurs in the flaming, rather than the smoldering, phase of combustion. In some states burning is regulated at a state level to avoid overloading airsheds with too much smoke at one time. A state-level coordinator approves individual burn units for a given day based on projected weather for dispersal, wind direction for avoidance, and estimated fuel to be consumed on each unit. As population grows and disperses to rural areas, either to live or recreate, smoke management will become even more important if prescribed fire is to persist.

In many fire environments there is a tradeoff between wildland fire smoke and prescribed fire smoke. In a sense it's similar to the old oil filter advertisement, "You can pay me now" (buy a new filter) or "pay me later" (the engine fails due to dirty oil). If prescribed fire is implemented more widely, particularly in areas where fire was historically quite frequent, it is likely that wildland fire emissions will decrease, as the wildland fires will be less intense and controlled at smaller sizes. This is an intuitively appealing argument as a justification for the production of smoke from prescribed fires. However, to date there is little convincing evidence that such a tradeoff in fact yet exists. If it does, there will be an inevitable transition period while sufficient prescribed fire area is burned to have an effect on size of wildland fires, during which time both wildland fire and prescribed fire smoke levels may be high.

The Escaped Fire
The threat of an escaped fire often is a major problem constraining the fire manager. It may result in an otherwise model managed wildland fire (pnf) being classified as a wildland fire, or a prescribed fire being ignited under conditions that may not be ecologically ideal. For example, a prescribed fire may be constrained by ignition to periods of "no-wind" because of the fear of escape. This can be a poor choice, as it gives the ignition crew little help on where to ignite the fire. If the wind does appear, it may blow the fire in unexpected directions. Too much caution in developing the prescription can then create escape problems. In the West spring burning often is desired because fires are easier to manage then with a lower probability of escape. Although fires are applied with a very high level of control, they are burning in an "unnatural" season, and may have deleterious effects on seeder-type shrubs, or the vigor of desirable trees. Obviously assuring that fires will not escape is a primary job of the burning team, and appropriate prescriptions are needed; otherwise, the burning program may have a very short life.

The issue of the escaped fire will increasingly become important if prescribed fire expands as a management tool and development in rural areas continues to increase, as average parcel size decreases. The paradox of the need for fire on such landscapes when the "windows" for burning are ever-narrowing is apparent, and there are no simple solutions in sight.

The Fine Filter for Special Resources
The need for fine filter approaches for conservation of biota were addressed in Chapter 2. The application of fire begins as a coarse filter approach to ecosystem management, as it addresses landscape-level issues much more clearly than individual features of the landscape. Fire may kill mature lodgepole pine, for example, but by doing so ensure that lodgepole pine remains a dominant landscape component over time. Nevertheless, there are good reasons for addressing the needs of individual species at risk, even in areas managed for natural values. Species such as the red-cockaded woodpecker in the South, butterflies in prairies across the continent, Kirtland's warbler in the upper Midwest, and the bull trout, northern spotted owl, and goshawk in the West are species at risk, and those living in forests are clearly indices to forest health. Although long-term presence of fire may

be compatible with the health of the species and the forest, short-term fire effects may need to be mitigated. Fire was historically a major coarse filter for each of these species across their ranges, but habitat alteration (grazing, farming, logging, fire exclusion) has put each at risk.

Mitigation of current fire effects may include burning only within certain seasons; burning only a certain proportion of a landscape in a given time period, so that some appropriate habitat is always available on a local basis; or avoiding burning in currently available habitat until other habitat has recovered. The risk of further habitat loss if areas are not burned (e.g., for fuel reduction) also must be weighed in the mitigation plan. In most low- and moderate-severity fire regimes, catastrophic habitat loss may occur if fire is not considered in planning.

REFERENCES

Agee, J. K. 1993. *Fire ecology of Pacific Northwest forests*. Island Press, Washington, D.C.

Andrews, P. L., and L. S. Bradshaw 1990. *RXWINDOW: defining windows of acceptable burning conditions based on desired fire behavior*. USDA Forest Service General Technical Report INT-273.

Bessie, W. C., and E. A. Johnson. 1995. The relative importance of fuels and weather on fire behavior in subalpine forests. *Ecology* 76:747–762.

Biswell, H. H. 1989. *Prescribed burning in California wildlands vegetation management*. University of California Press, Berkeley, Calif.

Brown, J. K., and S. F. Arno. 1991. The paradox of wildland fire. *Western Wildlands* (Spring):40–46.

Brown, J. K., R. W. Mutch, C. W. Spoon, and R. H. Wakimoto. 1995. *Proceedings: Symposium on fire in wilderness and park management*. USDA Forest Service General Technical Report INT-GTR-320.

Burgan, R. E., and R. C. Rothermel. 1984. *BEHAVE: fire behavior prediction and fuel modeling system —FUEL subsystem*. USDA Forest Service General Technical Report INT-167.

Christensen, N. L., J. K. Agee, P. F. Brussard, J. Hughes, D. H. Knight, G. W. Minshall, J. M. Peek, S. J. Pyne, F. J. Swanson, J. W. Thomas, S. Wells, S. E. Williams, and H. A. Wright. 1989. Interpreting the Yellowstone fires. *Bioscience* 39:678–685.

Fischer, W. C. 1984. *Wilderness fire management planning guide*. USDA Forest Service General Technical Report INT-171.

Green, L. R. 1977. *Fuelbreaks and other modifications for wildland fire control*. USDA Agriculture Handbook 499.

Hardy, C. C., and S. F. Arno. 1996. *The use of fire in forest restoration*. USDA Forest Service General Technical Report INT-GTR-341.

Helms, J. A. 1979. Positive effects of prescribed burning on wildfire intensities. *Fire Management Notes* 40:10–13.

Lotan, J. E., B. M. Kilgore, W. C. Fischer, and R. W. Mutch. 1985. *Proceedings: Symposium and workshop on wilderness fire*. USDA Forest Service General Technical Report INT-182.

Martin, R. E., and J. D. Dell. 1978. *Planning for prescribed burning in the inland Northwest*. USDA Forest Service General Technical Report PNW-76.

Moeur, M. 1985. *COVER: a user's guide to the CANOPY and SHRUBS extension of the Stand Prognosis model*. USDA Forest Service General Technical Report INT-190.

Ottmar, R. D, M. F. Burns, J. N. Hall, and A. D. Hanson. 1993. *CONSUME user's guide, version 1.0*. USDA Forest Service General Technical Report PNW-GTR-304.

Parsons, D. J., and J. Landres (1996). Restoring natural fire to wilderness: How are we doing? *Tall Timbers Fire Ecology Conference Proceedings*. Tallahassee, Fla.

Pyne, S. J., P. L. Andrews, and R. D. Laven (1996) *Introduction to wildland fire*. John Wiley and Sons, New York.

Romme, W. L. and D. Despain. 1989. Historical perspective on the Yellowstone fires of 1988. *Bioscience* 39:695–699.

Rothermel, R. C. 1991. *Predicting behavior and size of crown fires in the northern Rocky Mountains*. USDA Forest Service General Technical Report INT-438.

Sandberg, D. V., and F. N. Dost. 1990. Effects of prescribed fire on air quality, pp. 191–218. In J. D. Walstad, S. R. Radosevich, and D. V. Sandberg (eds.). 1990. *Natural and prescribed fire in Pacific Northwest forests*. Oregon State University Press, Corvallis, Oreg.

Swezy, D. M., and J. K. Agee. 1991. Prescribed fire effects on fine root and tree mortality in old growth ponderosa pine. *Canadian Journal of Forest Research* 21:626–634.

Tall Timbers Research Station. 1962-present. *Proceedings: Tall Timbers fire ecology conferences*, vols. 1–current. Tallahassee, Fla.

Thomas, T. L., and J. K. Agee. 1986. Prescribed fire effects on mixed conifer forest structure at Crater Lake, Oregon. *Canadian Journal of Forest Research* 16:1082–1087.

USDI, National Park Service. 1990. *National park service western region fire monitoring handbook*. NPS Western Region, San Francisco.

Van Wagner, C. E. 1977. Conditions for the start and spread of crown fire. *Canadian Journal of Forest Research* 7:23–34.

van Wagtendonk, J. W. 1995. Large fires in wilderness areas, pp. 113–116. In J. K. Brown, R. W. Mutch, C. W. Spoon, and R. H. Wakimoto (tech. coord.). *Proceedings: Symposium on fire in wilderness and park management*. USDA Forest Service General Technical Report INT-GTR-320.

van Wagtendonk, J. W. 1996. Use of a deterministic fire growth model to test fuel treatments, pp. 1155–1165. In *Sierra Nevada ecosystem project: final report to Congress, vol. ii, assessments and scientific basis for management options*. University of California, Davis, Centers for Water and Wildland Resources.

Wakimoto, R. W. 1990. National fire management policy. *Journal of Forestry* 88:22–26.

Walters, C. 1986. *Adaptive management of renewable natural resources*. Macmillan, New York.

Weatherspoon, C. P., and C. N. Skinner 1996. Landscape-level strategies for forest fuel management, pp. 1471–1492. In *Sierra Nevada ecosystem project: final report to Congress, vol. II, assessments and scientific basis for management options*. University of California, Davis, Centers for Water and Wildland Resources.

Woodley, S. 1995. Playing with fire: vegetation management in the Canadian Parks Service, pp. 30–33. In J. K. Brown, R. W. Mutch, C. W. Spoon, and R. H. Wakimoto. 1995. *Proceedings: symposium on fire in wilderness and park management*. USDA Forest Service General Technical Report INT-GTR-320.

7

WIND AND FOREST HEALTH

INTRODUCTION

Wind is a weather phenomenon that is largely out of the control of the forest manager. Wind can drive disturbances such as fires, but is a significant disturbance on its own, too. Although wind is an uncontrollable physical process, its effects in terms of damage by snapping trees or blowing them over depend on the species composition, density, and size of the trees, characteristics that can be managed. In addition to its direct effects, wind has synergistic interactions with itself and other forest disturbances. For example, windthrow may begin a pattern of subsequent wind susceptibility for residual trees in the stand. If wind injures trees, other disturbances such as bark beetle epidemics may begin. Wind is also a major vector of spread for many insects and biological disease agents.

There are a number of ecological roles played by wind that will not be covered in this chapter. For example, in the 1930s wind caused major desertification effects in the midwestern United States (the dustbowl) and continues to do so around the world. Wind is the major means of distributing heat energy around the earth, and carries many beneficial insects as well as seeds to once-barren landscapes. This chapter focuses on the interaction of wind with forest health and protection, treating wind as a disturbance while recognizing that vegetation affects, as well as is affected by, this powerful force of nature.

WIND AS A PHYSICAL PROCESS

Wind can be described as simply gentle, brisk, or strong, but a more accurate weather scale is the **Beaufort Scale** (Table 7.1). Categories of breezes, gales, storms, and hurricanes are recognized, with attendant ranges of windspeeds. Winds represent nature's attempt to equalize pressure differentials around the earth due to unequal heating between the equator and the poles, and between land and water surfaces. Because these processes repeat on an annual basis, there is some predictability to the general wind patterns in any region, although topography will

TABLE **7.1** | **The Beaufort Scale of Wind Velocity**

Beaufort description Number	Windspeed	
	kph (km/hr)	mph (mi/hr)
1 Calm	0	0
2 Light air	1–5	1–3
3 Light breeze	5–12	4–7
4 Gentle breeze	12–19	8–12
5 Moderate breeze	19–29	13–18
6 Fresh breeze	30–39	19–24
7 Strong breeze	40–50	25–31
8 Moderate gale	51–61	32–38
9 Fresh gale	62–74	39–46
10 Strong gale	75–87	47–54
11 Whole gale	88–101	55–63
12 Storm	102–116	64–72
13 Hurricane	>117	>72

create substantial subregional variation. For example, gale-force winds can occur any month of the year in coastal Alaska, but are much more common in the September to April period. Whole gales to hurricanes are rare, occurring about 2% as often as winds between 50 to 87 kph (31–55 mph), and are restricted from October to February. Over a 25-year period this has been a consistent pattern (Harris 1989).

Peak gusts can be significantly higher than the recorded windspeed, as an instantaneous gust can blow much stronger than the "sustained fastest mile" windspeed recorded by averaging the instantaneous windspeeds over 1 minute each hour of the day. In Alaska the average ratio of gust windspeed to "sustained fastest mile" windspeed is about 1.4, similar to that found in Scotland (1.4–1.5) and lower than that found in England (1.8–2.0) (Coutts and Grace 1995). Gust windspeeds are important even though they may be short-lived, because those gusts may be well above the threshold of wind damage. For example, a 80 kph (50 mph) recorded "sustained fastest mile" wind may have gusts close to 113 kph (70 mph).

Tornadoes are the most destructive weather phenomenon known. Violently swirling winds and extremely low atmospheric pressure associated with thunderstorms can create funnels that become classed as tornadoes if they touch the ground. Most tornadoes in North America occur east of the Rocky Mountains. A tornado path may be as narrow as 30 m (100 ft) or as wide as a km (0.6 mi), and will twist trees out of the ground or break them off, generally scattering the broken pieces over wide areas.

Prevailing wind direction also is predictable from general circulation patterns, and is described from the direction the wind is coming (e.g., the windward

Track of the eye of 1938 hurricane (arrow) through central New England, showing the distribution of forest damage. The Harvard Forest is located in north-central Massachusetts within the area of extensive damage.

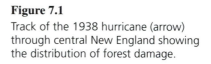

Figure 7.1

Track of the 1938 hurricane (arrow) through central New England showing the distribution of forest damage.

direction). A wind from the north is, therefore, a north wind. In the northern hemisphere wind direction is counterclockwise around low-pressure cells and clockwise around high-pressure cells. Those forests to the east of the track of the 1938 hurricane (a major low-pressure cell) that hit the northeastern United States (Figure 7.1) were first approached by winds coming from the southeast, and after the major damage had occurred, from the southwest (Foster 1988). The worst damage usually is on the eastern side of the storm track, where the direction of the storm is additive to the direction of the wind. The strong winter storms along the Pacific Coast usually have a strong south to southwest component (Gratkowski 1956). However, there may be substantial subregional variation in windstorm effects. In the Columbia Gorge area, a major air corridor in the Pacific Northwest between the west and east Cascades, most wind damage comes from the northeast due to strong, cold blasts of continental air in winter constrained through a narrow corridor.

Windstorms of sufficient force to damage forests occur all across North America. As temperate trees typically live for centuries, even infrequent windstorms (20- to 50-year intervals) can play significant roles in the dynamics of forest succession. The southeastern United States has a high frequency of hurricanes and tornadoes, averaging about one major storm every 3 years somewhere in the region from Texas through North Carolina (Gresham et al. 1991). The northeastern United States has fewer such storms, averaging about one every decade, some of which are extensions of southeastern storms and others that come more directly from the Atlantic Ocean. The 1938 hurricane (Figure 7.1) had major impact on northeastern forests, to be discussed later in this chapter. The Midwest and Lake states have an occasional tornado, which will uproot and shatter the trees in the narrow path of the storm. Wind damage in the Rocky Mountains is most commonly associated with foehn winds (see Chapter 3). Strong pressure gradients cause warm winds to descend out of the mountains toward the plains and can be associated with swaths of windthrown trees (Knight 1994). Along the northwestern Pacific Coast, large windstorms occur about three times a century, but appear to strike different parts of the coast. Along particular sections of the coast, storms strong enough to create individual tree gaps occur annually, and small-scale storms opening substantial canopy gaps occur about once a century; large-scale catastrophic events, such as the "1921 Blow" in the Olympic Mountains of Washington, occur about once every four centuries at this subregional scale.

WIND EFFECTS ON TREES

The ecological effect of "wind" often will incorporate extra loading of mass on the tree crown from ice and snow. Normal storm winds can then have a more catastrophic impact on the stand. The discussion of windsnap and windthrow implicitly incorporates the additional impacts that various forms of precipitation may have on susceptibility of tree crowns and boles to breakage.

Crown Morphology

Flagging is the most obvious morphological effect of wind on tree crowns (Figure 7.2). On the windward side of the tree, foliage is repeatedly killed by winds that desiccate the needles (Kimmins 1997). In summer this may be due to excessive transpiration and associated foliar moisture deficits. In winter, where soils become very cold, root uptake and movement of water will be restricted, and similar foliar moisture deficits may occur. Foliage can be abraded by soil or ice particles, and near the coast salt spray carried by wind can have a toxic effect on windward foliage of some tree species, while preventing other species from establishing at all. In species with limber branches, winds can "train" the branches by permanently bending them away from the wind over a period of years. Prevailing winds can be reconstructed from the patterns of flagged trees using the direction

Figure 7.2
A wind-flagged stand of trees. The damaging winds come from the left and cause mortality of needles on a regular basis on the windward side of the tree.

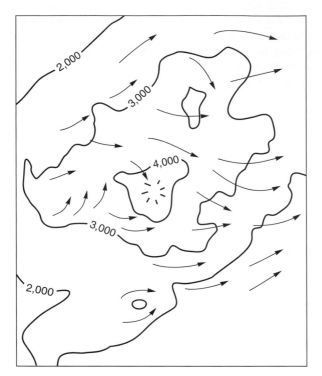

Figure 7.3

Wind patterns on White-face Mountain as reconstructed from direction of flagging on trees. Elevations are in feet.

(Source: Adapted from Holroyd 1970.)

of branch growth and reaction wood in the trunk tops (Figure 7.3). Flagging is most pronounced on isolated trees or those on the windward edge of stands; more protected trees seldom exhibit flagging (see Chapter 9).

The most severe form of flagging is where the leader is never able to rise above the protected zone, where the protection may be a rock, or the surface of the snow in winter. Individual trees will assume a shrublike morphology known as **krummholz**, and in continuous stands the forest will appear as a shrubfield. This is a common tree form at the ecotone between subalpine and alpine environments at the timberline. Abrasion by snow or soil particles removes the cuticle wax from leaves that are growing above the protected zone (e.g., above the snowline) and results in needle death by desiccation. Abrasion effects may be exacerbated by short growing seasons that do not allow adequate cuticle development before the onset of winter (Hadley and Smith 1986).

A subtle but significant impact of wind on tree crowns is the effect of adjacent crowns on one another by wind whipping the branch tips. Usually there is little overlap between adjacent crowns, and although competition for light may partially explain this lack of overlap, the physical impact of wind causing one branch tip to break off the branch tips from adjacent trees also is important. After a significant storm in fully stocked stands, it is common to see broken branch leaders scattered across the forest floor.

Stem and Root Shape

Conifers and angiosperms react differently to wind coming from a fairly constant direction. Conifers will develop **compression wood** on the lee side of the stem, enlarged annual rings that will help the tree remain upright against the tendency to compress the wood in that direction. Angiosperms react quite differently, developing **tension wood** on the windward side of the stem, resisting the tension (Coutts and Grace 1995). Compression and tension wood, both also known as reaction wood, are not solely related to the presence of constant winds, but where such winds occur this response is common, and results in elliptical stem shapes with the long axis in the direction of the wind.

Stem buttressing is another form of stem response to wind (Kimmins 1997). Particularly in tropical areas, swaying of tree stems results in augmented annual ring development near the tree base. Trees will form large buttresses in directions that provide the most support against the prevailing winds. Mechanical damage to the stem probably increases production of ethylene, with subsequent localized increases in growth. The appearance of buttressing may occur for other reasons, however. Trees that germinate on nurse logs typically have enlarged bases as the log decays and the roots merge above the soil surface. On large nurse logs the roots may become individual stilts, and such trees are very prone to windthrow.

In populations of western hemlock in southeast Alaska, **fluting**, or vertically oriented furrows that form above root buttresses and spiral into the crown, are common (Julin et al. 1993). Fluting makes logs difficult to debark and can reduce wood quality. Mechanical stress appears to be a significant factor in flute development, and is most common in dominant trees of coastal, young, even-aged stands exposed to strong winds. Eventually after 100-plus years, the stem becomes more cylindrical and the flutes are no longer obvious. Western hemlock appears genetically predisposed to fluting, whereas Sitka spruce growing nearby is much less affected.

Roots on trees exposed to wind will develop asymmetrical shapes, with growth allocated to the top and bottom at the expense of growth to the side, due to the same effect on localized growth caused by ethylene. In this manner the root becomes shaped much like an "I" beam, and has much more resistance to bending in the vertical direction than a root of the same volume but circular in shape (Harris 1989). Root grafting also can improve windfirmness. The effective root mass of the tree is extended by the physical bonding with roots of other trees, although these grafts can act as vectors for biological agents that cause root diseases.

Windsnap (Stem Breakage) and Windthrow (Uprooting)

Stem breakage and uprooting are more severe effects of wind than those mentioned above, and are most significant in terms of a forest health perspective. The windspeed, timing of gusts, and duration of wind are clearly initiating factors in windsnap and windthrow. In addition to the strength of the wind, there are several factors associated with the probability that a tree may break or be uprooted (Figure 7.4). Species characteristics, stand age and structure, and degree of decay all play important roles in evaluating the probability of severe wind damage.

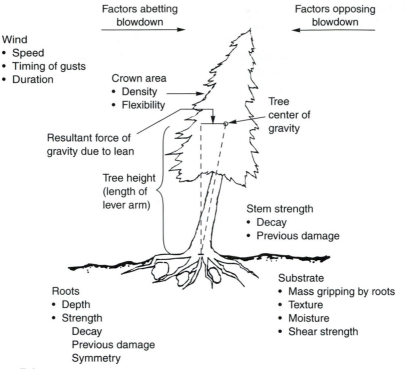

Figure 7.4
Factors affecting tree stability.
(Source: From Harris 1989.)

Crown area provides resistance to windflow, and acts as a "sail" tending to push over the tree (Coutts and Grace 1995). Species with flexible or low-density crowns, or deciduous species out of leaf, will have less of a sail effect, and may suffer only minor top damage while other species are blown over. Trees with large crowns also may have well-developed buttresses or root systems, so they may be stronger than trees with thinner crowns that may not be as well supported. On the Olympic Peninsula of Washington, the four largest western redcedar trees in the state are found, each with a deformed top due to repeated wind damage. Wind-snap and windthrow have repeatedly removed most of the associated spruce and hemlock with denser and/or less flexible crowns. In areas where snowstorms can occur, those species able to shed snow will be much less prone to windsnap and windthrow than species that will retain ice and snow in the crown. Trees with dwarf mistletoes that cause "brooms" (see Chapter 14) will accumulate snow and ice and will easily break in the wind.

Tree height acts as a lever so that tall trees are more susceptible to wind damage than shorter ones. Because height often is indexed to tree age, as stands mature they become more susceptible to wind damage. In a silvicultural sense, windthrow

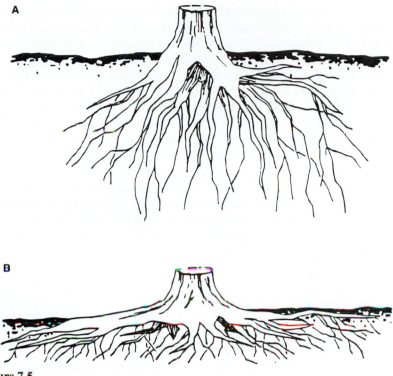

Figure 7.5
Typical rooting habit of Sitka spruce on two soils. **A,** Predominately vertical rooting on deep soils. **B,** Wide, shallow rooting on poorly drained soils.
(Source: From Harris 1989.)

and windsnap are equivalent to a thinning from above, such as a selection or crown thinning effect (Smith 1986).

The rooting medium of rocks and soil can greatly affect probability of wind damage. Extremely well-anchored trees, such as in rocks, are more likely to snap than be uprooted. Well-drained soils are more likely to provide good anchoring than soils with restricted rooting zones (Figure 7.5) due to seasonal flooding or cemented glacial till. When wet, soils are less cohesive than when dry, so that their shear strength declines during the wettest periods of the year. In coniferous forest areas like the Pacific Northwest, when the wet period also has the highest wind velocities, windthrow events are most common during the winter months.

Wind pushing against a tree crown will produce an overturning moment, the product of the force of wind at the center of pressure in the crown and the height of that point. A second overturning moment is produced when the tree is swayed to one side, caused by the displaced weight of the stem and crown. It is determined by the horizontal displacement of the center of gravity of the tree and crown and the height of the center. These turning moments, if they exceed the resistive force of the tree stem or its anchoring in the soil, will produce windsnap or windthrow.

Root and stem decays (see Chapters 12 and 14) can significantly weaken tree resistance to windsnap and windthrow. Stem decays may eliminate resistance to windsnap in the heartwood, leaving a thin rind of sapwood to provide stability. Root rots can remove resistance to windthrow around a portion of or the entire perimeter of the root system. Often these low-vigor trees will be killed by insect attack before they are damaged by wind. Some species, such as incense-cedar infected with Annosus root rot, caused by *Heterobasidion annosum,* will more likely show a declining crown vigor but remain alive with a progressively weakening root system until blown over. In densely developed portions of Yosemite Valley in the Sierra Nevada, a special hazard rating system was developed for incense-cedar (*Calocedrus decurrens*) because so many tourist cabins suffered damage from windthrow (Srago et al. 1978). The associated infected pines were killed by bark beetles and were easily identified as hazards, but the incense-cedars were not attacked by beetles so they remained upright until felled by wind. Declining crown vigor was apparent in the infected incense-cedars, and became the basis for the hazard rating system.

WIND EFFECTS ON FOREST SUCCESSION

Forest succession is enough of a site-specific process that a discussion of wind effects will be too general, so several case studies are included to illustrate more site-specific effects of wind. Wind will generally thin from above, so that fast-growing or older trees or stands are more likely to suffer damage than smaller, younger trees or stands. Often more trees in intermediate or suppressed crown classes will be affected, but more volume is lost in the larger-diameter classes. A typical result of windsnap or windthrow is the creation of a gap in the canopy. At the scale of a single tree, the adjacent trees may simply extend their crowns into the small gap and fill it. If the gap is at the scale of several trees, there is likely to be a significant environmental response at ground level with increased nutrients, light, moisture, and an altered microclimate. Existing understory herbs and shrubs, and existing understory trees that are likely to be shade-tolerant, late-successional species, will be released in growth and eventually fill the gap. There may be opportunity for shade-intolerant species to establish on the larger blowdowns, but large gaps may still not provide adequate microsites for shade-intolerant species, particularly if a shade-tolerant tree understory is released by the removal of the canopy dominants.

Windthrow will upturn a significant amount of soil with the root ball of the tree, creating a hummocky topography called **pit-and-mound** topography (Oliver and Larson 1990). In areas where pit-and-mound topography is ubiquitous, wind tends to have been, and be, a dominant disturbance process. This upturning can have significant effects on the microsite of the gap. Windthrow overturns the local soil profile, aerating the soil, preventing the continuation of the podzolization process, which in northern latitudes can lead to the formation of bogs (Ugolini and Mann 1979), while exposing mineral soil for species favoring such microsites. Mounds are generally better microsites than pits, which may be flooded in low-lying terrain or floodplains, and both microsites are poor until the microtopography has stabilized.

The creation of coarse woody debris from the uprooted or broken stems will have positive and negative consequences for wildlife and fish (Harmon et al. 1986). The stubs of broken stems, if large enough in diameter, will be feeding and nesting sites for woodpeckers and other cavity nesters. The downed material will increase fire hazard for a while, but also will create habitat favored by small mammals such as shrews, deer mice, and salamanders, and serve as perch and viewing platforms for grouse and squirrels. Coarse woody debris also may provide sites for nitrogen-fixing organisms, and may serve as nurse logs for tree regeneration. If the windthrow is large, some salvage of the windthrown timber can be done without negative consequences to fish and wildlife, as coarse woody debris above certain minimum levels is not a limiting habitat factor. Cleaning the forest floor of all debris is not recommended. Salvage of dead timber must generally be completed in 2 to 3 years, before decay lowers or eliminates the commercial value of the downed material. This is particularly true for hemlock and true fir species, and small trees with a high proportion of sapwood. Species such as redwoods or cedars, which have decay-resistant heartwood, are salvageable for much longer periods of time. Coarse woody debris in streams dissipates the erosive energy of water and creates pool habitat. During the 1970s it was common to see projects misdirected at removing logs from streams, until aquatic ecologists were able to demonstrate the benefits of logs in streams. Current projects now are often aimed at adding large pieces of woody debris to encourage stream habitat diversity (Naiman and Bilby 1998).

Case Study: Pacific Northwest Spruce-Hemlock Forests

The coastal forests of the Pacific Northwest comprise a strip 25 to 35 km (15–20 mi) wide at most, except where it extends up coastal river valleys. Sitka spruce is a dominant along with western hemlock and western redcedar. The climate is mild, with 200 to 300 cm (80–120 in.) of annual precipitation; summer fog and low clouds are common (Franklin and Dyrness 1973). Fires tend to be uncommon, and wind is the primary disturbance agent. Small-scale wind events are more common and over time have at least as much impact on successional processes as the large events, although the larger-scale events tend to be more impressive to the observer. At the stand level, one study in coastal Oregon suggested the small-scale wind event return interval at 119 years and the large-scale interval at 384 years (Harcombe 1986). At a regional scale, larger events occur more frequently, perhaps several times a century, but strike different areas, recurring in a particular stand only every few centuries. Although these intervals may seem very long, they are but a fraction of the life span (400–1,000 years) of species in these forests. Windthrow susceptibility in decreasing order is western hemlock, Sitka spruce, Douglas-fir, and western redcedar (Agee 1993).

The small-scale events maintain a mosaic of Sitka spruce and western hemlock, with western redcedar and Douglas-fir generally less common. Western redcedar usually is heavily browsed by elk, and gaps are seldom large enough to allow Douglas-fir to successfully regenerate. Periodic windthrow allows recruitment of the moderately shade-tolerant Sitka spruce, as well as the more shade-tolerant western hemlock. Competition on the forest floor from moss and shrubs can be significant. One of the common coastal shrubs, salmonberry (*Rubus spectabilis*) can have over 100 km ha^{-1} of rhizomes on treeless sites, so if it is present as an understory species it may fill the growing space rapidly (Tappeiner et al. 1991). Over time the pattern of repeated recruitment of understory trees by windsnap and windthrow of canopy dominants creates a multiaged stand. Coarse woody debris inputs in these forests tend to be episodic, and probably maintain on average higher log biomass than upland forests that have more of a U-shaped distribution of log biomass over centuries following major disturbances like fire (Agee 1993). Douglas-fir appears to be rarely recruited after windthrow events, although it regenerates well after fires.

On January 29, 1921, the largest windfall known in the world to that time occurred on the western Olympic Peninsula (Boyce 1929). Over 125 cm (50 in.) of rain had fallen in December and January, followed by a storm with south-westerly winds recorded over 225 kph (140 mph). In the relatively flat terrain of the coastal strip (Figures 7.6 and 7.7) blowdown was extreme. To the north and east, ridges reduced the power of the wind, and much of the light damage to the south was across mostly cutover land. Roughly 600,000 ha (1.5 million acres) experienced significant damage, with a core area of about 200,000 ha (0.5 million acres) having severe damage. Vistas of 2 km or more were created in stands where prewindthrow sight distances were 100 m. Few of the available photographs show extremely large timber, suggesting that such events may occur before old-growth conditions occur at a landscape level. The storm released the understory of the ragged forest stands left behind, and both western hemlock and Sitka spruce have continued to dominate these stands, with scattered western redcedar. In 1962 another regional windstorm of catastrophic proportions (145 kph [90 mph] winds with gusts to 217 kph [135 mph]) passed across the coastal region, but did more damage to coastal Oregon forests than those of Washington (Orr 1962).

Case Study: Northeast Hardwood–White Pine–Eastern Hemlock

New England forests experience periodic hurricanes. Some of the most destructive, in 1635, 1788, 1815, and 1938, extended through central New England (Figure 7.1); these infrequent but intense disturbances appear to be the most significant

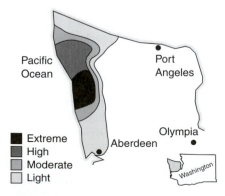

Figure 7.6

Severity of blowdown caused by the 1921 windstorm on the Olympic Peninsula, Washington.

Figure 7.7
A completely windthrown stand of young western hemlock from the 1921 windstorm.
(Source: Photo from Boyce 1929.)

disturbance events affecting the vegetation of the region (Foster 1988). A total of about four to five such storms per century occur in the area, so that forests are still adjusting to the effects of the last storm, in a successional sense, when the next large hurricane passes through. Storms have very strong southwest to east winds, so that more southerly aspects or flat areas exposed to the southeast experience the most damage. The importance of hurricanes decreases inland and to the north. Other events (fire, pathogens, etc.) are significant at a local scale but are not so regionally synchronized.

Human influence on northeastern forests has been extensive over the last 200 years, with land clearing and, more recently, reforestation of abandoned fields. The younger forests are more resistant to wind damage, being shorter and often with a more uniform canopy than older, natural forests of the region. These younger forests often have substantial edge effects at their border with fields, roads, and developments.

The 1938 hurricane brought 15 to 35 cm (6–14 in.) of precipitation with winds over 200 km hr^{-1} (120 mph). At the Harvard Forest over 70% of the timber volume was windthrown, and substantial pit-and-mound topography was created. Most of the windblown trees were blown over oriented to the northwest. In exposed locations conifer stands (eastern white pine [*Pinus strobus*] and red pine [*Pinus resinosa*]) were heavily damaged if more than 15 years old and were completely windthrown if 30 years of age or older. Hardwood stands, consisting of oak hickory pine (*Quercus-Carya–Pinus strobus*), red maple–oak (*Acer rubrum–Quercus rubra*), and northern hardwood–hemlock (*Tsuga canadensis*) forests, showed a gradual increase in damage with age (Foster 1988). Stands over 70 years old experienced complete windthrow, with those containing emergent eastern white pine showing the most damage. In older conifer and hardwood stands, dominants tended to be uprooted; younger stands showed more of a tendency for leaning (partial uprooting), particularly in the subdominant crown classes. All species showed increasing damage with stand age. Order of decreasing susceptibility to damage was pines > aspen > paper birch > red maple > [red oak and American ash] > [hickory, white oak (*Quercus alba*), black oak (*Quercus velutina*), and eastern hemlock]. The rapid-growing, shade-intolerant species that tended to be emergents ranked as most susceptible to damage (pine, aspen, and birch). At the other end of the spectrum, slow-growing, more shade-tolerant species tended to suffer the least damage because they were protected in the understory. Stands in protected locations showed less damage.

The primary successional changes initiated by the 1938 hurricane were conversion of eastern white pine and pine-hardwood forests to hardwood and hardwood-hemlock forests. These changes were due to higher damage in pine stands, selective removal of pines in mixed stands, and the ability of hardwoods to either release in growth as residual understory trees or vegetatively reproduce.

Case Study: Southern Pines

Wind is not the disturbance agent one usually thinks of when southern pines are mentioned. Historically forest fires (see Chapter 4) and southern pine beetles (*Dendroctonus frontalis*) (Chapter 21) might be thought of as more important than wind, but local wind gusts as well as hurricane-level winds do have significant effects on southern pines. Under normal weather without unusually high winds, loss from wind is usually not high. Measurement over several years of longleaf pine (*Pinus palustris*) stands in northern Louisiana indicated about 0.3% per year loss of trees to the combined effects of wind and insects. In Georgia wind followed only lightning as a cause of death of mature longleaf pine trees. In the highest-density size classes, total mortality was generally less than 5%, but 31% of that over a 4-year period was due to windthrow. In eastern Texas 29.6% of total mortality of loblolly, shortleaf (*Pinus echinata*), and longleaf pine trees used for red-cockaded woodpecker nest sites was due to windthrow, some attributed to adjacent cutting activities that funneled wind to these trees (Conner et al. 1990). Under normal conditions longleaf pine is more resistant to windthrow than loblolly pine (*Pinus taeda*) or slash pine (*Pinus elliottii*) (Gresham et al. 1991). In the early 20th century, much windthrow of longleaf pine was attributed to stem weakening from continued turpentining activities (Mattoon 1925). In more recent years stem diseases (see Chapter 13) such as fusiform rust (caused by *Cronartium fusiforme*), which will attack all three of the pines mentioned above but is most serious on loblolly pine, also has been implicated in windthrow under normal weather conditions.

Hurricane-level winds may occur infrequently but have much more severe effects on bottomland forests as well as southern pine stands. In 1989 Hurricane Hugo came ashore in South Carolina carrying windspeeds exceeding 220 km hr^{-1}. Return intervals for all hurricanes in this area average about one per 6 years, and those with intensities as high as Hugo about one in 25 years. With tree species that can live hundreds of years, a surviving old-growth tree will have withstood tens of hurricanes during its life. When hurricanes hit southern pine stands, the results can be almost a "clean sweep" of timber, but usually are not. However, early in the 20th century, one of the largest mills in the South operated for about a year on windthrown timber following a hurricane. More often there is substantial damage that varies by species.

Longleaf pine tends to suffer less damage than its associates because of taproot development and usually well-developed lateral rooting. Of 10 species

TABLE **7.2** | **Variation in Wind Damage by Percent of Stems in Hobcaw Forest, South Carolina During Hurricane Hugo**

Species[a]	No damage[b]	Light damage	Moderate damage	Heavy damage	Percent of total stems[c]
Longleaf pine	46.7	41.0	4.5	7.8	6.7
Loblolly pine	27.3	46.2	17.5	9.0	47.0
Pond pine	27.7	37.8	7.7	26.7	1.2
Bald cypress	11.2	73.0	12.2	3.6	1.8
Blackgum	15.3	50.1	30.9	3.5	4.7
Sweetgum	14.9	55.1	28.5	11.6	8.7
Swamp tupelo	5.4	86.6	6.6	1.5	2.1
Laurel oak	15.7	52.9	11.7	19.4	5.0
Live oak	27.1	47.7	21.2	4.3	2.8
Water oak	15.9	51.8	12.1	21.1	2.7
Other species	(42 others)				17.3

Note: *Light damage* is bent or limbs broken; *moderate damage* is defoliated or top broken; and *heavy damage* is bole broken, uprooted, and downed.
[a]Species not defined in text include blackgum (*Nyssa sylvatica*) and sweetgum (*Liquidambar styraciflua*)
[b]Sum across rows of Undamaged, Light, Moderate, Heavy Damage = 100%; may not be exact due to rounding. For sweetgum, mistake in original publication, sums to 110.1%
[c]Sum down column = 100%
Source: Adapted from Gresham et al. (1991).

surveyed in South Carolina after Hurricane Hugo (Gresham et al. 1991), only 13% of longleaf pines were damaged, with that damage comprising only 6.7% of all damaged trees (Table 7.2). Loblolly pine had 26.5% of its stems damaged; pond pine (*Pinus serotina*) had 34.5% of its stems damaged. The increased damage for the latter two pines was attributed to foliar and branch morphology differences, and to shallow rooting, particularly for pond pine. In terms of least damage by basal area removed, the top four species were swamp tupelo (*Nyssa sylvatica biflora*) , bald cypress (*Taxodium distichum*), longleaf pine, and live oak (*Quercus virginiana*). The first two species have buttressed boles, and the latter is a short species with very strong wood comprising the stem. Tall species with typically shallow rooting, such as water oak (*Quercus nigra*) and laurel oak (*Q. laurifolia*) and pond pine, had one-third or more of their basal area removed by the hurricane.

FOREST MANAGEMENT STRATEGIES

Shelterbelts

Trees are not only affected by wind but also have a large effect on wind. Although not a focus of this chapter, shelterbelts, or bands of trees oriented perpendicular to prevailing winds, have long been used to decrease winds to the lee side of the

T A B L E **7.3**	**The United Kingdom Windthrow Hazard Classification**

Regional Windiness[a]

Wind zone	A	B	C	D	E	F	G
Score	13.0	11.0	9.5	7.5	2.5	0.5	0

[a]Zones A–G indicate equal gradations of decreasing wind exposure, and generally decrease from west to east across Great Britain.

Elevation[b]

Elevation (m)	Score
0–60	0
286–315	5
541+	10

[b]Score increases with elevation in Great Britain. Not all categories are shown.

Topex[c]

Topex total (degrees)	Alternative classification	Score
0–9	Severely exposed	10
16–17	Very exposed	8
23–24	Very exposed	6
41–70	Moderately exposed	2
71–100	Moderately sheltered	1
100+	Very sheltered	0

[c]This is sum of angles (degrees) to the horizon at the eight major compass points. Not all categories are shown.

(continued)

barrier. Narrow shelterbelts tend to be more effective than wide barriers. The distance over which windspeed drops is proportional to the height of narrow barriers. At a lee distance five times the barrier height, windspeed is only one-third of the free windspeed, and at 10 tree heights it is about 50%, becoming 80% and more after 20 tree heights (Harris 1989). Shelterbelts are commonly used to protect agricultural crops from desiccating or otherwise damaging wind.

Windthrow Hazard Identification

Windthrow hazard classifications have been developed for plantations, particularly those where large-scale afforestation has occurred (Miller 1985, Quine et al. 1995). Such classifications identify the most likely locations for windthrow

T A B L E **7.3**	**The United Kingdom Windthrow Hazard Classification (*concluded*)**

Soil Rooting

Root development	Soil group	Score
Unrestricted rooting >45 cm	Brown earths, podzols	0
Restricted rooting but some >25 cm	Deep peats, loamy gleys	5
Very restricted rooting <25 cm	Peaty gleys, waterlogged soils	10

Windthrow Hazard Classes[d]

Score (total A–D)	Hazard class
Up to 8.0	1
8.0–13.5	2
14.0–19.0	3
19.5–24.5	4
25.0–30.0	5
>30.5	6

[d]Higher numbers indicate higher windthrow hazard.
Source: From Miller (1985).

damage on the basis of a rating system for important contributing factors. These factors will vary depending on site, as noted below. The system developed for the United Kingdom (Table 7.3) is based on regional windiness, elevation, topography, and soils. Regional windiness is based on meteorological data and the tatter, over several years, of small exposure flags placed across the landscape. Scores for wind are weighted slightly more heavily than the other factors. Elevation in this system is positively correlated with damage, as rainfall and mean windspeed increase with elevation. This factor may be rated quite differently elsewhere. For example, on the Olympic Peninsula of Washington, the strongest winds are along the western coastal plain near sea level, and the dissected topography of higher elevations, except for the very tops of ridges, tends to decrease mean windspeed. Topography in the United Kingdom system is measured by topex assessment. It is an index to exposure, based on the sum of angles to the horizon from the eight major compass points. Larger sums indicate more protected locations. An alternative in this system, in the absence of topex data, is to assign a score based on classifying the site in one of five categories ranging from "severely exposed" to "very sheltered." The soils factor is a simplified approach to a complex situation, and is based on rooting depth available to the trees. The scores for the four factors are summed to give a windthrow hazard class. These classes are then used to evaluate silvicultural activities.

Site Preparation and Planting

Site preparation and planting are normally not times when wind hazard is expected to be significant. On some sites where spaced furrow plowing is used to prepare sites on gley soils, the planting of trees on the raised rows between furrows can prevent adequate anchoring by roots, as they do not spread across the exposed hardpan in the furrows. Whole rows of trees may later be windthrown. Shallow plowing is advised on such sites.

Initial spacing of trees is not a windthrow concern, as they are too small to be blown over soon after planting. However, control of tree density is important to allow individual trees to develop anchoring root systems.

Thinning

Thinning can occur through a variety of methods (Smith 1986): low, crown, selection, and mechanical (Figure 7.8). The effect of thinning on wind damage potential is therefore complex. If the initial stand has very high tree densities and little crown stratification, all the trees in the stand are likely to have little windthrow resistance. The stand may not as yet have suffered significant wind damage because of the energy-dampening effect of closely spaced crowns on one another. Thinning of these stands will increase sway in the residual stems, and significant loss from wind may be expected. If root disease is present, losses from windthrow will be even greater. Several entries, each of which takes a small proportion of the total density, may be the safest way to build windfirmness in the residual trees.

Low thinnings, where trees of subordinate crown classes are removed, will leave codominant and dominant trees that have large crowns but are likely have well-developed root systems. Wind damage is likely to be minimal in low thinnings. Crown thinnings remove a range of crown classes and may expose at least some trees with poorly developed root systems near locations where dominant trees were removed. Selection thinnings remove the largest size classes of trees and often leave stems that have poorly developed root systems and have experienced little wind stress. In multispecies stands, these intermediate to suppressed trees are generally of shade-tolerant species, and are more likely to have root or butt rots. Mechanical thinnings are most common in even-aged plantations, and create linear gaps that may tend to funnel wind or expose an entire line of residual trees at once.

Pruning

The effects of pruning from below (butt log) on windthrow appear to be minor. Windspeed near the ground will increase slightly. On small plantation trees grown in short rotations, such pruning will increase the height of the center of gravity of the trees, and can increase the turning moment created by the wind. Evidence of windthrow between pruned and unpruned radiata (Monterey) pine (*Pinus radiata*) in New Zealand showed pruned stems sometimes less and sometimes more prone to windthrow (Somerville 1980). In North American forests, with the expectation of taller trees and longer rotations, and production of more logs per tree, butt log pruning is probably insignificant in affecting windthrow hazard; by the time the

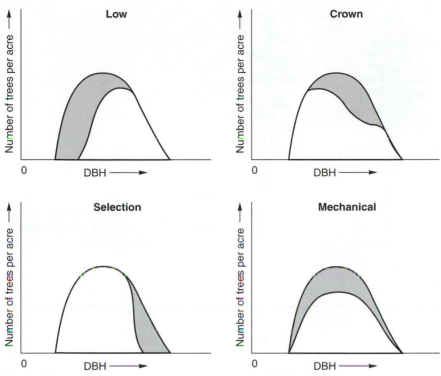

Figure 7.8

Diameter distribution of the same pure, even-aged stand that would be removed by four different thinning methods. Parts that would be removed are indicated by shading.

(Source: From Smith, D., *The Practice of Silviculture*, ©1986, John Wiley & Sons., Inc. Reprinted with permission.)

trees are tall enough to be seriously affected, the loss of some lower limbs has a negligible effect on center of gravity and turning moment. If the trees being pruned are part of a narrow shelterbelt, pruning may provide a funneling effect through the shelterbelt and result in winds higher in velocity than the free wind. Pruning may reduce damage to adjacent trees caused by branch abrasion during windy periods.

Another pruning treatment is to remove portions of the crown, such as alternate whorls in conifers. This practice would be unusual for commercial forests but may be more applicable in urban forests, where individual trees may be more valuable and in more fragmented and isolated situations. Such pruning can reduce the sail area of the tree and lower the center of gravity, reducing windthrow probability (Figure 7.9). Edges of urban forests, and next to new edges on adjacent properties, are places where such pruning is likely to be effective (Bradley 1995).

Clearcuts

Clearcutting is a common regenerating strategy in silviculture. In areas with strong winds, substantial wind-related expansion of clearcut edges and other artificial edges, such as roads, will occur. Edge expansion occurs primarily on the lee side

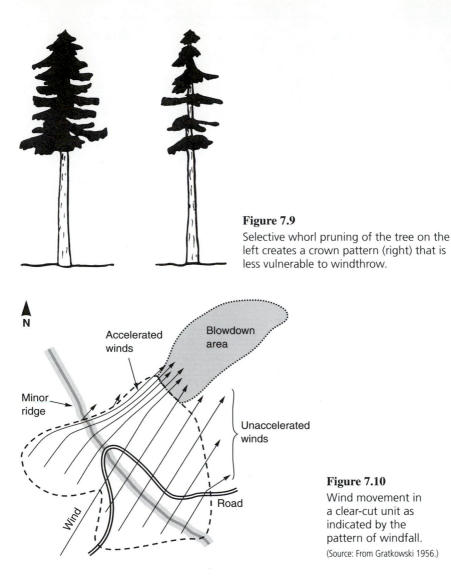

Figure 7.9
Selective whorl pruning of the tree on the left creates a crown pattern (right) that is less vulnerable to windthrow.

Figure 7.10
Wind movement in a clear-cut unit as indicated by the pattern of windfall.
(Source: From Gratkowski 1956.)

of the cut unit (Figure 7.10) or road, and will also occur on the lee sides of low ridges oriented perpendicular to the prevailing wind. Winds can be funneled along edges of cut units and blowdown will be concentrated at the timbered convergence zone of the funnel (Gratkowski 1956, Alexander 1964). Notches cut into a windward-facing edge also can act to concentrate wind. A number of stand and topographic conditions should be evaluated when planning clearcut or retention silvicultural activities (Table 7.4).

There is currently a trend away from large cutting units to more dispersed, smaller settings. This can have the unintended effect of increasing edge and decreasing the ability to locate cut lines along naturally windfirm boundaries. Small units

T A B L E **7.4**	**Factors Associated with Above- and Below-Average Windthrow Hazard**

Above Average

1. Leeward cutting boundaries
2. Narrow strips and small patches
3. Cutting boundaries at right angles to the contour
4. a. Saddles in ridges
   b. Ridgetops
   c. Lower and middle slopes
5. Slopes facing the wind
6. Moderate to steep slopes
7. Abrupt changes in direction, long straight lines and square corners, or indentations in boundaries.
8. Shallow or slowly drained soils
9. Stands of old trees, especially if defective
10. Trees grown in forest stands
11. Special topographic situations conducive to acceleration of the wind.

Below Average

1. Windward cutting boundaries
2. Wide strips and large patches
3. Cutting boundaries parallel to the contour and along roads
4. Stream bottoms and upper slopes
5. Slopes facing away from the wind
6. Flats and gentle slopes
7. Irregular or smoothly curved boundaries
8. Medium and deep soils
9. Moderate to rapidly drained soils
10. Young stands of sound trees
11. Open-grown trees

Source: Adapted from Alexander (1964).

have much more edge than large units. For example, consider a square kilometer of land undivided compared to the same section divided into 25 4 ha blocks. The undivided section, if cut as a large block, has a perimeter of 4 km, equaled by the perimeter of only 5 separated units of the 25 smaller units comprising the same square kilometer. There are a number of good reasons for smaller unit sizes and the associated slower cutting rates in watersheds, but in wind-prone areas the consequences are a continual need for salvage operations in areas with awkward access or the production of more coarse woody debris than is needed for any wildlife purposes.

At the lee edge of the clearcut, intermediate and suppressed trees are likely to be the first to be windthrown, and may be the most common windthrown crown class in terms of tree numbers. Larger trees also are very susceptible because of their sail area and exposure, and in terms of volume, loss due to windthrow around cut units is often 70% or more comprised of dominant trees.

The most common recommended cutting pattern in wind-prone areas is a modification of strip cutting into the wind (Figure 7.11). In these diagrams cutting begins at a stable leeward stand edge, which might be a rock outcrop, avalanche track, or natural clearing, and then progresses in a windward direction over time. Lee edges of the cut strips are mostly young, short stands less prone to wind damage. Such planning presupposes that damaging winds will come from only one direction. Although "normal" winds can be planned for by such cutting patterns, unusual storm winds from other directions can negate planning for unidirectional winds (Gordon 1973). This is likely why forest silvicultural plans in somewhat average wind conditions usually do not include contingencies for wind damage.

Stream Buffers, Interior Leave Patches, and Other Retention Strategies

Current forest management policies across public and private lands in North America are encouraging the retention of isolated or aggregated leave trees where even-aged regeneration strategies such as clearcutting were once dominant. These

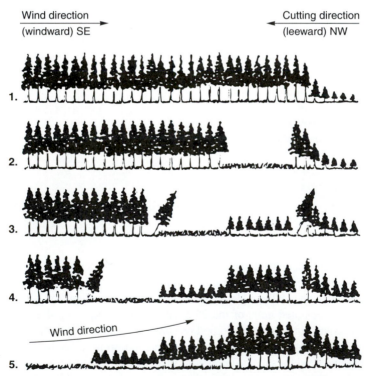

Figure 7.11
Distribution of age and size classes as a result of progressive strip cutting, showing how cutting progresses into the wind.
(Source: From Harris 1989.)

legacies of the previous stand may be important for hiding or thermal cover, or nesting for wildlife, or to create or maintain structural diversity provided by wood in streams, or to provide for a continuing supply of large-sized snags and logs through the next generation. Through the actions of disturbance agents like wind, however, green retention trees can be converted to snags and logs much more quickly than intended by the planners: rather than a gradual conversion over many decades, most retention trees can be blown over in less than a decade. Potential for windsnap and windthrow can be reduced by planning to leave trees in the most wind-sheltered locations.

Stream buffers consist of strips of trees either left uncut or selectively cut near streams. The stream environment often is sheltered from wind, as it is the lowest point on the landscape, but in headwater areas the riparian corridor also can be the location of funneled winds across saddles, and on plains or broad terraces the stream may not be incised enough to provide any protection from wind. Riparian zones are typically the wettest part of the landscape, too, so shear strength of soils may on average be less than elsewhere. Buffer strips have traditionally been narrow corridors of 10 to 25 m (30–80 ft) each side of the stream, so they are quite vulnerable to windthrow. Deeply incised canyons are less likely to suffer damage than broad floodplains.

Additional leave trees may be prescribed across units, and often are allowed to be either scattered or clustered. These usually are called *retention strategies*. Clustering in sheltered locations is preferable to scattering the same number of trees across the unit, from a windthrow perspective. Individual leave trees are more susceptible to damage than trees left in blocks. Feathering the edges of riparian buffers by concentrating additional leave trees adjacent to the stream buffers is another lower-risk strategy. The topographic buffering and that provided by the riparian trees can increase protection for both types of leave trees.

SYNERGISTIC EFFECTS

Disturbances damage and kill trees, creating situations where damaged trees may be killed by subsequent disturbances and other trees, untouched by the first disturbance, can be damaged or killed. One of the most important synergisms created by wind damage is the creation of susceptible loci for subsequent windthrow. Once a gap has been opened, it may expand through increased force of subsequent winds on residual trees. Two years after the 1921 Olympic Peninsula windstorm in Washington, another windstorm affected a 12 by 40 km area, "mopping up" damaged and exposed trees from the 1921 storm (Boyce 1929).

Recently downed trees are breeding sites for bark beetles, and the expanded populations of beetles can infest previously healthy trees. In the Douglas-fir region of the Pacific Northwest, the Douglas-fir beetle (*Dendroctonus pseudotsugae*) can reach epidemic proportions after major windthrow events, and subsequent mortality can extend for several more years. The synergistic effect is not linked to every windthrow, however, so there likely is a complex set of initiating factors besides windthrow.

Fire hazard obviously increases with wind-generated fuels on the ground. Whether fire danger increases, and for how long, depends on the risk of a fire starting, and how long the major fire spread potential exists. Where decomposition is rapid, lowering hazard, and risk is low (lightning is uncommon, human-caused risk is either low or intensively managed), increased fire danger may not persist for long. If the values-at-risk are low (see Chapter 6), reduction in hazard may not be justified.

Wind is a common carrier for insects and disease-causing organisms. Most insects are not large or strong enough to resist the power of the wind, and can be carried many kilometers before being deposited in another forest, range, agricultural, or urban community. This dispersal of insects can be beneficial or damaging to forest health, depending on the insect and where it is deposited. Defoliating insects, such as adult spruce budworms (*Choristoneura occidentalis*) (see Chapter 20), may be transported hundreds of kilometers by storms that "capture" them as they briefly fly above treetop level. Conversely, levels of outbreak in the source area may decline if large numbers of the female moths are removed. An excellent example of wind interactions with disease is provided by blister rust (caused by *Cronartium ribicola*). This rust, which attacks white pines (see Chapter 13), has an alternate host (currants or gooseberries, *Ribes* spp.). In an area near Lake Superior, both species were present but the rust was not present on the lakeside pines, although it was present on pines further inland. The rust spores were released at night from the currant bushes, and the land breeze (an offshore local wind at night, the counterpart of the sea breeze, an onshore wind, in the daytime) carried the spores out over the lake. They were then lofted by the relatively warm air currents over the lake to higher altitude, and transported inland again above the lakeside pines to areas further away from the lake, where they lost altitude and were deposited on these further inland trees (Van Arsdel 1965). These examples suggest that both gradient winds and local winds may have important synergistic implications for other disturbances (see Chapter 16).

Management for windthrow is essentially the management of risk. Some aspects of risk are unmanageable, such as windstorms, but the probability of large storms can be assessed. Management of species, age classes, and types and intensities of intermediate and regeneration treatments provide some ability to manage for minimal impact from wind. Some forest types or areas have so little risk of wind damage that incorporation of such strategies is not worthwhile; in other areas risk is high enough that wind is a primary concern in management. Rotation lengths may be predicated on the age at which trees reach a critical height. Wind risk can be evaluated to assess which, if any, portions of the managed forest need special treatment.

REFERENCES

Agee, J. K. 1993. *Fire ecology of Pacific Northwest forests*. Island Press, Washington, D.C.

Alexander, R. R. 1964. Minimizing windfall around clear cutting in spruce-fir forests. *Forest Science* 10:130–142.

Boyce, J. S. 1929. *Deterioration of wind-thrown timber on the Olympic Peninsula*. USDA Technical Bulletin 104, Washington, D.C.

Bradley, G. A. 1995. *Urban forest landscapes*. University of Washington Press, Seattle.

Conner, R. N., D. C. Rudolph, D. L. Kulhavy, and A. E. Snow. 1990. Causes of mortality of red-cockaded woodpecker cavity trees. *Journal of Wildlife Management* 55:531–537.

Coutts, M. P., and J. Grace (eds.). 1995. *Wind and trees*. Cambridge University Press, Cambridge, England.

Foster, D. R. 1988. Species and stand response to catastrophic wind in central New England, U.S.A. *Journal of Ecology* 76:135–151.

Franklin, J. F., and C. T. Dyrness. 1973. *Natural vegetation of Oregon and Washington*. USDA Forest Service General Technical Report PNW-8.

Gordon, D. T. 1973. Damage from wind and other causes in mixed white fir–red fir stands adjacent to clearcutting. USDA Forest Service Research Paper PSW-90.

Gratkowski, H. J. 1956. Windthrow around staggered settings in old-growth Douglas-fir. *Forest Science* 2:60–74.

Gresham, C. A., T. M. Williams, and D. J. Lipscomb. 1991. Hurricane Hugo wind damage to southeastern U.S. coastal forest tree species. *Biotropica* 23 (4a):420–426.

Hadley, J. L., and W. K. Smith. 1986. Wind effects on needles of timberline conifers: seasonal effects on mortality. *Ecology* 67:12–19.

Harcombe, P. A. 1986. Stand development in a 130-year-old spruce-hemlock forest based on age structure and 50 years of mortality data. *Forest Ecology and Management* 14:41–58.

Harmon, M. E., J. F. Franklin, F. J. Swanson, P. Sollins, S. V. Gregory, J. D. Lattin, N. H. Anderson, S. P. Cline, N. G. Aumen, J. R. Sedell, G. N. Lienkamper, K. Cromack, Jr., and K. W. Cummins. 1986. Ecology of coarse woody debris in temperate ecosystems. *Advances in Ecological Research* 15:133–202.

Harris, A. S. 1989. *Wind in the forests of southeast Alaska and guides for reducing damage*. USDA Forest Service General Technical Report PNW-GTR-244.

Holroyd, E. W. 1970. Prevailing winds on Whiteface Mountain as indicated by flag trees. *Forest Science* 16:222–229.

Julin, K. R., C. G. Shaw III, W. A. Farr, and T. M. Hinckley. 1993. The fluted western hemlock of Alaska. I. Morphological studies and experiments. *Forest Ecology and Management* 60:119–132.

Kimmins, J. P. 1997. *Forest Ecology*. Macmillan, New York.

Knight, D. L. 1994. *Mountains and plains: the ecology of Wyoming landscapes*. Yale University Press, New Haven, Conn.

Mattoon, W. R. 1925. *Longleaf pine*. USDA Department Bulletin 1061.

Miller, K. F. 1985. *Windthrow hazard classification*. Forestry Commission Leaflet 85. HMSO, London.)

Naiman, R. J., and R. E. Bilby. 1998. *River ecology and management*. Springer-Verlag, New York.

Oliver, C. D., and B. C. Larson. 1990. *Forest stand dynamics*. McGraw-Hill Book Company, New York.

Orr, P. W. 1962. *Windthrown timber survey in the Pacific Northwest*. Division of Timber Management. USDA Forest Service, Pacific Northwest Region, Portland, Oreg.

Quine, C., M. Coutts, B. Gardiner, and G. Pyatt. 1995. *Forests and wind: management to minimise damage*. Forestry Commission Bulletin 114, HMSO, London.

Smith, D. 1986. *The practice of silviculture,* 8th ed. John Wiley and Sons, New York.

Somerville, A. 1980. Wind stability: forest layout and structure. *New Zealand Journal of Forestry Science* 10:476–501.

Srago, M., J. R. Parmeter, Jr., J. Johnson, and L. West. 1978. *Determining early failure of root diseased incense-cedars in Yosemite Valley.* USDA Forest Service, USDI National Park Service, San Francisco.

Tappeiner, J., J. Zasada, P. Ryan, and M. Newton. 1991. Salmonberry clonal and population structure: the basis for a persistent cover. *Ecology* 72:609–618.

Ugolini, F. C., and D. H. Mann. 1979. Biopedological origin of peatlands in southeastern Alaska. *Nature* 281:366–368.

Van Arsdel, E. P. 1965. Micrometeorology and plant disease epidemiology. *Phytopathology* 55:945–950.

8

INTRODUCTION TO DISEASES

INTRODUCTION

Organisms causing diseases occur naturally in forests, but until recently they were considered to be important only because of the losses they caused to forest productivity and timber volumes. New views of forest management, however, have forced us to examine the role of diseases in forest ecosystems since disease-causing organisms play important roles in maintaining properly functioning ecosystems, particularly in providing wildlife habitat and biodiversity (Kohm and Franklin 1997). On the other hand, introduced disease organisms can cause great ecological change in areas where trees have little natural resistance (Gilbert and Hubbell 1996). All tree species suffer diseases, although some are more resistant than others. In addition, some disease-causing organisms attack a wide **variety** of host species; others have a very limited host range. Diseases thus strongly influence forest succession. Improper management practices also can alter the role of native disease-causing organisms in forest ecosystems.

Pathogens and abiotic agents have had a large impact on the world's forests with a long list of large-scale disturbances occurring since 1875 (Table 8.1). Many **epiphytotics** involving both introduced and native organisms occurred in North America between 1890 and the present. In particular, the introduced fungi causing chestnut blight (1904), white pine blister rust (1909 in the East and 1921 in the West), and Dutch elm disease (1930) have greatly influenced North American forests (Gibbs and Wainhouse 1986, Tainter and Baker 1996). A number of forest declines involving a multitude of causes and tree species have occurred in recent years, but many of these have abiotic causes or multiple causes. Introduction of quarantine measures has been generally successful in keeping out introduced organisms since the 1950s. However, the fungus causing dogwood anthracnose appears to have been introduced in the 1970s and now is causing a problem to native dogwoods in the eastern and western United States (Tainter and Baker 1996). With the recent movement of logs around the world, concern has been raised about potential new threats from exotic pathogens and pests and the efficacy of quarantines.

TABLE **8.1** | Incidence of Introduced or Native Pathogens and Abiotic Agents Causing Serious Tree Disease Epidemics, Epiphytotics, or Declines Between 1875 and the Present

Year introduced or first recognized	Disease	Cause	Introduced or native pathogen or insect
1875	Tannensterben (European silver fir decline)	Abiotic	
1890	Beech bark disease	Scale insect/ canker fungus	Introduced insect/ native pathogen?
1900	Alaska yellow cedar decline	Abiotic	
1904	Chestnut blight	Canker fungus	Introduced
1909	Blister rust (eastern North America)	Rust fungus	Introduced
1918	Elm yellows	Phytoplasma	Native
1920	White pine blister rust (eastern North America)	Rust fungus	Introduced
1921	white pine blister rust (western North America)	Rust fungus	Introduced
1921	Jarrah dieback (Australia)	Root disease fungus	Introduced
1925	Ash dieback	Abiotic	
1927	Larch canker	Canker fungus	Introduced
1929	Pole blight	Abiotic	
1930	Dutch elm disease	Vascular wilt fungus	Introduced
1930	Birch dieback	Abiotic	
1932	Littleleaf disease	Abiotic root rot fungus	Native
1940	Fusiform rust	Rust fungus	Native
1942	Oak wilt	Vascular wilt fungus	Native
1950s	Ponderosa pine decline	Abiotic	
1951	Sweetgum decline	Abiotic	
1951	Oak decline	Gypsy moth/wilt/ root rot	Introduced insect/ native fungi
1952	Port-Orford-cedar decline	Root root fungus	Introduced
1954	Annosus root disease in southern pines	Root rot fungus	Native
1955	Lethal yellowing of palms	Insect/phytoplasma	Native
1957	Maple blight	Abiotic	
1970s	Dogwood anthracnose	Foliage and twig fungus	Introduced
1979	Waldsterben (European forest decline)	Abiotic	
1980s	Pacific madrone decline	Canker fungus	Introduced?
1980s	Spruce-fir decline	Abiotic/balsam woolly adelgid	Introduced insect

Although we typically think of diseases as being caused by biological agents, such as fungi, abiotic agents, like air pollutants and adverse weather, also cause diseases, commonly called **abiotic diseases** (Sinclair et al. 1987, Manion 1991, Agrios 1997). Furthermore, symptoms resembling disease can be caused by insects and large animals such as deer, kangaroos, porcupines, and bears. All these factors also can interact. Thus it is important that the causes of damage to plants be established if correct management decisions are to be made. Forest pathologists generally do not study insects since they are the domain of entomologists, but relationships between insects and diseases are considered by pathologists. Many disease-causing organisms are spread by insects, such as those causing Dutch elm disease and black-stain root disease, and root-diseased trees are commonly attacked by bark beetles (Schowalter and Filip 1993). The role of insects in forest ecosystems is considered in detail in Chapters 17 to 25.

As mentioned previously forest management has changed the incidence of many disease organisms in their natural environments. Where clearcutting and intensive management for a single species like Douglas-fir have been practiced, root diseases have generally increased; others have decreased, especially stem decays and dwarf mistletoes in conifers in North America (Tainter and Baker 1996). This has occurred because rotations of managed stands are shorter than "natural rotations," and dwarf mistletoes and stem decays take considerable time to develop. On the other hand, dwarf mistletoes tended to increase with uneven-aged management and fire suppression. Understanding the behavior of both native and introduced organisms is extremely important in maintaining forest health.

The following topics are covered in this introductory chapter: definition and types of diseases; history of forest pathology; impact of diseases; the disease triangle; the concept of pathogenic, saprophytic, and symbiotic relationships; signs and symptoms of disease; linkages between diseases; insects, fire, and wind; and diseases in urban forests. Abiotic and animal-caused injuries and biological disease-causing agents are discussed in detail in Chapters 9 and 10, respectively. Descriptions of specific diseases are covered in Chapter 11 (Nursery Diseases and Mycorrhizae), Chapter 12 (Root Diseases), Chapter 13 (Foliage Diseases and Rusts), Chapter 14 (Stem and Branch Diseases), Chapter 15 (Forest Declines), and Chapter 16 (Management of Forest Diseases and the Deterioration of Wood Products).

DEFINITION OF DISEASE AND TYPES OF DISEASES

Tree disease is defined as sustained physiological and/or structural damage to tree tissues caused by biological agents (**fungi, bacteria, viruses, phytoplasmas, nematodes, parasitic plants** and **protozoans**) or nonbiological agents (such as air pollution, etc.) ending sometimes in tree death. A third category is caused by a combination of abiotic and biotic stresses, such as freeze damage predisposing trees to **infection** by canker or decay fungi. Examples of physiological damage are (1) impaired photosynthesis, which results in reduced food storage (e.g., starch) and reduced growth of shoots and roots, and (2) impaired water transport. An example of structural damage is decay. There are many definitions of disease and none is absolutely sacred. Disease also can be defined as a sustained impairment in function, structure, or form of an organism as provoked by biological, chemical, or physical factors of the environment.

Commonly disease and the biological agent causing that disease are used interchangeably, but this is incorrect and it is important for the reader to distinguish between them. For example, white pine blister rust is the disease, whereas the pathogen causing the disease is the fungus *Cronartium ribicola*. Thus the terms "infectious diseases" or "introduced diseases," are technically incorrect although they are commonly used.

All parts of trees are affected by diseases (Figure 8.1) such that diseases are commonly named after the tree components affected—for example, fine root diseases, coarse or structural root diseases, butt diseases, bole or stem diseases, branch diseases, and foliage (needle or leaf diseases). Note that we have included **mycorrhizae** in Figure 8.1 because they are an important component of tree health and interact strongly with disease organisms. The concept of pathogenic, saprophytic, and symbiotic relationships are covered later in this chapter. Mycorrhizae are covered in detail in Chapter 11.

A BRIEF HISTORY OF FOREST PATHOLOGY

Tree diseases were first studied in Europe following the work of the botanist Anton deBary (the "father" of modern plant pathology), who in 1853 demonstrated the cause of rusts and smuts in cereals. A German forester, Robert Hartig (1839–1901) is recognized as the "father" of forest pathology. He first associated fungal fruiting bodies with decay in trees and contributed the first text in the field in 1874 (Tainter and Baker 1996).

In the United States forest pathology began at the turn of the century with the establishment by the U.S. Department of Agriculture of the Mississippi Valley Laboratory at Washington University in St. Louis, Missouri. Dr. Hermann Von Schrenck, whose doctoral dissertation covered heart decay of trees, was the first field agent. Several other scientists were added in 1902: G. G. Hedgcock and Perley Spaulding, who were joined later by Ernst Bessey and Caroline Rumbold, who were interested in tree diseases. In 1907 the office was transferred to Washington, D.C., where it became the Laboratory of Forest Pathology and later the Office of Forest Pathology in the Bureau of Plant Industry. Additional scientists were hired and the scope of work included microbiological deterioration of wood, wood stains, a survey of diseases on National Forests, heartwood decay and salvage of infected stands, nursery and plantation diseases, and the discovery of the introduction of white pine blister rust fungus in the northeastern United States. In 1931 the name was changed to the Division of Forest Pathology. This division was transferred to the U.S. Department of Agriculture Forest Service in 1954 and its name was changed to the Division of Forest Disease Research. Research in the Forest Service regions is carried out in the Forest Experiment Stations.

The Forest Pest Control Act of 1947 eventually led to the transfer of pest control programs from the Forest Experiment Stations to the newly created branch known as Forest Pest Control in State and Private Forestry at the regional level.

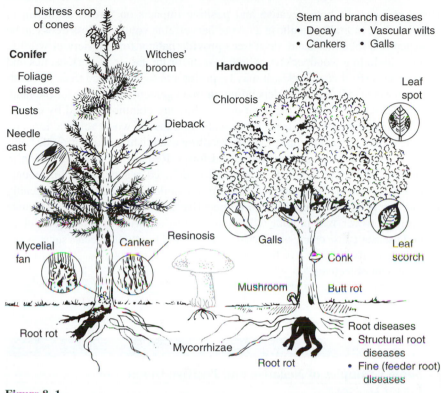

Figure 8. 1
Classification of tree diseases by the part of the tree that is affected.

This organization was renamed Forest Insect and Disease Control, then Forest Insect and Disease Management, followed by Forest Pest Management and now Forest Health. It is responsible for the prevention, detection, evaluation, and suppression of forest diseases, insects, and weeds on federal lands and indirectly through cooperative efforts, for the protection of state and private lands. Many states also employ forest pathologists in their departments of forestry, lands, or natural resources.

Other countries have similar structures for handling forest disease management and research. For example, forest pathology research in Canada is conducted by Forestry Canada, whereas provinces are largely involved with disease surveys and management. Worldwide forest pathology sections have been established in state and federal forestry departments, although many now are listed under Forest Health. Most of the research, however, is still being conducted in North America, Europe, Chile, New Zealand, and Australia. Marks et al. (1982) describe diseases in Australia. Much is still to be learned about forest diseases in Asia, Africa, and Central and South America, particularly in relation to the conservation of tropical forests (Gilbert and Hubbell 1996).

IMPACT OF DISEASES

Diseases can have both negative and positive impacts on forests as shown in Box 8.1. Tree mortality results in a loss of harvestable volume unless trees can be salvaged. On the other hand, dead trees provide habitat for a variety of wildlife species, including woodpeckers, which search for the larvae of bark beetles and other insects (Bull et al. 1997). It may be preferable to leave dead trees for wildlife rather than salvage them, depending on management objectives for the land. If trees have cavities resulting from decay, they are commonly used by cavity-nesting birds and bats, but decay also results in a loss of wood volume. Logs on the forest floor can act as nurse logs for seedling establishment and provide habitat for small mammals, invertebrates, and fungi. Reduced growth, reduction in pulp yield, stain and reduced wood quality, delayed regeneration, inadequate stocking, and site deterioration resulting from a buildup of pathogens, especially in soil, are generally considered to be negative impacts of disease. Inadequate stocking, however, may provide wildlife habitat by allowing understory development. Diseases cause changes in forest structure and function, species succession, and biodiversity, which may be considered negative or positive, depending on management objectives.

BOX 8.1

Some Examples of Negative and Positive Impacts of Forest Diseases

Negative impacts are considered to be related to timber and economic losses; positive impacts are ecological, especially related to wildlife habitat. These impacts are largely defined by human values and are subject to change.

Mortality: negative and positive

Reduced growth: negative

Decay of merchantable wood: negative and positive

Reduction in pulp yield: negative

Stain and reduced wood quality: negative

Delayed regeneration: negative

Inadequate stocking (low number of trees per hetare): negative and positive

Site deterioration—buildup of pathogens: negative

Change in forest structure and function (e.g., nutrient cycling): negative and positive

Change in species composition and species succession: negative and positive

Change in ecosystem biodiversity (generally increase biodiversity): positive

Creation of wildlife habitat: positive

The Disease Triangle, Square, or Tetrahedron

Plant pathologists have long recognized that diseases develop only in the presence of a pathogen and if environmental conditions are correct (Tainter and Baker 1996, Agrios 1997). This resulted in the development of the concept of the disease triangle, relating host, pathogen, and the environment (Figure 8.2). In recent years this concept has been expanded to the disease square (Figure 8.3) or tetrahedron (Figure 8.4) where the environment has been subdivided into the physical and biological environments. Time is also an important consideration in the relationship between pathogens, hosts, and the environment.

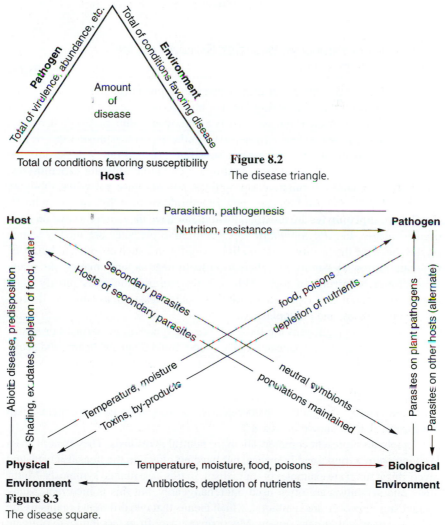

Figure 8.2

The disease triangle.

Figure 8.3

The disease square.

(Source: From Tainter, F.H. and F.A. Baker, *Principles of Forest Pathology*, ©1996 by John Wiley & Sons, Inc. Reprinted by permission of John Wiley & Sons, Inc.)

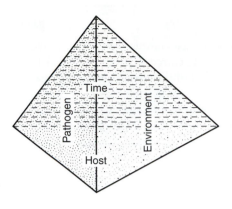

Figure 8.4

The disease tetrahedron.

(Source: From Agrios, G.N., *Plant Pathology,* 4th ed., ©1997 by Academic Press.)

THE CONCEPT OF PATHOGENS, PARASITES, SAPROPHYTES, AND SYMBIOTIC RELATIONSHIPS

A pathogen is an organism that causes disease; a **parasite** is an organism that grows on or within another living species and derives part of its food from it (Agrios 1997). A parasite is not necessarily a pathogen. Pathogenic organisms can be autotrophic (deriving energy from photosynthesis, such as mistletoes) or heterotrophic (deriving energy from organic substrates, such as fungi). Heterotrophic organisms can be classified along a continuum including **obligate saprophytes, facultative parasites,** facultative saprophytes, and **obligate parasites.** Obligate parasites, such as the rust fungi and powdery mildews, must live on living hosts. Facultative **saprophytes** are mostly parasitic, but have the faculty to live on dead organic matter, like *Phytophthora* spp. Facultative parasites are mostly saprophytes, but have the faculty to live on living organisms, such as root and butt rots. Obligate saprophytes must gain their food from dead organic matter like many decay fungi occurring on logs. Although this terminology has mostly been applied to fungi, it also applies to the obligate parasites, such as parasitic flowering plants, viruses, and phytoplasmas.

Parasites and pathogens are thus not exactly synonymous terms. For example, pathogens that are not parasites include sooty mold fungi and heartwood decay fungi. Heartwood decay fungi, because they grow in dead wood in tree boles, are considered to be saprophytes. However, we might consider them to be parasites if we consider that they grow on living hosts.

Pathogens also might be considered along a continuum from commensalism to saprophytism as shown in Box 8.2.

Epiphytes represent commensalism, or neutral **symbiosis**. They benefit from growing on tree trunks and branches, but have no effect on the trees. Lichens and mycorrhizal fungi are examples of mutualism, in this case mutual symbiosis. Mutualism and symbiosis are often used interchangeably, but this is incorrect. Many mutualisms, however, are symbiotic, which means that two different organisms are living together with organic union. Mycorrhizae vary from facultative to obligate in their associations with plants, and some associations between mycorrhizal fungi and trees can be near pathogenic. For example, Douglas-fir seedlings artificially

BOX 8.2

The Pathogen Continuum

		Effect on	
Type of interaction	Example	Species 1	Species 2
Commensalism (neutral symbiosis)	Epiphytes	+	0
Mutualism (mutualistic symbiosis)	Lichens, mycorrhizae	+	+
Parasitism (obligate parasite)	Rusts and mildew fungi	+	–
(facultative saprophyte)	*Phytophthora* spp.	+	–
Saprophytism (facultative parasite)	Root and butt rot fungi	+	–
(obligate saprophyte)	Wood decay fungi	+	0

Symbols: 0 = no effect; + = a positive effect of one species on another; – = a negative effect of one species on another.

infected with some mycorrhizal fungi have poorer growth than uninfected seedlings, suggesting that the fungus was robbing the plant of carbon. Rusts and mildew fungi (obligate parasites) and *Phytophthora* spp. (facultative saprophytes) are examples of parasitism where the fungi benefit from the relationship and the host species does not. A similar relationship occurs for saprophytes, especially facultative parasites, although it can be argued that there is no negative effect for species 2 (see Box 8.2) with obligate saprophytes because the host is already dead (e.g., logs). Pathogens also can be considered as symbiotic organisms in which the relationship is of net detriment to one partner.

SIGNS AND SYMPTOMS OF DISEASE

Diagnosis of disease depends largely on identification of **signs** and symptoms of disease. Signs are associated with the disease-causing organism, whereas symptoms are the plant response to the presence of a pathogen or abiotic agent. Some symptoms are general; others are specific to a particular disease. Resistance to pathogens varies from susceptible to tolerant to **immune,** where resistance is the ability of a tree to exclude or overcome the effect of a pathogen. A tolerant species can sustain the effects of a pathogen without dying or suffering serious injury or growth loss, whereas immune species are not affected. Typical signs and symptoms of pathogens are shown in Table 8.2. As mentioned previously signs are indicators of the causal agent, and for fungi they include large fruiting bodies such as **mushrooms, conks** of wood decay fungi, and small fruiting structures embedded in leaves. A variety of other signs for fungi include mycelia, **rhizomorphs,** and **spores.** Bacterial ooze from cankers is commonly seen. Parasitic plants such as mistletoes are seen with relative ease. Signs of viruses and phytoplasmas cannot be seen with the naked eye and an electron microscope is needed.

T A B L E **8.2** | **Typical Signs of Pathogens and Host Symptoms**

Signs   (indicating presence of disease causing organism)

Fungi

    Fruiting bodies: mushrooms, conks, etc.

    Mycelial fans, ectotrophic hyphae, setal hyphae

    Rhizomorphs (resembling roots)

    Spores: chlamydospores, conidia

Bacteria

    Ooze or slime

Parasitic plants: presence of plants growing from tree branches

Nematodes: presence of nematodes protruding from roots

Symptoms   (plant response)

Necrosis: death of tissues or cells

    Decay: decomposition of wood cells in roots and stem—sap, white and brown rots

    Cankers: death of cambium cells (target, diffuse, with ooze, mechanical)

    Leaf diseases: death of localized areas of cells—defoliation

    Vascular wilts: death and plugging of parenchyma cells in branches, stems, and roots

    Blights or diebacks: sudden death of all or part of tree

Hypertrophy: overgrowth of tissues

    Witches'-brooms: proliferation of adventitious buds

    Galls: overgrowths on stems and branches

    Leaf blisters: localized enlargements, puckering

Atrophy: lack of growth or failure of development

    Dwarfing

    General, interveinal, or marginal chlorosis

Physiological and other responses

    Resin flow on stems and roots

    Chlorosis

    Reddening

    Excess cone crops

Symptoms generally fall into four major categories (Table 8.2): **necrosis** (death of tissues or cells), **hypertrophy** (overgrowths), **atrophy** (underdevelopment), and physiological responses (resin flow). Biological agents generally tend to cause more random symptoms at the leaf, tree, stand, and landscape scales in contrast to abiotic agents, which generally cause more nonrandom or systematic symptoms. However, it needs to be pointed out that many diseases caused by biological agents may have nonrandom symptom patterns (e.g., casting of needles of only one age on a conifer and aggregated mortality in a root rot "pocket").

To link signs and symptoms it may be necessary to utilize Koch's postulates or rules of proof of **pathogenicity** as listed in Box 8.3.

Box	8.3

Koch's Postulates: Rules of Proof of Pathogenicity

1. Establish constant association of organism and disease symptoms.
2. Isolate organism and grow in pure culture.
3. Inoculate healthy plant and produce disease symptoms.
4. Reisolate organism

Koch's postulates now are rarely used because relationships between the signs and symptoms for most diseases now is well established. However, it may be necessary to conduct them for newly introduced pathogens.

LINKAGES BETWEEN DISEASES, INSECTS, FIRE, AND WIND

Diseases, insects, fire, and wind are strongly linked in ecosystems and this is strongly emphasized in this book. Insects and diseases interact directly and indirectly (Schowalter and Filip 1993). Many insects disperse fungal spores, particularly bark beetles and weevils that spread fungi, causing **vascular wilts.** Indirect effects involve bark beetle attacks on trees that are weakened by root diseases, such as Armillaria root disease. Defoliating insects, like the European gypsy moth, may weaken deciduous hardwoods, especially oaks, to the point where they are susceptible to *Armillaria*. In Japan and North America bark beetles spread the pine wood nematode that causes a wilt disease.

Interactions between fire, insects, and diseases are strongest in areas where fire frequency and moisture stress is high, such as the conifer ecosystems in inland western North America (Campbell and Liegel 1996). Fire scars at the base of stems provide entry points for decay fungi especially in areas where fire frequencies are low and fires are intense, such as the coastal areas of western North America and the productive cool temperate eucalypt forests in Victoria and Tasmania in Australia and in the forests of southern Chile.

In southern Brazil there are also numerous of examples of interactions among fire, insects, and diseases (personal communication, Ronaldo Viano Soares, Forestry School, Federal University of Paranas, Curitiba, PR, Brazil). The main concern involves introduced eucalypt and pine plantations and the native Brazilian *Araucaria angustifolia,* which occurs in natural and planted stands. Fire-injured tree stems are frequently attacked by fungi that cause stains, mainly bluestains, and decay, reducing the commercial value of the wood. On the other hand, fire has been suggested as a treatment for root diseases caused by *Armillaria* spp. and *Cylindrocladium clavatum.* Leaf cutting ants of the genera *Atta* and *Acromyrmex* are the most destructive insects in forestry plantations in Brazil. Ants cut leaves and bring them to underground nests where they feed a fungus that decomposes the leaves and provides food for the ants. Although fire generally does not affect underground nests, it does cause death of understory vegetation. Death of herbaceous vegetation, which is an ant food source, may result in ant attack and damage of large trees not normally fed upon. Scolytid beetles also may attack fire-damaged eucalypt trees.

Finally, wind causes trees damaged by fire, insects, and pathogens to fall, creating fuel for future fires. Wind also creates wounds through branch and top breakage, resulting in entry points for fungi and insects.

DISEASES IN URBAN FORESTS AND STREET TREES

Although this book focuses on forest ecosystems, it should be recognized that many of the diseases that occur in forests also occur in urban forests, parks, and street trees. Particular problems in the urban environment involve root diseases (especially Armillaria root disease, Annosus root and butt rot, Phytophthora root diseases, and laminated root rot) and wilt diseases, such as Dutch elm disease and Verticillium wilt. Decay of hardwood limbs and stems also is important because broken limbs and tree failures can cause considerable damage to property and cause loss of human life. Trees are deemed to be hazard trees if they have a target, such as houses, campgrounds, and people. Wind tends to exacerbate the influence of root diseases and stem decay by bringing down weakened trees and limbs. Publications concerning urban tree diseases and hazard trees are included in the references at the end of this chapter (e.g., Tattar 1978, Stipes and Campana 1981, Harvey and Hessberg 1992, Wallis et al. 1992, Blanchard and Tattar 1997).

FOREST PATHOLOGY BOOKS

The treatment of forest diseases in this book is not detailed in comparison to books devoted solely to forest pathology. Older, but still useful tree and forest pathology books include Hubert (1931), Boyce (1938, 1948, 1961), Baxter (1952, 1967), Pirone (1959), Peace (1962), Smith (1970), Hepting (1971), and Tattar (1978). More recent books include Manion (1981), Sinclair et al. (1987), Shigo (1989) Manion (1991), Tainter and Baker (1996), and Blanchard and Tattar (1997). A number of excellent field guide books are available including Foster and Wallis (1974), Hagle et al. (1987), Skelly et al. (1987), Scharpf (1993), and Allen et al. (1996). Hansen and Lewis (1997) provide an excellent compendium of North American conifer diseases. With the advent of the World Wide Web, a variety of sites devoted to forest pathology now exist, including the web version of the book by Allen et al. (1996), at the following address:

www.pfc.cfs.nrcan.gc.ca/health/td_web/index.html.

Teaching resources and links to other forest pathology sites throughout the world can be found at the following selected web sites:

Forest and shade tree pathology: College of Environmental Studies and Forestry, State University of New York, Syracuse, New York:

www.esf.edu/course/jworrall/default.htm

Tree pests and disease: Department of Forest Ecosystem Science, University of Maine, Orono, Maine:

www.ume.maine.edu/~nfa/fes/int256/home/tpd_home.htm.

Forest pathology: Department of Forest Sciences, University of British Columbia, Vancouver, B.C., Canada:

mrn.media-services.ubc.ca/disted/Courseout/Forestry/Frst309.html

Links to forest pathology sites:

www.esf.edu/course/jworrall/others.htm

REFERENCES

Agrios, G. N. 1997. *Plant pathology,* 4th ed. Academic Press, New York.

Allen, E., D. Morrison, and G. W. Wallis. 1996. *Tree diseases of British Columbia.* Forestry Canada, Pacific Forestry Center, Victoria, B.C., Canada.

Baxter, D. V. 1952. *Pathology in forest practice,* 2nd ed. John Wiley and Sons, New York.

Baxter, D. V. 1967. *Disease in forest plantations: thief of time.* Cranbrook Institute of Science Bulletin 51.

Blanchard, R. O., and T. A. Tattar. 1997. *Field and laboratory guide to tree pathology,* 2nd ed. Academic Press, New York.

Boyce, J. S. 1938. *Forest pathology,* 1st ed. McGraw-Hill, New York.

Boyce, J. S. 1948. *Forest pathology,* 2nd ed. McGraw-Hill, New York.

Boyce, J. S. 1961. *Forest pathology,* 3rd ed. McGraw-Hill, New York.

Bull, E. L., C. G. Parks, and T. R. Torgensen. 1997. *Trees and logs important to wildlife in the interior Columbia River Basin.* USDA Forest Service, Pacific Northwest Station, General Technical Report PNW-GTR-391, Portland, Oreg.

Callan, B., and A. Funk. 1994. *Introduction to forest diseases.* Forest Pest Leaflet 54. Forestry Canada, Pacific Forestry Centre, Victoria, B.C., Canada.

Campbell, S., and L. Liegel (technical coordinators). 1996. *Disturbance and forest health in Oregon and Washington.* USDA Forest Service, Pacific Northwest Station, General Technical Report PNW-GTR-381, Portland, Oreg.

Foster, R. E., and G. W. Wallis. 1974. *Common tree diseases of British Columbia.* Canadian Forestry Service Publication 1245, Ottawa, Canada.

Gibbs, J. N., and D. Wainhouse. 1986. Spread of pests and pathogens in the northern hemisphere. *Forestry* 59:141–153.

Gilbert, G. S., and S. P. Hubbell. 1996. Plant diseases and the conservation of tropical forests. *BioScience* 46:98–106.

Hagle, S., S. Tunnock, K. E. Gibson, and C. J. Gilligan. 1987. *Field guide to diseases and insect pests of Idaho and Montana forests.* USDA Forest Service, State and Private Forestry, Northern Region. Publication R1-89-54, Missoula, Mont.

Hansen, E. M., and K. J. Lewis (eds.). 1997. *Compendium of conifer diseases.* American Phytopathological Society Press, St. Paul, Minn.

Harvey, R. D., and P. F. Hessberg, Sr. 1992. *Long-range planning for developed sites in the Pacific Northwest: the context of hazard tree management.* USDA Forest Service, Pacific Northwest Region, FPM-TP039-92, Portland, Oreg.

Hepting, G. H. 1971. *Diseases of forest and shade trees of the United States.* USDA Agricultural Handbook 386.

Hubert, E. E. 1931. *An outline of forest pathology.* John Wiley and Sons, New York.

Kohm, K. A., and J. F. Franklin (eds.). 1997. *Creating a forestry for the twenty-first century: the science of ecosystem management.* Island Press, Washington, D.C.

Manion, P. D. 1981. *Tree disease concepts.* Prentice-Hall, Englewood Cliffs, N.J.

Manion, P. D. 1991. *Tree disease concepts,* 2nd ed. Prentice-Hall, Englewood Cliffs, N.J.

Marks, G. C., B. A. Fuhrer, and N. E. M. Walters. 1982. *Tree diseases in Victoria.* Forest Commission of Victoria, Handbook No. 1, Melbourne, Australia.

Peace, T. R. 1962. *Pathology of trees and shrubs with special reference to Britain.* Oxford University Press, London.

Pirone, P. P. 1959. *Tree maintenance,* 3rd ed. Oxford Press, New York.

Scharpf, R. T. 1993. *Diseases of Pacific Coast conifers.* USDA Forest Service, Agricultural Handbook 521, Washington, D.C.

Schowalter, T. D., and G. M. Filip (eds.). 1993. *Beetle-pathogen interactions in conifer forests.* Academic Press, New York.

Shigo, A L. 1989. *A new tree biology: facts, photos and philosophies on trees and proper care,* 2nd ed. Shigo and Trees Associates, Durham, N.Y.

Sinclair, W. A., H. H. Lyon, and W. T. Johnson. 1987. *Diseases of trees and shrub.* Cornell University Press, Ithaca, New York.

Skelly, J. M., D. D. Davis, W. Merrill, A. E. Cameron, H. D. Brown, D. B. Drummond, and L. S. Dochinger (eds.). 1987. *Diagnosing injury to eastern forest trees: a manual for identifying damage by air pollution, pathogens, insects and abiotic stresses.* National Acid Precipitation Assessment Program. Vegetation Survey Cooperative. USDA Forest Service, Forest Pest Management and Pennsylvania State University, University Park, Pa.

Smith, W. H. 1970. *Tree pathology: a short introduction.* Academic Press, New York.

Stipes, R. J., and R. J. Campana (eds.). 1981. *Compendium of elm diseases.* American Phytopathological Society, St. Paul, Minn.

Tainter, F. H., and F. A. Baker. 1996. *Principles of forest pathology.* John Wiley and Sons, New York.

Tattar, T. A. 1978. *Diseases of shade trees.* Academic Press, New York.

Wallis, G. W., D. J. Morrison, and D. W. Ross. 1992. *Tree hazards in recreation sites in British Columbia.* Joint Report 13, British Columbia Ministry of the Environment and Parks and Canadian Forestry Service, Victoria, B.C., Canada.

9

Abiotic (Nonbiological) and Animal-Caused Injuries

Introduction

Abiotic or nonbiological factors, like extremes of temperature and moisture, lack of soil oxygen, air pollution, chemicals such as herbicides, nutrient deficiencies, elemental toxicities, salt, and lack or excess of light, can cause injuries and effects that resemble those caused by biological agents like fungi (Tainter and Baker 1996). Abiotic injuries also may make trees more susceptible to infection by fungi. It also is useful to consider the effects of mechanical injuries and fire, although these act differently than the other factors, and may not be considered as agents of disease because they occur in a discrete period of time. Likewise, extremes of temperature may not be considered as agents of disease, although they certainly affect tree health. Animal injuries (Black 1992) (e.g., bear damage) may resemble disease, so we also will consider this type of injury here.

The Pattern of Abiotic Injuries

Abiotic injuries and effects can be distinguished from those caused by biotic agents. Diseases caused by pathogens tend to be more random, whereas abiotic injuries usually are more nonrandom or uniform, but not always. As mentioned in Chapter 8, some diseases caused by biological agents may have nonrandom symptom patterns (e.g., casting of needles of only one age on a conifer and aggregated mortality in a root rot "pocket"). An example of a systematic pattern caused by an abiotic agent is injury at the edges and between the veins of beech leaves caused by high sulfur dioxide (SO_2) concentrations in the atmosphere (Skelly et al. 1987) (Figure 9.1). Very cold temperatures also can cause necrosis at the tips and/or edges of leaves and needles (Sinclair et al. 1987). This contrasts with hardwood foliage diseases, which typically have more random infection patterns (see Figure 13.1). At the landscape level, cold-temperature injuries usually are related to cold air drainage patterns and may occur along valley bottoms (Scharpf 1993).

199

Figure 9.1

A typical damage pattern for abiotic injury; inter-veinal and leave edge and tip necrosis on SO_2 damaged American beech leaves.

(Source: From Skelly et al. 1987.)

The micrometeorology of forests is extremely important to consider in interpreting abiotic injury patterns. Insects, like defoliators and bark beetles, sometimes cause widespread damage at the landscape scale, but the presence of insects usually is easy to detect. Defoliators and bark beetles are covered in detail in Chapters 20 and 21, respectively.

INJURIES CAUSED BY TEMPERATURE EXTREMES

Very high or very low temperatures can cause plant injuries. High-temperature injuries include heat defoliation, sugar exudation, and pole blight, which is exhibited as resin flow in pole-sized western white pine (Sinclair et al. 1987, Tainter and Baker 1996). Sunscald also is common. For example, young Douglas-fir trees are very susceptible to sunscald after thinning (Foster and Wallis 1974). Hardwood trees also are susceptible to sunscald. In the northern hemisphere most damage is in the south and southwest quadrants of stems where radiation loads are highest. In contrast, most damage is in the north and northwest quadrants in the southern hemisphere. Young trees are particularly sensitive because of their thin bark. Thinned and pruned trees may be more extensively damaged than thinned trees alone because the cambium is damaged. In ornamental plantings young trees growing close to south-facing walls in the northern hemisphere and north-facing walls in the southern hemisphere are particularly sensitive to high-temperature injuries because of reradiation from the walls.

Cedar **flagging** in western redcedar (*Thuja plicata*) occurs even in relatively old trees after a hot, dry summer (Foster and Wallis 1974, Tainter and Baker 1996). Older foliage turns red/orange in later summer or fall and eventually is shed. Trees generally are not permanently injured and typically recover. Cedar flagging results from the combined effects of both warm temperatures and low soil moisture.

Low-temperature injuries fall into two categories: frost injuries and winter damage (Tainter and Baker 1996). Both early and late frosts can cause extensive damage. Early frosts in autumn cause injury because tree tissues are not hardened off and buds may not be set. Young trees, particularly seedlings in open areas or in frost pockets, are more susceptible than older trees. Late frosts occurring after

Figure 9.2
Calloused vertical frost crack in a Douglas-fir stem.

budburst cause extensive damage because new tissues are very susceptible. Vertical frost cracks can occur along the stem of susceptible species due to differential swelling and shrinkage of tissues (Figure 9.2). The mechanism of freeze damage is not completely understood. Plants have natural frost protection because water in the cell vacuole contains solutes that depress the freezing point of water below $0^{\circ}C$ (Larcher 1995). In **intercellular** spaces freezing draws water from cells, increasing concentrations of solutes within cells, which further reduces the freezing point and protects the plant. Very rapid freezing, however, may not allow water to move out of the cells quickly enough, causing ice formation in **intracellular** spaces and membrane rupture. Intracellular freezing usually is lethal, but plants often recover from intercellular freezing. Exposed wood in frost cracks also is susceptible to infection by spores of wood decay fungi.

Winter damage is generally not as much of a problem as frost damage for species growing in their natural range because trees are hardened off and buds are set, but it does occur. Generally conifers are more susceptible to winter damage than deciduous hardwoods because deciduous trees are dormant and do not have foliage in winter (Sinclair et al. 1987). Conifer winter damage can be extensive in areas that typically have mild winter climates, such as the Puget Sound area of Washington. Every so often very cold air arrives from inland areas during extended periods of high pressure occurring over a wide area. Lowland Douglas-fir is particularly susceptible to foliage damage, although it may not be expressed until spring when older needles turn brown. Typically the buds are not killed and the trees survive. On the other hand, exotic species from warmer climates may die under these conditions. For example, eucalypts from Australia are particularly susceptible to winter damage when planted in Oregon and Washington.

On the slopes of broad mountain valleys in cold northerly environments, such as the Rocky Mountains, winter damage known as "red belt" sometimes occurs in conifers (Scharpf 1993) (Figure 9.3). Red belt occurs in a band midway to

Figure 9.3
"Red belt" shown as white stripe in the midslope area in a broad mountain valley in the Rocky Mountains near Banff, Alberta.

two-thirds up the slope because this is usually the warmest location on the slope (Lowry 1969). Frozen soils in combination with warming temperatures create a situation where trees attempt to transpire, but cannot move water to the foliage. They subsequently dehydrate and the foliage turns red. Because the buds are hardened trees usually do not die and they recover in the summer. Although this symptom occurs as a result of cold temperatures, it is really drought injury. If the soil freezes plants cannot take up enough water and desiccation damage results. Winter desiccation is more common on sites with little snow cover. However, although snow affords protection from very low temperatures and winter desiccation, plants may still suffer due to low oxygen levels, resulting in symptoms comparable to flood-induced hypoxia (Larcher 1995). Dead tops on Douglas-fir sometimes develop in the Puget Sound of Washington and the Willamette Valley of Oregon after cold, dry winters. Dry easterly winds blowing across mountain passes warm as they descend, causing winter desiccation.

Occasionally very cold winter conditions will affect trees over extremely wide areas—for example, temperature extremes during the winter of 1950–1951 damaged trees in the southeastern United States (Tainter and Baker 1996). The autumn of 1950 was unusually warm, but in November temperatures suddenly dropped to as low as -17°C and remained below freezing for 5 days. The winter of 1951 was extremely cold, and many trees died or developed injuries that had a high incidence of bacterial slime flux the following spring. A similar situation occurred in western Oregon and Washington in 1955. Young trees were particularly susceptible, and many were killed; multiple leaders or stem crooks developed in surviving trees. Decay fungi entered many of the damaged tissues.

LOW AND HIGH MOISTURE AND LOW SOIL OXYGEN

Extremely dry conditions commonly cause drought distress symptoms in trees including wilting, loss of foliage, dead branches, and even tree death (Tainter and Baker 1996). Drought stress also causes trees to be more susceptible to root diseases and bark beetles. In areas where rainfall is unreliable, such as the eastern slopes of the Cascade Range in Oregon and Washington and in the Sierra Nevada

Mountains in California, drought stress is common. Fire suppression in this region has resulted in increased tree density, which has exacerbated drought symptoms during times of low rainfall. Drought stress also occurs in other areas of the world, especially in Australia.

Winter drying and leaf scorch also can be considered to be drought symptoms. At the northern edges of species ranges in the northern hemisphere, winter drying is common. Leaf scorch occurs either due to prolonged drought or if roots are diseased and exposed to a continuous warm wind. Scorched conifers have needles with dead tips or drooping needles (Sinclair et al. 1987).

Excesses of moisture or flooding also cause tree stress and wilting symptoms and death (Larcher 1995). Soils may go anaerobic, placing the roots under oxygen stress. Douglas-fir trees treated with biosolids (sewage sludge) in Washington State developed oxygen stress and defoliated in areas where the water table was close to the surface and biosolids effectively sealed the soil surface. Root weevils attacked the weakened trees, resulting in considerable mortality.

Few tree species are well adapted to withstand long periods of flooding. Swamp cypress in the southern United States is an exception, being adapted for growing in flooded areas by having aboveground knees, enabling oxygen exchange to roots. Hardwoods generally grow better in wetter soils than conifers. Heavy textured conifer nursery soils are easily waterlogged, resulting in poor seedling growth. Wet soils also favor development of root diseases caused by *Phytophthora* spp. (Tainter and Baker 1996).

GLOBAL CHANGE

Trees are very responsive to weather conditions, and the current distributions of tree species are largely determined by climate. In recent years there has been great concern that human activities are causing global change, particularly global warming due to increased atmospheric CO_2 (Houghton 1997). This is a result of increasing atmospheric CO_2 concentrations because of fossil fuel combustion, forest burning, and increased decomposition as a result of forest cutting. Carbon dioxide is a greenhouse gas that traps longwave radiation in the atmosphere. Atmospheric CO_2 has increased from about 275 ppmv in the 1850s to over 350 ppmv today. Other greenhouse gases like nitrous oxides, chlorofluorocarbons, methane, and tropospheric ozone also are increasing. Chlorofluorocarbons also reduce stratospheric ozone, allowing more radiation to reach the earth's surface, contributing to warming. Global temperatures are predicted to rise 1.5 to 3.5°C with greater increases at higher than lower latitudes.

Rainfall patterns also will change as a result of global warming, and some computer models predict a decrease in summer moisture over middle and high latitudes. Such a change may cause this region to be drier and have longer growing seasons. Global change has occurred in the past, but the rate of change caused by recent human activities is thought to be more rapid than natural rates of change, and this has great implications for trees (Houghton 1997). Trees adapt to change slowly because of their long generation time. Many of the diebacks discussed in Chapter 15 might be related to climate change. Disease-causing organisms and insects, because of their short life cycles, are able to adapt more quickly to climate change than trees and might pose a more serious threat to forest health than global change alone. Plantation forestry might become more important in the future if we are forced to switch to species better adapted to new climates.

AIR POLLUTION

Types of Air Pollutants

Air pollution has caused injuries to trees ever since the industrial revolution (Treshow and Anderson 1989). The main types of air pollutants are gases, particulates, and acid precipitation. The air pollutants affecting plants and their sources are shown in Table 9.1. Air pollutants can be classified as primary or secondary pollutants. Primary pollutants are those that originate at the source in a form toxic to plants, like sulfur dioxide (SO_2); secondary pollutants develop as a result of reaction among pollutants, like ozone (O_3) and other oxidants and acid precipitation (Legge and Krupa 1986). The major gases causing plant injuries are SO_2 and O_3. Minor gases are nitrogen oxides (NO_x), peroxyacetyl nitrate (PAN), hydrogen fluoride (HF), silicon tetrafluoride (SiF_x), ammonia (NH_3), chlorine (Cl_2) and Freon (Tainter and Baker 1996). Carbon monoxide (CO), a major pollutant gas affecting humans and animals, is not considered to be a problem to plants in the open atmosphere.

Primary Pollutants   Sulfur dioxide is the main primary air pollutant gas, but other gases, such as NO_x, HF, SiF_4, NH_3, Cl_2, and Freon also cause problems. Nitrogen oxides, the precursors to ozone, come from automobile and truck exhausts (Treshow and Anderson 1989). The source of HF is aluminum smelters, and NH_3 comes from urea fertilizer plants, tanker spills, and agricultural activities such as feed lots, particularly from pigs. Chlorine gas usually results from industrial leaks or from swimming pools and Freon comes from refrigeration and air-conditioning leaks.

Sulfur dioxide occurs naturally in the atmosphere as a result of volcanic activity, but industrial sources such as copper smelters and the burning of coal for power generation are more important (Tainter and Baker 1996). Burning of high-sulfur coal produces copious quantities of SO_2. Around copper smelters at Sudbury, Ontario; Copper River, Tennessee; Trail, British Columbia; and Kellogg, Idaho, vegetation was killed outright for kilometers around due to SO_2 exposure. In Australia vegetation was killed around a copper smelter near Mount Lyell in Tasmania, and there are many other examples around the world. The very high concentrations that occurred around smelters and polluting power plants in the past generally no longer occur because of pollution abatement measures, although some high-emission sources still exist, especially in Eastern Europe. Many older smelters have been closed, and new technology has improved the quality of emissions. Scrubbers have been installed to reduce SO_2 at the source, and smoke stacks have been raised to reduce pollutant levels near the ground. The micrometeorology of air pollution is discussed later in this section.

T A B L E **9.1** │ **The Major Air Pollutants Affecting Trees and Their Sources**

Pollutant	Source(s)
Primary pollutants	
Gases	
Sulphur dioxide (SO_2)	Combustion of coal, refining and use of petroleum and natural gas, smelting and refining of ores: copper, lead, zinc, nickel; volcanic eruptions
Hydrogen fluoride (HF) and silicon tetrafluoride (SiF_4)	Aluminum reduction, manufacture of phosphate fertilizer or steel, brick plants, ferroenamel works, petroleum refineries
Nitrogen oxides (NO_x)	Internal combustion engine exhaust; petroleum refining; combustion of natural gas, fuel oil, and coal; incineration of organic wastes
Ammonia	Urea fertilizer plants, accidental spillage: tanker truck, train accidents
HCl	Refineries, glass making, incineration, scarp burning, accidental spillage, combustion of PVC
Chlorine (Cl_2)	Same sources as HCl
Polyvinyl chloride (PVC)	Manufacture of packaging materials, and wire insulation
Hydrogen sulfide (H_2S)	Anaerobic soils, cesspools, sewers
Ethylene	Chemical industry
Freon	Air conditioners, refrigerators
Particulates	Combustion of coal, gasoline, fuel oil; cement production; lime kilns; incineration; agricultural and forest burning; road surfaces, particularly freeways
Secondary pollutants	
Ozone (O_3)	Photochemical oxidation of NO_x in the presence of hydrocarbons, including natural plant terpenes; injection of stratospheric ozone by thunderstorms
Peroxyacetyl nitrate (PAN) and aldehydes	Photochemical oxidation of NO_x and olefin-type hydrocarbons
Acid rain and excess nitrogen	Reaction of SO_2 and NO_x with water in the atmosphere to form sulphuric acid (H_2SO_4) and nitric acid (HNO_3). Fertilizers and animal feed lots also produce excess atmospheric nitrogen

Source: From Legge and Krupa (1986).

Particulates are the second main type of primary pollutants, but particulates generally do not cause serious problems to plants (Legge and Krupa 1986, Treshow and Anderson 1989). Particulates range in size from 0.001 to 100 μm or so and usually are derived from industrial activities such as cement production and coal-burning power plants. Close to expressways plant leaves are commonly coated with particulates. Particles also form from interaction of pollutants in the atmosphere.

Secondary Pollutants   The major secondary pollutants are ozone (Legge and Krupa 1986, Tainter and Baker 1996) and acid precipitation and associated excess nitrogen (Aber et al. 1995, 1998). Ozone is a major air pollutant gas throughout the world. There is some confusion with respect to the beneficial and negative influences of ozone in the atmosphere. In the upper atmosphere, or stratosphere, ozone serves to protect the biosphere from exposure to ultraviolet radiation and is beneficial. Near the ground surface ozone is a pollutant gas and is largely a result of photochemical oxidation of nitrogen oxides from the exhaust of internal combustion engines in the presence of hydrocarbons (Legge and Krupa 1986) (see Table 9.1). Some stratospheric ozone is injected into the lower atmosphere by thunderstorms, but this is a minor source. Concentrations of O_3 are particularly high in areas where solar radiation and vehicular traffic are high and where topography confines the movement of pollutants (Olson et al. 1992). Levels of ozone are particularly high in southern California because of these conditions (Miller and McBride 1999). Other areas with high O_3 concentrations occur in the southern and eastern United States, Europe, Mexico, Central and South America, and Asia. Ozone also is high near population centers in the southern hemisphere, such as Sydney and Melbourne in Australia, Sao Paulo in Brazil, and Santiago in Chile. As populations and automobile use increase, O_3 steadily rises unless emission controls are used.

The other major secondary air pollutant is acid precipitation (Schulze et al. 1989, Johnson and Lindberg 1992). Acid precipitation is formed by interaction of sulfur dioxide and nitrous oxides in the atmosphere with water vapor, forming sulfuric and nitric acids, respectively. Normal rainfall pH is 5.3. Acid precipitation in rain, snow, and fog can have pHs as low as 2.5; fog pH usually is lower than rain or snow pH. Low-pH fog has been implicated in forest decline problems at higher elevations in the Great Smoky and Appalachian Ranges in the eastern United States at locations such as Camel's Hump in Vermont, and Clingman's Dome and Mount Mitchell in North Carolina (Johnson and Lindberg 1992). These areas are downwind of industrial and power generation sources in the eastern United States, particularly the Ohio Valley, which are sources of sulfur dioxide. Automobiles are largely the source of nitrous oxides. Acid precipitation has been implicated in forest decline in Europe (Schulze et al. 1989).

Excess N from both dry and wet deposition now is considered an air pollution problem affecting the functioning of forest ecosystems in eastern North America and Europe (Vitousek et al. 1997). The hypothesized influence of N saturation on ecosystems is shown in Figure 9.4, especially the reduction in net primary production (Aber et al. 1995, 98). Threshold levels for N saturation appear to be lower in Europe (>10 kg ha^{-1} yr^{-1}) than North America (>15–20 ha^{-1} yr^{-1}), perhaps because of longer exposure or different cultural practices in forests. There is currently a large-scale research effort being conducted in Europe sponsored by the European Union called NITREX (Nitrogen Saturation Experiments on Forest Ecosystems) to determine the influence of nitrogen saturation on European forests (Gundersen et al. 1997).

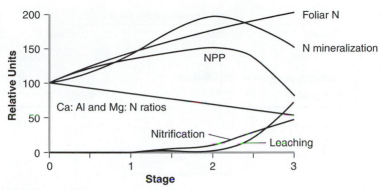

Figure 9.4

Hypothesized time course of response to chronic nitrogen additions of relative foliar N concentrations, N mineralization rates, net primary production (NPP), foliar Ca:Al and Mg:N ratios, nitrification and nitrate leaching in temperate forest ecosystems . The four stages of N saturation are: Stage 0—unimpacted ecosystem; Stage 1—increased N deposition along with increased N mineralization, foliar N and NPP, and reduced Ca:Al and Mg:N ratios, but no effect on nitrification and nitrate leaching; Stage 2—continued increased foliar N and NPP—N mineralization is starting to decrease and nitrification and nitrate leaching beginning to increase; Stage 3—foliar N and nitrification and nitrate leaching continue to increase, but NPP and N mineralization drop dramatically. Visible effects apparent in Stage 3.

Dispersion of Air Pollutants

The concentration of pollutants downwind from a source depends on the type of source, concentration at the source, height of the source above the ground surface, the terrain, and meteorological conditions, such as windspeed and atmospheric stability (Lowry 1969). Sulfur dioxide is typically emitted in a plume from a point source such as smokestack. The relationship between atmospheric stability and dispersion from a stack is shown in Figure 9.5. Maximum dispersion and lowest concentrations at the ground occur when the atmosphere is unstable and windspeeds are high, a condition known as "looping." Dispersion also is good under "coning" conditions. Minimum dispersion occurs when the atmosphere is stable and windspeeds are low. Temperature decreases with height in the atmosphere under unstable conditions, but under stable conditions the reverse occurs, a condition called a *temperature inversion.* Air pollutants trapped below inversions can cause health problems to plants, animals, and humans. If the air is very stable, then wind speeds are low and "fanning" occurs. Under these conditions pollutants do not reach the ground close to the source. However, if there is heating at the ground, then pollutants will reach the ground in high concentrations because of stable conditions above and unstable conditions below the top of the smokestack; this condition is known as "fumigation." Increasing the height of a smokestack is one way to prevent fumigation so that the top of the smokestack is above the inversion. The

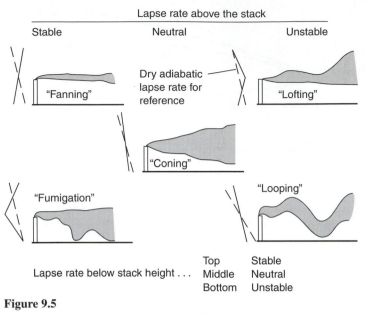

Figure 9.5

Relationships between atmospheric stability and dispersion from a smokestack.

heat from the plume can also push through the inversion before it is dispersed, resulting in a condition known as "lofting" (Lowry 1969).

Temperature inversions occur as a result of rapid radiative cooling of the ground surface or as a result of air moving down from the upper atmosphere and heating as it falls. Radiation inversions typically occur in mountain valleys, whereas subsidence inversions usually occur at a regional scale. For example, in the northern hemisphere air rising at the equator moves north and west, finally moving back toward the land surface about 25 to 35° N, or the latitude of Los Angeles, trapping pollutants below the inversion.

Point sources of air pollution are generally easy to identify and control. Area sources, such as those associated with ozone pollution, are much more difficult to manage. In areas such as the Los Angeles basin, morning traffic produces NO_x and under strong radiation it is converted to O_3 in the presence of hydrocarbons (Olson et al. 1992). Highest O_3 concentrations can thus occur at some distance from the source, since the morning sea breezes moving off the ocean to the land move the precursors to the east. There may be a second O_3 peak in the afternoon related to afternoon rush-hour traffic. In the evening, air may move back to the west as the land cools and air moves from the land to the ocean. The presence of the San Gabriel and San Bernadino Mountains contributes to the trapping of O_3 in the Los Angeles basin. Because O_3 is such a reactive gas, it is scavenged out of the atmosphere by NO_x in the evening hours and may even drop to near zero overnight. Only during westerly Santa Ana conditions does the air clear in Los Angeles.

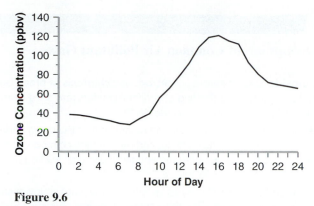

Figure 9.6

A typical diurnal curve for O_3 in the Cedar River watershed in the Puget Sound area of Washington on May 31, 1986.

A similar situation occurs in many other areas of the world. In the Puget Sound region of Washington, an O_3 plume drifts southeast from the Seattle metropolitan area toward Mount Rainier National Park on warm spring and summer days. The plume is confined between the Olympic and Cascade Mountains and is trapped by an inversion which typically occurs at about 1,500 m elevation. A typical diurnal curve for O_3 in the Puget Sound area is shown in Figure 9.6 (Edmonds and Basabe 1989). Note that there is only one peak about 3 to 4 P.M. because less O_3 is produced after the evening peak traffic, and it is rapidly scavenged. Concentrations drop to a minimum at about 6 to 7 A.M. At high elevations average daily ozone concentrations may be similar to those at low elevations, but diurnal fluctuations are less.

Damage Caused by Air Pollutants

Air pollution damage is classified as either acute or chronic (Tainter and Baker 1996). Acute damage implies necrotic symptoms are present, whereas chronic damage (e.g., reduced growth) is more subtle. Air pollutants can affect cell membrane permeability, accelerate respiration, decrease photosynthesis and growth, and inhibit enzymes (Larcher 1995). Chronic damage can occur at the level of the individual leaf or needle, the tree, or the ecosystem. Trees may be weakened so that they are more susceptible to root diseases, bark beetles, and insect defoliators. Chronic air pollution injury now is more common than acute air pollution injury. Symptoms of common air pollutant gases are shown in Box 9.1.

The degree of damage due to gaseous pollutants is related to the following factors: (1) the concentration of the gas, (2) the exposure or dose, (3) the species of plants involved, and (4) the macro- and micrometeorology of the location (Treshow and Anderson 1989).

Where SO_2 concentrations and dosages are very high—for example, in the vicinity of copper smelters—acute damage resulting in mortality can occur. Sulfur dioxide enters leaves or needles through stomates, and even if stomates are closed it can enter through the cuticle (Larcher 1995). With increased SO_2 uptake

Box **9.1**

Typical Symptoms of Common Air Pollutant Gases

Sulfur dioxide Causes interveinal necrosis and **chlorosis** of broad-leaved trees and tip necrosis and banding of conifers; conifers may shed needles prematurely; high concentrations cause death.

Ozone On broad-leaved plants flecking or stippling of the upper surface occurs; tip necrosis, chlorotic mottling, and banding are common on conifers; older needles and leaves are more susceptible to damage, leading to premature defoliation of older foliage.

Peroxyacetylnitrate (PAN) Causes symptoms (bronzing) on lower surfaces of leaves, particularly vegetables; also causes interveinal necrosis on trees such as white ash; young leaves are sensitive.

HF Causes marginal necrosis and chlorosis of leaves of broadleaves and tip necrosis and chlorosis on conifers.

Cl$_2$ Causes similar symptoms to HF.

Source: Adapted from Tainter and Baker (1996).

---

internal acidification occurs, sulfite levels in the chloroplasts rise, and chlorophyll is destroyed. High concentrations of SO$_2$ result in interveinal necrosis (Figure 9.1) and eventually plant death. Most of the smelters in the world near where mortality occurred have either closed or reduced their emissions such that dramatic examples of acute damage now only rarely occur. Recovery in the devastated areas has been slow, and in many cases tree reestablishment is poor and sparse even after 20 years (Figure 9.7).

Acute damage caused by ozone was first determined to occur in the early 1950s in the San Bernadino Mountains near Los Angeles (Olson et al. 1992, Miller and McBride 1999). Tree mortality was observed in ponderosa pine because this species is particularly sensitive to ozone. Typically visible damage to vegetation is thought to occur at ozone concentrations greater than 80 ppbv. At background, or clean sites, O$_3$ concentrations typically vary between 30 and 40 ppbv with little change during the day. Ozone typically causes chlorotic mottling of leaves and needles due to oxidation and death of tissues (Figure 9.8). Ozone enters leaves through the stomates like SO$_2$, where it disassociates into oxygen and peroxides. These peroxides affect the plasmalemma and other membranes, impairing transfer processes. Eventually necrosis develops, spreading out from where O$_3$ enters the leaf or needle (Larcher 1995). Older needles are affected more than younger needles, resulting in premature defoliation of the older needles and a thin crown condition in surviving trees (Figure 9.9). Average hourly concentrations of more than 400 ppbv were commonly observed in the Los Angeles area in the 1960s. With vehicle emission controls these maximum concentrations have been reduced, but they still commonly exceed 80 ppbv.

Figure 9.7
Slow vegetation recovery near a smelter where vegetation was killed by high sulfur dioxide and heavy metal concentrations in the atmosphere near Smelterville, Idaho.

Figure 9.8
Ozone damage showing chlorotic mottling on ponderosa pine in California; motttling is more intense on older needles.
(Source: From Scharpf 1993.)

Figure 9.9
During winter only shortened current-year needles remain on this ponderosa pine severely damaged by ozone in California. The crown is very thin and transparent.
(Source: From Scharpf 1993.)

Ozone concentrations capable of causing chronic damage occur widely throughout the United States (Johnson and Lindberg 1992). Not all tree species are equally affected, however, and susceptibility varies greatly (Sinclair et al. 1987) (Table 9.2). As mentioned previously ponderosa pine is extremely susceptible as well as the white pines. Mortality in ponderosa pine in southern California initially averaged about 8% per year, but this rate has been much reduced in recent years due to a combination of reduced emissions and survival of resistant individuals. Species succession and composition has been affected in southern California (Miller and McBride 1999). There is now a higher proportion of oaks and less ponderosa pine. Ecosystem functioning also has been affected in that forest floor decomposition and nutrient-cycling rates have been changed (Olson et al. 1992). It is suspected that ozone is influencing forest growth and composition in the southeastern United States and Europe, particularly in Germany.

Other gaseous pollutants also cause damage to trees, such as PAN, HF, and Cl_2, but to much lesser extent than SO_2 or O_3 (Legge and Krupa 1986). Emissions of ammonia gas in high concentrations, especially resulting from tanker truck spills, have also caused tree mortality. White spruce mortality also was noted in the vicinity of a urea fertilizer plant in Kenai, Alaska (Whytemare et al. 1997).

Effects of Acid Precipitation

A massive research program into the effects of acid precipitation on forest ecosystems in the United States and Europe was mounted in the 1980s. The main conclusions were that (1) aquatic systems were being adversely affected in areas where streams and lakes had low acid-neutralizing capacity; (2) fish populations were lowered or even decimated, particularly in the northeastern United States, eastern Canada, and Scandinavia where the bedrock is low in bases such as Ca and Mg; (3) sulfur and nitrogen deposition caused adverse impacts on certain highly sensitive ecosystems, especially high-elevation spruce-fir ecosystems; and (4) gradual leaching of soil nutrients, especially Ca, Mg, and K, from sustained inputs of acid rain could eventually affect the nutrition and growth of trees (NAPAP 1993, 1998). Original hypotheses on forest effects were that increased acidity in rain and fog would (1) decrease soil pH and the availability of nutrient ions; (2) increase soil Al concentrations, resulting in the killing of fine roots and mycorrhizae and result in forest decline; and (3) leach cations from foliage and soils, causing poor growth and increased susceptibility to Armillaria root disease (Schulze et al. 1989). It has, however, proven to be extremely difficult to verify all these hypotheses in the field because the observed forest decline problems in Europe and North America are due to multiple causes: acid precipitation; other air pollutants such as ozone and excess N; insects, such as the balsam woolly adelgid; weather; forest age; and global change. Forest decline is treated in detail in Chapter 15.

Management of Air Pollution Problems

Air pollution injuries of forests can be managed by

1. Controlling emissions at the source
2. Using resistant species or varieties

T A B L E **9.2** | **Susceptibility of Selected Tree Species to Ozone**

Gymnosperms	Angiosperms
Tolerant	
Abies balsamea (balsam fir)	Acer rubrum (red maple)
Gingko biloba (gingko)	Acer saccharinum (sugar maple)
Picea abies (Norway spruce)	Acer platanoides (Norway maple)
Picea pungens (Colorado blue spruce)	Fagus sylvatica (European beech)
Pinus aristata (bristlecone pine)	Pinus sabiniana (digger pine)
Pinus lambertiana (sugar pine)	Juglans nigra (eastern black walnut)
Pinus rubra (red pine)	Quercus rubra (northern red oak)
Pseudotsuga menziesii (Douglas-fir)	Robinia pseudoacacia (black locust)
Sequoiadendron giganteum (giant sequoia)	Ulmus spp. (hybrid elms)
Sequoia sempervirens (redwood)	
Tsuga canadensis (eastern hemlock)	
Taxus spp. (yew)	
Intermediate	
Abies concolor (white fir)	Acer negundo (box elder)
Pinus contorta (lodgepole pine)	Acer saccharinum (silver maple)
Pinus rigida (pitch pine)	Fraxinus pennsylvanica (green ash)
Pinus sylvestris (Scots pine)	Platanus x acerifolia (London plane)
Pinus echinata (shortleaf pine)	Tilia spp. (basswood)
Pinus torreyana (Torrey pine)	
Pinus nigra (Austrian pine)	
Sensitive	
Larix decidua (European larch)	Fraxinus americana (white ash)
Pinus banksiana (jack pine)	Liriodendron tulipifera (tulip tree)
Pinus coulteri (Coulter pine)	Platanus racemosa (California sycamore)
Pinus jeffreyi (Jeffrey pine)	Populus spp. (poplars)
Pinus ponderosa (Ponderosa pine)	Salix babylonica (weeping willlow)
Pinus radiata (Monterey pine)	
Pinus monticola (western white pine)	
Pinus strobus (eastern white pine)	
Pinus virginiana (Virginia pine)	

Source: From Sinclair et al. (1987).

3.   Using chemical protectants
4.   Forecasting air pollution episodes

Implementation of legislation in the United States, particularly the different iterations of the Clean Air Act, has helped to reduce pollutants at the source. Point source pollutants are the easiest to manage. Scrubbers have been placed on stacks emitting SO_2, and new technology is being developed. Use of low-sulfur coal also has helped reduce emissions of SO_2. Some countries now have active programs to curtail SO_2 emissions—for example, Canada and the United States. Use of catalytic converters has helped reduce ozone concentrations in southern California, but at the same time the number of vehicles has increased as the population of California has increased. In many areas in the United States, flexible covers have been placed over the end of gas pump hoses to prevent gasoline fumes (i.e., hydrocarbons, or volatile organic carbon) from entering the atmosphere. Ozone is one of the more difficult pollutants to control because automobiles are mostly responsible for this problem. Global warming is likely to increase O_3 concentrations (Houghton 1997).

Resistant species can be planted in areas where pollutants are likely to be a problem, and breeding techniques can be used to develop resistant **strains.** Strains of resistant species are likely to remain resistant in comparison to strains developed against fungal pathogens, since the pollutants cannot change with time like pathogens.

Chemicals can be sprayed on surfaces or used as soil drenches for protection against O_3. A soil drench of benomyl-protected pinto beans at O_3 exposures of 25 ppm for 4 hours. In Germany lime has been added to many forests to counteract the effects of soil acidification due to acid precipitation (Huettl and Zoettl 1993). It also has been added to lakes in Sweden and the northeastern United States.

Forecasting ground-level concentrations of SO_2 can be used to reduce pollutant emissions. Power plants then might be able to curtail emissions during sensitive periods. Ozone concentrations also can be forecast. For example, in the Puget Sound region ozone concentrations can be forecast using maximum air temperatures and solar radiation (Edmonds and Basabe 1989).

HERBICIDES

Herbicides are used in forestry to control competing vegetation during stand establishment, especially for conifers. They fall into two groups: those used for broad-leaved vegetation such as red alder and big leaf maple in the Pacific Northwest and those used for grasses and herbs (Helgerson et al. 1992). Triclopyr, 2,4-D and dicamba are in the first group and act as plant hormones that disrupt normal growth forms. In sublethal doses they cause aberrant growth forms (Figure 9.10), but conifers may outgrow the problem within 1 or 2 years. Other symptoms include cupped leaves, curled leaf margins, swollen buds, and many others. These herbicides are sprayed aerially or from the roadside and can cause injury to nontarget plants, because they can drift away from the target. The amount of drift depends on release height, droplet size, and wind velocity. For herbaceous vegetation, soil-active residual herbicides such as atrazine, simazine, hexazinone, and sulfometuron, and foliar-active glyphosate, are effective. Simazine is a much used preemergence herbicide that may cause problems after application to nursery soils because it persists in the soil.

Conifer seedlings are sensitive to many herbicides, but this depends on the species. Severe injury can result when shoot growth is in progress. Foliar-active

Figure 9.10
Abnormal growth patterns in Douglas-fir caused by a
herbicide. Normal foliage is on the right. Herbicides may
act as plant hormones, disrupting normal growth forms.
Note that the tree may outgrow the injury in the year
after application.
(Photo provided by Ken Russell, Washington Department of Natural Resources, Retired.)

herbicides should not be applied over planted seedlings in the first growing season. Soil-active herbicides can be safely broadcast over dormant conifers in the
autumn.

MECHANICAL INJURIES

Mechanical injuries to trees are caused by wind, snow and ice, lightning, hail, and
equipment used during thinning operations, but should not strictly be considered
as causing disease because of their short time action. Pruning also may cause
mechanical injuries.

The influence of wind is covered in detail in Chapter 7. Wind damage results
in blowdown of individual trees, particularly along stand edges. Occasionally large
areas are blown down by hurricanes in the eastern United States, by cyclones in
Australia and Southeast Asia, and by windstorms in the coastal Pacific Northwest,
such as the "1921 blow" on the Olympic Peninsula of Washington. Hurricane
Camille, on September 21, 1989, destroyed more than 1.6 million ha of trees in
South Carolina (Tainter and Baker 1996). Tornadoes also damage swaths of forests
in the United States every year. Uprooted trees and trees with stem breakage usually are attacked by stain and decay fungi, and insects such as *Ips* bark beetles,
wood borers, powder post beetles, and ambrosia beetles. Rapid drying also
causes checking, which reduces the value of logs for lumber. Hardwoods frequently survive breakage longer than conifers, but hardwoods are particularly susceptible to decay fungi.

Trees that are severely bent by wind may not be attacked by insects or fungi,
but will have reduced wood quality (Tainter and Baker 1996). Leaning trees produce reaction wood, known as compression wood in conifers and tension wood

in hardwoods, which reduces timber values. Wind also can cause strange stem deformities, such as fluting in western hemlock.

Although many tree species are adapted for handling heavy snow loads (e.g., high-elevation or high-latitude spruces and firs, which have a pyramidal shape that reduces snow accumulation), snow can cause tree damage, particularly in low-elevation species. For example, wet, heavy snow accumulates on Douglas-fir branches in western North America, causing considerable breakage. Heavy snow accumulations often can bend young conifers, especially in the Rocky Mountains and the Cascade Mountains. However, bent stems usually straighten out with no permanent damage. Moving deep snowpacks in steep terrain, however, can result in more permanent deformities, including butt and stem sweep, doglegs, S-curves, and even stem failure (Tainter and Baker 1996).

Ice buildup on branches can cause breakage because of the increase in weight. Glaze occurs frequently in the southeastern United States when rain contacts conifer and hardwood branches that have surface temperatures below freezing (Tainter and Baker 1996). Long needle conifers are especially susceptible to glaze injury, and trees may have their branches stripped, or they may be uprooted, bent over, or broken off. The poor form and condition of many hardwood stands in the southern Appalachians may be due to glaze. Repeated limb and top injury deforms trees, reduces growth, and provides entry points for insects and rot fungi. Tree damage from hail is common in some areas, particularly where thunderstorms are common. Hail strips leaves and needles from branches, but damage is usually not extensive (Sinclair et al. 1987).

Like hail, lightning damage is relatively localized and causes less damage than snow and ice. Lightning-struck trees may be shattered, or a strip of bark from the top to the bottom of the tree may be blown off as the lightning is discharged into the soil. Trees may recover from this injury or they may slowly die. Trees shatter because heating of the wood literally causes it to explode as sap is converted to steam. Sometimes struck trees may have little or no visible injury yet unexpectedly die a short time later. Bark beetles often are attracted to lightning-struck conifers, and many isolated southern pine beetle attacks are related to lightning strikes (Tainter and Baker 1996).

Wounding of stems and roots by machinery or logs being pulled through the forest during thinning operations is of great concern. Trees are particularly susceptible to damage during the spring when the sap begins to flow and the bark is loose. Young trees and thin-barked species are very susceptible to damage. Large open wounds can easily be infected by decay fungi. Laying of green branches on the forest floor protects roots from damage from tracked machinery. Rubber-tired skidders are better.

Ornamental trees are particularly susceptible to mechanical damage at the base of the stem from "weed whackers" used to remove grass (Tainter and Baker 1996). They also are susceptible to damage from excavation, resulting in root damage or severance. The placement of earth around the base of trees causes soil compaction and reduction of oxygen to roots. Even people walking around the base of trees can cause compaction, damage fine roots, and even result in tree death.

| TABLE **9.3** | **Typical Concentrations of Essential Elements in Foliage Necessary for Healthy Plants** |

	Concentration in dry mass
Element	**µg/g (ppm)**
Macronutrients	
Nitrogen (N)	15,000 (1.5%)
Phosphorus (P)	2,000
Potassium (K)	10,000
Magnesium (Mg)	2,000
Calcium (Ca)	5,000
Sulfur (S)	1,000
Oxygen (O)	450,000
Carbon (C)	450,000
Hydrogen (H)	60,000
Micronutrients	
Iron (Fe)	100
Zinc (Zn)	100
Copper (Cu)	6
Manganese (Mn)	50
Molybdenum (Mo)	0.1
Boron (B)	20
Chlorine (Cl)	100

Source: Adapted from Tainter and Baker (1996).

Many of the large famous trees throughout the world are protected by fences or boardwalks to prevent this type of damage.

NUTRIENT DEFICIENCIES

Sixteen nutrient elements are required for plant growth (Tainter and Baker 1996). One can remember the 10 major elements required through the mnemonic CHOP-KNSCaFeMg (pronounced "see Hopkin's cafe mighty good"), which of course stands for carbon, hydrogen, oxygen, phosphorus, potassium, nitrogen, sulfur, calcium, iron, and magnesium. All 16 elements are shown in Table 9.3 along with concentrations in foliage typical of most higher plants. It should be recognized, however, that there is considerable variability among species. Note that some are classified as macronutrients, like N, P, K, Ca, Mg, and S; others are micronutrients required in much lower concentrations, like Fe, Cu, Zn, B, Mn, Mo, and Cl. Nickel, Si, Na, Co, and Se may also be beneficial.

The major functions of essential elements excluding C, H, and O are shown in Box 9.2.

Box 9.2

Major Functions of Essential Elements in Plants

Macronutrients

Nitrogen (N)   Important component of proteins, enzymes, nucleic acids, and chlorophyll. Required in relatively large amounts.

Phosphorus (P)   Component of high-energy phosphate bonds of adenosine triphosphate (ATP) responsible for energy transfer in biochemical systems; also involved with RNA and DNA, which mediate protein synthesis and transfer of genetic information.

Potassium (K)   Activator of many enzymes involved with photosynthesis and respiration and maintaining internal osmotic pressure in cells; a very mobile nutrient and is present in high concentrations in plant tissues.

Calcium (Ca)   Occurs in middle lamella as calcium pectate; needed during cell division and functioning of cell membranes.

Magnesium (Mg)   Important enzyme activator and essential in photosynthesis, respiration, and formation of nucleic acids.

Sulfur (S)   Important component of proteins and occurs in the amino acids cysteine, cystine, and methionine.

Micronutrients

Iron (Fe)   Involved with enzymes, chlorophyll formation, and respiration.

Copper (Cu)   Associated with chloroplasts and proteins and oxidation/reduction enzymes.

Zinc (Zn)   Important in the functioning of many enzymes and required to produce the growth regulator indoleacetic acid.

Manganese (Mn)   Involved with chloroplast formation and evolution of oxygen in photosynthesis; an enzyme activator.

Boron (B)   Involved with nucleic acid synthesis, sugar translocation, respiration, reproduction, and water relations.

Molybdenum (Mo)   Involved with nitrate reductase enzyme in nitrogen fixation.

Chlorine (Cl)   Chloride ion plays a role in cell division and photosynthesis.

Source: Adapted from Tainter and Baker (1996).

Nutrient elements may be deficient, optimal, or toxic in plant tissues (Figure 9.11) and each element has its own typical range. For current Douglas-fir foliage, for example, N concentrations range from 0.8 to 3.0% with an optimum at 1.5% (Carter 1992). Foliar analysis may be required to diagnose problems, especially if more than one nutrient is involved. In the southeast United States and Australia, P deficiency must be corrected before trees will respond to N fertilization.

Most nutrient deficiencies show typical symptoms that can be used for initial diagnosis. Some symptoms may be confused with injuries caused by fungi or other biotic disease agents, but with experience one usually can separate the two.

Box 9.3

Typical Nutrient Deficiency Symptoms in Foliage

Macronutrients

Nitrogen (N)   Foliage chlorotic or yellow; deficiency first in older needles since N is mobile and moves from older to younger tissues because of demand; in severe deficiency needles small and yellow; most limiting element in soils.

Phosphorus (P)   Foliage often red or purple; small, fused needles; deficiency first in older needles; younger foliage green; second only to N as most limiting element in soils.

Potassium (K)   Dead areas on edges of leaves or tips of needles; symptoms develop quickly in older tissues because K is so mobile.

Calcium (Ca)   Death of terminal bud; deficiency symptoms in youngest tissue because Ca is very immobile.

Magnesium (Mg)   Symptoms first in younger tissues; leaves may be cupped, mottled, or have chlorotic spots; old needles on conifer may be yellow with brown tips; basal portion of older needles may be green; youngest needles may be green.

Sulfur (S)   Veins on leaves light green; upper needles yellowish; older needles still green; usually not a major problem.

Manganese (Mn)   Spots of dead tissue may develop; leaves become increasingly chlorotic with veins remaining darker green; new leaves are stunted and growth may stop; young needles pale green or yellowish; in severe deficiency shoot tip may die.

Micronutrients

Iron (Fe)   Young leaves chlorotic, but veins remain green; young needles bright yellow; older needles green.

Copper (Cu)   Young leaves wilted, but no spotting or chlorosis.

Zinc (Zn)   Young needles pale green, then become chlorotic or necrotic.

Boron (B)   Terminal twigs and root tips fail to elongate normally; stunting; youngest needles bunched resembling **rosettes;** foliage gray/green or bronze; stubby roots.

Molybdenum (Mo)   Young needles turn pale green or yellow, then brown; rare.

Chlorine (Cl)   None—chloride ion absorbed in far greater quantities than needed.

Source: Adapted from Tainter and Baker (1996).

Typical nutrient deficiency symptoms are shown in Box 9.3. Once again it should be pointed out that there is considerable variability among plant species.

Nitrogen deficiency is typically exhibited on poor, young soils. It is very apparent in the Pacific Northwest (Walker and Gessel 1990) and northern Europe, particularly on soils formed after recent deglaciation. Phosphorus deficiency typically

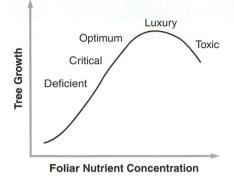

Figure 9.11
Nutrient elements may be deficient, optimal, or toxic in plant tissues.
(Source: Adapted from Carter 1992.)

occurs in areas with very old soils, such as the southeastern United States and many areas in Australia and soils developed on sand dunes in New Zealand. Potassium, Ca, and Mg deficiencies are rarer. Trace element deficiencies are common on older soils. In Australia Cu, Zn, and B deficiencies are common (Chappell et al. 1992).

Most elemental deficiencies can be easily corrected with application of fertilizers. Nitrogen is applied in the form of urea (NH_2CONH_2), NH_4NO_3 or $(NH_4)_2SO_4$; phosphorus is applied as rock phosphate or superphosphate; Mg is applied as Epsom salts of $MgSO_4$ (Tainter and Baker 1996).

The nutritional status of a tree can alter its response to pathogens, particularly those causing root diseases, or insect attack. Seedlings grown with too much N often are spindly, with very succulent tissue that is very susceptible to **damping-off** fungi and frost damage. Nursery managers may limit fertilization in early growth to reduce susceptibility to damping-off and cause seedlings harden-off earlier. Phosphorus fertilization in nurseries also may be adjusted to allow colonization of roots by mycorrhizal fungi. If soil P levels are too high, seedlings will not form mycorrhizal associations. There also is some evidence that nutrient deficiency in aspen increases susceptibility to Hypoxylon canker caused by *Hypoxylon mammatum*.

NUTRIENT IMBALANCES AND ELEMENTAL TOXICITIES

Nutrient imbalances can result from overfertilization of elements, particularly N. For example, Mg deficiency resulted from application of biosolids (sewage sludge) to Douglas-fir and grand fir Christmas trees in western Washington. High NH_4 uptake interfered with Mg uptake, resulting in Mg deficiency in needles. This also was observed in the Kenai Peninsula, Alaska, in seedlings near a urea fertilizer plant. Emissions from the plant have built up N levels in the soil (Whytemare et al. 1997), resulting in Mg deficiency in young white spruce seedlings. Magnesium deficiencies also have been observed in conifers in areas of Europe receiving high atmospheric N inputs as a result of air pollution.

As indicated in Figure 9.11, very high concentrations of nutrients in foliage can cause toxicities. This usually does not happen in natural situations, except on odd soil types, such as serpentine soils. Areas of serpentine soils in the Oregon and Washington Cascades, for example, have unusually high Mg concentrations, resulting in high Mg uptake and stunted growth because of interference with the uptake of other nutrients.

In Australia rapidly growing radiata pine trees planted on old agricultural land heavily fertilized with N have deformities in the lower stem known as "speed wobbles," particularly in soils where NO_3 concentrations are high. The upper portion of the tree is often straight. This occurs only in trees grown from seedlings. If cuttings are used, the stems grow straight. Apparently the high NO_3 levels have a greater effect on younger tissues in seedlings, and since cuttings are physiologically older they grow straighter. "Speed wobbles" also have been observed in rapidly growing Douglas-fir trees treated with biosolids (sewage sludge).

Nonnutrient elements such as heavy metals, like Cd and Al, can also cause toxicity. High Al concentrations typically cause root deformations, including club roots. Heavy metals are more mobile in acid soils, and soil acidification due to acid rain is thought to lead to Al toxicity. Heavy metals also can cause root damage in acid mine spoils.

FIRE

Fire commonly causes scars on trees, although it cannot really be considered to be a disease agent because of its mode of action; fire ecology is covered extensively in Chapter 4. Scars caused by fires are easy to identify because of the presence of charcoal. Fires scars also are important entry points for basal stem decay fungi such as *Phaeolus schweinitzii* in conifers (Tainter and Baker 1996). Prescribed burning in stands where fire suppression has been practiced over the years sometimes results in injury to fine roots and tree decline and mortality. This usually occurs because the fine roots were occupying the burned litter layer. Soil heating also resulted in fine root death in the mineral soil (Agee 1993).

SALT INJURY

There are a number of sources of toxic salts, including highway deicing salts, ocean salt spray, flooding with ocean salt water, soil salinity, irrigation with saline water, and cooling tower drift. Injury due to salt comes from direct spray on the foliage and uptake through roots. The major sources are from the ocean and the application of salt to roads during the winter months in North America. Sodium chloride is the major salt causing problems (Tainter and Baker 1996).

Trees growing close to the ocean usually are adapted to exposure to salt; for example, shore pine (*Pinus contorta* var. *contorta*) in western north America. Windstorms, however, especially hurricanes along the Atlantic and Gulf coasts, can carry salt water as much as 80 km inland. During hurricanes, storm surges can trap salt water that cannot drain back to the ocean, and this is generally more damaging than salt spray. Salt burn occurs as result of salt uptake by trees and mortality may result. Generally, salts are leached beyond the rooting zone relatively rapidly so the damage is not long lasting. Species vary in sensitivity to salts, and in the southeast United States loblolly pine is more sensitive than slash pine and other conifers (Tainter and Baker 1996).

In northern latitudes in North America, especially in the eastern midcontinent regions, sodium chloride and sometimes calcium chloride is used on roads to prevent ice formation on road surfaces. Adjacent to highways trees often are covered with salt spray, especially conifers. During the spring foliage may turn brown, especially in the midportion of the crown. Lower crowns usually are protected from the salt because of snow accumulation. Buds are not usually killed so that the new foliage that emerges is healthy. Older needles will drop off. Vigor of these trees is low and mortality may result in the long run.

High salt concentrations influence plants due to osmotic retention of water and to specific ion effects on the protoplasm. As salt concentrations increase, water becomes less available to the plant and high concentrations change the normal ratio of K and Ca to Na. This influences proteins and membranes. Photosynthesis is impaired, by both stomatal closure and salt effects on chloroplasts. Dwarfism and necrosis of roots, buds, leaf margins, and shoot tips results, and growth rates and biomass production are impaired (Larcher 1995).

Just as tree species have varying **tolerances** to ocean salt, they also have varying tolerances to road salt (Tainter and Baker 1996). Conifers tend to be more damaged than deciduous hardwoods. They have foliage in the winter and their foliage and branches are more efficient in intercepting airborne particles and spray. Eastern white and red pines are severely damaged; other pines are more tolerant. Eastern hemlock also is severely damaged; eastern redcedar and spruce seem to be tolerant. Hardwoods also vary in their susceptibility. Red and white oak and birch species seem to be tolerant; red and sugar maples are susceptible. Apparently tolerant species are able to withstand above-normal concentrations of chloride in their foliage. Salt damage due to irrigation is more common in agriculture where irrigation is practiced in areas where evaporation is high, bringing salts to the surface—for example, in inland western North America and Australia.

ANIMAL DAMAGE

Ever since humans began growing trees in plantations, animal damage has been recognized as a problem to forest regeneration. Seeds, seedlings, and older trees are subject to damage by animals (Black 1992). Feeding by birds and mammals is primarily responsible for seed destruction, cone severing, browsing, clipping, bud injury, seed pulling, tree cutting, and bark damage. The most common animal damage problems in North America are shown in Table 9.4 in relation to stage of forest development. All stages of forest development have animal damage problems. Animal damage management strategies are shown in Table 9.4 and usually

TABLE **9.4** | **Animal Damage Problems in North America and Control Measures**

Problem animals	Stage of forest development			Control measures	
	Seed to plantation establishment	Established plantation (seedling to sapling)	Poles to mature timber	Silviculture[a]	Direct[b]
Seed eaters	X			None	
Voles	X			X	
Deer and elk	X	X		X	X
Black bears			X		X
Porcupines	X	X	X	X	X
Pocket gophers	X	X	X	X	X
Mountain beavers	X	X	X	X	X
Tree squirrels		X	X	X	
Snowshoe hares	X	X		X	X
Livestock	X	X		Herd management	

[a]Silviculutral control is management of habitat conditions with silviculutral practices that reduce the carrying capacity of the habitat for problem animals species.
[b]Direct control is the protection with nonsilviclcultural techniques of control, such as toxic baits, trapping, foliar repellents, plastic-mesh tubing, drift fences, etc.
Source: Adapted from Black (1992).

involve either silvicultural treatments to manipulate habitat conditions or direct control such as mesh tubing, trapping, or toxic baits (Black 1994). The leading problems in North America are caused by seed eaters, voles, deer, elk, tree squirrels, porcupines, black bears, pocket gophers, mountain beavers, snowshoe hares, and livestock (Black 1992). Flooding by beaver dams also causes tree mortality. Interestingly, this has become a major problem in areas in the southeastern United States after attempts to reintroduce beavers.

Similar problems occur throughout the world. For example, in the southern hemisphere radiata pine seedlings are heavily browsed by rabbits and kangaroos in Australia and by exotic deer in New Zealand. White cockatoos in Australia also damage cones and treetops in Australia and are particularly difficult to control.

Seed Eaters
Many bird and mammal species consume tree seeds (Black 1992). Examples of birds include nuthatches, sparrows, and grosbeaks; common seed-eating mammals are shrews, deer mice, chipmunks, and ground squirrels. Deer mice are particularly difficult to manage. Seed eaters are more of a problem in the early stages of regeneration in areas where natural regeneration is relied upon rather than planting of seedlings. No control for seed eaters is currently practiced.

Voles

Species of *Microtus* can cause a large amount of damage to young plantations (Black 1992). Natural *Microtus* habitats include grassy openings in the forest, herbaceous stream margins, and understocked brushfields. Voles feed on grass roots and forbs during the growing season, but move to rhizomes, bark, and cached food during the winter. Debarking of young seedlings can result in considerable mortality. In Douglas-fir plantations established on abandoned pastureland or on land treated with municipal biosolids (sewage sludge) seedling mortality can be extremely high. High grass cover in treated areas allows vole populations to expand, and herbicide treatment of the grass is usually needed for successful vole management.

Deer and Elk

The most widespread damage to plantations in North America is caused by deer and elk feeding on the leader of seedlings, particularly the new growth in spring (Black 1992). Repeated browsing converts seedlings into brushy forms and suppresses height growth. Deer and elk browsing results in ragged breaks in terminal shoots in contrast to the clean-angled cuts produced by hare and mountain beaver. In addition, deer and elk may cause bark damage to larger trees by rubbing. The larger droppings of elk in comparison to deer helps identify them as the cause of damage. Elk also may cause trampling because they are larger than deer and travel in herds. Planting of large, robust seedlings able to withstand moderate browsing is recommended. Drift fences to deflect elk from plantations, natural-product big-game repellents, and plastic-mesh tubing are commonly used. Some tree species are particularly vulnerable to deer browse, such as western redcedar, and it is very difficult to establish this species in the western United States because of intense browsing. Douglas-fir seedlings also are susceptible to browsing but not to the same extent as cedar.

Tree Squirrels

Tree squirrels feed on foliage buds, elongating shoots, cones, and phloem and xylem in the spring, stripping short strips of bark (2.5–7.6 cm in length). Their presence is indicated by discarded tree parts littering the ground under the crown (Black 1992). Debarking by red squirrels on the branches and upper boles of young lodgepole pine trees in south-central British Columbia has resulted in deformed, spike-topped trees. Reduction in stand density by precommercial thinning reduces tree squirrel damage.

Porcupines

Porcupines are found throughout North American forests (Black 1992). They feed on the bark and sapwood of all age classes of conifers from the ground upward, but rarely in old-growth forests (British Columbia Ministry of Forestry and Lands 1988). Repeated feeding in young trees results in crown death and poor form and seedlings can be killed by basal girdling. Horizontal and oblique teeth marks in the sapwood (0.3 cm wide) are characteristic of porcupine feeding, along with the

Figure 9.12
Girdling caused by a black bear on a Douglas-fir tree in western Washington. Note copious resin flow where bark has been stripped.

presence of quills and oblong droppings up to 2.5 cm in length. Removal of large, overstory roost-trees and control of ground vegetation reduces porcupine populations. New forestry practices may encourage porcupines by leaving nest sites in trees and in coarse woody debris. Hunting also is practiced to manage porcupines.

Black Bears

Black bears cause considerable damage in stands that are 15 to 50 years of age regardless of forest type (Black 1992). Douglas-fir, western larch, Engelmann spruce, and lodgepole pine are preferred species. In spring bears damage trees by striping bark from trunks of vigorously growing trees (Figure 9.12). Sugars and carbohydrates are utilized by bears after emergence from hibernation. Initial stripping may only partially girdle the tree, but repeated visits may completely kill the tree. Teeth marks are evident in the exposed sapwood. Silvicultural practices to manage bears are currently not used, but thinning and fertilization may increase bear damage. Hunting, snaring, and supplemental feeding using pelletized food are commonly used where bear damage is high in plantations. Supplemental feeding appears to reduce bear damage, but is controversial. Porcupine damage can sometimes be confused with bear damage. However, bears strip bark at the base of trees; porcupines usually cause damage higher in the bole.

Pocket Gophers

Pocket gophers are a major problem in the drier mixed-conifer types east of the Cascade Mountains, causing root pruning, basal stem injury, basal stem debarking, and stem clipping in plantations (Black 1992). Larger trees also may sustain injury, particularly under snow. Vegetation management and toxic baits have been used to control pocket gophers.

Mountain Beavers

Mountain beavers cause considerable damage in forests west of the Cascade crest, clipping the main stem and lateral shoots of seedlings and debarking small Douglas-fir trees up to 40 cm in diameter (Black 1992). They construct elaborate networks of tunnels and also can damage roots. Mountain beaver root wounds in young coastal western hemlock stands can allow infection by airborne spores of *Heterobasidion annosum,* the cause of Annosus root and butt rot. Site preparation typically reduces mountain beaver populations, but trapping and protection of seedlings with plastic-mesh tubing also are used.

Snowshoe Hares

Snowshoe hares occur widely and when populations are high they cause considerable seedling mortality by clipping (Black 1992). Damage resembles that of mountain beavers, but snowshoe hares seldom clip stems greater than 0.6 cm in diameter, whereas mountain beavers can clip stems up to 2.5 cm in diameter.

Livestock

Livestock includes both sheep and cattle. Grazing on public lands is common in the western Unites States. Cattle cause damage by trampling, browsing, and rubbing against sapling-sized trees. They also can cause soil compaction, degrading the site, particularly in sensitive riparian zones. On the other hand, both cattle and sheep can be used to control competing vegetation. They can be managed by fences and the placement of salting areas.

REFERENCES

Aber, J. D. A. Magill, S. G. McNulty, R. D. Boone, K. J. Nadelhoffer, M. Downs, and R. Hallett. 1995. Forest biogeochemistry and primary production altered by nitrogen saturation. *Water, Air and Soil Pollution* 85:1665–1670.

Aber, J., W. McDowell, K. Nadelhoffer, A. Magill, G. Bernston, M. Kamakea, S. McNulty, W. Currie, L. Rustad, and I. Fernandez. 1998. Nitrogen saturation in temperate forest ecosystems—hypotheses revisited. *BioScience* 48:921–934.

Agee, J. K. 1993. *Fire ecology of Pacific Northwest forests.* Island Press, Washington, D.C.

Black, H. C. 1992. *Silvicultural approaches to animal damage management in Pacific Northwest forests.* USDA Forest Service, Pacific Northwest Research Station, General Technical Report PNW-GTR-287, Portland, Oreg.

Black, H. C. (ed.). 1994. *Animal damage management handbook.* USDA Forest Service, Pacific Northwest Station, General Technical Report PNW-GTR-332, Portland, Oreg.

British Columbia Ministry of Forests and Lands. 1988. Porcupines as forest pests. Pestopics 25. Pest Management, Protection Branch, British Columbia Ministry of Forests and Lands, Victoria, B.C., Canada

Carter, R. 1992. Diagnosis and interpretation of forest stand nutrient status, pp. 90–97. In H. N. Chappell, G. F. Weetman, and R. E. Miller (eds.). *Forest fertilization: sustaining and improving nutrition and growth of western forests.* Contribution 73, Institute of Forest Resources, College of Forest Resources, University of Washington, Seattle.

Chappell, H. N., G. F. Weetman, and R. E. Miller (eds.). 1992. *Forest fertilization: sustaining and improving nutrition and growth of western forests.* Contribution 73, Institute of Forest Resources, College of Forest Resources, University of Washington, Seattle.

Edmonds, R. L. and F. A. Basabe. 1989. Ozone concentrations above a Douglas-fir forest canopy in western Washington, USA. *Atmospheric Environment.* 23:625–629.

Foster, R. E., and G. W. Wallis. 1974. *Common tree diseases of British Columbia.* Canadian Forestry Service Publication 1245, Ottawa, Canada

Gundersen, P., B. A. Emmett, O. J. Kjonaas, C. J. Koopmans, and A. Tietema. 1997. Impact of nitrogen deposition on nitrogen cycling in forests: a synthesis of NITREX data. *Forest Ecology and Management* 101:37–55.

Helgerson, O. T., M. Newton, D. deCalesta, T. Schowalter, and E. Hansen. 1992. Protecting young regeneration, pp. 384–420. In S. D. Hobbs, S. D. Tesch, P. W. Owsten, R. E. Stewart, J. C. Tappeiner II, and G. E. Wells (eds.). *Reforestation practices in south western Oregon and northern California.* Oregon State University, Corvallis, Oreg.

Houghton, J. T. 1997. *Global warming: the complete briefing.* 2nd ed. Cambridge University Press, New York.

Huettl, R. F., and H. W. Zoettl. 1993. Liming as a mitigation tool in Germany's declining forests—reviewing results from former and recent trials. *Forest Ecology and Management* 61:325–338.

Johnson, D. W. and S. E. Lindberg (eds.). 1992. *Atmospheric deposition and forest nutrient cycling: a synthesis of the integrated forest study.* Ecological Studies 91. Springer-Verlag, New York.

Larcher, W. 1995. *Physiological plant ecology: ecophysiology and stress physiology of functional groups.* Springer-Verlag, New York.

Legge, A. H., and S. V. Krupa (eds.). 1986. *Air pollutants and their effects on terrestrial ecosystems.* John Wiley and Sons, New York.

Lowry, W. P. 1969. *Weather and life.* Academic Press, New York.

Miller, P., and J. R. McBride (eds.). 1999. *Oxidant air pollutant impacts in the montane forests of southern Calfornia: a case study of the San Bernadino Mountains.* Ecological Studies 134. Springer, New York.

NAPAP (National Atmospheric Precipitation Assessment Program). 1993. 1992 report to Congress, Washington, D.C.

NAPAP (National Atmospheric Assessment Program). 1998. Biennial report to Congress: an integrated assessment. NOAA, Silver Spring, Md.

Olson, R. K., D. Binkley, and M. Bohm (eds.). 1992. *The response of western forests to air pollution.* Ecological Studies 97. Springer-Verlag, New York.

Scharpf, R. F. 1993. *Diseases of Pacific Coast conifers.* USDA Forest Service, Agricultural Handbook 521, Washington, D.C.

Schulze, E. D., O. L. Lange, and R. Oren. 1989. *Forest decline and air pollution: a study of spruce (Picea abies) in acid soils.* Ecological Studies 77. Springer-Verlag, Heidelberg.

Sinclair, W. A., H. H. Lyon, and W. T. Johnson. 1987. *Diseases of trees and shrub.* Cornell University Press, Ithaca, New York.

Skelly, J. M., D. D. Davis, W. Merrill, A. E. Cameron, H. D. Brown, D. B. Drummond, and L. S. Dochinger (eds.). 1987. *Diagnosing injury to eastern forest trees: a manual for identifying damage by air pollution, pathogens, insects and abiotic stresses.* National Acid Precipitation Assessment Program. Vegetation Survey Cooperative. USDA Forest Service, Forest Pest Management and Pennsylvania State University, University Park, Pa.

Tainter, F. H., and F. A. Baker. 1996. *Principles of forest pathology.* John Wiley and Sons, New York.

Treshow, M., and F. K. Anderson. 1989. *Plant stress from air pollution,* John Wiley, New York.

Vitousek, P. M., J. Aber, R. W. Howarth, G. E. Likens, P. A. Matson, D. W. Schindler, W. H. Schlesinger, and G. D. Tilman. 1997. *Human alteration of the global nitrogen cycle: causes and consequences.* Issues in Ecology 1. Ecological Society of America, Washington, D.C.

Walker, R. B., and S. P. Gessel. 1990. Mineral deficiency symptoms in Pacific northwest conifers. *Western Journal of Applied Forestry* 5(3):96–98.

Whytemare, A. B., R. L. Edmonds, J. D. Aber, and K. Latjtha. 1997. Influence of excess nitrogen deposition on a white spruce (*Picea glauca*) ecosystem in southeastern Alaska. *Biogeochemistry* 38:173–187.

10

DISEASE-CAUSING ORGANISMS

INTRODUCTION

Tree diseases are caused by a wide variety of organisms including fungi, bacteria, viruses, phytoplasmas, parasitic plants (especially true and dwarf mistletoes), and small animals such as nematodes and protozoans (Agrios 1997). Because of the woody nature of trees, fungi and dwarf mistletoes tend to be the most important causes of forest diseases (Manion 1991, Tainter and Baker 1996). Bacteria, viruses, phytoplasmas, and nematodes are important causes of diseases in soft-tissued agricultural plants along with the fungi. Flagellate protozoans also have been implicated in plant disease (e.g., phloem necrosis of coffee trees) but are not important causes of forest diseases (Agrios 1997). Protozoans also attack insects (see Chapter 19). In this chapter we consider the features of disease-causing organisms; the classification, life cycles, and diseases caused by fungi; and the spread and ecology of fungi.

FEATURES OF DISEASE-CAUSING ORGANISMS

Living organisms now are classified into three domains or superkingdoms: Eukarya (eukaryotic organisms), Bacteria (prokaryotic true bacteria), and Archae (prokaryotic Archaebacteria—halophiles, methanogens, and extreme thermophiles). The Eukarya include the following kingdoms: Plantae, Animalia, Fungi, Protozoa, and Stramenopila or Chromista (fungal and algal groups with **cellulose** in cell walls). Reproduction, morphology, dispersal, and the type of diseases caused by fungi, parasitic plants, bacteria, viruses, phytoplasmas, and nematodes are discussed below. Morphological and reproductive features for each group are shown in Figure 10.1.

Fungi

It has been estimated that more than 1 million species of fungi exist on earth, but only about 100,000 of these species are known (Alexopolous et al. 1996). About 10% of the known species cause plant diseases and a smaller number cause forest diseases.

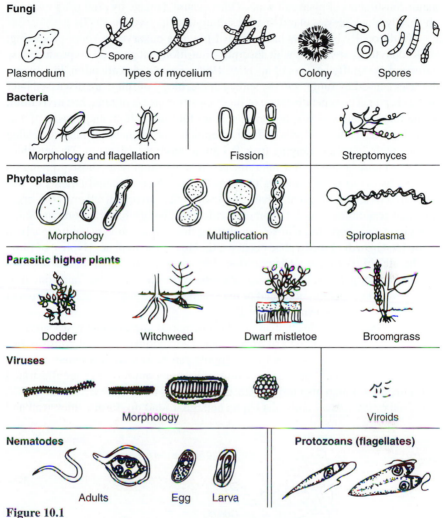

Figure 10.1

Morphological and reproductive features of fungi, bacteria, phytoplasmas, parasitic higher plants, viruses, nematodes, and protozoans.

Some are obligate parasites (i.e., they can grow only on living hosts), whereas others can live on dead organic matter as well as living hosts and are either facultative parasites or facultative saprophytes. The vast majority of fungi, however, are obligate saprophytes or decomposers. Many are intimately associated with insects (see Chapters 19 and 21). Some fungi form beneficial symbiotic associations with plant roots known as mycorrhizae, which assist plants in nutrient and water uptake and protect them against disease organisms (see Chapter 11). Most fungi cause disease by enzymatically breaking down the cellulose and/or lignin in plants, which are the major constituents of plant cell walls. Others cause damage by producing toxins or physically blocking vascular tissues, resulting in wilts (Agrios 1997).

Traditionally biologists have defined *fungi* as eukaryotic, spore-producing, achlorophyllous organisms with absorptive nutrition that generally reproduce sexually and **asexually** (Alexopolous et al. 1996). Fungal classification, life cycles, diseases caused by fungi, and the spread and ecology of fungi are discussed later in the chapter. The vegetative state of fungi mostly consists of branched **mycelium.** Some fungi such as yeasts, however, seldom form mycelium and consist of single cells or collections of cells. Individual branches or filaments of fungi are called *hyphae* with diameters ranging from 0.5 μm to more than 100 μm. The length of mycelium may be only a few micrometers in some fungi (e.g., those growing on rock surfaces) to several meters in fungi that produce mycelial strands or rhizomorphs (Alexopolous et al. 1996). A rhizomorph is an organized mycelial structure that resembles a root. Commonly rhizomorphs are black or brown. Some fungi have rhizomycelium, or strands that are dissimilar and vary continuously in diameter. Growth of the hyphae occurs at the tips of the hyphae. Hyphae of fungi can be made up of many cells partitioned by cross walls (septa) containing one or more nuclei per cell, or they may be **coenocytic,** have no septa, and consist of one continuous, tubular, branched or unbranched, multinucleate cell. Features of septate hyphae are shown in Figure 10.2.

Fungi produce spores that are used for dispersal and in some cases survival. Some spores can have flagella like bacteria and are motile in wet soil. Most spores, however, are airborne (Agrios 1997). Spores can be sexual or asexual, and can be produced externally or in sacs. Airborne asexual conidiospores are illustrated in Figure 10.2. Fungi also form **chlamydospores** or thick-walled asexual spores. Terminal or intercalary cells of a hypha enlarge, round up, form a thick wall and

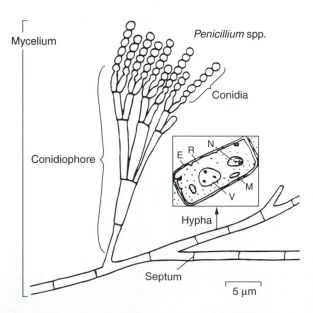

Figure 10.2

Characteristics of a septate fungal hyphae and asexual conidia. N = nucleus, V = vacuole with volutin granules, R = ribosome, E = endoplasmic reticulum, M = mitochondrion.

TABLE **10.1** | **Important Families and Genera of Parasitic Plants Causing Tree Diseases**

Family	Genus	Common name
Cuscutaceae	*Cuscuta*	Dodders
Viscaceae	*Arceuthobium*	Dwarf mistletoes of conifers
	Phoradendron	True mistletoes of broad leaved trees
	Viscum	European true mistletoes

separate. Other fungi produce **arthrospores** or **oidia** where hyphae break into their component cells and behave as spores. Survival structures like sclerotia or microsclerotia also are produced. These are hard resting bodies resistant to unfavorable conditions that may remain dormant for long periods of time and germinate when favorable conditions return (Alexopolous et al. 1996). Fungi are treated in more detail later in this chapter.

Parasitic Plants

Plants that live parasitically on other plants are common and more than 2,500 species are known (Agrios 1997). Like plants rooted in soil, they produce flowers and seeds. They belong to several widely separated botanical families. Only a few of the known parasitic plants, however, cause important plant diseases. The most common families and genera causing tree diseases are shown in Table 10.1; dwarf and true mistletoes are the most important. Some parasitic plants have chlorophyll like the mistletoes, but others have little or no chlorophyll, such as dodders, which depend entirely on their hosts for their existence (Agrios 1997).

A dwarf mistletoe plant of *Arceuthobium tsugense* subsp. *tsugense* occurring on lodgepole pine is shown in Figure 10.3. Dwarf mistletoes are small yellowish to brownish-green or olive-green plants with simple or branched jointed stems, which are generally less than 10 cm long (Hawksworth and Wiens 1996). Leaves are inconspicuous, scalelike, in opposite pairs, and are the same color as the stem. Inside the branches or stems of the host plant is a ramifying system of **haustorial** strands and sinkers located in the phloem and xylem from which the plant obtains water and nutrients. They typically produce swellings (Figure 10.3) and brooming on branches on the host plant (see Figure 14.1) and retard growth. Very severe infections can cause tree death. Short-distance spread is mostly by sticky airborne seeds that are shot out from the plant, but birds are thought to be responsible for long-distance spread (Hawksworth and Wiens 1996). Dwarf mistletoes are discussed in more detail in Chapter 14.

True or leafy mistletoes (Figure 10.4) generally occur on broad-leaved plants throughout the world, particularly in warmer climates. They attack not only forest trees but also occur on apple, cherry, citrus, rubber, cacao, and coffee trees (Agrios 1997). Some conifers such as juniper and cypress also have true mistletoes (Scharpf 1993). The plants are evergreens with well-developed stems and

Figure 10.3

Dwarf mistletoe plant (*Arceuthobium tsugense* subsp. *tsugense*) on lodgepole pine (*Pinus contorta* var. *contorta*). Shoots of female plant with mature fruit. Note branch swelling at the point of infection.

(Source: From Hawksworth and Wiens 1996.)

Figure 10.4

True mistletoe plant, *Phoradendron densum,* on western juniper. True mistletoes are mostly on broad-leaved plants, however.

(Source: From Scharpf 1993.)

leaves typically 1 to 2 cm in diameter. Maximum stem diameter, however, can reach 30 cm and plant height varies from a few centimeters to more than 1 m. Mistletoes plants are sometimes so numerous they can make up almost half the foliage on a tree. Like dwarf mistletoes they produce haustoria and sinkers rather than roots and cause swollen branches and brooming. They are typically spread by birds that eat the seed-containing berries (Tainter and Baker 1996). The plants often are used for Christmas decorations.

Bacteria

Bacteria now are classified in the domain Bacteria and are simple one-celled, prokaryotic organisms with no nuclear membrane. Shape and size of cells; being gram positive or negative; the presence, absence, and position of flagella; substrates they grow on; colony characteristics; and molecular techniques are used to identify bacteria (Agrios 1997). Bacteria generally have three common shapes: cocci or round, rod-shaped, and spiral. However, some bacteria have a mycelial habit, like Actinomycetes; others have pleomorphic shapes or are springlike. Typical bacterial cell diameters are around 0.5 μm; some rod-shaped bacteria can be as long as 10 μm (Agrios 1997). Some cells have flagella that are anchored in the cytoplasm and extend out through the cell membrane. Flagella are used for swimming and may occur in one specific location or all around the cells (peritrichous flagella).

Bacteria have no specialized structures for dispersion. They are dispersed passively in the atmosphere by being blown off plant or soil surfaces, or they can move through soil water. Some bacteria may be spread by insects. Some bacteria form thick-walled specialized survival spores that are particularly resistant to adverse conditions, and some can survive being boiled in water. Relatively few bacterial genera cause plant diseases: *Agrobacterium, Clavibacter, Erwinia, Pseudomonas, Xanthomonas, Streptomyces,* and *Xylella* are the most common. *Erwinia amylovora* is the cause of fire blight of apple and pear (Agrios 1997). Many bacteria are associated with insects and some are used for biological control, such as *Bacillus thuringiensis* (see Chapter 19).

Viruses and Viroids

Viruses are the simplest form of life and live in the cells of other living things, although they also might be considered to be nonliving. Considered alone, a virus is a lifeless particle that cannot reproduce outside a cell. However, inside a cell it becomes an active organism that can multiply hundreds of times using the cell's material to live and reproduce. The number of known viruses is more than 2,000 and about one-fourth cause plant diseases (Agrios 1997); they also cause insect diseases (see Chapter 19). Viruses cause disease not by decomposing cells or killing them with toxins, but by disrupting their metabolism. Characteristic symptoms, serological tests such as ELISA (enzyme-linked immunosorbent assay), and genetic techniques can be used to identify viruses.

Viruses have many shapes; they may be elongate (shaped like rods or threads), spherical (isometric or polyhedrals), or rhabdoviruses (bacilluslike). They vary in size from 0.01 to 0.3 μm (10–300 nm), although some flexible threads can be as long as 2,000 nm. Most viruses can be seen only with the scanning electron microscope and are much smaller than bacteria. Viruses have two basic parts: a nucleic acid core consisting of either RNA and DNA and an outer protective protein coat or capsid. Some viruses have more than one size of nucleic acid and proteins, and some are even made up of additional chemical compounds, such as polyamines, lipids, or specific enzymes. There also are a number of viruslike pathogens, such as viroids (small, naked, circular RNAs), capable of causing diseases (Agrios 1997).

Dispersal of viruses is by (1) parasitic animals, such as insects, mites and nematodes; (2) parasitic plants, such as dodder; (3) parasitic fungi; (4) seeds; (5) pollen; (6) grafting; (7) contact; and (8) vegetative propagation (Agrios 1997). Plant cells have tough cell walls that viruses cannot penetrate and thus wounding is generally involved in their transmission. Most plant viruses enter through wounds made mechanically by sapsucking vectoring insects, such as aphids or leafhoppers, and by deposition into an ovule by an infected pollen grain. Insects penetrate the cell walls while feeding on a plant, enabling viruses to enter. Some viruses move from cell to cell through parenchyma cells in a relatively slow manner (e.g., 1 mm/day), whereas others can move rapidly through the phloem (Agrios 1997).

Symptoms of virus infection are most obvious on leaves, but can occur on other plant parts. Ring spots and **mosaics** are the most common symptoms. Mosaics are characterized by yellow, light green, or white areas intermingled with normal green. Other symptoms include stunting, dwarfing, leaf roll, yellows, streak, pox, tumors, pitting, and stem distortion or flattening. Agricultural crops such as tobacco, tomato, cucumbers, wheat, barley, maize, soybean, cucurbits, crucifers, and legumes are particularly susceptible to virus diseases (Agrios 1997). Fruit trees also are susceptible. Although more than 100 virus species have been noted on forest trees, few are currently known to be economically important (Cooper 1993). Interestingly, many of the viruses occurring on trees also occur on agricultural crops. For example, ash ring spot is caused by tobacco mosaic virus and elm mosaic is caused by cherry leaf roll virus. One of the main reasons why viruses appear to cause few forest diseases is that trees are propagated from seed in nature and are not often grafted, an important means of virus spread in fruit trees (Cooper 1993).

Phytoplasmas

About 200 distinct plant diseases have been shown to be caused by phytoplasmas (formerly known as mycoplasma-like organisms [MLOs]), but few are currently known to occur in trees (Hiruki 1988). Elms and ashes are most susceptible. Phytoplasmas are prokaryotic organisms and have no **nucleus** or true cell wall. Cells contain cytoplasm, ribosomes, and strands of nuclear material and are bounded by a single triple-layered membrane. Phytoplasmas can be round to filamentous and range in diameter from 175 to 250 nm. They reproduce by budding and binary fission, have no flagella, produce no spores, and are gram negative. Most are transmitted by sucking insects like leafhoppers and planthoppers.

Elm phloem necrosis or yellows has killed thousands of trees in the central and southern United States (Agrios 1997). Leaves droop and curl, turn bright yellow, then brown, and finally fall. Trees usually are killed quickly. The pathogen is transmitted from diseased to healthy trees by leafhoppers. Discoloration of the inner bark or phloem commonly occurs in the latter stages of the disease.

Ash yellows is spread by leafhoppers and planthoppers. Symptoms include slow apical and radial growth, diminished apical dominance, suppressed root development, precocious flowering and witches'-brooms. Foliar deformities and chlorosis often occur. Another phytoplasma disease is lethal yellowing of coconut palms. The disease has killed thousands of palms in the Caribbean, West Africa, and Florida. All the fronds are killed and finally drop off, leaving a "telephone pole." Planthoppers have been implicated in its spread.

Nematodes

Nematodes are wormlike animals, but are quite distinct from the true worms. There are several thousand species of nematodes and most live in fresh water, salt water, and soil, feeding on microbes, plants, and other animals. Many cause human, animal, and plant diseases as well as diseases of insects (see Chapter 19). Plant parasitic nematodes are small (15–35 µm in diameter and 30 µm to 4 mm long) (Jenkins and Taylor 1967, Agrios 1997). They are generally not visible to the naked eye, but are easily observed under a microscope. Sometimes, however, root nematodes may be so numerous that they resemble many little needles projecting out of the root.

Nematodes are more or less transparent, generally eel-shaped, round in cross section, with smooth unsegmented bodies, and without legs or other appendages. Females of some species may becomes swollen at maturity and have pear-shaped or spherical bodies (Tainter and Baker 1996). Characteristics of a parasitic nematode are shown in Figure 10.5. The nematode body has a colorless cuticle which has striations or other markings, and the digestive system is a hollow tube extending from the mouth to the esophagus, intestine, rectum, and anus. Surrounding the mouth are lips, usually six. In contrast to other nematodes, plant-parasitic nematodes have a hollow stylet or spear, which punctures plant cells. Reproduction is through eggs, but many species lack males.

The life cycles of most nematodes are very similar. Eggs first hatch into larvae; there are four larval stages terminated by a molt. The first molt usually occurs in the egg, and in the final molt larvae differentiate into adult males and females. Females can produce fertile eggs by mating, parthenogenetically, or hermaphroditically by producing sperm. Life cycles can be completed within 3 to 4 weeks. Part of the nematode life cycle usually is spent in the top 15 cm of soil, although some may go deeper. Soil temperature, moisture, and aeration affect survival and movement. Under their own power nematodes travel very slowly, probably no more than a meter, but they move more easily in moist soils where there are films of moisture in the soil pores. They also can move considerable distances in irrigation or floodwaters, on equipment and animal feet, in dust storms, and on farm produce and nursery plants. Although most plant-parasitic nematodes are in the soil attacking roots, *Aphelenchoides* spp. (bud and leaf nematodes) are spread by rain splash and overhead watering, and can move along wet stem surfaces. *Bursaphelenchus* (pine wilt nematode) and *Rhadinaphelenchus* (coconut red-ring nematode) are spread by insects.

Plant-pathogenic nematodes are either ectoparasitic (not entering the host feeding on cells near the root surfaces) or endoparasitic (entering the host). They can be migratory or sedentary (i.e., once within a root they do not move about).

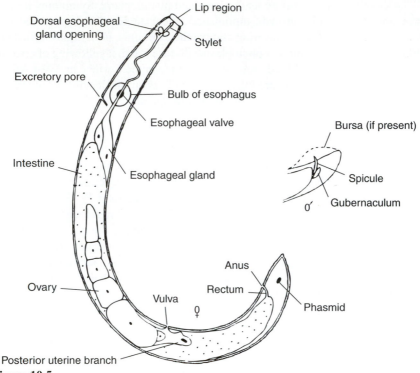

Figure 10.5

Morphological features of a fictitious plant pathogenic nematode.

(Source: From Tainter, F.H. and F.A. Baker, *Principles of Forest Pathology*, ©1996 by John Wiley & Sons, Inc. Reprinted by permission of John Wiley & Sons, Inc.)

Nematodes cause symptoms in roots and aboveground plant parts. Root symptoms include root knots or galls, root **lesions,** excessive branching, and injured root tips. *Meloidogyne* spp. cause root knots and *Pratylenchus* spp. cause lesions.

These root symptoms usually are accompanied by aboveground symptoms including yellowing of foliage and excessive wilting, stunting, and reduced yields. Aboveground nematodes cause galls, necrotic lesions, distortions of leaves and stems, and abnormal flower development. Most of the damage seems to be caused by secretion of saliva rather than mechanical injury inflicted by the nematodes. Some feed rapidly, puncturing the cell wall, inject saliva, suck part of the cell contents, and move on within a few seconds. Others feed slowly and remain at the same puncture site for hours or even days. Plant tissues can be dissolved by nematodes' enzymes. Others cause plants to have hypertrophic growth.

Nematodes do not always act alone and in many cases they form complexes with fungi. For example, Fusarium wilt increases in incidence and severity when plants are infected with root knot or stunt nematodes. Similar effects have been noted with Verticillium wilt, Pythium damping-off, and Rhizoctonia and Pythium root rots. The fungus is not transmitted by the nematodes; they damage roots and lower resistance, allowing fungi to more easily enter roots. Relatively few cases of nematode-bacteria disease complexes are known. Viruses, on the other hand, may be transmitted by nematodes (e.g., tobacco and tomato ring spot viruses).

The most damaging nematode problem in forests is caused by the pine wood nematode (*Bursaphelenchus xylophilus*) (Fielding and Evans 1996). Affected trees wilt and die within 40 to 50 days. It is a particularly serious problem in Japan in black and red pines. It was apparently introduced from North America where injury to native conifers is minor. It is not known to occur in Europe. Beetles in the genus *Monochamus* (Cerambycidae) are largely responsible for spread of the disease—*M. alternatus* in Japan and *M. carolinensis, M. scutellatus,* and *M. titillator* in the United States. Losses are greatest where mean summer temperatures exceed 20°C for 6 to 8 weeks. Emerging beetles carry the nematode in their tracheae. They feed on the bark of young twigs of healthy trees, introducing the nematode in the feeding wound. The nematodes then molt to become adults, mate and reproduce rapidly in resin canals, and feed on parenchyma and epithelial cells. They may also secrete a toxin that kills cells. In stressed, dying, and recently dead trees, emerging beetles may deposit nematodes during oviposition. In this case the nematodes feed on fungi in wood, including blue-stain fungi introduced by bark beetles. As pupae mature in these trees, the dispersal stage of the nematode (dauerlarvae) enters the insect through the spiracles and leaves the tree with the emerging adults. Millions of nematodes can reside in one tree.

CLASSIFICATION, LIFE CYCLES, AND DISEASES CAUSED BY FUNGI

Fungi have always been difficult to classify, and there have been many classification schemes. Originally fungi were considered to be plants and were classified using vegetative structures and sexual or asexual spores. For example, the Oomycetes and Zygomycetes used to be grouped under the Phycomycetes, which was distinguished from other fungi by having **aseptate** hyphae or at least very few septa. It was then discovered that Oomycetes have cellulose in their cell walls, whereas Zygomycetes do not, and because of the different chemical composition of their hyphae and differences in spores they were classified in different subdivisions of the kingdom Fungi (Alexopoulos et al. 1996).

Recent applications of molecular techniques, especially analysis of nuclear-encoded ribosomal RNA genes, has enabled further reclassification of fungi. Use of the polymerase chain reaction (PCR) and restriction fragment length polymorphism (RFLP) techniques now are commonly used for fungal classification. In one scheme recognizing monophyletic groups (i.e., those sharing a close common ancestor), the organisms once classified as fungi are considered in three separate groups: the monophyletic kingdoms Fungi and Stramenopila and four protozoan phyla. The kingdom Fungi then includes true fungi in the phyla Chytriodomycota, Zygomycota, Ascomycota, and Basidiomycota. Only the latter three phyla cause tree diseases and their features, roles, types of disease they cause, and representative genera are shown in Table 10.2. In Table 10.2 the

TABLE 10.2 | Characteristics of Fungi Causing Forest Diseases

Phylum	Features	Class, order, and representative genera[a]	Type of disease or role
Kingdom Fungi (True Fungi)			
Zygomycota	Coenocytic mycelium containing glucans and chitin. Sexual reproduction by zygospores (thick-walled resting spores), conidia, arthrospores, chlamydospores. Absence of flagellate cells. Airborne or animal-dispersed spores. Saprophytes; dung fungi; bread molds; animal, plant, and fungal pathogens; endomycorrhizal fungi; storage molds of seedlings.	Zygomycetes (the bread molds) Mucorales (*Rhizopus*) Glomales (*Glomus, Gigaspora*)	Storage molds Endomycorrhizal vesicular arbuscular mycorrhizae (VAM) fungi
Ascomycota (Ascomycetes, the sac fungi)	Unicellular (yeasts) or mycelia with septa. Mannans and glucans in cell walls. Dikaryon stage in life cycle. Sexual reproduction by ascospores. Asexual reproduction by conidia, budding yeasts), sclerotia and chlamydospores. Airborne spores. Saprophytes, foliar pathogens, cause cankers, wilts, diebacks, fine root diseases, stain fungi, galls fungi, nematode-trapping fungi, endophytes, ecto- and ericoid mycorrhizal fungi, lichen fungi.	Archiascomycetes (*Taphrina*) Saccharomycetes (yeasts) Filamentous Ascomycetes Erisyphales (*Erisphye*) Pyrenomycetes (with perithecia) Hyocreales (*Nectria*) Microascales (*Ceratocystis*) Phyllachorales Ophiostomatales (*Ophiostoma*) Diaporthales (*Gnomonia, Cryphonectria*) Xylariales (*Hypoxylon, Xylaria*) Loculoascomycetes (with ascostroma) Dothideales Capnodiales (*Capnodium*) Pleosporales (*Venturia*) Discomycetes (with apothecia) Rhytsimales (*Lophodermium, Rhabdocline, Rhytisma*) Helotiales (*Monilinia,Diplocarpon*)	Leaf curls and blisters Powdery mildews Twig and stem cankers Oak wilt, sap and blue stains Dutch elm disease, stains Anthracnose, leaf spots Chestnut blight Cankers, wood decay Sooty molds Apple scab Conifer and hardwood foliage diseases Cherry brown rot Rose black spot

Phylum	Features	Class, order, and representative genera[a]	Type of disease or role
Kingdom Fungi (True Fungi) (continued)			
	Imperfect fungi have septate mycelia. Sexual reproduction rare or unknown. Asexual spores (conidia) formed on conidiophores singly, grouped in synnemata or sporodochia or in structures (acervuli and pycnidia). Most are Ascomycetes.	Deuteromycetes (Imperfect or asexual fungi) *Aspergillus* *Botrytis* *Fusarium* *Verticillium* *Dothistroma* *Rhizoctonia* (Basidiomycete)	Molds Gray mold Root diseases Root diseases Foliage pathogen Root diseases
Basidiomycota (Basidiomycetes, the club or mushroom fungi)	Mycelia with septate hyphae and clamp connections. Mannans and glucans main cell wall components. Rhizomorphs, sexual reproduction by basidiospores. Asexual reproduction by conidia, fragmentation of hyphae into arthrospores and oidia, and urediniospores. Saprophytes; decay fungi; root diseases; rusts; smuts; and ecto, arbutiod, monotropoid, and orchid mycorrhizal fungi; puffballs; jelly fungi; mushrooms; conks; bird's nest fungi. Airborne spores.	Ustilaginales (*Ustilago*) Uredinales (*Cronartium, Gymnosporangium, Melampsora*) Exobasidiales (*Exobasidium*) Ceratobasidiales (*Thanatephorus*) Agaricales (*Armillaria, Agaricus, Pleurotus, Pholiota*) Aphyllophorales (*Heterobasidion, Phellinus, Ganoderma, Inonotus*)	Smut fungi Rusts Galls Root diseases (*Rhizoctonia*) Root diseases, wood decays, ectomycorrhizal fungi Root and butt decays, stem decays
Kingdom Stramenopila (Chromista)			
Oomycota (water molds, downy mildews)	Aseptate mycelium except at base of reproductive organs. Cell walls comprised mainly of glucans and cellulose. Sexual reproduction by Oospores. Asexual reproduction by motile, biflagellate zoospores. Occur in fresh and salt water and soil. Most terrestrial Oomycetes are highly specialized parasites of vascular plants	Oomycetes (*Pythium, Phytophthora, Peronospora*)	Damping-off of seedlings, seed decay Fine root diseases Molds

[a]Representative genera are shown in parentheses.

Deuteromycetes (Fungi Imperfecti) are included in the Ascomycota (Agrios 1997), although some people still consider them as a separate phyla (Deuteromycota), since a few Basidiomycota are included. The Stramenopila includes three phyla of algal groups, based on **flagellum** hair structure. Only the Oomycota cause forest diseases, and the features of this phylum also are shown in Table 10.2. The remaining four phyla are considered as protists or **slime molds.** None cause forest diseases and are not considered further here. In an alternative classification scheme, the fungi are classified in the kingdoms Chromista (Oomycetes) and Mycota (Zygomycetes and Dikaryomycetes) (Paul and Clark 1996). Beyond phylum, fungi are classified by class, order, family, genus, species, and **race** (plant pathogenic fungi with similar **virulence** on particular host plants).

The life cycles and features of the phyla that cause forest diseases (i.e., the Zygomycota, Ascomycota and Deuteromycetes, Basidiomycota, and Oomycota) are discussed below.Although the life cycles of the fungi of the different groups vary greatly, the great majority go through a series of steps that are quite similar. Some pathogens have yearly life cycles, such as many of the leaf pathogens, which typically overwinter in fallen leaves on the soil surface. Others produce spores almost continuously throughout the year, such as the perennial conk fungi that cause decay in trees, although it takes some time to develop the conks on tree trunks after initial infection. Some fungi have long cycles—for example, 2 to 3 years in *Cronartium ribicola*, the cause of white pine blister rust (Tainter and Baker 1996).

The mycelium of many fungi is spent in the haploid (1*N*) condition, and the diploid (2*N*) occurs only in the zygote (Alexopolous et al. 1996). In some fungi like the Basidiomycetes, vegetative mycelia may contain two haploid nuclei that remain separate in the cell-forming **dikaryotic** cells that behave like diploids. The true diploid condition, however, occurs only in the basidium before production of haploid **basidiospores.** There are two basic mating systems in fungi: homothallic and heterothallic (Alexopolous et al. 1996). Fungi are homothallic when male **gametes** fertilize female ones on the same mycelium. If male gametes can fertilize female gametes only on another compatible mycelium, the fungus is heterothallic.

Zygomycota

The life cycle of a typical Zygomycete is shown in Figure 10.6. Asexual **sporangiospores** are produced internally in a sporangium produced on aseptate hyphae. In the sexual stage compatible hyphae fuse to form **gametangia** and zygotes, usually at the end of the growing season or after the food supply is exhausted (Agrios 1997). After meiosis and mitosis zygospores form that typically overwinter in decaying tissues. Zygospores then germinate and produce a sporangium, sporangiospores, and eventually aseptate mycelium. Spores are usually airborne.

Diseases Caused by Zygomycetes   The Zygomycetes are strictly terrestrial fungi with airborne spores (Alexopolous et al. 1996). Most are saprophytes or weak pathogens and none cause forest diseases. Two genera cause disease in other plants or fruits: *Choanephora,* which causes soft rots of squash and similar plants,

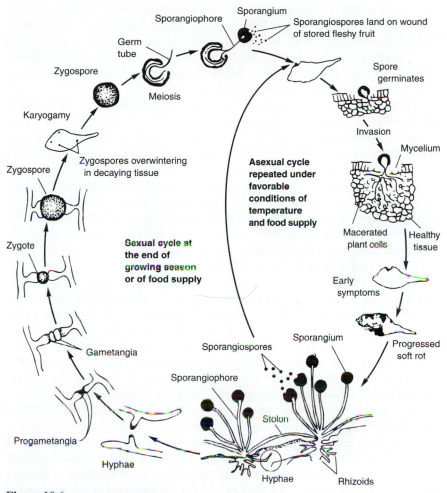

Figure 10.6
Life cycle of a typical Zygomycete.

and *Rhizopus*, the common bread mold fungus, which also causes soft rots of fleshy fruits, vegetables, flowers, bulbs and corms, and storage molds of tree seedlings. *Mucor* also causes bread mold. Some are endomycorrhizal fungi in the genera *Glomus* and *Gigaspora*.

Ascomycota and Deuteromycetes (Imperfect Fungi)

If sexual spores are produced in a sac or **ascus** (usually eight per sac), the spores are called **ascospores** and the fungi are classified in the phylum Ascomycota (Alexopolous et al. 1996). The Ascomycota are classified into Archiascomycetes, Saccharomycetes (yeasts), and filamentous ascomycetes. The Archiascomycetes have no **ascocarp** or **fruiting body** and the ascospores are forcibly discharged from the ascus; no species cause forest diseases. Yeasts can be single cells or have mycelial form and reproduce asexually by budding. Generally, they do not cause forest diseases, but they do occur in slime fluxes in wounds (Tainter and Baker 1996). They also produce alcohol by fermentation, and this is important in attracting insects such as bark beetles to trees.

Filamentous Ascomycetes, on the other hand, cause many forest diseases and are classified with respect to the type of ascocarp they form: **apothecia** (open, saucerlike), **perithecia** (closed, flask, or elongated with an opening at the top), **cleistothecia** (closed), ascostroma or pseudothecia (no distinct structure) and no ascocarp (Manion 1991). These shapes are illustrated in Figure 10.7. Hysterothecia are variations of perithecia and are long and elongated. Pyrenomycetes, including the genera *Cryphonectria, Nectria, Ophiostoma,* and *Ceratocystis*, produce perithecia. Apothecia can be large and fleshy and occur on the forest floor or soil, sometimes resembling orange peels, or smaller and embedded in leaves as they are with foliar pathogens. **Discomycetes** produce apothecia and include the foliar pathogens in the genera *Rhytisma, Rhabdocline,* and *Elytroderma* (Tainter and Baker 1996). This group of fungi also form ectomycorrhiza (including the truffle fungi), ericoid and monotropoid mycorrhizae, lichens, **endophytes,** and burned substrate fungi like the root disease fungus *Rhizina undulata* (a pyrophilous Discomycete). Cleistothecia are produced by the obligate powdery mildew parasites like *Erisyphe graminis*. Fungi in the Loculoascomycetes include *Venturia inequalis* (the cause of apple scab) and many lichen fungi produce ascostroma (Agrios 1997).

The life cycle of a typical filamentous Ascomycete is shown in Figure 10.8. Haploid sexual ascospores are produced in asci and usually are airborne, but can also be spread by insects. After spores germinate, an ascocarp will form if hyphae are compatible. Nuclei will fuse, meoisis and mitosis will occur, and new ascospores will form. Most Ascomycetes also have an asexual or **conidial** cycle, which is also illustrated in Figure 10.8. Like ascocarps the fruiting structures of the Imperfect Fungi are used for classification (Manion 1991). They are **synnemata** or coremia (a group of upright hyphae that produce spores), **sporodochia** (cushion-shaped), **pycnidia** (spherical) or acervuli (open saucerlike) and some are illustrated in Figure 10.7. Some fungi produce more than one type of conidia. In many cases the sexual and asexual stages have not been connected and thus some fungi have only Deuteromycete names or **anamorphs** (Alexopolous et al. 1996). Some have both anamorph and teleomorph (sexual stage) names. For example, the teleomorph of the Dutch elm disease fungus is *Ophiostoma ulmi,* while there are two anamorphs (*Pesotum ulmi* and *Sporothrix* spp). The International Code of Botanical Nomenclature allows for a system of separate names for the teleomorph and anamorph. Not all Imperfect Fungi are Ascomycetes; some are also Basidiomycetes (e.g., *Spiniger* spp. produces conidia and is the anamorph for *Heterobasidion* annosum and *Rhizoctonia solani* is the anamorph for *Thanatephorus* spp.

Diseases Caused by Ascomycetes and Imperfect Fungi   The most common forest diseases caused by Ascomycetes or Imperfect Fungi are shown in Table 10.3.

TABLE **10.3**	**Some Common Forest Diseases Caused by Species of Ascomycetes or Deuteromycetes (Fungi Imperfecti)**

Fungus species	Disease
Foliar Diseases	
Elytroderma deformans	Leaf spot and witches'-broom of pines
Rhabdocline pseudotsugae	Needle cast of Douglas-fir
Rhytisma spp.	Tar spot of maples and willows
Scirrhia acicola (Septoria)	Brown spot needle disease of pine
Scirrhia pini (Dothistroma)	Needle blight of pine
Lophodermium pinastri	Lophodermium needle blight
Phaeocryptopus gaeumannii	Swiss needle cast
Vascular Wilt Diseases	
Ophiostoma ulmi	Dutch elm disease
Ceratocystis fagacearum	Oak wilt
Verticillium albro-atrum	Verticillium wilt
Fusarium oxysporum f. sp. perniciosum	Mimosa wilt
Root Diseases	
Leptographium wageneri	Black-stain root disease
Leptographium procerum	Procerum root disease
Rhizina undulata	Rhizina root rot, teapot fungus
Fusarium oxysporum	Nursery root disease
Canker Diseases	
Hypoxylon mammatum	Hypoxylon canker
Ceratocystis fimbriata	Ceratocystis canker
Fusarium subglutinans	Pitch canker
Nectria coccinea var. faginata	Beech bark disease
Nectria galligena	Nectria canker
Eutypella parasitica	Eutypella canker
Cryphonectria parasitica	Chestnut blight
Gremmeniella abietina	
(Scleroderris lagerbergii)	Scleroderris canker on conifers
Dothichiza populina	Dothichiza canker of poplar
Anthracnose Diseases	
Gnomonia platani	Sycamore anthracnose
Discula destructiva	Dogwood anthracnose
Molds	
Botrytis cinerea	Gray mold disease of seedlings
Stains	
Ceratocystis coerulescens	Blue stain of pines
C. virescens	
C. piliferum	

Ascomycete (Sexual) Fruiting Structures (Ascocarps)

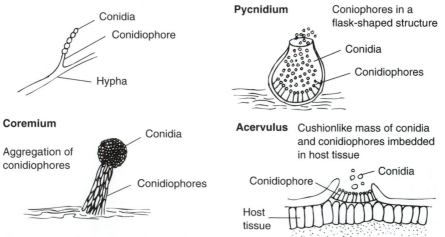

No ascomata

Ascospores
Asci
Host leaf

Apothecium

(cuplike ascomata)

Ascospores
Asci

Pseudothecium

Host woody tissue
and fungus stroma

Ascospore

Locule cavities

Asci

Perithecium   Ascospores

Ascus

(flasklike ascomata)

8 ascospores

Asci

Cleistothecium

(spherical
ascomata)

Ascospores

Asci

Perithecia embedded in fungus
stroma on host tissue

Fungi Imperfecti (Asexual) Fruiting Structures

Conidia
Conidiophore
Hypha

Pycnidium   Coniophores in a
flask-shaped structure

Conidia
Conidiophores

Coremium

Aggregation of
conidiophores

Conidia

Conidiophores

Acervulus   Cushionlike mass of conidia
and conidiophores imbedded
in host tissue

Conidiophore   Conidia

Host
tissue

Figure 10.7

Shapes of ascocarps and Fungi Imperfecti fruiting structures.

(Source: *Tree Disease Concepts*, 2/e by Manion, Paul D., ©1991. Reprinted by permission of Prentice-Hall, Inc., Upper Saddle River, NJ.)

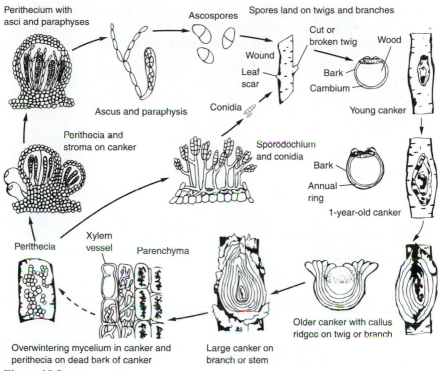

Figure 10.8

Life cycle of a typical filamentous Ascomycete, *Nectria galligena.*

Most Imperfect Fungi represent the asexual or conidial stage of Ascomycetes so they are considered together here. Ascomycetes cause many important forest diseases, including leaf diseases in broad-leaved trees, conifer needle diseases or needle casts, blights, cankers, anthracnoses, vascular wilts, soft rots, gray mold, and some stem and root rots (Tainter and Baker 1997). They also cause sooty molds, leaf curls, powdery mildews, ergots, scabs, and postharvest decays of fruits and vegetables (Agrios 1997). Furthermore, they are common causes of seedling storage molds and cause molds on green lumber, particularly in the sapwood (e.g., *Aspergillus* and *Penicillium* spp.) Some Ascomycete fungi are mycorrhizal, especially with forest trees—for example, *Elaphomyces* spp.

Basidiomycota

Basidiomycetes have many different forms and consist of the following traditional groups. A more advanced classification is given in Alexopolous et al. (1996).

Hymenomycetes

Agaricales (mushrooms and boletes)

Aphyllophorales: Polyporaceae (pore fungi—the major wood decay organisms), Clavariaceae (coral fungi), Cantherellaceae (chanterelles), Hydnaceae (tooth fungi), Meruliaceae and Thelephoraceae.

Dacrymycetales (some jelly fungi)

Exobasidiales

Teliomycetes

Uredinales (rusts)

Ustilaginales (smuts)

Gasteromycetes

Lycoperdales (puff balls and earth stars)

Nidulariales (bird's nest fungi)

Phallales (stink horns)

Sclerodermatales

Heterobasidiomycetes

Tremellales (some jelly fungi)

Auriculariales

Basidiospores, which are usually airborne, are produced externally on a club or basidium (usually four per basidium). The Hymenomycetes bear their basidiospores on short stalks called *sterigmata* on one-celled, club-shaped basidia on gills, in pores, or on the surface of the hymenium; rust and smut fungi bear basidiospores on segmented basidia (Figure 10.9). Typical life cycles of Hymenomycetes and rusts are shown Figures 10.10 and 10.11, respectively. After spores of Hymenomycetes germinate, dikaryons result if strains are compatible and new fruiting bodies are produced (Alexopolous et al. 1996). Life cycles of rusts are more complicated than that of Hymenomycetes and involve more spore stages and **alternate hosts.** The life cycle of smuts is not illustrated because they do not cause tree diseases.

Diseases Caused by Basidiomycetes   Basidiomycetes cause many of the most important forest diseases as shown in Table 10.4. They are the major causes of wood decay and cause structural root rots (e.g., Armillaria root disease), stem decays, and rusts (Tainter and Baker 1996). They also are the most important group of fungi involved with decay in wooden structures, such as buildings, boats, and decks. A few Basidiomycetes are members of the Fungi Imperfecti, such as *Rhizoctonia solani,* which causes seedling root diseases and *Spiniger* spp., the imperfect form of *H. annosum* (Tainter and Baker 1996). Basidiomycetes form ectomycorrhizal associations with economically important forest trees including those in the families Pinaceae, Betulaceae, Fagaceaea, and Myrtaceae (Paul and Clark 1996). Many mushrooms on the forest floor are mycorrhizal fungi, such as *Amanita, Boletus, Laccaria,* and *Russula* spp. (Alexopolous et al. 1996).

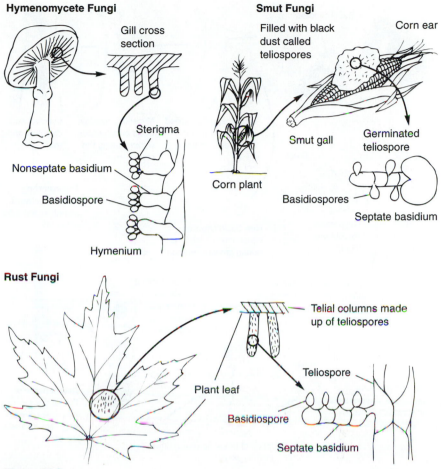

Hymenomycete Fungi

Gill cross section

Sterigma

Nonseptate basidium

Basidiospore

Hymenium

Smut Fungi

Filled with black dust called teliospores

Corn ear

Corn plant

Smut gall

Germinated teliospore

Basidiospores

Septate basidium

Rust Fungi

Telial columns made up of teliospores

Plant leaf

Teliospore

Basidiospore

Septate basidium

Figure 10.9

Structures bearing basidiospores.

(Source: *Tree Disease Concepts*, 2/e by Manion, Paul D., ©1991. Reprinted by permission of Prentice-Hall, Inc., Upper Saddle River, NJ.)

Oomycota

Oomycetes are commonly found in water, wet soils, or on the upper portion of plants and in very wet conditions. The life cycle of a typical Oomycete is shown in Figure 10.12. The aseptate hyphae give rise to **sporangia** that rupture and release swimming biflagellate zoospores in water or wet soils that germinate and encyst. Cysts then germinate if the environment is favorable, and the hyphae invade substrates, particularly fine roots. When conditions are favorable to sexual reproduction, hyphae give rise to **oogonia** and **antheridia.** Antheridia are much smaller than oogonia. After fertilization, a diploid zygote develops forming an oospore. A sporangium develops from the oospore, releasing zoospores. Zoospores are produced in a sac called a *sporangium* and are released when the sac ruptures.

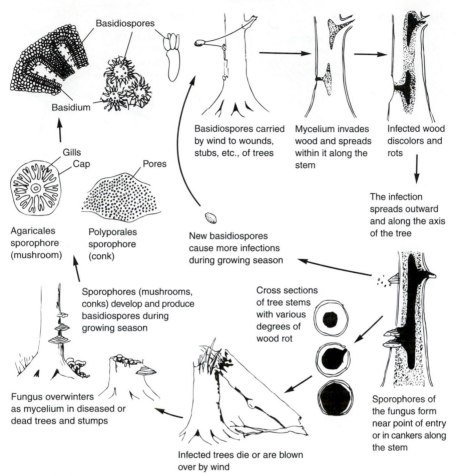

Figure 10.10
Life cycle of a typical hymenial wood-decaying Basidiomycete.

Diseases Caused by Oomycetes Oomycetes cause some very important plant diseases in forest nurseries, forests, and agricultural fields throughout the world (Agrios 1997). They often are called *water molds* because their impacts are greater in damp or wet soils, and they possess swimming zoospores that are particularly suited for dispersal in water. Plant diseases caused by Oomycetes are of two types: type 1—those affecting plant parts present or in contact with the soil, such as fine roots, lower stems, tubers, and seeds; and type 2—those affecting the aboveground plant parts, such as leaves, young stems, and fruits. Fungi in the genera *Pythium* and *Phytophthora* cause the major problems. *Pythium* and *Phytophthora* spp. cause type 1 diseases. *Phytophthora* also causes type 2 diseases, but *Pythium* seems to be confined to the soil.

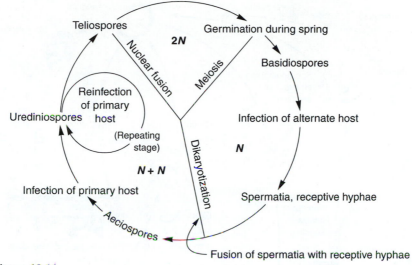

Figure 10.11

Life cycle of a rust Basidiomycete.

(Source: From, Alexopoulos, C.J., C.W. Mims and M. Blackwell, *Introductory Mycology,* ©1996 by John Wiley & Sons, Inc. Reprinted with permission of John Wiley & Sons, Inc.)

T A B L E 10.4	Some Common Tree Diseases Caused by Species of Basidiomycetes
Species	**Disease**
Root and Butt Diseases	
Armillaria spp.	Armillaria root disease
Heterobasidion annosum	Annosus root and butt rot
Phellinus weirii	Laminated root rot
Phaeolus schweinitzii	Schweinitzii butt rot
Stem Decays	
Echinodontium tinctorium	Indian paint fungus
Ganoderma applanatum	Artist's conk
Phellinus pini	Red ring rot
Phellinus igniarius	False tinder conk
Rusts	
Cronartium ribicola	White pine blister rust
Cronartium quercuum f. sp. *fusiforme*	Fusiform rust
Endocronartium harknessii	Western gall rust
Melampsora occidentalis	Poplar rust

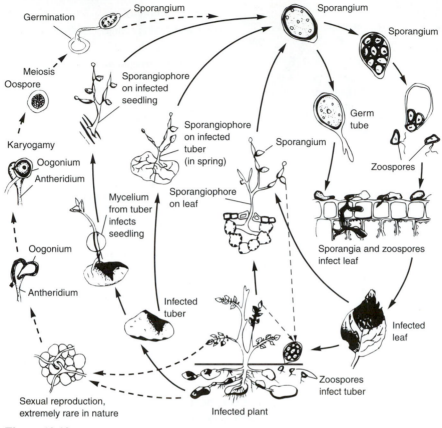

Figure 10.12
Life cycle of a typical Oomycete, *Phytophthora infestans.*

Pythium spp. typically cause damping-off disease of seedlings. If a seedling is attacked near the soil line and subsequently falls over, it is known as postemergence damping-off (Tainter and Baker 1996). Seeds and seedling roots during germination also may be attacked, a condition known as preemergence damping-off. Several species of *Pythium* are involved: *P. aphanidermatum*, *P. debaryanum*, *P. irregulare*, and *P. ultimatum*. These species attack both agricultural crops and tree seedlings and are widely distributed in soils throughout the world (Agrios 1997).

Phytophthora spp. cause a variety of diseases of agricultural crops, shrubs, tree seedlings, and forest trees (Agrios 1997). They cause damping-off; root rots of fine roots; rots of lower stems, tubers, and corms; rots of buds or fruits; and blights of foliage, young twigs, and fruits. Late blight of potatoes caused by *P. infestans* caused considerable social disruption in Ireland in the early 1850s by destroying the potato crop several years in a row (Agrios 1997). Potatoes were the staple diet of the Irish and rather than face starvation many Irish emigrated to the eastern United States, explaining why there are so many people of Irish descent in this area today.

Some species of *Phytophthora* are host specific, such as *Phytophthora lateralis,* which causes Port-Orford-cedar root disease. Others such as *P. cinnamomi* have a very broad host range including oaks, pines, chestnut, and eucalypts (Tainter and Baker 1996, Agrios 1997). Specifically, *P. cinnamomi* causes two very important forest diseases: littleleaf disease of pine in North America and dieback in eucalypt forests in Australia. Phytophthora root disease occurs in nearly every part of the world where soil becomes too wet for good growth of susceptible plants and the temperature remains fairly low (i.e., between 15–23°C). Small plants usually are killed quickly, whereas forest trees may die rapidly or take years to die, showing gradual decline or dieback symptoms.

Downy mildews also are caused by Oomycetes and are mostly a problem in agriculture, causing foliage blights (Agrios 1997). The fungi spread rapidly in young tender green tissues including leaves, twigs, and fruit. No important downy mildew diseases occur in forests, but downy mildew fungi occur on a wide variety of plants including most of the cultivated grain and vegetable crops, ornamentals, shrubs and vines, particularly grapes. The fungi involved are obligate parasites belonging to the family Perononsporaceae. *Peronospora* is a common genus involving downy mildew of onion, spinach and soybeans, and mildew (blue mold) of tobacco. The development of the disease depends on the presence of a film of water on the plant surfaces and high relative humidity (Agrios 1997).

SPREAD AND ECOLOGY OF FUNGI

Fungi spread long and short distances by means of spores and locally by vegetative mycelia (Alexopolous et al. 1996). Spores are spread in the air, in soil, in water, and by animals, insects, and birds. Humans also can spread spores on equipment, boots, and so on. Some airborne spores can travel hundreds of kilometers in viable condition. Most of the fungi capable of traveling long distances are heavily pigmented to give protection against damaging ultraviolet radiation (e.g., conidia and many rust spores). Many of these are passively released by wind blowing across surfaces of leaves, and so on, and dispersal occurs during daylight hours. Nonpigmented (**hyaline**) basidiospores are commonly actively released at night. Other airborne spores may be actively released after rain (e.g., many ascospores). Both basidiospores and ascospores are passively spread by wind after their initial active release.

In the soil spores can be spread by animals—for example, many small mammals eat truffles and spread mycorrhizal spores. Mites and other small soil animals also spread spores. Swimming spores, such as zoospores, can spread in soil water in wet soils and free running water in overland flow. Zoospores also can spread in streams. Insects spread many spores, especially those of blue-stain fungi and fungi causing vascular wilts such as Dutch elm disease (Tainter and Baker

1996). Spores spread by insects line the surface of insect galleries where they contact insect bodies. Some insects have specialized structures, called *mycangia,* for transporting fungi and inoculating hosts. Birds may spread the spores of branch canker fungi.

To spread disease, spores must be in a viable condition. Although millions of airborne spores are dispersed, survival is generally low because of exposure to radiation, dessicating conditions, temperature extremes, smoke, air pollutant gases, and so on (Alexopolous et al. 1996). Once spores land on a suitable substrate, environmental conditions must be favorable for germination and survival of the mycelium. Spores can generally withstand broader ranges of temperature and moisture than free mycelium which typically survives and grows between -5 and 45°C (Paul and Clark 1996). Many fungi produce survival structures such as sclerotia, a compact mass of hyphae, usually with a darkened rind, which are capable of surviving in unfavorable environments. Sclerotia can also be disseminated, but to a much lesser extent than spores. Some fungi also produce rhizomorphs.

If conditions are too hot, too cold, too wet, or too dry the infection process will not occur. Some fungi grow only on plant surfaces, but plant pathogens generally need to enter the host to cause disease. Pathogens do not enter plant tissues easily, however. Although some pathogens can penetrate directly, leaf pathogens and rust fungi typically enter leaves through the stomates or other openings such as hydathodes or lenticels (Tainter and Baker 1996). Some have specialized structures for penetration called haustoria or **appressoria** (Agrios 1997). Other pathogens require wounds in the bark or damage caused by insects to enter. Many forest management activities, such as thinning and pruning, create wounds in the stem and roots can be easily wounded by logging equipment. Once inside plant tissues environmental conditions are generally more favorable for fungal pathogens, especially moisture conditions. Many plant pathogenic fungi have an optimum temperature for growth near 25°C (Agrios 1997).

Even though temperature and moisture conditions may be favorable for the development of mycelium, most hosts have resistance mechanisms that may be inherent, such as chemicals like phenolics, which give wood its color, or induced in response to infection (Agrios 1997). We are still not sure how pathogens recognize their hosts or how hosts recognize pathogens. There is usually a wide variety of resistance among the genotypes of individual species. Most plants have evolved in the presence of pathogens, and it is not in the best interest of a pathogen to be able to easily kill hosts and reduce its food base. Introduced pathogens to North America, such as those causing chestnut blight, Dutch elm disease, and white pine blister rust have not evolved with local hosts and can have a devastating effects on tree species. For example, American chestnuts were almost eradicated by chestnut blight (Tainter and Baker 1996).

Forest management also serves to reduce the genetic base for resistance to tree pathogens because genotypes may be selected for superior growth. Of course they can also be selected for disease resistance, but this further narrows the genetic base. The concern with genetic selection is new virulent mutant strains of the fungus can overcome host resistance, and this is a particular problem for trees grown on 50- to 80-year rotations. Continuous production of new strains of pathogens has been a constant problem in agriculture, particularly with wheat rust (Agrios 1997). Wheat breeders are constantly having to produce new resistant strains to combat new wheat rust strains, and the genetic basis for resistance has been narrowed considerably. Fortunately, most tree species have a wide genetic base and there has been tendency with forest tree breeding to look for multiple genes for resistance as opposed to the single gene resistance used with many agricultural crops.

REFERENCES

Agrios, G. N. 1997. *Plant pathology,* 4th ed. Academic Press, New York.

Alexopoulos, C. J., C. W. Mims, and M. Blackwell. 1996. *Introductory mycology.* John Wiley and Sons, New York.

Cooper, J. I. 1993. *Virus diseases of trees and shrubs.* Chapman and Hall, London.

Fielding, N. J., and H. J. Evans. 1996. The pine wood nematode *Bursaphelenchus xylophilus* (Steiner and Buhrer) (=*B. lignicolus* Mamiya and Kiyohara): an assessment of the current position. *Forestry* 69:35–46.

Hawksworth, F. G., and D. Wiens. 1996. *Dwarf mistletoes: biology, pathology, and systemics.* USDA Forest Service Agrcultural Handbook 709, Washington, D.C.

Hiruki, C. (ed.). 1988. *Tree mycoplasmas and mycoplasmas diseases.* University of Alberta Press, Edmonton.

Jenkins, W. R., and D. P. Taylor. 1967. *Plant nematology.* Reinhold Publishing, New York.

Killham, K. 1994. *Soil ecology.* Cambridge University Press, Cambridge, United Kingdom.

Manion, P. D. 1991. *Tree disease concepts.* Prentice-Hall, Englewood Cliffs, N.J.

Paul, E. A., and F. E. Clark. 1996. *Soil microbiology and biochemistry,* 2nd ed. Academic Press, New York.

Scharpf, P. F. 1993. *Diseases of Pacific Coast conifers.* USDA Forest Service Agricultural Handbook 521, Washington, D.C.

Tainter, F. H., and F. A. Baker. 1996. *Principles of forest pathology.* John Wiley and Sons, New York.

11

NURSERY DISEASES AND MYCORRHIZAE

INTRODUCTION

Many forests throughout the world are not established by natural regeneration, but through the planting of seedlings that are raised in nurseries (Hamm et al. 1990, Aldhous and Mason 1994). Even when natural regeneration is used, seedlings often are interplanted to ensure that sites are regenerated in an adequate time frame. Sometimes poor weather or lack of a seed crop makes it difficult to regenerate a site relying entirely on natural regeneration.

There are economic and ecological reasons for rapid site regeneration, and sites that are regenerated quickly are less susceptible to runoff and erosion (Perry et al. 1987). Although most readers of this book may not be involved directly with forest nurseries, it is important to understand how nurseries operate and how healthy, disease-free seedlings are produced. It is extremely important that seedlings are in healthy condition at the time of outplanting if they are to have a good chance of survival. Understanding the conditions under which seedlings are grown in the nursery also helps in interpreting problems in the field after outplanting.

Around the world forest tree nurseries provide millions of seedlings every year for reforestation efforts. This operation is very intensive and is actually more akin to agriculture than forestry. As a result considerably more money is spent on controlling nursery diseases than diseases in the forest because the crop has such high value and chemicals are commonly used.

Tree seedlings may be grown in "normal soil" in bare root nurseries, but they also are grown in containers in synthetic soil containing peat or vermiculite, which may be steamed or fumigated to reduce pathogen populations. Bare root nursery soil also may be fumigated as we will see later. Containerized seedling production often is done in greenhouses with overhead sprinkler systems (Hamm et al. 1990). Heavy fertilization is common, and moist soil conditions typically favor the development of root pathogens, as well as mold fungi on the leaves and needles (Sutherland et al. 1989). Steamed or fumigated soil mixes usually are devoid or have very low populations of soil fungi, bacteria, and animals—and pathogenic microbes can rapidly colonize a medium in which competition and antagonism from other soil organisms is missing.

Although bare root seedlings may be mycorrhizal, containerized seedlings usually are not, since the growing medium usually has been sterilized. Mycorrhizae, or mutualistic fungus/root associations, play important roles in forest nurseries as well as in forests, and it is appropriate to consider the role of mycorrhizae in tree health in this chapter. The high-fertilization regimes in nurseries may tend to inhibit the development of mycorrhizae, especially in containerized seedlings. There also is usually a completely different suite of mycorrhizal fungi on nursery seedlings than on forest seedlings (Robson et al. 1994).

Like the diseases that occur in forests, nursery disease problems can be classified into biotic diseases and abiotic injuries. Insects also cause problems in nurseries and these are covered in Chapter 24. In this chapter we cover the selection of nursery sites, including soil types, causes of noninfectious injuries, such as frost damage, and the common infectious diseases that are typically seed-borne diseases, root rots, and diseases affecting stems, leaves, and needles, and disease management.

SELECTION OF NURSERY SITES

Correct selection of nursery sites, particularly with respect to soil type, is extremely important in management of nursery diseases (Sutherland et al. 1989, Aldhous and Mason 1994). As you will see later, many of the diseases, particularly damping-off, are worsened by the type of soil, specifically heavy clay soils. An ideal nursery soil should be well drained with no more than 10% clay and 10 to 15% silt and be stone-free (Hamm et al. 1990). It is best to err on the sandy/gravelly side than to use a soil with too much clay. You can always water the soil, but it is much more difficult to improve the drainage in the soil. Compacted layers in the soil also will impede drainage. Soil organic matter and pH are important to consider. Most conifer seedlings grow best in soil with a pH between 4.5 and 5.5. Most hardwoods, such as poplars, also do well in this range, although some may require slightly higher pH. When a soil is too acid, it can be readily brought to its optimum by adding lime. Fertilizers also can change soil pH. Ammonium sulfate and ammonium nitrate will lower soil pH, whereas urea will raise soil pH. Adequate levels of P, K, and Mg also are important and can be added with fertilization.

It also is a good idea to avoid frost hollows, since seedlings are very susceptible to frost injury, as well as areas that are susceptible to high winds and flooding. Sites with potential or known disease problems also should be avoided, such as those near forests with a high incidence of western gall rust or fusiform rust.

ABIOTIC INJURIES

Most abiotic injuries in the nursery are a result of adverse weather conditions (e.g., frost), poor nutrition, over- or underwatering, and chemical injuries particularly from herbicides used to control weed competition (Sutherland et al. 1989).

Because seedlings have few woody tissues, they are particularly susceptible to weather damage. Weather damage falls into two main categories: (1) low-temperature injury—spring or "late" frosts, autumn or "early" frosts—winter cold, frost heave, and secondary infection following low-temperature injury and (2) heat and moisture stress.

Low-Temperature Injuries

Tree species, provenances, and varieties differ greatly in their ability to withstand low-temperature injury (Aldhous and Mason 1994). Species from warmer climates and lowland provenances tend to be more susceptible than species from higher latitudes and higher elevations. Most tree species are susceptible to spring frosts, but some species are less susceptible—for example, lodgepole pine and Scots pine (Aldhous and Mason 1994). Death of buds, new shoots, and leaves is the commonest form of damage and the injured tissues typically shrivel and turn black or brown. Occasionally bark on older shoots may be killed. Trees may die, but they also may recover by producing new shoots from surviving or adventitious buds. These plants, however, may have poor form, because they often have multiple leaders.

Autumn frosts may cause death of unhardened buds and shoots and foliage browning. Some species and provenances may grow late into the season and are very susceptible to early frosts. Certain conditions also make plants more susceptible to early frost damage, such as very warm weather just before the first frost, defoliation by diseases or pests, and nutritional status. Fertilizing with too much nitrogen may increase the length of the growing season and delay budset. Douglas-fir is very susceptible to autumn frosts (Hamm et al. 1990).

Winter cold can cause injury, especially if mild temperatures occur in periods before low temperatures occur. High winds during periods of low temperature cause further injury by desiccation. Buds may not be killed, because they are hardened. However, browning of foliage may occur when soils are frozen or near-frozen and seedlings are unable to take up water, thereby suffering drought damage. Douglas-fir, western hemlock, and western redcedar are susceptible to winter injury. Repeated freezing and thawing of soil, particularly in heavier soils, can eventually push seedlings out of the ground, resulting in frost heave. Seedlings may be replaced in the soil when conditions are suitable, but in many cases this is impractical and the frost-heaved seedlings may be lost. Needles damaged by cold may be invaded by pathogens such as *Botrytis cinerea,* which causes gray mold. Other fungi may invade damaged shoots.

High-Temperature Injuries

High-temperature injuries and moisture stress may occur in areas with warm climates (Aldhous and Mason 1994). Symptoms vary from wilting, fading, and death of foliage to shoot dieback. Browning of foliage usually commences on leaf margins or needle tips. Drying winds can exacerbate the damage. Reradiation from light-colored soils also may damage the root collar of small, thin-barked seedlings.

BIOTIC DISEASES

Biotic diseases can be divided into two main classes: diseases affecting seeds, roots, or root collar areas, and diseases affecting shoots. Many of the diseases affecting forest nurseries are common throughout the world, although some are specific to a certain region, particularly the rusts.

Diseases Affecting Seeds, Roots, or Root Collar Areas

A flow chart for identifying diseases affecting seeds, roots, or root collar areas is shown in Box 11.1 (Sutherland et al. 1989). Tree seeds, if not stored under low temperature and moisture conditions, will be attacked by seed-inhabiting fungi, which can markedly reduce their viability. These storage fungi typically are molds in the Fungi Imperfecti genera *Aspergillus* and *Penicillium,* although other genera such as *Alternaria* and *Cladosporium* also can be found. The presence or absence of particular combinations of fungi may affect the early survival of seedlings, and some may actually protect them from damping-off fungi. Some fungi develop on seeds in cones, like the Ascomycete *Caloscyphe fulgens,* which inhabits forest duff. Cones picked from the ground or from squirrel caches usually are infected. In British Columbia the pathogen occurs in about 25% of all stored Sitka, Engelmann, and white spruces and in some Douglas-fir and grand fir seedlots. The fungus spreads during stratification (Sutherland et al. 1989).

Damping-off describes a set of diseases resulting in the rotting of seeds, germinants, and succulent seedlings. There are two types: preemergence and postemergence. Preemergence damping-off kills seeds and germinants before they are visible; postemergence damping-off kills newly emerged succulent seedlings. They both affect bareroot and containerized seedlings (Sutherland et al. 1989, Tainter and Baker 1996).

The fungi causing damping-off are unspecialized pathogens and can survive for long periods in the soil in the absence of hosts, because of resistant resting stages. The common genera involved worldwide are the Oomycetes *Phytophthora* and *Pythium,* the Fungi Imperfecti *Fusarium* and *Cylindrocladium,* and the Basidiomycete *Rhizoctonia*. These fungi are all common on soft-tissued agricultural crop plants. The life cycle of *Phytophthora* is illustrated in Figure 10.12. Failure of seedlings to emerge is the only aboveground evidence of preemergence damping-off. Heavy, cool, wet, compacted soils tend to favor disease, particularly if it is caused by Oomycetes, which have zoospores that swim in free water and can easily spread in nurseries. *Pythium* and *Phytophthora* also can be spread by running water, rain splash, wind, and in soil adhering to machinery, boots, and tools. Sclerotia of *Rhizoctonia* also can be physically carried by running water, wind, and soil adhering to seedlings when transplanted. Dense sowing also favors the disease. Temperature, moisture, light, soil composition, pH, nutrition, and time of sowing also influence damping-off (Sutherland et al. 1989).

The main symptoms of normal postemergence damping-off are stem rotting near the groundline and subsequent toppling of the shoot. Until they become woody, usually 4 to 6 weeks after emergence, seedlings are susceptible to this disease. Late

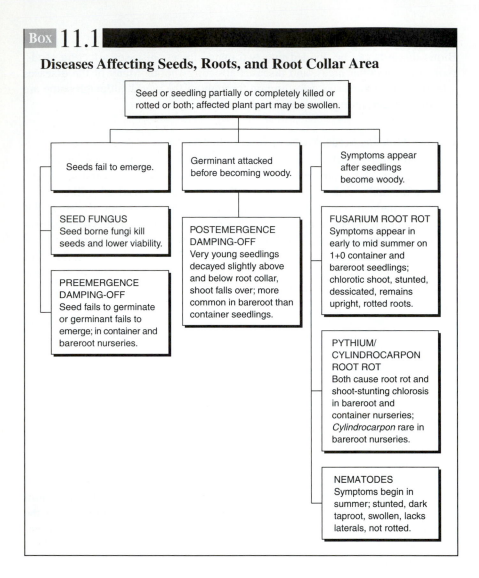

Box **11.1**

Diseases Affecting Seeds, Roots, and Root Collar Area

Seed or seedling partially or completely killed or rotted or both; affected plant part may be swollen.

Seeds fail to emerge.

Germinant attacked before becoming woody.

Symptoms appear after seedlings become woody.

SEED FUNGUS
Seed borne fungi kill seeds and lower viability.

PREEMERGENCE DAMPING-OFF
Seed fails to germinate or germinant fails to emerge; in container and bareroot nurseries.

POSTEMERGENCE DAMPING-OFF
Very young seedlings decayed slightly above and below root collar, shoot falls over; more common in bareroot than container seedlings.

FUSARIUM ROOT ROT
Symptoms appear in early to mid summer on 1+0 container and bareroot seedlings; chlorotic shoot, stunted, dessicated, remains upright, rotted roots.

PYTHIUM/ CYLINDROCARPON ROOT ROT
Both cause root rot and shoot-stunting chlorosis in bareroot and container nurseries; *Cylindrocarpon* rare in bareroot nurseries.

NEMATODES
Symptoms begin in summer; stunted, dark taproot, swollen, lacks laterals, not rotted.

damping-off sometimes occurs after stems and some of the roots have begun to become woody. New roots of these seedlings are susceptible as long as they remain tender. Top damping-off also occurs, but rarely. In this case cotyledons which are held in the seed coat after emergence from the soil are attacked. In bareroot nurseries the disease may kill scattered seedlings randomly, but expanding patches of disease may develop (Sutherland et al. 1989).

Fusarium, Pythium, and *Cylindrocarpon* spp. cause root rots in woody seedlings in both bare root and container nurseries. Fusarium root rot becomes evident in midsummer when postemergence damping-off is declining. Chlorosis of terminal needles is followed by the needles becoming flaccid, turning purple,

Pathogen overwinters, usually as chlamydospores produced in small pieces of diseased roots or in organic matter.

Needle chlorosis, browning, desiccation, and crozier-leader tip symptoms then develop.

In early summer chlamydospores germinate and infect when seedling roots grow nearby.

Later in growing season when seedlings are larger, more crowded, and stressed, fungus ramifies, destroying root system.

Figure 11.1
Life cycle of *Fusarium* spp. in forest nurseries.

and then turning brown and dry. The shoot tip frequently bends into a shepherd's crook and because the stem is woody it remains upright. This distinguishes it from postemergence damping-off. The life cycle of this fungus is shown in Figure 11.1. High temperatures and drought predispose seedlings to disease and contribute to rapid disease progress. Infection can be seed-borne or may result from airborne spores (Sutherland et al. 1989).

Pythium spp. have a wide host range, but usually attack only succulent tissues or root tips or lateral roots of woody seedlings. Symptoms are poorly developed root systems, shoot stunting, and chlorosis (Figure 11.2). The life cycle of this fungus is illustrated in Figure 11.3. Like *Phytophthora*, *Pythium* spp. can survive for long periods as thick-walled resting spores (oospores) or as saprophytes on dead organic matter.

Cylindrocarpon occurs on a variety of conifer and hardwood seedlings in the United States and Canada (Tainter and Baker 1996). Diseased roots are dark brown and stunted. It is a saprophyte or weak pathogen and spreads by mycelium in the soil to new substrates. Chlamydospores and conidia also are produced. It is common in alkaline soils.

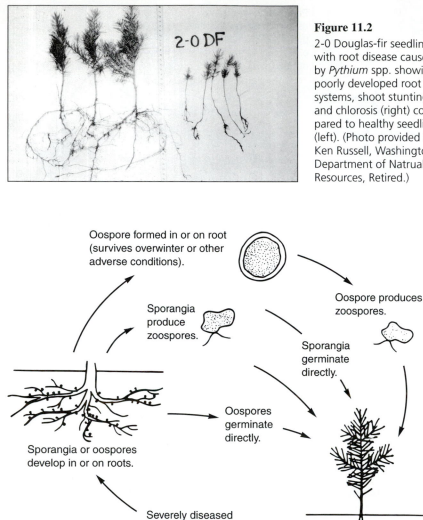

Figure 11.2
2-0 Douglas-fir seedlings with root disease caused by *Pythium* spp. showing poorly developed root systems, shoot stunting, and chlorosis (right) compared to healthy seedlings (left). (Photo provided by Ken Russell, Washington Department of Natrual Resources, Retired.)

Oospore formed in or on root (survives overwinter or other adverse conditions).

Oospore produces zoospores.

Sporangia produce zoospores.

Sporangia germinate directly.

Sporangia or oospores develop in or on roots.

Oospores germinate directly.

Severely diseased seedlings become stunted and chlorotic.

Root infection; disease results.

Figure 11.3
Life cycle of *Pythium* spp.

Black root disease is a problem in the southeastern United States. It is caused by the fungus *Macrophomina phaseoli*. *Fusarium* and nematodes may increase the intensity of the disease. Shoot stunting occurs late in the growing season and the tap root becomes enlarged, blackened, and roughened. Black microsclerotia are produced on dead roots and the lower stem. All of the commercially important

Box 11.2

Root-knot nematode: *Meloidogyne* spp.—causes galls on roots

Pine-cystoid nematode: *Meoloidodera* spp.

Root-lesion nematode: *Pratylenchus* spp.

Dagger nematode: *Xiphinema* spp.

Lance nematode: *Hoplolaimus* spp.

Sheath nematode: *Hemicyciophora* spp.

Ring nematode: *Macroposthonia* spp.

Pin nematode: *Pratylenchus* spp.

Sting nematode: *Belonolaimus* spp.

southern pines are susceptible. It is enhanced by high soil temperatures, moisture stress, and N fertilization.

Although nematodes do not cause many problems in forests, they can cause major problems in nurseries and can attack nearly all plant species (Sutherland et al. 1989). There are two main groups of species: ectoparasitic species that remain outside roots and feed on epidermal or cortical cells, but occasionally enter roots, and endoparasitic species that enter, feed, and reproduce inside roots. It usually takes a large population of nematodes to cause disease. Plant parasitic nematodes feed with a stylet. Nematodes may predispose trees to other pathogens such as *Fusarium* spp. Nine genera of nematodes attack tree roots and are listed in Box 11.2.

Corky root disease is a problem caused by the nematode *Xiphinema bakeri* in western North America. Douglas-fir is particularly sensitive to *Xiphinema*, although it also attacks spruces and western hemlock (Sutherland et al. 1989). On Douglas-fir the first symptoms are slightly chlorotic secondary needles and stunted growth on random first-year seedlings. Examination of the roots reveals that the taproot has few laterals and the dark and swollen tips are club-shaped. The life cycle of this nematode is shown in Figure 11.4. Disease is worse in nurseries where forests have recently been converted to nurseries and where soils are sandy and low in nutrition. They can move quickly in sandy soils.

Diseases Affecting Shoots

A flow chart for identifying seedling diseases affecting shoots is shown in Box 11.3 (Sutherland et al. 1989). A variety of diseases affect shoots including gray mold, Rosellinia blight, Sirococcus blight, molding of stored seedlings, Phomopsis canker and foliage blight, Colletotrichum blight, Fusarium top blight, Hyocotyl rot, Phoma blight, smothering fungus, and a number of rusts.

Gray mold is caused by *Botrytis cinerea,* which occurs worldwide and has a broad host range. Pines seem to be less affected than other conifers. The disease seldom is a problem in bareroot nurseries, but it is a major problem in container nurseries. Initial symptoms include formation of watery lesions and killing of lower seedling needles. The fungus then moves upward and can kill the whole

Eggs laid in summer;
populations peak in early fall.

Nematode overwinters in
soil as eggs and juveniles.

Eggs hatch in spring.

Feeding results in stunted and
chlorotic shoots and dark, swollen,
club-tipped roots with few laterals.

Last three juvenile stages
and adults feed on root tips.

Figure 11.4
Life cycle of the nematode, *Xiphinema bakeri,* cause of corky root disease.

plant. Commonly masses of gray-brown mycelium grow on dead tissues, giving it the name "gray mold." It can be introduced on seeds, by airborne conidiospores, spores in irrigation water, or even on workers' gloves. Disease spread is favored by moderate temperatures, high moisture, dense foliage, and crowded seedlings. Tissues damaged by frost in bareroot nurseries are susceptible to infection. *Rosellinia minor,* which causes Rosellinia blight, also occurs on container-grown seedlings and resembles gray mold, but its mycelium is white-brown (Sutherland et al. 1989).

Sirococcus strobilinus causes Sirococcus blight and infects conifers in bareroot and container nurseries throughout the northern temperate zone (Tainter and Baker 1996). Seedlings are killed from the base upward and dead seedlings remain upright. Needles are colored light to reddish-brown with pycnidia that darken with age. *Phomopsis occulata* and *P. lokoyae* cause Phomopsis canker and blight. These common saprophytes can form cankers on stems and branches of winter-dormant conifer seedlings and girdle the stem. Other diseases of less importance are Colletotrichum blight on container seedlings, Fusarium top blight caused by *F. oxysporum,* resulting in severe losses in 1-0 Douglas-fir and bareroot and container ponderosa pine seedlings, and hypocotyl rot caused by *Fusarium* and *Phoma* spp. which affects Douglas-fir, ponderosa pine and lodgepole pines in container

Box 11.3

Diseases Affecting Shoots

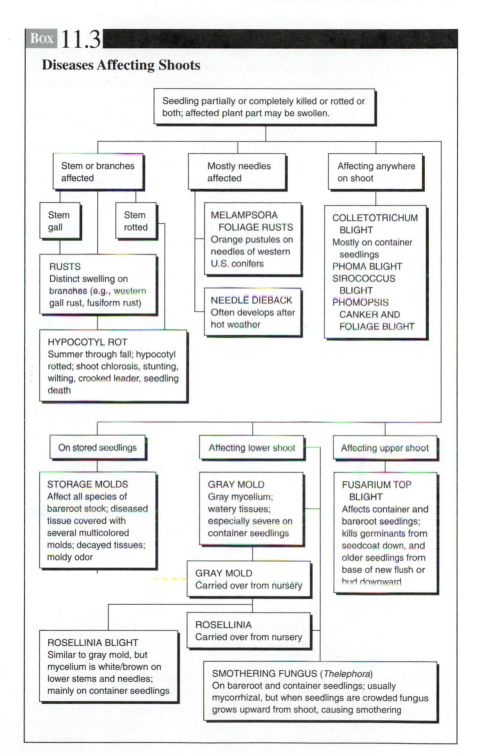

Seedling partially or completely killed or rotted or both; affected plant part may be swollen.

Stem or branches affected

Mostly needles affected

Affecting anywhere on shoot

Stem gall

Stem rotted

MELAMPSORA FOLIAGE RUSTS
Orange pustules on needles of western U.S. conifers

COLLETOTRICHUM BLIGHT
Mostly on container seedlings
PHOMA BLIGHT
SIROCOCCUS BLIGHT
PHOMOPSIS CANKER AND FOLIAGE BLIGHT

RUSTS
Distinct swelling on branches (e.g., western gall rust, fusiform rust)

NEEDLE DIEBACK
Often develops after hot weather

HYPOCOTYL ROT
Summer through fall; hypocotyl rotted; shoot chlorosis, stunting, wilting, crooked leader, seedling death

On stored seedlings

Affecting lower shoot

Affecting upper shoot

STORAGE MOLDS
Affect all species of bareroot stock; diseased tissue covered with several multicolored molds; decayed tissues; moldy odor

GRAY MOLD
Gray mycelium; watery tissues; especially severe on container seedlings

FUSARIUM TOP BLIGHT
Affects container and bareroot seedlings; kills germinants from seedcoat down, and older seedlings from base of new flush or bud downward

GRAY MOLD
Carried over from nursery

ROSELLINIA
Carried over from nursery

ROSELLINIA BLIGHT
Similar to gray mold, but mycelium is white/brown on lower stems and needles; mainly on container seedlings

SMOTHERING FUNGUS (*Thelephora*)
On bareroot and container seedlings; usually mycorrhizal, but when seedlings are crowded fungus grows upward from shoot, causing smothering

and bareroot nurseries. *Thelephora terrestris,* a mycorrhizal fungus associated with many conifer species, readily invades fumigated soils (Allen 1991, Zak 1971). With little competition its mycelium and fruiting bodies can grow up the stems of bareroot and container seedlings covering stems and needles, resulting in "smothering" of the plant and sometimes causing mortality.

Molds on stored bareroot seedlings can cause huge losses if not controlled. Seedlings are stored because the times for lifting and outplanting rarely coincide. Seedlings are placed in storage packages, usually large paper bags with special linings, and stored under refrigerated conditions just above freezing (1°C) until needed. Numerous mold fungi are involved including *Fusarium, Rhizopus, Aspergillus, Penicillium, Epicoccum,* and *Cylindrocarpon.* They naturally occur on foliage or in soil particles as saprophytes and develop in storage containers if conditions are too warm and moist. Cottony mycelium develops and can eventually kill the seedlings if prolonged. Only two fungi are involved with molding of stored container seedlings, *Botrytis cinerea* and *Rosellinia* spp. (Sutherland et al. 1989).

A variety of rusts cause problems in forest nurseries, including conifer–aspen rust (caused by *Melampsora medusae*), conifer–cottonwood rust (caused by *M. occidentalis*), and western gall rust (caused by *Endocronartium harknessii*) in western North America and fusiform rust (caused by *Cronartium quercuum f. sp. fusiforme*) in the southeast United States (Tainter and Baker 1996). The life cycles of rusts are covered in detail in Chapter 13. Douglas-fir, western larch, tamarack, and ponderosa and lodgepole pines are the conifer hosts for *M. medusae;* Douglas-fir is the host for *M. occidentalis.* Symptoms are confined to the current year's foliage on conifers, which may be killed and shed in the autumn. Shoots of severely infected seedlings may die. Aecia form on conifer hosts. Western gall rust infects lodgepole and ponderosa pine seedlings (Sutherland et al. 1989). Fusiform rust alternates between oaks and pines. Loblolly, slash and longleaf pines are susceptible, although the latter is more resistant (Tainter and Baker 1996).

MANAGEMENT OF SEEDLING DISEASES

There are many techniques available to protect seedlings against abiotic diseases (Hamm et al. 1990, Aldhous and Mason 1994). Seedlings can be protected against low-temperature injury by overhead shelter or spraying with water. Protection by irrigation results from the release of latent heat when water freezes on plant surfaces. Irrigation also can prevent heat and moisture stress. Use of shade cloths or cover crops is common in tropical environments. Containerized seedlings in well-watered greenhouses usually do not suffer moisture stress, but provision of adequate ventilation is important. Maintenance of correct soil pH is important; 5.5 to 6.2 is best for Douglas-fir, for example.

Management of infectious diseases is more difficult, but usually involves an understanding of the life cycles of individual disease organisms and how environmental factors affect their development. Regulating irrigation, fertilization, and cultural regimes so that they do not favor diseases, using sanitation and exclusion practices, establishing disease-monitoring schemes, and using chemicals judiciously usually results in production of healthy seedlings (Hamm et al. 1990). Sanitation and exclusion practices include not allowing movement of infected plant material or contaminated equipment within or among nurseries, weed control, removing infected seedlings, checking water supplies periodically for pathogens, and removing diseased trees and alternative hosts for rusts from areas around nurseries.

Chemicals have been used widely to control diseases in nurseries; these include seed treatment chemicals, chemical sprays for foliage diseases, and soil fumigation for root diseases. In some cases chlorination of irrigation water can be carried to reduce *Pythium* and *Phytophthora* populations. Captan, thiram, and benomyl are fungicides commonly used to control gray mold and some foliage diseases.

Soil fumigants have been used widely to control damping-off and root diseases in forest nurseries (Aldhous and Mason 1994). Typically gaseous fumigants such as methyl bromide/chloropicrin or vapam are used. Soil fumigation is expensive and typically reduces or eliminates the population of beneficial soil organisms, including mycorrhizae. Once a fumigation cycle is started, it usually has to be refumigated every few years because pathogens can readily build up their populations. Many other fungicides, such as Banrot, Bravo, Captan, and Maneb, also decrease ectomycorrhizal development .

Methyl bromide is a broad-spectrum pesticide used in control of pathogens, nematodes, insects, weeds, and rodents. In the United States it is primarily used in agriculture and forestry for soil fumigation, but there is some use for commodity and quarantine treatment and structural fumigation. North America has the highest use (41%), followed by Europe (26%), Asia (24%), and South America and Africa (9%). It is typically injected into the soil at a depth of 40 to 80 cm before seeds are sown. Immediately after injection, the soil is covered with plastic tarps, which are removed after 24 to 72 hours. As much as 95% of the methyl bromide enters the atmosphere, however, and acts as a potent ozone-depleting substance. This is contributing to the reduction of the ozone layer in the upper atmosphere that protects life on earth from harmful ultraviolet radiation. Because of this, under the Clean Air Act of 1990, the Environmental Protection Agency (EPA) has prohibited the production and importation of methyl bromide starting January 1, 2002. This deadline was recently extended. Alternatives are now being investigated (EPA 1997). There are a number of chemical alternatives: 1, 3-dichloropropone, dazomet, chloropicrin, and metam sodium, and Basamid. Basamid seems to hold promise for forest nurseries When applied to moist soils, it breaks down into methyl isothiocyanate, which is effective against fungi (particularly *Phytophthora*), nematodes, and weeds. It is a solid material and can be stored safely and easily and tarping after application is not necessary.

Nonchemical alternatives are (1) soil pasteurization by solar irradiation or steam/hot water treatment, (2) biological controls that result in pest suppression, and (3) cultural practices that reduce plant stresses. In warm environments, such as the southeast United States, soil temperatures under the clear plastic tarps get warm enough to accomplish partial solar sterilization and reduction of pathogen populations. In cooler areas, such as the Pacific Northwest United States and

Canada, this technique does not work, and steam treatment may have to be used or a combination steam pasteurization with solarization. Cultural management is another alternative. This includes production of high-quality vigorous seed, monitoring seeds to reduce the likelihood of unwanted pest introduction, achieving uniform irrigation, reducing the need for N fertilization during the summer months, and modifying root morphology through root pruning, undercutting, and bed wrenching. Biological control also is possible. Use of plants, such as mustard (*Brassica* spp.), which produce natural fungicides also has been proposed. Fallowing also reduces the populations of some pathogens, but some may survive on the roots of fallowing plants. Monitoring of populations of *Pythium, Phytophthora,* and *Fusarium* in nursery beds allows nursery managers to follow the buildup of pathogen populations. In most cases, whole nursery beds may not need treatment since the pathogens tend to occur in "hot spots."

An integrated approach to managing nursery diseases is preferable to a single approach like methyl bromide fumigation. Managing diseases in nurseries takes a broad knowledge of pathogens, cultural techniques, and the specific requirements of individual species.

MYCORRHIZAE

Mycorrhizae (fungus/root associations) are extremely important in forest nurseries and in maintaining healthy trees in forests. All of the commercially important forest trees have mycorrhizae (Richards 1987, Smith and Reed 1997). Not all plants, however, have mycorrhizae. Aquatic plants typically do not possess mycorrhizae, and even certain terrestrial plants such as those in the mustard family do not have them. Many early successional plants, especially those involved with primary succession on sand dunes or after glacial retreat, are nonmycorrhizal or facultatively mycorrhizal (Allen 1991). Thus there is a gradient in mycorrhizal associations from plants that are nonmycorrhizal to those that obligately mycorrhizal.

There are basically two types of mycorrhizae: **ectomycorrhizae** and **endomycorrhizae** (Smith and Reed 1997). Both produce hyphal networks that extend out into the soil. The features of the different types of mycorrhizae are presented in Table 11.1 along with the major fungal and host taxa. Ectomycorrhizae are characterized by (1) a sheath or mantle of fungal tissue that encloses the tips of fine roots and (2) the Hartig net or penetration of fungal hyphae between the epidermal and cortical cells (Figure 11.5). Hyphae of ectomycorrhizal fungi usually do not penetrate cell walls of the host plant. In contrast, hyphae of endomycorrhizal fungi do penetrate into cortical cells (Figure 11.6). Most endomycorrhizae don't have an external mantle, although some do. A few fungi are considered to form ectendo mycorrhizae (Richards 1987).

There are five distinct types of endomycorrhizae: vesicular-arbuscular, arbutoid, monotropoid, ericoid, and orchid (Table 11.1). Vesicular-arbuscular mycorrhizae (VAM) are named because they produce arbuscules, branched hyphae-like trees, and vesicles, which probably serve as storage organs (Figure 11.6). However, many produce only arbuscules and probably should be called arbuscular

T A B L E 11.1 | Characteristics of Types of Mycorrhizae

Character	Ectomycorrhizae	Endomycorrhizae				
		Vesicular-arbuscular	Ericoid	Arbutoid	Monotropoid	Orchid
Fungi septate	+	−	+	+	+	+
Aseptate	(+)[a]	+	−	−	−	−
Hyphae enter cells	−	+	+	+	+	+
Fungal sheath (mantle) present	+	−	−	+	+	−
Hartig net formed	+	−	−	+	+	−
Hyphal coils in cells	−	+	+	+	−	+
Haustoria dichotomous	−	+	−	−	−	−
Not dichotomous	−	−	−	−	+	+ or −
Vesicles in cells or tissues	−	+(or −)	−	−	+	−
Achlorophylly	−	+(or −)	−	−(or +)	+	+
Fungal phyla	Basidiomycota Ascomycota Zygomycota	Zygomycota	Ascomycota (Basidiomycota)	Basidiomycota	Basidiomycota	Basidiomycota
Host taxon	Gymnosperm	Bryophyte Gymnosperm Angiosperm	Ericales	Ericales	Monotropaceae	Orchidaceae

[a]Entries in parentheses are rare.

267

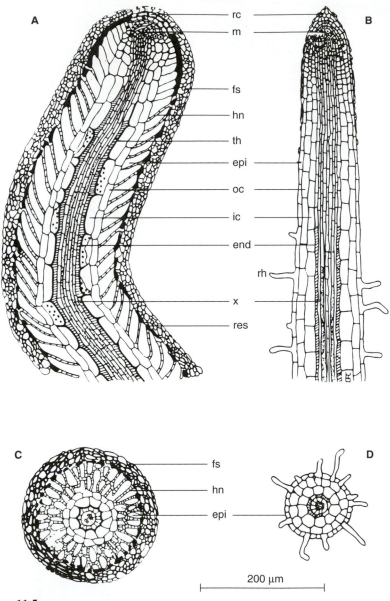

Figure 11.5

Anatomy of ectomycorrhizae of *Eucalyptus* compared to uninfected roots. **A, B,** Median longitudinal sections of mycorrhiza and nonmycorrhizal root, respectively. **C, D,** Transverse sections of mycorrhiza and uninfected root, respectively. rc = root cap, m = meristematic region, fs = fungal sheath or mantle, hn = Hartig net, th = thickened walls of inner cortex, epi = epidermis, oc = outer cortex, ic = inner cortex, end = endodermis, rh = root hair, x = lignified protoxylem, res = collapsed residues of cap cells.

Figure 11.6

Schematic diagram of the association of arbuscular mycorrhizal fungi and soil aggregate of a plant root. The external mycelium bears chlamydospores (CH). Infection of the plant can occur through root hairs or between epidermal cells. Arbuscules in progessives stages of development and senescence are shown (A–F) as is a vesicle (V), and a macroaggregate (M)

mycorrhizae. As well as having mycorrhizal associations, some plants have tri-partite associations, such as N-fixing plants (e.g., legumes and actinorrhizal plants like red alder).

Most of the commercially important genera of trees in the temperate zones are ectomycorrhizal—for example, those in the families Pinaceae, Betulaceae, Fagaceae, Dipterocarpaceae, and Myrtaceae (Richards 1987). Ectomycorrhizae usually are visible with the naked eye (Figure 11.7). Many are dichotomously

Figure 11.7
Western hemlock ectomycorrhizae. Note fungal mantles on short roots, mycelium, and rhizomorphs.
(Source: From Zak 1971.)

branched; others are coralloid, or pinnate (Smith and Reed 1997). Mantle morphotypes also have been used to classify mycorrhizae. Some fungi have loose mantles, and others have compact mantles. A number of mycorrhizal fungi also produce hyphal strands or rhizomorphs. However, since many fungi produce similar mantles, and as many as 2,000 species of fungi are thought to be mycorrhizal with one tree species like Douglas-fir, it has been difficult to determine which fungi are actually associated with tree roots in the field. There is not always a strong association between fungal fruiting bodies and the fungi associated with fine roots, but recent use of molecular techniques has allowed better exploration of this interesting topic.

Although ectomycorrhizae dominate many of the commercially important tree genera, VAM are the most abundant and are found in practically every taxonomic group of plants, including important agricultural plants (Smith and Reed 1997). Those gymnosperms that do not form ectomycorrhizae form VAM, such as plants in the Cupressaceae (e.g., western redcedar), and in the Araucariaceae and Podocarpaceae in the southern hemisphere. Some species in genera like *Acacia, Casuarina,* and *Populus* have both VAM and ectomycorrhizae. There are far fewer genera of fungi that form VAM than ectomycorrhizae; the dominant genera are *Glomus, Endogone, Gigaspora, Acaulospora,* and *Sclerocystis.* All are Zygomycota.

Roles of Mycorrhizae

Mycorrhizal fungi form mutualistic associations with plants. The plant provides carbohydrates to the fungus; the fungus provides nutrients and water to the plant, extends the effective root surface area and life of fine roots, and protects roots against pathogens, especially those that attack fine roots (Allen 1991). Some mycorrhizal fungi possess limited ability to decompose simple substrates, and they also are important in improving soil structure by binding soils particles together through hyphal networks (Perry et al. 1989).

Ectomycorrhizae develop most strongly on acid, podsolized soils with well-developed litter layers (Allen 1991). They appear to be more important for nutrient uptake in plants that have poorly branched root systems than in plants that have highly branched root systems. Mycorrhizae may be more important for protection against pathogens and enhancing fecundity (i.e., seed production) in plants with highly branched roots. Nutrients, such as N and P, and carbon also are able to move from one plant to another via an interconnected hyphal network, and this may aid in plant survival. There is an inverse relationship between nutrients and mycorrhizae. The greater nutrient availability, particularly for N and P, the less mycorrhizal development—and this has great implications for forest nurseries where fertilizers are routinely applied and in fertilized plantations.

Plants that have well-developed mycorrhizae may be more drought-resistant, and ectomycorrhizal fungi that produce rhizomorphs are thought to give more drought resistance to conifer seedlings than those that do not. In addition, VAM may enhance water uptake by lowering root resistance or increasing stomatal conductance.

Ectomycorrhizae seem to protect roots against pathogens by (1) producing antibiotics, (2) competing for available nutrients in the rhizosphere, (3) creating a physical barrier to infection with the fungal mantle, and (4) allowing the development of antagonistic bacteria and fungi in the mycorrhizosphere (Smith and Reed 1997). Fine roots are thought to be protected by ectomycorrhizal fungi against infection by *Phytophthora cinnamomi*. *Leucopaxillus cerealis* var. *picceina* on shortleaf and loblolly pines under controlled conditions has been shown to produce antibiotics that protect the host from infection. Some VAM fungi also confer protection against pathogens. Mycorrhizae also may reduce the incidence of Verticillium wilt, perhaps due to the improved P status of mycorrhizal plants.

Management of Ectomycorrhizae

There is substantial evidence from field studies that there is considerable potential for improving tree survival and production by inoculating seedlings with specific ectomycorrhizal fungi (Robson et al. 1994, Smith and Reed 1997). Early in this century attempts to establish northern hemisphere conifers in the southern hemisphere, especially in Australia and Rhodesia, failed until soil or organic matter was introduced containing compatible mycorrhizal fungi from the species' natural environment. A similar situation occurred in the Caribbean region.

Many significant responses also have been noted in nursery and field experiments with mycorrhizae in the southeastern United States. Southern pines inoculated with *Pitholithus tinctorius* (Pt) performed better in the nursery and in outplanting studies than noninoculated control seedlings (Cordell 1985). Survival of conifer seedlings inoculated with Pt is particularly enhanced on eroded and degraded sites such as mine spoils, where nutrients are low. Pt does not universally give a positive growth response, and other fungi may give better responses because they are adapted better to that environment. For example, in dry southwestern Oregon, *Rhizopogon* spp. are better suited to the soils and environment and are more effective in promoting early growth than Pt, even though Pt does occur in western North America. Responses to **inoculation** with ectomycorrhizal fungi are not confined to harsh sites, however, and some good responses have been observed where ectomycorrhizal **inoculum** is abundant. For example, inoculation of fast-growing eucalypts with specific fungi increased early tree growth. More than 500 specific fungal–host tree associations have now been studied and many appear to have a positive effect on growth of the host. Some fungi have a neutral effect; others cause growth loss and could even be considered to be parasitic.

Mycorrhizal inoculum can be obtained from (1) fungi in culture or (2) basidiocarps and spores. Not all species can be grown in culture or fruit profusely, however. *Pitholithus tinctorius* is easily grown and is commercially available for addition to nursery soils. Mycelial forms of inoculum often are more effective than spores in colonizing roots and are commonly contained in peat-vermiculite mixes, liquid media, or alginate beads.

Although nursery seedling roots may be well inoculated with mycorrhizal fungi that are capable of enhancing growth, these fungi may be replaced by natural mycorrhizal fungi in the field. Some mycorrhizal fungi are restricted to a single host or a few hosts; others have a broad host range. In addition, many mycorrhizal fungi may be associated with one tree species, such as Douglas-fir. Understanding the natural ecology of mycorrhizae is important in interpreting the success of inoculated seedlings in the field. Mycorrhizal fungi change with succession and stand age, and the fungi associated with seedlings are likely to be different than those associated with older stands. Early in primary succession the first plants are likely to be nonmycorrhizal or facultatively mycorrhizal. This certainly seems to be the case on sand dunes, pyroclastic flows at Mount St. Helens, and many mine spoils. Here there is little or no organic matter and mycorrhizal inoculum is usually absent. As succession proceeds obligately mycorrhizal species increase until in late succession they dominate and nonmycorrhizal species are virtually absent. Inoculum of obligately mycorrhizal species arrives on the site via wind-borne spores or is carried by animals on their hooves or by burrowing animals such as gophers. In secondary succession mycorrhizal inoculum usually is not reduced to the same extent, although clearcutting on some dry pumice soil types in southwest Oregon caused ectomycorrhizal inoculum to be reduced drastically because the hyphae holding soil particles together died after the trees were severed and the soil lost its structure (Perry et al. 1987, 1989). Attempts at replanting these sites resulted in failure. However, seedlings planted near or under surviving *Ceanothus* plants were able to survive because of the higher organic matter in the soil and the longer survival of mycelium. Animal activity also was higher in this area.

Early successional arbutoid shrubs appear to have the same fungal associates as later successional ectomycorrhizal trees. These fungi can survive fire and clearcutting because the shrubs can resprout. Observing this, Dave Perry and colleagues at Oregon State University proposed his "bootstrapping" hypothesis (Perry et al. 1989). The major concept is that disturbed ecosystems rarely reestablish "from scratch" (i.e., they do not go back to primary succession), but mycorrhizal fungi survive in patches, and thus pull the system back up "by their bootstraps." This probably also happened on Mount St. Helens even in the pyroclastic flow area. Logs in forests also may act as patches for survival of mycorrhizal mycelium during harsh dry conditions, since logs tend to hold moisture. Mycorrhizal mats often are associated with logs.

The ability to identify sites where inoculated trees are likely to succeed is of great consequence. Two factors are very important: (1) soil nutrient status (particularly for P, although this is not always directly related to growth improvement resulting from inoculation) and (2) the abundance of inoculum. Responses are likely to be greatest where there are low populations of indigenous fungi. The largest effects of inoculation are expressed when there is high nutrient requirement—that is, when trees are young. Inoculated Douglas-fir seedlings in Oregon and eucalyptus seedlings in southwestern Australia had significantly increased growth on recently clearcut sites (Robson et al. 1994). This suggests that indigenous mycorrhizal fungi may be dominated by late successional species that are less effective than those from the nursery in these cases.

Forest management and human activities have had a large influence on mycorrhizae, particularly clearcutting, fertilization, fire, and the use of exotic species (Perry et al. 1987). Air pollution and acid precipitation also influence mycorrhizae. In forested areas impacted by high air pollution in Eastern Europe, the activity and diversity of mycorrhizal fungi were dramatically reduced compared to areas with less pollution (Fellner and Peskova 1995). Mycorrhizal fungi also are likely to be affected by increasing atmospheric CO_2 and global change.

REFERENCES

Aldhous, J. R., and W. L. Mason. 1994. *Forest nursery practice.* Forestry Commission Bulletin 111, London.

Allen, M. F. 1991. *The ecology of mycorrhizae.* Cambridge University Press, Cambridge, United Kingdom.

Cordell, C. E. 1985. The application of *Pitholithus tinctorius* ectomycorrhizae in forest land management, pp. 69–71. In R. Molina (ed.). *Proceedings of the sixth North American conference on mycorrhizae.* Forest Research Laboratory, Oregon State University, Corvallis, Oreg.

EPA (Enivronmental Protection Agency). 1997. *Stratospheric ozone protection. Methyl bromide alternatives. 10 case studies. Soil, community and structural use,* vol. 3. Office of Air and Radiation, USEPA, Washington, D.C.

Fellner, R., and V. Peskova. 1995. Effects of industrial pollution on ectomycorrhizal relationships in temperate forests. *Canadian Journal of Botany* 73:1310–1315.

Hamm, P. B., S. J. Campbell, and E. M. Hansen. 1990. *Growing healthy seedlings: identification and management of pests in Northwest forest nurseries.* Forest Research Laboratory Special Publication 19, Oregon State University, Corvallis, Oreg.

Paul, E. A., and F. E. Clark. 1996. *Soil biology and biochemistry,* 2nd ed. Academic Press, New York.

Perry, D. A., R. Molina, and M. P. Amaranthus. 1987. Mycorrhizae, mycorrhizospheres and reforestation: current knowledge and research needs. *Canadian Journal of Forest Research* 17:929–940.

Perry, D. A., M. P. Amaranthus, J. G. Borchers, S. L. Borchers, and R. E. Brainerd. 1989. Bootstrapping in ecosystems. *BioScience* 39:230–237.

Richards, B. N. 1987. *The microbiology of terrestrial ecosystems.* Longman Scientific, Essex, England.

Robson, A. D., L. K. Abbott, and N. Malajczuk (eds.). 1994. *Management of mycorrhizas in agriculture, horticulture and forestry. Developments in plant and soil sciences,* vol. 56. Kluwer Academic Publishers, Dordrecht, The Netherlands.

Smith, S. E., and D. J. Reed. 1997. *Mycorrhizal symbiosis,* 2nd ed. Academic Press, New York.

Sutherland, J. R., G. M. Shrimpton, and R. N. Sturrock. 1989. *Diseases and insects in British Columbia forest seedling nurseries.* Forestry Canada and B.C. Ministry of Forests, FRDA Report 065, Victoria, B.C., Canada.

Tainter, F. H., and F. A. Baker. 1996. *Principles of forest pathology.* John Wiley and Sons, New York.

Zak, B. 1971. Characterization and identification of Douglas-fir mycorrhizae, pp. 38–53. In E. Hacskaylo (ed.). Mycorrhizae. *Proceedings of the first North American conference on mycorrhizae,* Superintendent of Documents, Washington, D.C.

ROOT DISEASES

INTRODUCTION

Root diseases have important influences on forest ecosystems (Sinclair et al. 1987, Morrison 1989, Manion 1991, Johannson and Stenlid 1994, Tainter and Baker 1996). Not only do they cause loss of timber production, but they also change stand structure by causing mortality, and influence forest succession, species biodiversity, decomposition, and nutrient cycling processes. Fungi causing root diseases interact with insects, particularly bark beetles and defoliators, fire, wind, drought, and air pollutants in forests (Schowalter et al. 1997).

Forest management practices have generally increased the incidence of root diseases throughout the world, resulting in mortality, growth loss, reduced stand densities, site deterioration, and loss of long-term productivity (Shaw and Kile 1991, Tainter and Baker 1996, Woodward et al. 1998). Until recently root diseases in managed forests were considered to have only negative impacts. Forest managers typically viewed root diseases as nuisances interfering with timber management objectives. However, new views of forest management involving ecosystems have forced us to examine beneficial roles of root diseases in managed forests, particularly those involving wildlife.

Management of root diseases in forests is extremely limited by economic and ecological constraints. Thus managers must learn to manage forests within the constraints imposed by root diseases. In some cases it may be best to leave patches of forest with root disease to provide wildlife habitat and biodiversity. However, these diseases cannot be completely ignored since they will cause increasing damage to timber resources in the future. Fire suppression in areas with high natural fire frequency has had a major impact on forest health (see Chapter 4), by changing tree species, and increasing the incidence of root diseases, insects, and large intense wildfires. In these areas active management of root diseases may be necessary to return ecosystems to a healthy condition. The following topics are covered in this chapter: types and causes of root diseases; signs, symptoms, and mechanisms of spread; interactions of root diseases, insects and fire; management of root diseases;

and the use of root disease models. Root diseases occurring in forests throughout the world are discussed, but emphasis is on North America. Root diseases in forest nurseries are covered in Chapter 11.

TYPES AND CAUSES OF ROOT DISEASES

There are basically three types of root diseases: structural, feeder, and vascular root diseases. All important root diseases in forest ecosystems are caused by fungi. Table 12.1 shows common names, species of fungi involved, main hosts, and geographic distribution of the most important root diseases. Hosts and distributions are discussed in more detail in the following section.

Five dominant genera of fungi are involved: *Armillaria, Heterobasidion*, and *Phellinus*, which cause structural root diseases, *Phytophthora*, which causes feeder root diseases, and *Leptographium*, which causes vascular root diseases (Hadfield et al. 1986, Morrison 1989, Tainter and Baker 1996). Pathogenic species of these genera may have races or strains that vary in virulence—that is, the ability to cause severe symptoms on a particular host. Virulent strains also can vary in aggressiveness. Hosts may range from very resistant to very susceptible.

Structural Root Diseases

Structural root rots are caused by Basidiomycete fungi, which decay the larger woody roots. Some also cause butt rots as well as root rots. *Armillaria* has a worldwide distribution with a huge host range, attacking most woody plants including hardwoods, conifers, and shrubs (Shaw and Kile 1991). It was originally thought to be only one species in the northern hemisphere (*Armillaria mellea*). However, recent examination has revealed that it is not one species, but rather a species complex involving at least 30 species (Table 12.1). Within those species there is considerable variation in virulence; some appear to have only strains that are low in virulence such as *A. gallica, A. borealis*, and *A. sinapina*. Other species such as *A. mellea, A. ostoyea,* and *A. luteobubalina* have highly virulent strains. *Armillaria mellea* occurs in Europe and in the eastern United States, but not in the western United States, except in California (Shaw and Kile 1991). There are at least nine interfertility groups in North America; all but two are classified as species. Five species occur in Europe and three are common to both continents: *A. ostoyae, A. mellea,* and *A. gallica. Armillaria ostoyae* is the dominant pathogen in western North American forests (Shaw and Kile 1991).

Heterobasidion annosum also is widely distributed throughout the temperate zone in the northern hemisphere and has even been reported from Australia (Morrison 1989, Woodward et al. 1998). It mostly occurs on conifers but has been reported on hardwoods. There at least two strains of *H. annosum* in northern hemisphere conifer forests; the S or spruce strain, which occurs on spruces, firs, and hemlocks, and the P strain, which typically occurs on pines, but also on S-type hosts. A third group, the F group, has recently been recognized on *Abies* in Italy. In the western United States both the S and P strains occur; only the P strain occurs in the eastern United States. In Europe the S type mainly causes butt rot on Norway spruce; the P type causes root rot in Scots pine and some butt rot in Norway spruce and other tree species.

Phellinus weirii has a narrower distribution and occurs in the Pacific Northwest United States, British Columbia, and Alaska (Thies and Sturrock 1995). It also has been reported from Japan and Manchuria. Two groups are recognized in *P. weirii:* a cedar form (that includes the type specimen from Idaho) and a fir form. The host range of *P. weirii* is relatively broad, but the fir form rarely attacks western redcedar, and the cedar form occurs primarily on western redcedar. Although both forms are still called *P. weirii,* some taxonomists believe that the fir form should be correctly named *Phellinus sulphurascens.* There also may be a third Asian form.

Species names have continually changed over time as taxonomic studies have become more sophisticated. This sometimes leads to confusion in teaching forest pathology, since many different scientific names for the same fungus appear in the literature. *Heterobasidion annosum* was formerly known as *Fomes annosus* and *Fomitopsis annosa. Poria weirii* was changed to *Phellinus weirii.* Molecular techniques now are being used to determine species and strains of root disease fungi.

A number of other root diseases deserve mention, including the structural root diseases caused by the Basidiomycetes *Phaeolus schweinitzii, Inonotus tomentosus,* and *Perenniporia subacida* (common in North American conifer forests) and *Phellinus noxius, Rigidiporus lignosus,* and *Ganoderma lucidum* (important in lowland tropical forests) (Morrison 1989) (Table 12.1). The impact of *Inonotus tomentosus* in Canada is second only to *Armillaria* (Hansen and Lewis 1997).

Feeder and Vascular Root Diseases

Feeder root diseases are typically caused by Oomycete fungi, particularly in the genus *Phytophthora,* which has many species. The Oomycetes have recently been placed in another kingdom than the true fungi (see Chapter 10). *Phytophthora cinnamomi* is the main species causing problems worldwide, and it even attacks many species of eucalyptus in Australia (Old 1979). *Phytophthora lateralis* attacks Port-Orford-cedar in the Pacific Northwest United States (Roth et al. 1987). *Phytophthora* spp. cause most damage in moist soil conditions, since the fungus spreads by swimming spores.

Vascular root diseases are generally caused by Ascomycetes. These fungi typically infect the vascular system in the roots and lower stem, causing wilting *Leptographium* spp. (anamorph) or *Ophiostoma* spp. (teleomorph) are the most common causes of these diseases (Harrington and Cobb 1988). *Leptographium wageneri* (formerly *Verticicladiella wageneri*) attacks conifers in western North America. This species has at least three forms: one attacking Douglas-fir, one attacking ponderosa and Jeffrey pines, and one attacking pinyon pine. Several other *Leptographium* species are pathogenic, including *L. procerum,* although it is not as pathogenic as *L. wageneri.* This fungus has been reported from the eastern United States, Eastern Europe, Yugoslavia, New Zealand, and Canada. The feeder root disease caused by *Rhizina undulata,* an Ascomycete, which has been reported from North America and Europe, is interesting because *R. undulata* is

TABLE **12.1** | **Common Root Diseases, Causal Organisms, Main Hosts, and Geographic Distribution**

Common name	Causal organism	Main hosts	Geographic distribution
		Structural Root Diseases	
Armillaria root disease	*Armillaria* species (at least 30)	Conifers and hardwoods	Worldwide (Africa, Asia, Australia, Central and South America, India, New Zealand)
North America and Europe			
	A. ostoyae	Conifers and hardwoods (e.g., Douglas-fir and ponderosa pine, birch, and maple)	North America (especially western)
	A. mellea	Hardwood trees and shrubs	Europe
	A. gallica = A. bulbosa	Hardwoods and conifers (butt rot)	Eastern North America, California, Europe (low elevations)
	A. cepistipes	Conifers	North America, Europe (low elevations)
	A. borealis	Conifers	Europe (high elevations), British Columbia, Washington
	A. sinapina	Conifers and hardwoods	Europe and Russia (high latitude)
	A. tabescens	Conifers and hardwoods	Northern coniferous—conifers west, hardwoods east
	A. calvescens	Hardwoods	Southeast United States, also reported from Northeast and Midwest
	A. nabsoma	Hardwoods	East Coast, Midwest, Canada
	A. gemina	Hardwoods	Western North America
			Eastern North America
Australia and New Zealand			
	A. luteobubalina	Dry and wet sclerophyll eucalypt	Australia
	A. hinnulea	Mixed forest	Australia, New Zealand
	A. novae-zelandie	Cool temperate rainforest, mixed forest, wet sclerophyll eucalypt	Australia, New Zealand, South America
	A. fumosa	Dry sclerophyll eucalypt	Australia

Common name	Causal organism	Main hosts	Geographic distribution
Structural Root Diseases (continued)			
Annosus root and butt rot	*Heterobasidion annosum*	Conifers and hardwoods (rarely)	Worldwide (including Australia, New Guinea)
	S strain	Spruces, firs, hemlocks	
	P strain	Pines and S-type hosts	
	F strain	Firs	Europe
Laminated root rot	*Phellinus weirii*		Western North America, Asia
	Cedar strain	Western redcedar	Western North America
	Fir strain	Douglas-fir, grand fir, Mountain hemlock	Western North America
Brown root and butt rot	*Phaeolus schweinitzii*	Douglas-fir, firs, larches, spruces eastern and western white pines	Worldwide
Tomentosus root rot	*Inonotus tomentosus*	Conifers	Europe and North America
Yellow root rot	*Perenniporia subacida*	Conifers	Western North America
	Phellinus noxius, Rigidiporus licnopus, Ganoderma lucidum	Tropical hardwoods and conifers	Australia, New Guinea, Vanuatu, West Africa
Feeder Root Diseases			
Phytophthora root diseases	*Phytophthora* spp.	Conifers and hardwoods	Worldwide
Phytophthora root rot	*Phytophthora cinnamomi*	Conifers and hardwoods	Worldwide including Australia
Port-Orford-cedar root disease	*Phytophthora lateralis*	Port-Orford-cedar	Oregon and California
Rhizina root disease	*Rhizina undulata*	Conifers	Northern hemisphere
Vascular Root Diseases			
Black stain root disease	*Leptographium* spp.		
	L. wageneri		
	Race 1—*pseudotsugae*	Douglas-fir	U.S. Pacific Coast
	Race 2—*ponderosum*	Ponderosa and Jeffrey pines	Western United States
	Race 3	Pinyon pine	California and southwest United States
Procera root disease	*L. procerum*	Eastern white pine and other conifers	Worldwide including United States and Canada

279

a pyrophilous, or fire-loving, fungus (Hansen and Lewis 1997). Dormant spores in the soil germinate after the soil is heated by fire. The fungus then attacks the fine root system of trees. It is named the "tea pot" fungus because the disease was observed at locations in England where woods workers lit fires to boil water for their morning and afternoon tea. It has been observed in western Washington, killing Douglas-fir seedlings planted in recently slash-burned clearcuts. Currently slash burning is not practiced widely in this area because of air quality restrictions, so there are few problems from this fungus. However, if fire is reintroduced to forests as a management tool, incidence of this disease could increase.

SYMPTOMS, SIGNS, HOSTS, DISTRIBUTION, AND SPREAD OF ROOT DISEASES

Structural Root Diseases

Structural root diseases are relatively easy to recognize. A number of general aboveground symptoms common to all may be expressed to varying degrees (Hadfield et al. 1986). In the early stages of infection, there usually is a loss of older needles in conifers resulting in a thinning crown. Hardwoods also exhibit thin crowns. Chlorosis is common. A loss of leader growth results in a flat top on the tree. Young trees generally die more quickly than older trees. Trees old enough to produce a cone crop usually produce a large number of cones or a "distress" cone crop. Crown symptoms, however, are not always reliable indicators of root rot, since insects or animals such as bears and porcupines also may produce these symptoms. Trees may die standing up and slowly deteriorate from the top down, or they may be windthrown, revealing well-rotted roots. Vascular wilt and feeder root diseases usually kill trees more rapidly than structural root rots, and even large trees may quickly decline after infection, turn orange and die without losing a lot of foliage. Table 12.2 shows symptoms and signs of the three most important structural root diseases (Armillaria root disease, Annosus root and butt rot, and laminated root rot), in contrast to those for black stain and Phytophthora root diseases. Each disease has specific signs, symptoms, hosts, distributions, and mechanisms of spread as follows.

Armillaria Root Disease   *Armillaria* occurs worldwide in boreal, temperate, and tropical forests (Table 12.1), although not in hot lowland tropical forests (Shaw and Kile 1991). In the tropics it occurs in cooler higher-elevation forests. *Armillaria* has a wide host range, including conifers, hardwoods, shrubs, and many agricultural plants such as citrus and grapes. It causes root disease and butt rot in natural forests as well as plantations (Shaw and Kile 1991).

The general symptoms, signs, and mechanisms of spread of Armillaria root disease are shown in Figure 12.1. Spread occurs primarily by vegetative growth through mycelial growth and rhizomorphs (Wargo and Shaw 1985). Rhizomorphs are small-diameter (1 to 2 mm), dark-colored, stringlike structures that grow through the soil and forest floor, but are attached to the food source (e.g., stumps). They penetrate the roots of new hosts and expand the distribution of successful

TABLE 12.2 | **Symptoms and Signs of the Five Most Important Root Diseases**

	Armillaria root disease	Annosus root and butt rot	Laminated root rot	Black stain root disease	Phytoph-thora root disease
Symptoms					
Reduced height growth	X	X	X	X	
Yellow foliage	X	X	X	X	X
Slow loss of foliage	X	X	X	X	
Distress cones	X	X	X	X	
Slow crown decline	X	X	X	X	
Rapid tree death	X			X	X
Dead tree, no foliage loss	X				
Abundant basal resin flow	X				
Cinnamon stain in inner bark					X
Black stain in sapwood				X	
Roots rotted	X	X	X		
Windthrown live trees	X		X		
Insect galleries under bark	X	X	X	X	X
Signs					
Fleshy yellow-golden mushrooms on tree base	X				
Mycelial fans	X				
Rhizomorphs	X				
Leathery conks		X			
Setal hyphae			X		
Ectotrophic mycelium			X		
Creamy leathery pustules on roots		X			
Advanced decay					
Laminated decay with pits on both sides of sheets			X		
Laminated decay with pits on only one side of sheets		X			
Yellow, stringy decay with black zone lines	X				
White, stringy decay with black specks		X			

Source: Adapted from Hadfield et al. (1986).

genotypes. The disease also spreads through mycelium when roots of an uninfected tree come in contact with an old infected root or stump. Rates of spread usually are a maximum of 1 to 2 m/year. Basidiospore spread is limited, although new infection centers are thought to be initiated this way (Shaw and Kile 1991).

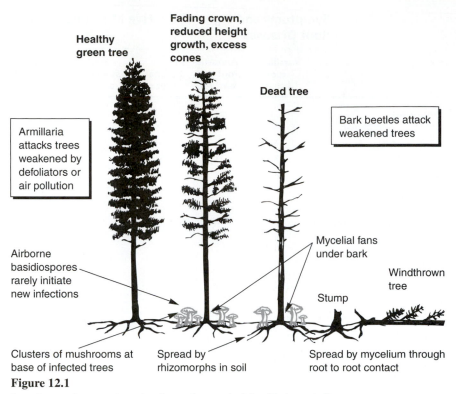

Figure 12.1

Symptoms, signs, and mechanisms of spread of Armillaria root disease.

Pathogenic species of *Armillaria* cause classical root rot symptoms including reduced leader growth, tree decline, thinning foliage, and excess cone crops (Table 12.2). Very young trees, however, may die very rapidly with few decline symptoms. *Armillaria* also causes resin flow near the base of the tree (Figure 12.2) and on roots. Signs include the presence of white mycelial fans under the bark (Figure 12.3) and as mentioned previously, rhizomorphs (Figure 12.4) Rhizomorph forms vary in different *Armillaria* spp; some have monopodial branching, like *A. gallica;* others, such as *A. ostoyae,* have dichotomous branching (Morrison et al. 1991). Incipient decay in roots has a "water-soaked" appearance and advanced decay is a white, spongy rot. Black **zone lines** are common in the wood (Table 12.2). A further sign is the presence of light brown to honey-colored mushrooms with a central stalk 8 to 25 cm long, with an annulus, appearing in clusters near the base of trees in the autumn (Figure 12.5). The cap is 5 to 13 cm across and may have brown scales. Gills are white to pale yellow and attached to the stalk (Shaw and Kile 1991, Scharpf 1993).

Let us now consider the role of *Armillaria* in natural forests, where it is widely distributed, but may cause little disease (Shaw and Kile 1991). A relatively stable balance appears to exist between hosts and pathogens in natural forests such that in absence of stress, minor infections have little effect on tree or forest health.

Figure 12.2
Resin flow near the base of a young Douglas-fir stem resulting from infection by *Armillaria ostoyae.*

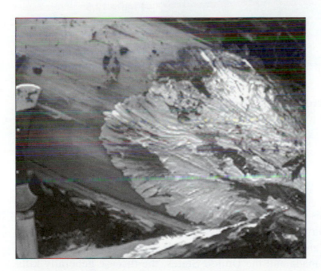

Figure 12.3
White mycelial fans of *Armillaria ostoyae* under conifer bark.

Management of native forests, however, usually changes this balance. The nature of the disease in native forests is shown in Box 12.1. Primary disease is a continuum from minor to major root infection, but it is not necessarily lethal or even visible. Secondary disease results from increased stress.

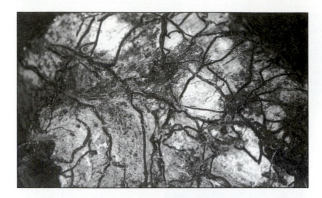

Figure 12.4
Rhizomorphs of
Armillaria.
(Source: From Skelly et al. 1987.)

Figure 12.5
Clusters of *Armillaria*
mushrooms commonly
occur near the base
of trees.

In the moist coniferous forests of North America and northern Europe, *Armillaria* does not have a dramatic effect and causes mortality of single or small groups of seedlings or saplings early in stand development. Here *Armillaria ostoyae* is the major pathogen, although *Armillaria sinapina,* which has less virulent strains, occurs widely in southeastern Alaskan forests. *Armillaria borealis* and *A. cepistipes* cause minor mortality and butt rot in natural Scandinavian forests (Shaw and Kile 1991).

Most conifer species in the western United States, except western larch, incense-cedar, and western redcedar, are severely or moderately susceptible. In this region hardwoods are usually less susceptible than conifers. In coastal Douglas-fir forests in Canada and the United States, the disease generally kills less than 1% of the trees and rarely more than 2 or 3%. Coastal Douglas-fir is only moderately susceptible to damage, and resistance to infection appears to increase as trees age. Mortality after age 25 seldom occurs unless the trees have been stressed.

Box **12.1**

Features of Disease Caused by *Armillaria* Species in Native Forests

Primary Disease

Root lesions or root rot, basal cankers, butt rot

Killing of natural regeneration, mortality decreasing with stand age

Killing of trees of all ages and sizes or singly in patches throughout the life of a stand

Secondary Disease

Killing of trees—weakened by stress, singly or on a standwide basis—by preexisting or new infections

Source: Adapted from Shaw and Kile (1991).

In contrast, in the drier inland forests of western North America, *Armillaria* causes a killing primary disease that affects hundreds of thousands of hectares in eastern Oregon and Washington, northern Idaho, western Montana, southern British Columbia, and the central and southern Rocky Mountains. Here *Armillaria ostoyae* is an aggressive pathogen on interior Douglas-fir, true firs, and pines. Trees of all sizes and ages are attacked, although damage tends to diminish with stand age. When these forests are stressed by drought, the severity of *Armillaria* increases. Past forest management practices have increased stress. Partial cutting of selected species like ponderosa pine has changed the forest to include more susceptible species such as Douglas-fir and true firs, and fire suppression has increased stem densities, resulting in concentrations of drought-stressed trees (McDonald et al. 1987). Environmental conditions strongly influence the development of the fungus, especially soil moisture, temperature, aeration, pH, nutrients, and organic matter.

Active disease centers can be several hectares in size and may involve more than 25% of the trees, causing large annual volume losses. Single clones of the fungus can occupy large areas—for example, several hundred square kilometers in southwest Washington and Michigan.

Armillaria also frequently forms complexes with other root disease pathogens, particularly *Phellinus weirii* and *Leptographium wageneri* and bark beetles in the western North America, and these are discussed later in this chapter (Schowalter and Filip 1993). *Armillaria* also has been recognized as a secondary pathogen in many of the diebacks and declines discussed in Chapter 15.

There are few natural forests remaining in Europe, but in the eastern Pyrenees in France mortality from *A. ostoyae* in relatively undisturbed pine forests has caused a disease called *ring disease of mountain pine.* Here ring diameters can reach 120 m, and they expand at a rate of 1 m/year. Rainfall is low in this region and bark beetles may be acting as a stress agent. *Armillaria* also is a primary pathogen in eucalyptus forests in Australia. For example, *A. luteobubilina* affects many dry sclerophyll forests in central Victoria as well as karri and jarrah forests in southwestern Australia where it occurs in expanding pockets. At least 81 eucalypt and understory species are susceptible.

Armillaria root disease is particularly important in plantations and it is widespread on planted hosts, both hardwoods and conifers, throughout temperate and tropical regions (Shaw and Kile 1991). Figure 12.6 shows the distribution of *Armillaria* attacks in planted hosts throughout the world. As well as attacking forest trees, it attacks other plantation species, such as apples, citrus, nuts, grapes, and berries. The disease tends to be very common in areas with moderate temperatures. In the tropics it tends to occur at higher elevations and appears to be absent in inland regions with very cold continental climates, for example, parts of Russia.

In Europe pure conifer plantations are attacked mainly by *A. ostoyae*, whereas *A. mellea* tends to attack mostly planted hardwoods. Interestingly, the most severe attacks in conifers appear to occur on former natural hardwood sites. On stressed sites several other less pathogenic species, such as *A. gallica,* are involved. *Armillaria* also attacks introduced species in Europe, such as Douglas-fir in Germany. Its impact, however, on forest plantations in Europe is less than that of *H. annosum*.

In North America *Armillaria* also attacks planted species (Figure 12.6). *Armillaria ostoyae* is the major pathogen on conifers; *A. mellea* and *A. gallica* occur on hardwoods. In the western United States and Canada, its importance in plantations pretty much mirrors its importance in native forests. In coastal areas of Oregon, Washington, and British Columbia it is relatively unimportant in conifer plantations because it causes low mortality and acts as a thinning agent in young stands. It also is of little consequence in California, but is extremely important as a mortality agent in plantations in inland western North America. *Armillaria* is present in eastern Canada and causes mortality in plantations of native and introduced fir, spruce, and pine species. It also causes mortality in young spruce and pine plantations and hardwoods in the northeastern United States, especially oaks, and in conifers in midwestern and southeastern states.

In Central and South America reports from planted hosts are less common, but it does cause mortality in exotic pines such as radiata pine in Mexico, Ecuador, Chile, and southern Brazil and slash pine in Cuba, Honduras, Jamaica, Columbia, and Peru. It has not been reported from the extensive plantations of eucalypt, pine, and *Gmelina* in the Amazon of Brazil, perhaps because this region is too warm.

Losses from *Armillaria* have been reported from the fast-growing exotic conifer and hardwood plantations in many African countries (Morrison 1989, Shaw and Kile 1991). The tree genera involved are *Auracaria, Pinus, Widdringtonia, Acacia, Albizzia, Cassia, Cedrela, Cupressus, Eucalyptus, Gmelina, Grevillea, Khaya, Tectona, Terminali, Toona* and *Vitis*. The most common species on planted hosts is *Armillaria heimii*.

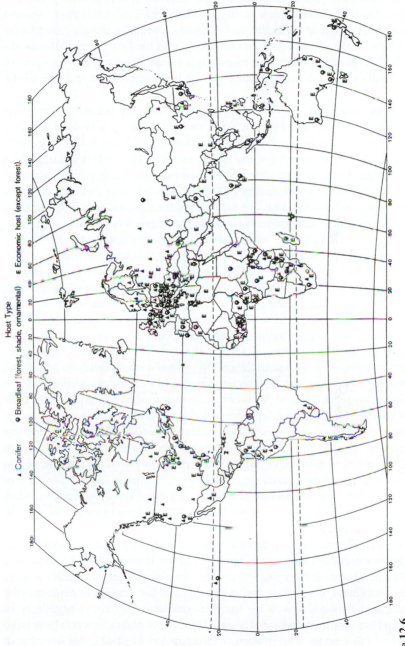

Figure 12.6

Distribution of recorded Armillaria attacks in conifer, broadleaf tree plantations, and other economic hosts throughout the world.

(Source: From Shaw and Kile 1991.)

Armillaria also is widespread in China and Japan and at higher elevations in India, Vietnam, Malaysia, Indonesia, Fiji, Hawaii, and Papua New Guinea in exotic plantations of pines, eucalypts, acacia, and teak. It occurs in all plantation species in Australia (exotic pines, eucalypts, *Auracaria*), but the impact seems to be minor. Both *A. novae-zelandiae* and *A. limonea* occur in radiata pine plantations in New Zealand, and losses can be dramatic in the first 5 years after establishment after clearing of the native forests. It is known to occur in second-growth stands of radiata pine, but losses do not appear to be as great (Shaw and Kile 1991).

Annosus Root and Butt Rot   The fungus causing Annosus root and butt rot (*Heterobasidion annosum*) occurs on a wide variety of living and dead conifers including spruces, firs, hemlocks, and pines (Table 12.1), but only rarely on hardwoods. It occurs throughout the world and is considered to be the most important forest pathogen in Europe, attacking Norway spruce and Scots pine (Woodward et al. 1998). In Asia it is replaced by *Heterobasidion insulare.* In North America the main hosts are loblolly, slash and red pines in the east, and true firs, mountain and western hemlock, and ponderosa and lodgepole pines in the west. It is common in old-growth forests in North America on true firs and hemlocks. Typical symptoms and signs are shown in Table 12.2, and spread mechanisms are shown in Figure 12.7*A* for pines and in Figure 12.7*B* for true firs, hemlocks, and spruces.

Heterobasidion annosum spreads by vegetative mycelia as well as airborne spores. Perennial fruiting bodies or conks (Figure 12.8) are common, although sometimes they are difficult to find. They can be found on the underside of roots of windthrown trees, under the duff in the root collar area, on the underside of downed stems, and inside rotting stumps, particularly in California. The pore surface of the fruiting body is cream-colored, has a sterile margin with no pores, and a distinctive smell. They can grow as large as 25 cm in diameter. Millions of airborne spores are produced, and this is one of the major mechanisms of spread. During favorable periods, the fungus produces numerous basidiospores. In the Pacific Northwest United States and England, spores are produced year-round, but have low production in the winter in Scandinavia and in the summer in the southeastern United States. Spores land on cut stump surfaces, germinate if the environment is favorable, and spread throughout the root system, crossing to living trees via root contacts and grafts. The fungus can grow at least 1 m/year. Basidiospores also infect wounds in non-pine species. The asexual stage of this fungus (*Spiniger meineckellus*) produces asexual conidiospores, but the role of conidia in spread of the disease is not known. *Heterobasidion annosum* also is spread by root infection where expanding roots contact the roots of infected trees or stumps where it can survive for as long as 50 years in large root systems, but for a shorter time in smaller stumps.

Like Armillaria root disease, forest management has caused an increase in the incidence and damage caused by Annosus root and butt rot, particularly in Europe and the southeastern United States. Thinning of plantations results in large numbers of cut stumps. The impact of thinning has been of great concern in Europe and North America since it has great potential to increase disease incidence and losses. Care also must be taken not to wound remaining trees during thinning.

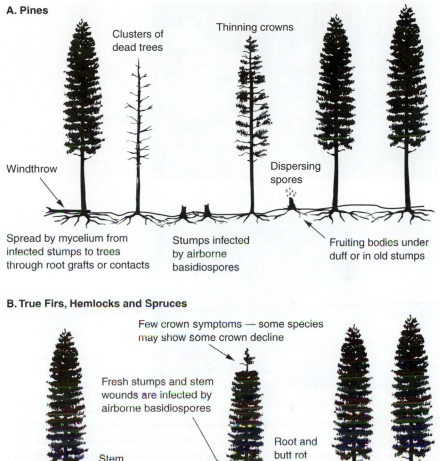

A. Pines

Clusters of dead trees

Thinning crowns

Windthrow

Dispersing spores

Spread by mycelium from infected stumps to trees through root grafts or contacts

Stumps infected by airborne basidiospores

Fruiting bodies under duff or in old stumps

B. True Firs, Hemlocks and Spruces

Few crown symptoms — some species may show some crown decline

Fresh stumps and stem wounds are infected by airborne basidiospores

Root and butt rot in trees

Stem breakage

Windthrow

Dispersing spores

Fruiting bodies in root crotches of downed logs, on rotted trees and old stumps

Mycelium spreads from infected stumps to trees via root grafts and contacts

Figure 12.7

Symptoms, signs, and spread mechanisms for Annosus root and butt rot in (**A**) pines and (**B**) true firs, hemlocks, and spruces.

In resinous trees such as pines, *H. annosum* tends to kill trees outright by girdling the root collar, whereas in whitewood genera such as spruces, firs, and hemlocks it causes both root and butt rot (Tainter and Baker 1996). Pines trees may die standing or be windthrown and are killed relatively rapidly. Weakened trees

Figure 12.8
Fruiting body (conk) of
Heterobasidion annosum
from western hemlock
(S strain).

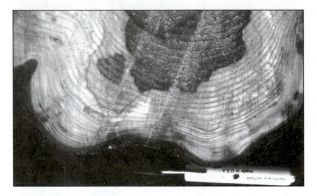

Figure 12.9
Incipient decay and
advanced decay of *Heter-
obasidion annosum* in a
western hemlock stem
section. Wetwood stain
is also present.

often are killed by bark beetles. High ground-level ozone concentrations in California are thought to make pines more susceptible to *H. annosum* and bark beetles. In spruces, firs, and hemlocks the fungus can cause extensive decay columns with few or no visible symptoms because the decay is in the xylem and does not influence the phloem until it is very well developed. Windthrow and stem breakage sometimes occur. Classical symptoms of root rot—including thinning foliage, declining height growth, and excess cone crops—are not always observed, particularly in species like western hemlock. Losses in whitewood species are mainly the result of butt rot, increased susceptibility to insect attack, and windthrow. In California and Oregon decay losses due to *H. annosum* in true firs are high (Filip and Goheen 1984).

Heterobasidion annosum in the roots and butts of spruces, firs, and hemlocks typically has a dark brown incipient decay stain with stringy advanced decay, which may include black specks (Figure 12.9). Sometimes the wood will laminate at the annual rings, but pits occur only on one side in contrast to *Phellinus weirii,* where pits form on both sides (Table 12.2) The incipient dark brown stain can sometimes be confused with wetwood, which is wood with a high moisture content with stains that usually are not brown, but pinkish (Shaw et al. 1995). Bacteria are commonly associated with wetwood, and there is no apparent degradation of the wood. In fact, wetwood may suppress the development of *H. annosum* and may be a natural biological control mechanism, particularly in western hemlock, where it appears to be associated with dead branch stubs. It was thought that thinning of western hemlock stands in coastal western North America was going to result in a major increase in disease severity like it did in Europe and the southeast United States. This has not occurred, although disease incidence has increased.

Laminated Root Rot   Laminated root rot has a much narrower geographic distribution than Armillaria root disease and Annosus root and butt rot (Table 12.1), but it is extremely important in the western United States and Canada, occurring on nearly all commercial conifers (Thies and Sturrock 1995). Douglas-fir, grand fir, and mountain hemlock are highly susceptible; pines and cedars are tolerant or resistant and hardwoods are immune.

Laminated root rot is one of the most destructive diseases and one of the most difficult root pathogens to manage. In Douglas-fir forests of western Oregon and Washington, at least 5% of the area is infected, and the average annual timber volume loss has been estimated to be 1 million cubic meters. *Phellinus weirii* is not distributed uniformly, but we are not sure how it is influenced by topography, climate, or soil conditions. Recent evidence from coastal Oregon suggests that trees growing on mid to upper slopes appear to have a higher incidence of infection than those on lower slopes, implying that moisture stress is factor in host susceptibility (Kastner et al. 1994). It occurs in old-growth stands, but the disease is most damaging in stands from about 25 to 125 years old (Thies and Sturrock 1995).

Signs and symptoms are shown in Table 12.2 and mechanisms of spread are illustrated in Figure 12.10. Laminated root rot tends to occur in infection centers

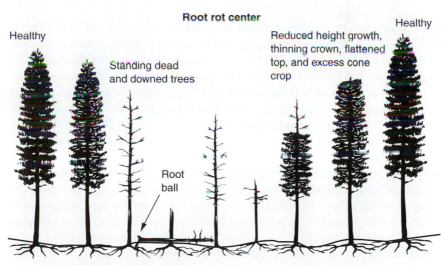

Figure 12.10

Symptoms, signs, and spread mechanisms of *Phellinus weirii*.

Figure 12.11
Windthrown Douglas-fir in a laminated root rot center with characteristic root-balls.

or "pockets" characterized by patches of dead and dying trees (Figure 12.11). Dying trees are prone to attack by bark beetles. Trees can die standing or be windthrown. Roots of infected trees commonly break off close to the root collar, forming characteristic "root balls." There is usually little successful regeneration of susceptible host species in the pockets. In western Oregon, Washington, and British Columbia infection centers usually are occupied by brush and herb species, immune hardwoods such as red alder and vine maple, and resistant conifers (Thies and Sturrock 1995). Douglas-fir regeneration usually is killed, thus speeding up the rate of succession to more shade-tolerant species such as western hemlock. In higher-elevation mountain hemlock stands in the Cascades in central Oregon, doughnut-shaped patterns occur when viewed from the air because conifer regeneration occurs in the center of large pockets. Infection centers west of the Cascade crest are generally less than 0.4 ha but are larger elsewhere. The fungus spreads mainly via ectotrophic mycelium. Pockets expand at an average rate of only 0.3 m/year radially. Some pockets expand rapidly and may coalesce with neighboring ones, while others appear inactive with very little spread. The fungus can remain alive for 50 or more years in very large old-growth stumps and root systems after trees have been cut. Survival time is less in smaller stumps. Susceptible trees become infected as their roots contact inoculum in stumps and roots, allowing the disease to continue into

Figure 12.12
Laminated advanced decay of *Phellinus weirii* in Douglas-fir. Wood delaminates at the annual rings with pits on both sides of sheets.
(Source: From Allen, et al. 1996.)

the subsequent stand. Long-term survival of the fungus is assisted by disease-caused resin saturation of wood surrounding infections and the ability of the fungus to enclose itself by producing a hyphal sheath.

Crown symptoms include reduced leader growth, thinning and yellowing foliage, and in Douglas-fir, distress cone crops (Hadfield et al. 1986, Thies and Sturrock 1995). Obvious crown symptoms may not occur until 50% or more of the root has been killed, which may take 5 to 15 years in larger trees. Only about half of the trees in an infection center produce crown symptoms, and old-growth trees may not exhibit distinct crown symptoms. For survival some trees may confine the fungus to a few roots or the interior of the butt; others are able to produce new roots. All trees within 5 m of a dead tree are infected, but infection is rare beyond 15 m. On freshly cut stump surfaces incipient decay is a reddish-brown crescent or circular stain that tends to fade shortly after exposure to light. In advanced decay wood in the roots and tree butts separates at the annual rings (Figure 12.12) with pits on both sides. The fungus rarely travels more than 1.5 to 2.4 m above ground level before the tree dies. It produces buff-colored mycelium with reddish setae (needle-shaped) **hyphae,** and grayish-white ectotrophic mycelia may be visible on the outside of roots.

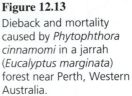

Figure 12.13
Dieback and mortality caused by *Phytophthora cinnamomi* in a jarrah (*Eucalyptus marginata*) forest near Perth, Western Australia.

Spread by spores is thought to be unimportant compared to vegetative spread by mycelia. Fruiting bodies are rarely produced and are of little value in diagnosing infection and usually form on the underside of downed trees, uprooted stumps, and sometimes on the boles of standing trees. In Douglas-fir stands fruiting bodies of the fir strain are annual and appear as **resupinate,** pore-covered crusts. Young fruiting bodies are light gray-brown, containing white sterile margins that change to chocolate-brown with time. The fruiting body of the cedar strain also has a brown pore surface, but is perennial.

Feeder Root Diseases

Feeder root diseases caused by *Phytophthora* spp. disrupt nutrient and water uptake. Once infected even very large trees wilt, turn red-brown, and die within a year or two. The most damaging species is *Phytophthora cinnamomi,* which is thought to have originated in southeast Asia or northeastern Australia and now occurs in tropical and temperate areas around the world, attacking more than 900 species and varieties of plants. Its major hosts are chestnut, eucalyptus, fir, pine, and ericaceous ornamentals, particularly azalea and rhododendron. *Phytophthora cinnamomi* is a soil-borne fungus that attacks the small absorbing roots and also may invade the large roots and the mainstem, growing in the inner bark and cambium. It causes littleleaf diseases of pines in the southeastern United States, and decline and death of *Eucalyptus* spp. in eastern and western Australia (Old 1979). Figure 12.13 shows mortality in a jarrah (*Eucalyptus marginata*) stand in western Australia near Perth.

The life cycle of *Phytophthora,* in this case illustrated by *P. infestans,* the cause of potato late blight is shown in Figure 10.12. Oospores, chlamydospores, sporangia, and zoospores are produced. Oospores and chlamydospores allow the fungus to survive adverse environmental conditions, and they can persist in the soil for several years in the absence of hosts. When conditions are favorable and in the presence of roots, oospores or chlamydospores give rise to mycelium or to sporangia (hyphal swellings) and zoospores. Zoospores are swimming spores and require free water for dispersal. *Phytophthora cinnamomi* requires warm, wet soil for significant pathogenic activity. It is unimportant in western North American forests, perhaps because soils are too dry when warm enough to favor the fungus. *Phytophthora lateralis,* on the other hand, is a problem in western North America, but it attacks only valuable Port-Orford-cedar which is native to a small region of southwestern Oregon and northern California, and associated Pacific yew (Roth et al. 1987). Port-Orford-cedar has also been planted widely as an ornamental and *P. lateralis* now has spread widely throughout the planted range. It is suspected of being of Eurasian origin. The fungus infects the small feeder roots and moves progressively to inner bark and cambial region where a brown stain typically develops. It kills small seedlings in weeks and large trees within 2 to 4 years. Forest management tends to hasten the spread of *Phytophthora* because spores or soil-containing spores can be transported on workers' boots and equipment, such as shovels, and by heavy machinery and vehicles such as bulldozers and road graders. As a result it is not uncommon for the disease to spread along road corridors. It also can be transported on the feet of animals and in infected plants. Transport in infected plants is a particular problem for the nursery and ornamental plant industry. *Phytophthora* root disease is a major problem in bareroot forest nurseries and care must be taken not to introduce it into nurseries. Once it is in nursery soil, it is extremely difficult to eradicate.

Vascular Root Diseases

The most common vascular root diseases are caused by *Leptographium* species (Harrington and Cobb 1988, Tainter and Baker 1996). The fungus does not decay root wood and the disease is very similar to oak wilt and Dutch elm disease, which cause vascular wilts in hardwoods. Black stain root disease is caused by three different host-specific races of *L. wageneri:* var. *pseudotsugae* infects Douglas-fir along the Pacific coast; var. *ponderosum* attacks hard pines (mostly ponderosa and Jeffrey pines); and a third variety occurs on pinyon pines in California and the southwestern United States. *Leptographium procerum* is found primarily in the eastern United States, and occurs on eastern white, Scots, loblolly, red, Virginia, slash, and Austrian pines. The majority of infected trees are on heavy, wet soils. It is not as pathogenic as *L. wageneri* and does not cause as much staining. This root disease has been associated with ozone-sensitive white pines in the Blue Ridge Mountains of Virginia (Skelly et al. 1987).

Leptographium wageneri attacks trees of all ages. It infects the roots and spreads throughout the sapwood to the lower boles, resulting in a loss of tree vigor. Growth of the terminal is reduced, needles are shorter and chlorotic, and the tree finally dies. Bark beetles may attack weakened trees. The disease is easy to identify in the field. It produces a distinctive sapwood stain in the roots, root crown, and lower bole (Figure 12.14). The stain is black to dark brown, often is infiltrated with resin, and occurs in arcs rather than the wedge shapes of blue stains. Long-distance spread is via insects; the root weevils *Pissodes fasciatus* and *Steremnius carinatus* and the root bark beetle *Hylastes nigrinus* are involved (Schowalter and Filip 1993). New infection centers commonly are associated with site disturbance and often are found along roads. Local spread is by mycelia through root contacts.

Figure 12.14
Black stain root disease
in the lower bole of a
Douglas-fir tree.

INTERACTIONS BETWEEN ROOT DISEASES, INSECTS, AND FIRE

Root pathogens, insects, and fire commonly interact and root pathogens may even interact with each other. It is quite common for more than one root pathogen to occur in a single location or even a single tree. Common pathogen combinations in western North America are (1) *P. weirii* with *A. ostoyae, H. annosum,* or *L. wageneri* and (2) *A. ostoyae* with *H. annosum, L. wageneri,* or *P. schweinitzii.* Occasionally *P. weirii, A. ostoyae,* and *H. annosum* occur together (Hadfield et al. 1986).

Pathogens and insects usually are closely associated (Schowalter and Filip 1993, Schowalter et al. 1997). In some cases there is a very intimate association between root disease fungi and insects—for example, *Leptographium wageneri* is vectored by root-feeding bark beetles and weevils. In most cases, however, insects do not vector the spores of root disease fungi, and the relationship is more indirect and related to the creation of ecological niches. Table 12.3 shows common bark beetle/root rot pathogen associations in western North America. Root diseases are particularly important as agents predisposing trees to bark beetles, especially mountain pine beetle (*Dendroctonus ponderosae*) in inland areas. Even in the moister coastal areas, bark beetles and pathogens form complexes. For example, Douglas-fir beetles may hasten tree death in *Phellinus weirii* pockets in Douglas-fir stands, and Port-Orford-cedar trees attacked by *P. lateralis* in southwestern Oregon and adjacent California often are attacked by bark beetles.

In some cases insects can weaken trees and make them susceptible to root diseases. For example, bark beetles in the Pyrenees in France act as a stress agent, predisposing pine forests to *Armillaria* infection. Defoliation of oaks by the gypsy moth in the eastern United States has made them very susceptible to attack by *Armillaria* (Wargo and Shaw 1985). In most cases trees do not die from defoliation, but from Armillaria root disease.

Fire interacts widely with both pathogens and insects. Fire suppression in dry interior forests in western North America along with selective cutting of larch and pines has resulted in the development of dense conifer forests of shade-tolerant species, such as true firs and Douglas-fir, that are susceptible to bark beetles, defoliators, root pathogens, dwarf mistletoes, drought, and inevitably catastrophic

T A B L E **12.3**	Root Pathogen–Bark Beetle Complexes in Western North America	

Pathogen	Host	Bark beetles
Armillaria ostoyae	Pines	Western pine beetle (Dendroctonus brevicomis), mountain pine beetle (D. ponderosae), pine engraver (Ips spp.)
	Douglas-fir	Douglas-fir pole beetle, Douglas-fir engraver
	True firs	Fir engraver (Scolytus ventralis)
	Western hemlock	Hemlock engraver
Heterobasidion annosum	True firs	Fir engraver (Scolytus ventralis)
Phellinus weirii	Douglas-fir	Douglas-fir pole beetle, Douglas-fir beetle (Dendroctonus pseudotsugae), Douglas-fir engraver
	True firs	Fir engraver (Scolytus ventralis), western balsam bark beetle
	Western hemlock	Hemlock engraver
Leptographium wageneri	Douglas-fir	Douglas-fir beetle (Dendroctonus pseudotsugae), Douglas-fir pole beetle, Douglas-fir engraver, Hylastes root bark beetle (Hylastes nigrinus), root weevil (Pissodes fasciatus), root weevil (Steremnius carinatus).
	Ponderosa pine	Mountain pine beetle (Dendroctonus ponderosae), western pine beetle (D. brevicomis), red turpentine beetle (D. valens), Hylastes macer
Phytophthora lateralis	Port-Orford-cedar	Cedar bark beetle

Source: Adapted from Hadfield et al. (1986).

crown fires. Fire is being reintroduced into these forests to keep them healthy. Like the situation in the western North America, fire suppression in the southern United States contributes to eventual replacement of pines by shade-tolerant hardwoods, largely because of pine mortality caused by the bark beetle *Dendroctonus frontalis*.

ROOT DISEASE MANAGEMENT

Root diseases can be managed using a variety of techniques, including inoculum reduction, use of resistant species, genetic resistance, chemical stump treatment, biological control, thinning, nitrogen fertilization, avoiding hazardous sites, natural climatic controls, prescribed burning, and quarantine. Doing nothing also may be a viable management strategy as we will explain later.

Inoculum Reduction

There are two approaches to inoculum reduction: stump removal and fumigation. Fumigation is discussed later in the section on chemical treatment of stumps. Removal of infected stump and root material (Figure 12.15) was pioneered in the western United States and Canada and can be utilized for aggressive killing strains

Figure 12.15
Backhoe removing stumps to reduce *Armillaria* inoculum in British Columbia.

of *Armillaria* spp. and *Phellinus weirii* (Morrison et al. 1991). Stump removal has also been effectively used for control of *H. annosum* in England (Woodward et al. 1998). Armillaria root disease often is severe in the first few years after conifer plantation establishment on sites that had other species—for example, native forests in New Zealand and hardwoods in Europe, or in the inland western United States. In this case it may be economically feasible to use this technique. Large tractors or backhoes are typically used. This technique removes vegetative inoculum so that roots from subsequently planted trees will not be infected. It is not necessary to completely remove stumps from the site. In many cases all that is necessary is to push the stump out of the ground, exposing it to drying conditions. It is difficult to remove all the roots from the soil, and many infected small-diameter roots may remain. However, the survival time of the fungus is short in these roots, and the fungus usually is dead before new tree roots come in contact with dead roots. Other organisms also quickly colonize broken root pieces in the soil. This technique can cause considerable site disturbance including compaction and increased erosion on slopes. It usually is practiced only on flat land or on relatively mild slopes with light (relatively low clay content) soils. Soil trenching to reduce inoculum and the spread of *Armillaria* and *Phytophthora* in urban areas is possible, but this is generally impractical in forested areas.

In Canada tree pushing using a backhoe with a thumb to harvest whole trees has recently been employed (Thies and Sturrock 1995). This tends to remove most of the infected root material and no stumps remain to harbor inoculum. The downed trees are cut into lengths at the site for transportation to mills.

Use of Resistant Species

Using resistant conifers or immune species in root rot centers is a viable management technique for many root diseases, and Table 12.4 shows the relative susceptibility of western North American conifers and hardwoods to the four most

T A B L E **12.4** | **Relative Susceptibility[a] of Conifers in Western North America to Damage by the Four Common Root Diseases**

Species	Armillaria root disease	Annosus root and butt rot	Laminated root rot	Black-stain root disease
Westside Douglas-fir[b]	2[c]	3	1	1
Eastside Douglas-fir[d]	1	3	1	3
Ponderosa pine	2	2	3	1
Lodgepole pine	2	2	3	3
Western white pine	2	3	3	4
Sugar pine	2	3	3	4
Grand fir	1	1	1	4
White fir	1	1	1	4
Pacific silver fir	2	1	2	4
Noble fir	2	2	2	4
Subalpine fir	2	2	2	4
California red fir	2	2	2	4
Western hemlock	2	2[e]	2	3
Mountain hemlock	2	1	1	3
Western larch	3	3	2	4
Engelmann spruce	2	3	2	4
Sitka spruce	2	3	3	4
Western redcedar	2	3	4[f]	4
Incense cedar	3	3	4	4
Port-Orford-cedar	3	3	4	4
Hardwoods	2	3	4	4

[a]Susceptibility = 1 = severely damaged 2 = moderately damaged 3 = seldom damaged 4 = not damaged (immune).
[b]West side of the Cascade Mountains.
[c]Westside Douglas-fir is moderately damaged up to age 25; susceptibility then decreases.
[d]East side of the Cascade Mountains.
[e]Western hemlock is not severely damaged until it exceeds 150 years.
[f]Western redcedar east of the Cascade Range may have butt rot.
Source: Adapted from Hadfield et al. (1986).

common root diseases. To manage the fir strain of the laminated root rot fungus, an immune hardwood species such as red alder could be planted in Douglas-fir infection centers. Since the primary mode of spread of this disease is by mycelia through root contacts, this gives time for the fungus to die off or be replaced by other fungi in the stump and root systems. During a single 50-year rotation of red alder, the inoculum is essentially eliminated and it may not take even that long if woody residues are small. There is some economic justification for doing this. High-quality red alder has high value as furniture wood, and a rotation of red alder would fertilize the site by adding nitrogen through natural N fixation as well as adding additional organic matter. It also would add species diversity on the landscape level.

Resistant conifer species, such as western redcedar or western white pine, also can be grown on laminated root rot infested sites to reduce the amount of inoculum over time or keep the damage within acceptable limits. The best opportunity to do this is at the time of final harvest. All hosts in an infection center should be cut plus a surrounding 15 m buffer zone. Western white pine is a good alternative species, although it is susceptible to white pine blister rust, so blister rust–resistant varieties are recommended. Western redcedar also is a viable species to use in some areas, but it is extremely susceptible to animal browse damage. In some instances heavily infected sites could be seeded to grass and other browse species for wildlife forage enhancement.

Alternative species are not very effective against Armillaria root rot because the host range is so wide, although western larch is seldom damaged (Table 12.4). The use of combinations of species has been successful in Canada where fewer and smaller diseases patches were observed when rows of susceptible Douglas-fir were alternated with less susceptible western redcedar or paper birch compared to disease development in pure Douglas-fir. Alternative species can be used for Annosus root and butt rot and black stain root disease (Table 12.4).

Genetic Resistance

Although it is recognized that root disease–resistant individuals occur in stands, there has been little attempt to breed or clone genetically resistant strains for planting in root rot centers. However, some attempts to select resistant strains of Port-Orford-cedar to *P. lateralis* have been attempted because it is a highly desirable and valuable tree for ornamental plantings in the Pacific Northwest (Hansen and Lewis 1997).

Chemical Treatment of Stumps

A variety of chemicals have been employed to control root diseases. Because *H. annosum* spreads by airborne spores, chemical stump treatment has been practiced (Woodward et al. 1998). Stump treatments to control spread were first suggested in the 1950s in England. Creosote was the first chemical used but, although it is effective, it is not really practical to use. A search for better chemicals resulted in the testing of urea, sodium nitrate, and powdered borax. Urea and sodium nitrate applied to stump surfaces promote other organisms to outcompete *H. annosum*. Powdered borax (tradename Sporax) works well on resinous species such as pines. It also is effective on nonresinous species such as western hemlock, and spruces or firs. However, it does not adhere well to sloping stump surfaces. Borax is used operationally in the southeastern United States and in high-hazard areas such as campgrounds in the western United States, especially on large stumps of susceptible pine, fir, and hemlock species where it can be applied carefully.

Fumigant chemicals have been used to reduce *Phellinus weirii* and *Heterobasidion annosum* inoculum in woody substrates. Methyl bromide has been used to fumigate soil and woody residues containing *Armillaria*. Placing liquid fumigant (allyl alcohol, chloropicrin, Vapam, or Vorlex) in holes drilled in Douglas-fir stumps infected with *Phellinus weirii* and then plugging the holes with wooden dowels has been suggested to reduce inoculum in high-hazard areas such as campgrounds (Thies and Sturrock 1995). After 1 year this treatment successfully eliminated *P. weirii* from stumps and most of the roots. It also has been tested on infected living Douglas-fir trees with some success. Such treatment is dangerous and requires extreme caution, however. Soil fumigant chemicals also have been used to control *Armillaria* and *Phytophthora* spp., but mostly in ornamental situations.

Biological Control

There are a number of examples of biological control of root diseases. *Phlebia gigantea*, a Basidiomycete decay organism, is applied to cut stump surfaces in England and Finland (Woodward et al. 1998). It has been tested in the southeastern United States and eastern Canada to prevent stump infection by *H. annosum* (Tainter and Baker 1996). In Finland spores of *P. gigantea* (tradename Rotstop) are applied in a liquid spray as Norway spruce or Scots pine trees are machine-felled during thinning operations. *Trichoderma* spp. also have been tested on stumps containing *Phellinus weirii* in Oregon with some success (Thies and Sturrock 1995).

Thinning

Thinning stands may reduce the spread rate of *Phellinus weirii* (Thies and Sturrock 1995). Removal of trees in a zone 15 m from the last tree in a pocket showing symptoms reduced the rate of spread because the fungus does not spread as rapidly through a dead root system. On the other hand the rate of spread by *Armillaria* may increase as a result of thinning because the fungus can occupy the stumps and roots of thinned trees, especially of large trees (Morrison et al. 1991). Precommercial thinning, however, does not increase *Armillaria* spread or mortality in the Oregon and Washington Cascades because the stumps are too small. Thinning also may increase the incidence of *H. annosum* because of the infection of stumps by airborne spores and subsequent infection of the remaining trees. Avoidance of wounding when thinning nonresinous species also reduces the incidence of infection by *H. annosum*. Generally the larger the wound, the greater the chance of infection. Thinning should be avoided in the spring when the bark is loose and the chance of damage is high, especially in thin-bark species. Black stain root disease also may increase with precommercial thinning.

Cutting high stumps may reduce the chance of residual tree infection in precommercially thinned western hemlock stands because the fungus is replaced or outcompeted by other organisms including nonpathogenic *Armillaria* strains. Pruning of infected roots can be used to control *Armillaria* but usually only in the early stages of infection and where only a small proportion of the roots are infected. This technique might be applied in parks or urban areas, but it would not be usual in forests. It also may increase other root diseases such as Annosus root and butt rot.

Fertilization

Nitrogen fertilization has been reported to cause both decreases and increases in root diseases. It is thought that laminated root rot can be reduced by N fertilization (Thies and Sturrock 1995). Prior to reforestation, sites infected with *P. weirii* can be given a heavy application of N fertilizer (usually urea) with the expectation of reducing the survival time of *P. weirii* in stumps and roots. The mechanism of action involves stimulation of soil organisms such as *Trichoderma* spp. that are antagonistic to *P. weirii*. The incidence of Armillaria root rot, however, may be increased by N fertilization, and there is evidence for this in Idaho. However, Douglas-fir stands in Idaho showed less loss to Armillaria root disease when fertilized with both urea and potassium (Mandzak and Moore 1994). It is not known how fertilization influences *H. annosum*.

Feeder root diseases also respond to fertilization. The effects of *P. cinnamomi* on *Pinus echinata* in the southeastern United States can be alleviated by application of N fertilizer. Similarly, phosphate application to infected *Pinus radiata* stands in New Zealand also reduced the disease. Attempts to control *P. cinnamomi* in eucalypts in Australia using fertilizers, however, have failed.

Avoidance of High-Hazard Sites

A considerable amount of effort has been expended in trying to identify hazardous sites where susceptible species should be avoided. For example, *Phellinus weirii* tends to be more prevalent on mid and upper slopes (Kastner et al. 1994); sites with poor drainage and high clay content favor *Phytophthora* spp. (Tainter and Baker 1996); sites with high pH and low soil aluminum may favor Armillaria (Browning and Edmonds 1993); recently burned sites may favor *Rhizina undulata*; and sandy soils favor development of *H. annosum* in southern pines (Tainter and Baker 1996). In the northern Rocky Mountains, it appears that *A. ostoyae* can be predicted by habitat type (McDonald et al. 1987). Hot-dry, cold-dry, and frost pocket habitat types appeared not to support *Armillaria* spp. Incidence was higher in cool-dry/moderate, cool-moderate/wet and warm-moderate/wet habitat types. In undisturbed sites the probability of *A. ostoyae* infection decreased as productivity increased. High productivity sites that had been disturbed, however, had more *Armillaria* compared to undisturbed high-productivity plots.

Natural Climatic Controls

Natural climatic controls have been used to manage root diseases. For example, high and low temperatures may prevent spore production and/or stump infection by *H. annosum* (Woodward et al. 1998). In the southeast United States, thinning of southern pines can be conducted in the warm summer months because stump temperatures are high and it is difficult for *H. annosum* to establish because it is outcompeted by another fungus, *Trichoderma viride,* which is favored by high temperatures (Driver and Ginns 1969). Thinning without borax stump treatment, however, should still be minimal. Conversely, in Scandinavia thinning is practiced during cold winter months when spores are not released and the possibility of stump infection is very low. In areas with more moderate temperatures, spores are released year round and stump temperatures are favorable for infection at most times of the year, particularly in coastal areas in North America, the British Isles, and other areas in Europe.

Prescribed Fire

With the current emphasis on ecosystem management, there has been considerable interest in using prescribed fire as a management tool. In forest ecosystems that typically have high fire frequencies, fire suppression has resulted in increased incidence of root disease and insect problems. Thus the reintroduction of fire in these ecosystems in the form of understory burning should result in a healthier forest with fewer root disease and insect problems. Smoke management, however, is a problem.

Slash burning, however, has generally had little or no impact on structural root fungi like *Phellinus weirii, Armillaria* spp., and *Heterobasidion annosum* (Hadfield et al. 1986). Most slash burns do not get hot enough to sterilize the soil and although stumps and roots may burn, the inoculum is generally not eliminated.

On the other hand, prescribed fire might increase the incidence of some root diseases. As mentioned previously it increases the incidence of *Rhizina undulata* because this is a fire-loving fungus requiring fire and soil heating for spore germination. In Australia prescribed fire is used for the purpose of hazard reduction and regeneration of desirable tree species. Hazard reduction burns are used in dry sclerophyll forests at relatively low intensities at intervals between 7 and 10 years. Periodic removal of the litter layer for the purpose of hazard reduction could significantly increase the susceptibility of these forests to *P. cinnamomi*. Removal of the litter layer in jarrah forests in western Australia has been shown to increase soil temperatures in spring, thus prolonging the period when the soil moisture and temperature regimes are suitable for spore production. Hazard reduction burning also changes the understory to favor *Banksia grandis*, which harbors the fungus.

Quarantine

Quarantine is particularly useful in preventing new species or strains of root diseases from spreading. It is most effective against exotic introductions, which can be particularly devastating. With the current increase in log movement from the southern hemisphere to the northern hemisphere, and from Russia to the U.S. west coast, there is a possibility of introducing a new species or strain to North America that could be devastating to the local population of trees. The Scandinavians also are concerned about introducing diseases and insects from North America. Thus it is important that serious quarantine measures are undertaken should this occur on a large scale.

Use of quarantine is particularly important for the feeder root diseases caused by *Phytophthora*. In western Australia quarantine and hygiene procedures are practiced seriously since *P. cinnamomi* has the capacity to build up to a very high population level from just a few propagules. Marked reductions in the rate of extension of the disease can be achieved with even moderate attempts at hygiene. For example, within the jarrah forest gravel from dead and dying forests was used for roads. When this practice was stopped, there was a marked reduction in disease extension. Contamination from forest nurseries also was reduced by imposition of stringent hygiene involving fungicidal treatment of vehicles, tools, boots, and so on, that could carry contaminated soil from nurseries to the forest. The same principles apply for Port-Orford-cedar in the western United States. Transportation on infected seedlings should particularly be avoided.

Doing Nothing or Passive Management

Doing nothing or passively managing root diseases is a viable option, particularly where infection is relatively low and the rate of spread is slow. Infection centers involving *Phellinus weirii* will naturally have a different array of species than other areas of the forest. West of the Cascade crest in Oregon, Washington, and British Columbia hardwoods such as red alder and vine maple occur in infection centers. During harvesting operations leaving small patches of forest that are infected, but in which the fungus is spreading slowly, promotes biodiversity and wildlife populations (Thies and Sturrock 1995). On the other hand, if infection centers are large and expanding rapidly, then the site productivity can be greatly impacted for many years. If a landowner owns only a small, heavily infected area, then active management to reduce inoculum and change species is more appropriate.

MODELING OF ROOT DISEASES AND DISEASE/BARK BEETLE INTERACTIONS

Computer models can be used to model disease behavior and spread and interactions with insects as well as to test disease management strategies. A good example of this modeling approach is provided by the Western Root Disease Model (WRDM), Version 3.0 (Frankel 1998). The WRDM is an extension for the Forest Vegetation Simulator (FVS), an individual-conifer-tree growth and yield model (Figure 12.16). The model simulates the effects of *Armillaria ostoyae*, *Phellinus weirii,* and *Heterobasidion annosum* and their interaction with beetles in the western United States. The WRDM is calibrated for each root disease, tree species, and location and can be used by forest managers to evaluate outcomes of different silvicultural prescriptions and direct pest management decisions. It also can be used to assess what happens to a stand when it is allowed to develop naturally with and without root disease and beetles (Figure 12.17). This model was developed using the knowledge of various experts in disease recognition, biology, impact, and management, as well as the potential users of the model.

To use the model an inventory of stand conditions must be available and the extent of the pest problem defined. New techniques have been developed for pinpointing individual root diseases, trees, and centers using aerial photography, computer-assisted technology, and global positioning systems (GPS). The major relationships captured in the model are the susceptibility of trees to infection by root pathogens, the vulnerability of trees to death once infected, reductions in stem growth due to disease, disease spread to previously uninfected areas, inoculum

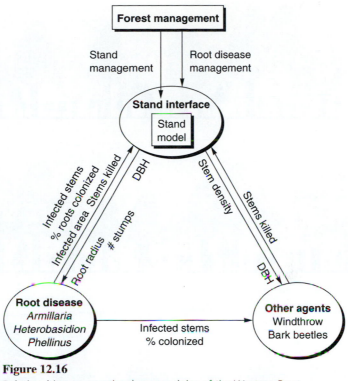

Figure 12.16
Relationships among the three modules of the Western Root
Disease Model.
(Source: Adapted from Shaw et al. 1991.)

dynamics in infected trees and stumps, and the effects of windthrow and attack
by bark beetles. Root disease dynamics are modeled in six major parts:

1. Size and distribution of root disease centers at the start of the simulation
2. Dynamics of tree infection inside centers
3. Tree mortality and growth inside the centers
4. Spread and enlargement of centers
5. Persistence of root disease following tree harvest, through a newly
   regenerated stand
6. Interaction of "other agents" (windthrow and four types of bark beetles)

The model is continually being developed and analyzed for sensitivity to changes
in the various parameters that control it. However, fire is currently not included,
and it has not been adapted for conditions in coastal areas in Oregon, Washington, and British Columbia. A root disease model involving *Phellinus weirii* in
Douglas-fir in western Canada, however, has been developed by scientists at the
Forestry Research Centre in Victoria, British Columbia. Models are discussed
further in Chapter 16.

A

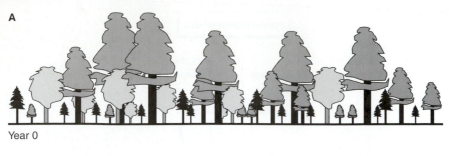

Year 0

B

Year 50: no root disease

C

Year 50: with root disease

Figure 12.17

Changes in a pine stand in the inland western United States predicted by the Western Root Disease Model over a 50-year period: **A,** initial conditions; **B,** 50 years later, without considering bark beetles and root disease; **C,** 50 years later, including the influence of bark beetles and root disease.

(Source: From Frankel 1998.)

CONCLUSIONS

Root diseases, especially Armillaria root rot and Annosus root and butt rot, have a great influence on forests throughout the world. Forest management has generally increased timber losses to these diseases. However, these losses can be minimized if forest managers incorporate disease management principles into their operations and if they aware of the ecological relationships between root pathogens and their hosts and the interactions of diseases, fire, insects, and wind. Many techniques are available for managing root diseases, such as inoculum reduction, use of resistant species, genetic resistance, chemical stump treatment, biological control, thinning, nitrogen fertilization, avoiding hazardous sites, natural climatic controls, prescribed burning, and quarantine.

Forest ecosystems can be negatively or positively impacted by root diseases. Root diseases are important disturbance agents in natural ecosystems, changing successional patterns, promoting biodiversity, and providing habitat for birds and other animals by changing the structure of the forest. A clear set of objectives for land management is needed in order to choose the appropriate management strategy. In some areas doing nothing may be appropriate. However, in other areas, such as dry, moisture-stressed and fire-dominated ecosystems, or in areas subject to air pollution, active management of root diseases may be necessary to restore ecosystems to a healthy condition. Use of root disease/insect models will help us to better understand the behavior of root diseases and will enable managers to predict the impact of management strategies over long time periods.

REFERENCES

Allen, E., D. Morrison, and G. W. Wallis. 1996. *Tree diseases of British Columbia.* Forestry Canada, Pacific Forestry Centre, Victoria, B.C., Canada.

Browning, J. E., and R. L. Edmonds. 1993. Influence of soil aluminum and pH on Armillaria root rot in Douglas-fir in western Washington. *Northwest Science* 67:37–43.

Driver, C. H., and J. H. Ginns, Jr. 1969. Ecology of slash pine stumps: Fungal colonization and infection by Fomes annosus. *Forest Science* 15:1–10.

Filip, G. M., and D. M. Goheen. 1984. Root disease causes severe mortality in white fir and grand fir of the Pacific Northwest. *Forest Science* 30:138–142.

Frankel, S. J. (tech. coord.). 1998. *User's guide to the western root disease model, version 3.0.* USDA Forest Service, Pacific Southwest Research Station, General Technical Report, PSW-GTR-165, Berkeley, Calif.

Hadfield, F. S., D. J. Goheen, G. M. Filip, C. L. Schmitt, and R. D. Harvey. 1986. *Root diseases in Oregon and Washington conifers.* USDA Forest Service, State and Private Forestry, Forest Pest Management, R6-FPM-250-86, Portland, Oreg.

Hansen, E. M., and K. J. Lewis (eds.). 1997. *Compendium of conifer diseases.* American Phytopathological Society Press, St. Paul, Minn.

Harrington, T. C., and F. W. Cobb, Jr. (eds.). 1988. *Leptographium root diseases of conifers.* American Phytopathological Society Press, St. Paul, Minn.

Johannson, M., and J. Stenlid. 1994. *Proceedings of the eigth international conference on root and butt rots.* Parts 1 and 2. IUFRO Working Party S2.06.01, Swedish University of Agricultural Sciences, Uppsala, Sweden

Kastner, W. W., D. J. Goheen, and R. L. Edmonds. 1994. Relationship between occurrence of laminated root rot and site characteristics in Douglas-fir forests of the northern Oregon Coast Range. *Western Journal of Applied Forestry* 9:14–17.

Mandzak, J. M., and J. A. Moore. 1994. The role of nutrition in the health of inland western forests, pp. 191–210. In R. N. Sampson and D. L. Adams (eds.). *Assessing forest ecosystem health in the inland west.* Haworth Press, New York.

Manion, P. D. 1991. *Tree disease concepts,* 2nd ed. Prentice-Hall, Englewood Cliffs, N.J.

McDonald, G. I., N. E. Martin, and A. E. Harvey. 1987. Armillaria in the northern Rockies: pathogenicity and susceptibility on pristine and disturbed sites. USDA Forest Service, Intermountain Research Station, Research Note INT-371, Ogden, Utah.

Morrison, D. J. (ed.). 1989. *Proceedings of the seventh international conference on root and butt rots.* IUFRO Working Party S2.06.01, Forestry Canada, Victoria, B.C., Canada.

Morrison, D. J., H. Merler, and D. Norris. 1991. *Detection, recognition and management of Armillaria and Phellinus root diseases in the southern interor of British Columbia.* Joint Publication of Forestry Canada and the British Columbia Ministry of Forests, FRDA Report 179, Forestry Canada, Victoria, B.C., Canada.

Old, K. M. 1979. Phytophthora *and forest management in Australia.* CSIRO, Melbourne, Australia.

Roth, L. F., R. D. Harvey, Jr., and J. T. Kliejunas. 1987. *Port-Orford-cedar root disease.* USDA Forest Service R6, FPM, PR 0101 91, Portland, Oreg.

Scharpf, R. F. 1993. *Diseases of Pacific Coast conifers.* USDA Forest Service Agricultural Handbook 521, Washington, D.C.

Schowalter, T. D., and G. M. Filip (eds.). 1993. *Beetle-pathogen interactions in conifer forests.* Academic Press, New York.

Schowalter, T., E. Hansen, R. Molina, and Y. Zhang. 1997. Integrating the roles of phytophagous insects, plant pathogens and mycorrhizae in managed forests, pp. 171–189. In K. A. Kohm, and J. F. Franklin (eds.). *Creating a forestry for the twenty-first century: the science of ecosystem management.* Island Press, Washington, D.C.

Shaw, C. G., III, and G. A. Kile. 1991. Armillaria *root disease.* USDA Forest Service Agricultural Handbook 691, Washington, D.C.

Shaw, C. G. III, A. R. Stage, and P. McNamee. 1991. Modeling the dynamics, behavior, and impact of Armillaria root disease, pp. 150–156. In C. G. Shaw III and G. A. Kile (eds.). *Armillaria root disease.* USDA Forest Service Agricultural Handbook 691, Washington, D.C.

Shaw, D. C., R. L. Edmonds, W. R. Littke, J. E. Browning, and K. W. Russell. 1995. Incidence of wetwood and decay in precommercially thinned western hemlock stands. *Canadian Journal of Forest Research* 25:1269–1277.

Sinclair, W. A., H. T. Lyon, and W. T. Johnson. 1987. *Diseases of trees and shrubs.* Cornell University Press, Ithaca, New York.

Skelly, J. M., D. D. Davis, W. Merrill, A. E. Cameron, H. D. Brown, D. B. Drummond, and L. S. Dochinger (eds.). 1987. *Diagnosing injury to eastern forest trees: a manual for identifying damage by air pollution, pathogens, insects and abiotic streses.* National Acid Precipitation Program, Vegetation Survey Cooperative, USDA Forest Service, Forest Pest Management, and Pennsylvania State University, University Park, Penna.

Tainter, F. H., and F. A. Baker. 1996. *Principles of forest pathology.* John Wiley and Sons, New York.

Thies, W. G., and R. N. Sturrock. 1995. *Laminated root rot in western North America.* USDA Forest Service Resource Bulletin PNW-GTR-349, Portland, Oreg.

Wargo, P. M., and C. G. Shaw, III. 1985. *Armillaria* root rot: the puzzle is being solved. *Plant Disease* 69:826–832.

Woodward, S., J. Stenlid, R. Karjalainen, and A. Huttermann (tech. eds.). 1998. Heterobasidion annosum: *biology, ecology impact and control.* CAB International, Wallingford, United Kingdom.

FOLIAGE DISEASES AND RUSTS

INTRODUCTION

Many diseases occur on the foliage of trees and include needle casts of conifers and leaf spots, blights, and mildews on hardwoods (Tainter and Baker 1996). Some diseases occur both on the foliage and branches such as anthracnose and rusts. Although rust fungi cause foliage diseases, we will consider them under the separate category of rust diseases in this chapter. Rust fungi can infect the foliage, stems, and branches of trees.

FOLIAGE DISEASES

Foliage diseases of trees are caused primarily by Ascomycetes or Fungi Imperfecti. Injuries to foliage also are caused by abiotic agents such as low and high temperatures and air pollutants, and other biotic agents like viruses, phytoplasmas, bacteria, insects, mites, and nematodes (Agrios 1997). Foliage diseases caused by Oomycetes are almost unknown in temperate forests, and no foliage diseases are caused by Basidiomycetes, except those caused by rust fungi (Manion 1991). Excellent descriptions and photographs of foliage diseases are given in Funk (1985), Sinclair et al. (1987), Skelly et al. (1987), Scharpf (1993), Allen et al. (1996), and Hansen and Lewis (1997).

Fungi growing in or on leaf surfaces are of concern because they reduce plant photosynthesis and growth, and reduce plant vigor making them more susceptible to attack by insects and other fungi. Foliar pathogens are considered to be weak parasites and generally do not kill trees, but can do so under extreme circumstances (Agrios 1997). Foliage diseases on conifers are thought to be more serious than those on hardwoods, although anthracnose diseases result in considerable damage. Damage to conifers is more serious because conifers cannot refoliate like hardwoods, and if they lose several years of foliage, growth can almost cease. The most damaging situations for conifers occur when trees are planted (1) "off site" (wrong type of site for species or out of its native range, e.g., radiata pine in New Zealand, Ecuador, or Chile) and (2) in pure, dense stands. Most damage occurs in stands less than 30 years old and in Christmas tree plantations.

Fungi are common on leaf surfaces (the phylloplane). Many are not pathogens, but rather are saprophytes living on leaf exudates and other substances on the leaf surface. There are hundreds of foliage diseases, but most do not cause serious damage (Tainter and Baker 1996). A list of the most serious foliage diseases of hardwoods and conifers is shown in Table 13.1, along with the causal fungi and host species. These diseases have the most potential for defoliation. Some of the fungi involved are very host specific; whereas others have a broad host range. You will note that the disease common names are very descriptive of typical symptoms you see on leaves and needles—for example, leaf and tar spots, shot-holes, blights, anthracnose, needle casts, downy mildews, and sooty molds. A leaf spot is a small lesion on a leaf. Tar spots are similar, but usually larger and dark in color due to the color of the fungi involved, thus resembling tar. Shot-holes occur when small diseased leaf fragments fall out, leaving small holes. A blight is a disease characterized by general and rapid killing of leaves and stems; anthracnose is a term coined in 1876 to describe numerous plant diseases characterized by ugly dark sunken lesions and blisters. Needle casts are characterized by premature shedding of conifer needles. In downy mildews the fungus appears as a downy growth on the lower surface of leaves and stems. A sooty mold involves a sooty coating formed by the dark hyphae of fungi growing on the secretions of leaf-sucking insects (honeydew), such as scale insects and aphids, and is managed by controlling the insects.

Most foliar pathogens parasitize leaves by extracting carbon and nutrients through intracellular haustoria or by the direct penetration of mesophyll and parenchyma cells (Agrios 1997). Haustoria are specialized hyphae of fungi that enter host cells and absorb food. Fungi producing haustoria usually do not kill host cells immediately, whereas those that directly penetrate cells usually cause rapid cell death. The host typically reacts to both types of infections by producing defensive chemicals.

Typical Symptoms

Symptoms vary markedly from simple localized necrotic spots on leaves to total necrosis and shriveling of leaves (Sinclair et al. 1987). Very heavy infections may cause premature defoliation of deciduous hardwoods and production of a second crop in the same growing season. New shoots and foliage may be produced from adventitious buds and sometimes witches'-brooms are formed. Terminal growth reduction and tip dieback may occur. Anthracnose fungi invade the stem as well as the leaves, causing twig death. In conifers premature defoliation and a thin crown are typical symptoms of heavy infection. Fungal fruiting bodies usually are common in necrotic areas and are typically pycnidia. If the leafspot is sharply delimited, dry, and necrotic, it may tend to fall out, causing "shot-holes." Typical symptoms and signs of hardwood and conifer foliage diseases are shown in Figures 13.1 and 13.2, respectively.

T A B L E **13.1** | **Serious Foliage Diseases of Hardwoods and Conifers**

Common name	Fungal species	Main hosts
Hardwoods		
Septoria leaf spot and canker	*Septoria musiva*	Eastern cottonwood, black cottonwood/black poplar hybrids
Phyllosticta leaf spot	*Phyllosticta negundinis*	Maple
	P. hamamelidis	Witch hazel
	P. cornicola	Dogwood
	P. platani	Sycamore
Anthracnose		
Sycamore	*Apignomonia venata*	Sycamore
Oak	*A. guercina*	Oak
Ash	*A. errabunda*	Ash
Dogwood	*Discula destructiva*	Dogwood
Leaf blister	*Taphrina caerulescens*	Red oak
	T. populina	Poplars, oaks
Powdery mildew	*Sphaerotheca* spp.	Oaks, and many other
	Erisyphe spp.	hardwoods
	Podosphaeria spp.	
	Microsphaera spp.	
	Uncinula spp.	
	Phyllactina spp.	
Sooty molds	*Capnodium* spp.	Many hardwood species
Tar spot of maples		
Large tar spot	*Rhytisma acerinum*	Maples
Small tar spot	*R. punctatum*	Maples
Conifers		
Brown spot needle blight	*Mycosphaerella dearnessii*	*Pinus echinata* and 25 other species
Elytrodema needle blight	*Elytroderma deformans*	Ponderosa and Jeffrey pines
Larch needle blight	*Hypodermella laricis*	Western larch
Larch needle cast	*Meria laricis*	Western and alpine larch, tamarack
Lophodermium needle cast	*Lophodermium seditiosum*	Pines
Diplodia tip blight	*Sphaeropsis sapinea*	Pines, Douglas-fir rarely
Dothistroma needle blight (red band needle blight)	*Mycosphaerella pini*	Pines (30 species, especially *P. radiata* and *P. nigra*), hybrids or varieties
Rhabdocline needle cast of Douglas-fir	*Rhabdocline pseudotsugae* *R. weirii*	Douglas-fir
Swiss needle cast of Douglas-fir	*Phaeocryptopus gaeumanii*	Douglas-fir
Snow blight	*Phacidium abietis*	True firs and Douglas-fir
Snow molds and brown felt blights	*Herpotrichia coulteri* *H. juniperi*	Pines Other conifes
Sphaeropsis tip blight	*Sphaeropsis sapinea*	Pines
Cedar leaf blight	*Didymascella thujina*	Cedars and incense-cedars

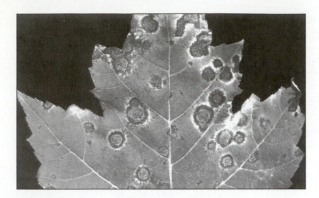

Figure 13.1
Typical hardwood foliage disease; Phyllosticta leaf spot on red maple.
(Source: From Skelly et al. 1987.)

Figure 13.2
Typical signs and symptoms of a conifer foliage disease caused by *Hypodermella laricis* on western larch.

Foliage diseases caused by fungi can be confused with foliage damage caused by viruses, bacteria, insects, mites, and abiotic causes. Observations on patterns of necrotic tissue usually enables one to tell the cause. Symptoms produced by fungi usually are located randomly on the leaf or in the crown, whereas frost or air pollution injuries are more uniform and the transition from healthy to damaged tissue is more abrupt. Saprophytic fungi may fruit on dead areas resulting from abiotic damage, and this can sometimes confuse the issue. Natural phenomena such as fall coloration and conifer needle drop, especially cedar flagging, are sometimes confused with foliage diseases caused by biotic agents (Manion 1991, Tainter and Baker 1996). Insect problems usually are easily identified by distinctive feeding patterns, and the presence of insects, particularly larvae and frass. Mites can cause distinctive morphological and color changes, such as needle curling and chlorosis. Viruses may cause flecking or ring spots and phytoplasmas may induce chlorosis and witches'-brooms. Bacteria may cause leaf spots and blights, such as fire blight of cherry caused by *Erwinia*, and can be isolated on agar media.

The Life Cycle of Foliage Pathogens

The life cycle of many Ascomycete foliage pathogens is remarkably similar (Figure 13.3). Production of ascospores usually occurs in the spring about the time that new foliage is emerging (Agrios 1997). A few fungi produce ascospores in late summer and early fall and the fungi involved infect buds. Conidiospores are produced during the summer growing season; they usually are produced in large numbers and can rapidly spread a new pathogenic strain. Spores alighting on a leaf or needle surface germinate if the environment is favorable and the hyphae either directly penetrate the cuticle and epidermal cells by enzymatic action or physical force or they go through stomates. Many fungi are host specific because the structure and chemical composition of the leaf surface layers varies from host to host, preventing penetration by all but a few fungi.

Foliar pathogens typically overwinter as mycelium in diseased leaves or needles that remain on trees or in dead foliage on the forest floor. Some overwinter as mycelium in infected shoots or buds. Most foliage diseases have both a parasitic and saprophytic phase to their life cycle. In contrast, the mildew fungi are obligate parasites and survive as cleistothecia rather than hyphae in dead host material (Agrios 1997).

Weather conditions are particularly important in the development of foliage diseases, and moist conditions usually are necessary. Moisture is required for ascospore or conidia release as well as for spore germination and infection. Thus foliage diseases are most serious in cool, moist years.

There is some confusion in the naming of foliar pathogens, which sometimes leads to difficulty when teaching students the names of these organisms. Some produce the sexual phase infrequently and are commonly recognized by their conidial stage. Thus they are better known by their anamorph names rather than their teleomorph names. This means the fungus has two names, and classification has been further complicated by the fact that mycologists have changed many of these names over the years. On the other hand, some rarely produce asexual spores and are known only by their sexual or Ascomycete name.

Specific Hardwood Foliage Diseases

Anthracnose of Sycamores, Oaks, and Other Hardwoods   Sycamore anthracnose, caused by *Apiognomonia veneta*, Ascomycota (Pyrenomycetes) or *Discula platani*, Fungi Imperfecti, occurs widely in the United States in forest, park, and street trees (Tainter and Baker 1996). American, Arizona and California sycamores are susceptible; London plane has some resistance, and Oriental and European

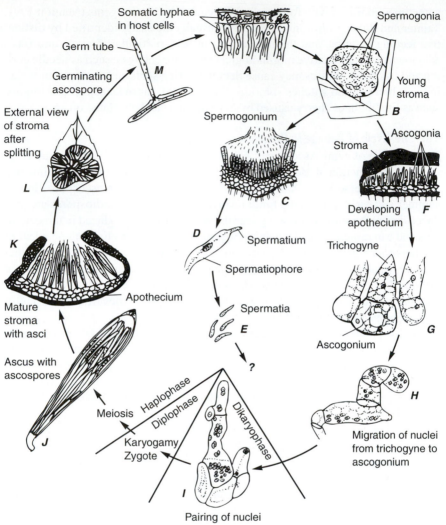

Figure 13.3
General life cycle of a *Rhytisma acerinum,* cause of large maple tar spot.

planes are resistant. Typical symptoms include leaf, bud, shoot, and twig blight, defoliation (Figure 13.4), and cankers. Severe infections weaken trees and predispose them to borers and winter injury, resulting in mortality. Oak anthracnose is caused by a different fungus, but symptoms are similar. Black and white oaks are severely infected.

The life cycle of the fungus causing sycamore anthracnose is as follows. Anthracnose is initiated in the spring by ascospores in perithecia and conidia in pycnidia that are produced from overwintering mycelium in infected twigs on the

Figure 13.4
Sycamore anthracnose symptoms including defoliation and crooked branches.

tree. Mycelium in infected buds will also infect emerging leaves. Secondary infection by conidia also occurs in the summer. The conidial state (**acervuli**), however, looks different in leaves than cankers. Sycamore anthracnose is very severe in cool-wet weather. If temperatures are above 16°C, shoot blight is slight, but if temperatures for a 2-week period after leaf emergence are below 10°C, shoot blight is severe. The shoot blight stage is not affected by rain, but the leaf blight stage and secondary cycles are favored by moisture since spores are dispersed by both wind and rain.

No control is practiced for forest trees, although wide spacing is recommended in plantations to reduce leaf moisture. Pure stands of sycamore should be avoided. Chemicals are commonly used for control on ornamental and shade trees. Fungicides are applied during bud break and 14 days later. Pruning infected twigs also reduces inoculum, but removal of fallen leaves does little because infection usually is initiated from diseased tissues on the tree. Fertilization to promote shoot growth appears to reduce infection. Resistant trees such as London and Oriental plane or hybrids should be selected for planting in parks and as street trees.

Dogwood Anthracnose   Dogwood anthracnose is an important disease of both forest and ornamental species in North America (Tainter and Baker 1996). It was first reported in the 1970s in New England and in the western United States now is widespread (Daughtrey et al. 1996). It increased dramatically in the southeastern

United States from 0.2 million ha in 1988 to 5.1 million ha in 1992, and dogwoods have been eliminated from the understory in some areas. Its origin is unknown, but it was probably introduced on *Cornus kousa*. The major hosts are flowering dogwood (*Cornus florida*) in eastern North America and Pacific dogwood (*Cornus nuttalli*) in western North America, including British Columbia.

The Ascomycete fungus *Discula destructiva* is responsible. There appears to be two types (Type 1 and Type 2). Type 2 may be a different species of *Discula*. Symptoms are similar to sycamore anthracnose with infections first developing on leaves in spring due to spore infection. Bracts also may be infected. The fungus then progresses down the petioles into shoots in winter and eventually produce cankers. A proliferation of epicormic shoots may form on large branches and the stem. Infection of these shoots allows direct entry to the main stem. Twig and branch death is a common symptom.

Again cool, moist conditions favor infection. In the eastern United States, high-hazard sites are close to water, on north-facing slopes, and above 1,000 m in elevation where foggy conditions are common. Control in forest stands usually involves thinning stands to lower the risk of infection. Ornamental dogwoods respond well to cultural treatment including pruning of dead twigs and branches, removal of infected leaves in the fall, mulching, watering (without wetting foliage), and fertilization. Fungicide sprays also can be used as well as resistant species and varieties. *Cornus kousa* is more resistant than *C. florida.*

Powdery Mildews of Oaks and Other Hardwoods   Infection by powdery mildew fungi results in a powdery whitish material on the foliage, shoots, flowers, and fruits (Agrios 1997). The mycelium is almost entirely superficial, except for the haustoria in the epidermal cells. Powdery mildews are widely distributed throughout the world and are common and obvious on foliage. They occur on thousands of species of plants and cause considerable economic damage on grapes, cereals, apples, and roses. However, they usually cause little damage on trees except on young sprouts and seedlings in nurseries. Deciduous hardwoods and at least 30 species of oaks are susceptible, but conifers are not susceptible (Tainter and Baker 1996).

Seven genera of fungi are involved—*Erisyphe, Phyllactina, Microsphaera, Podosphaera, Sphaerotheca, Cystotheca*, and *Uncinula*. All are Pyrenomycetes, and are obligate parasites that form cleistothecia, usually in the late summer and early fall, overwinter and discharge airborne ascospores in the spring. In summer they produce conidia in chains on short **conidiophores.** Powdery mildew fungi are somewhat less dependent on wet conditions than many other foliage pathogens and are very host specific; one species is narrowly restricted to the female catkins of alder.

Microsphaera penicillata and *Cystotheca lanestris* are the most common species on oak. The disease is initiated in spring by ascospores formed in overwintering black cleistothecia on fallen leaves, or by conidia from overwintering mycelium. Hyphae then penetrate the leaf surface and form haustoria in the epidermal and subepidermal cells. Superficial mycelium covers the plant surface, giving the powdery white or tan appearance. Leaf surfaces may be blistered and distorted. Secondary spread is by conidia in the summer, and high relative humidity and temperatures favor infections. Thus this problem typically is prevalent in greenhouses and nurseries. No control is necessary in forests, but wide spacing helps reduce incidence of powdery mildews. Fungicide dusts and sprays are used on nursery and ornamental trees, and the disease can be successfully controlled even after it is established because of its superficial nature.

Sooty Molds   Like powdery mildews, sooty molds are superficial on leaf and bark surfaces and cause little damage to trees, although they are unsightly (Hughes 1976, Tainter and Baker 1996). The associated sap-sucking scale insects and aphids, however, may cause damage and can retard growth and kill small trees. The disease is widespread in North America and commonly occurs mostly on hardwoods and shrubs such as *Camelia*. It also can occur on conifers. A number of genera of Ascomycetes and Fungi Imperfecti are involved (e.g., *Capnodium, Leptoxyphium, Fumago, Auerobasidium,* and *Cladosporium*). *Dimerosporidium* and *Adelopus* spp. occur on grand fir; *Dimeriella* and *Capnodium* spp. occur on Douglas-fir and eastern white pine, respectively.

Most sooty mold fungi are saprobes and have black mycelium growing on excrement from aphids and scale insects called *honeydew.* Spores also are usually dark colored. Plant exudates may be the substrate in some cases. Sooty mold fungi appear to be most abundant in mild, moist climates. No control is necessary in the forest, and insecticides are commonly used against aphids or scale insects in ornamental or nursery plants which indirectly controls the fungus.

Tar Spots of Maples   Although tar spots are common on maples, they are of little economic importance. *Rhytisma acerinum* causes large tar spot (Figure 13.1); *R. punctatum* causes small tar spot (Sinclair et al. 1987, Tainter and Baker 1996). This disease, when severe, can result in defoliation and is unsightly on shade ornamental and nursery trees. Large tar spot is common on red and silver maples and also occurs on Norway, sycamore, sugar, and bigleaf maples. Small tar spot also occurs on many maples and box elder. Tar spots also occur in madrone, yellow poplar, and willow. No control is practiced in forests. Sanitation, particularly raking of leaves and burning or removal of leaves in the fall, is a good method of control for ornamental trees because the fungus is an obligate parasite and does not produce conidia for secondary infection cycles.

Leaf Blisters and Other Diseases Caused by *Taphrina* spp.   *Taphrina* spp. are obligate parasites and cause hypertrophy in the host (i.e., overgrowth in infected areas), leading to blistering, puckering, and curling (Sinclair et al. 1987). Ascospores are produced in naked asci (no ascoma) on the leaf surface, and because ascospores keep dividing, the asci usually have more than the typical eight spores. Peach leaf curl is an important disease in orchards caused by *Taphrina deformans*. There is a curious disease caused by *Taphrina* on the female catkins of alder, causing the bracts to grow much longer than normal.

Specific Conifer Foliage Diseases

Needle casts or blights are general terms to describe situations where needles are lost, or cast, prematurely (Sinclair et al. 1987, Hansen and Lewis 1997). However, there are some strange situations where needles are kept longer than normal (e.g., on larch). At least 40 species of Ascomycete fungi cause needle casts and blights in North America. They produce **apothecia** on needle surfaces that are often dark and glossy. If fruiting bodies are elongated, they are called *hysterothecia* and have special "lip cells" that respond to wet conditions by opening a slit to expose the asci. When the weather is dry, they close again. Some produce pseudothecia in which the asci are formed directly in cavities, such as stomates. Pycnidia are produced by some needle cast fungi. Pines, spruces, firs, larches, cedars, and Douglas-fir all suffer from needle casts. Descriptions of the most common needle cast diseases follow.

Rhabdocline Needle Cast of Douglas-Fir   Rhabdocline needle cast is a problem in Christmas trees and young plantations (Hansen and Lewis 1997). It has severely affected the success of Douglas-fir plantations in Western Europe and occurs wherever exotic plantations of Douglas-fir occur in North America. It is caused by *Rhabdocline pseudotsugae,* which has subspecies *epiphylla,* and *Rhabdocline weirii,* which has subspecies *oblonga* and *ovata.* Conidial stages are unknown. Ascospores produced in apothecia are released in spring and infect newly emerging needles. First symptoms occur in autumn as yellow spots or bands. By late winter the yellowing has changed to purple-brown mottling. Rupturing of the epidermis by maturing apothecia then occurs followed by defoliation. Symptoms are shown in Figure 13.5. No control usually is necessary in plantations, but Christmas trees are commonly sprayed with fungicides beginning at budbreak and repeated at 2 to 3-week intervals until shoots and foliage are fully grown.

Swiss Needle Cast of Douglas-Fir   Like Rhabdocline needle cast, Swiss needle cast is a problem in Christmas trees and young plantations (Hansen and Lewis 1997). It was discovered in Switzerland in 1925, giving it the name. It occurs throughout the natural range of Douglas-fir, and wherever it has been planted in

Figure 13.5
Rhabdocline needle cast symptoms caused by *Rhabdocline pseudotsugae.*
(Source: From Skelly et al. 1987.)

Western Europe, including Scandinavia, in Japan and in New Zealand. It is currently causing extreme defoliation, growth loss, and mortality in young Douglas-fir plantations in north coastal Oregon, where it appears to have increased in recent years for reasons unknown. The disease is caused by *Phaeocryptopus gaeumannii* and only the sexual stage is known. Ascospores are released from stomatal pseudothecia in late spring that infect the newly emerging needles. Cool, wet springs favor development of Swiss needle cast. Appearance of slight yellow-green discoloration at needle tips is the first symptom in the fall, and ventral sootlike bands of stomatal perithecia begin to develop as early as November (Figure 13.6). The time between infection and defoliation is variable.

Forest plantations usually are not treated for this disease, although chemical sprays have been applied in cases of severe defoliation. Losses in Christmas tree plantations are minimized by early detection and fungicide spraying. The first application is typically carried out when new shoots are 1 to 5 cm long. A second spray 2 to 3 weeks later usually is needed and in wet years a third spray may be needed.

Lophodermium Needle Cast   Lophodermium needle cast has been a problem in pines in Europe for a century, but has been a problem in nurseries in North America only since the mid-1960s (Tainter and Baker 1996, Hansen and Lewis 1997). It has been reported from 27 states in the United States as well as British Columbia, Nova Scotia, and Ontario. The most common mode of long-distance spread is by infected seedlings, and it is a particular problem for the Christmas tree industry. The greatest injury is to red pine and the short-needled varieties of Scots pine, but as many as 26 species and varieties of pines are known to be susceptible.

Figure 13.6
Swiss needle cast caused by *Phaeocrytopus gaeumannii* on Douglas-fir, showing pseudothecia in stomates.
(Source: From Skelly et al. 1987.)

The perfect stage is *Lophodermium pinastri* and the pycnidial stage is *Leptostroma pinastri*. Ascospores are released from hysterothecia in late summer, but foliage spots are not observed until the following spring. Needles yellow then brown and defoliation results. Other *Lophodermium* spp. attack a variety of other conifers (i.e., true firs, spruces, and cedars)

Cultural methods for management usually are sufficient, but occasionally fungicidal sprays are needed during the spore dispersal period (July–September). Long-leafed varieties of Scots pine should be favored over the more susceptible short-leafed varieties.

Elytroderma Needle Cast   The major hosts of this disease are ponderosa and Jeffrey pines, but jack, lodgepole, and shortleaf pines also are susceptible (Scharpf 1993). The disease occurs from California to British Columbia and in Arizona and New Mexico with isolated examples in Colorado and Ontario. The fungus involved is *Elytroderma deformans,* and although it produces conidia in pycnidia, this stage has not been named. Airborne ascospores produced in hysterothecia (Figure 13.7) are dispersed in late spring and early summer and infect the new needles. The fungus colonizes the phloem of needles and twigs and infects needles in buds. In May and June pycnidia develop as blisters and during wet weather conidia are exuded in tendrils. As well as causing defoliation, brooming and shoot deformation are common symptoms, and the fungus is unique in that it can persist in tree tissues (Hansen and Lewis 1997). The combined effects of premature defoliation and perennial branch dieback lead to drastic growth reduction and sometimes death, especially when associated with attack by bark beetles. The disease tends to be worse at high elevations and is favored by high stand density. Trees of all ages can be attacked.

Elytroderma needle cast can be managed using silvicultural techniques. Damage can be reduced by maintaining healthy stands through thinning and removal of trees with high crown flagging.

Dothistroma Needle Blight   This disease, also known as red band needle blight, is a major problem in exotic pine plantations in the northern and southern hemispheres (Hansen and Lewis 1997). In the southern hemisphere it is a major problem in radiata pine plantations, particularly in New Zealand, Chile, and Ecuador. It also occurs in Australia. In the northern hemisphere it is mostly a problem in

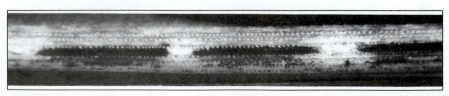

Figure 13.7
Slitlike fruiting bodies of *Elytroderma deformans,* on ponderosa pine.

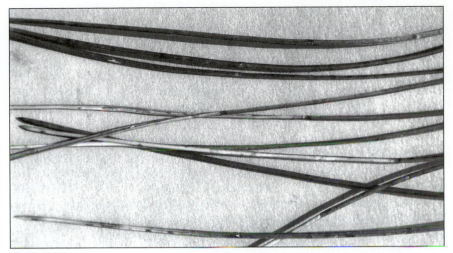

Figure 13.8
Dothistroma needle cast, showing banding on Austrian pine needles.

ornamental and Christmas tree plantations, particularly with Austrian pines. About 30 species of two-, three- and five-needle pines have been recorded as hosts (Tainter and Baker 1996), and the fungus was recently reported to be causing mortality of limber pine in Montana (Taylor and Schwandt 1998).

The fungus involved is *Dothistroma septospora,* which is the asexual stage. The perfect stage (*Mycosphaerella pini*) is rarely found. Conidia are produced in acervuli from late spring to early fall and are dispersed by rain splash. Both old and new foliage are susceptible, and dark green bands that develop into brown/reddish bands commonly appear first in early fall. The banding is due to the production of the toxin dothistromin. Distal necrosis of the needles usually is followed by defoliation. Infection is typically most severe in the lower canopy, but the whole crown can be involved. Symptoms are shown in Figure 13.8. In New Zealand young radiata pine plantations are commonly sprayed by crop-dusting aircraft using copper fungicides. Selection of blight-resistant provenances also holds promise in New Zealand.

Diplodia Tip Blight   This disease, like Dothistroma needle blight, is mostly a problem in exotic plantations of radiata pine and other exotic pines in New Zealand, Australia, South Africa, and Swaziland (Hansen and Lewis 1997). In Swaziland this pathogen causes a root stain disease of stressed loblolly and slash pines. It typically occurs in the northern and southern hemispheres between 40 and 50° north and south latitude, although it has not been reported from Asia. Thirty-three species of pines are known to be hosts and in the northern hemisphere Austrian, ponderosa, red, Scots, and mugo pines are the most important hosts.

The causal fungus is *Sphaeropsis sapinea* (formerly *Diplodia pinea*) the coni-dial stage. The perfect stage is unknown. Spores are produced in pycnidia in dead shoots or stunted needles from late spring to the end of the growing season and are dispersed primarily in wet weather. Infection occurs through stomates, on devel-oping shoots, and on umbos of second-year cones. New growth is mostly affected, and infected branches can be killed back to the main stem and successive attacks result in dieback of branches and tops. Tip blight usually increases with time and injury tends to be most severe in plantations as they approach 30 years. Wounds are not necessary for entry, but wounding created by hail and spittle bugs favors infec-tions. In New Zealand the fungus enters radiata pine trees through pruning wounds.

Fungicidal protection of new shoots during a critical 2-week period after bud break is the principal control. Two applications of protectant or **systemic** fungi-cide at 1-week intervals has proven effective. Avoidance of wounds immediately before and during the growing season reduces infection. This is particularly impor-tant with Christmas tree shearing.

Brown Spot Needle Blight   This is the most serious disease of grass-stage long-leaf pine in the southeastern United States (Hansen and Lewis 1997). Twenty-five other species of pine in the eastern and midwestern United States, however, are known to be susceptible, and the disease can be important on Christmas trees or in nurseries. It also has been reported from the western United States and Canada.

The pathogen is *Mycosphaerella dearnessii* (*Schirrhia acicola*) and the coni-dial stage is *Lecanosticta acicola*. Ascospores are wind dispersed from pseudothe-cia. Conidia are produced from acervuli on dead fallen needles and have rain-splash dispersal. Hyphae from germinating spores penetrate through stomates. In warm areas both types of spores are produced throughout the year during wet periods. Peak sporulation is in late summer in cooler climates. The first symptoms are yel-low spots on needles that change to brown bands and distal necrosis. Toward the end of the growing season defoliation begins.

This disease can be controlled in grass-stage plantations of longleaf pine by pre-scribed winter burning. Fungicidal sprays can be used in nurseries at 10- to 13-day intervals, depending on rain frequency, and root-dip treatment with benomyl before outplanting also gives excellent control. There is also potential for genetic resistance.

Larch Needle Diseases   Larch needle blight is caused by *Hypodermella laricis,* and it occurs on western and alpine larch as well as tamarack (Hansen and Lewis 1997). Infected needles turn yellow and then red-brown in spring and early sum-mer. Elliptical black spots (hysterothecia) form soon after the needles are killed (Figure 13.2). Diseased needles are retained in the fall after normal needle drop. This is in contrast to larch needle cast (caused by *Meria laricis*) where needles also turn prematurely yellow, but are shed soon after they turn brown. *Meria lar-icis* appears to attack only western larch. Neither of these diseases kill large trees, but repeated infections can result in growth reductions. Infected nursery seedlings can be killed by *Meria laricis*. The rapid onset of symptoms may resemble frost damage, but no fruiting bodies are formed in frost-damaged trees.

Snow Molds   Snow molds are sometimes called *brown felt blights* (Tainter and Baker 1996). They grow under snow in the spring and cover the needles of seedlings and lower branches of larger trees with a thick brown or black mycelium that can impede photosynthesis. In North America the fungi causing them are *Herpotrichia coulteri* on pines and *H. juniperi* on other conifers. Since these two fungi develop only under snow, they are only in the snow belt at higher elevations. Once foliage extends above the snowpack, snow molds do not develop.

RUSTS

Rusts occur on both hardwood and conifer trees (Ziller 1974, Tainter and Baker 1996). The name "rust" arises because of the rust-colored appearance of one of the spore stages (uredinial stage) in rusts that attack cereal crops, particularly wheat. On a worldwide basis the damage and economic losses caused by cereal rusts are immense and have resulted in the development of an international breeding program to produce crop strains that are rust resistant (Agrios 1997). The name rust now applies to any fungus in the Urediniales of the Basidiomycota.

Not only do rusts cause cankers on the branches and stems of trees, they also occur on leaves and needles and cones. Because of this they are discussed as a separate group of diseases rather than being classed as foliage, canker, or gall diseases. They also are unique in that they are obligate parasites and have an extremely complex life cycle not found in other fungi. The life cycle often involves five spore stages and two different plant hosts. A list of common North America rusts is shown in Table 13.2 including alternate hosts. There are hundreds of species, but only a few cause serious economic damage. The two most damaging rusts in North America are white pine blister rust on five-needle pines (Figure 13.9) and fusiform rust on southern pines (Tainter and Baker 1996) (Figure 13.10). Poplar rust, caused by several species of *Melampsora* (Figure 13.11), has become economically important on hybrid poplars, which now are grown for paper pulp. Another rust of some importance is western gall rust on lodgepole pine in western North America (Figure 13.12).

White pine blister rust was introduced from Europe into eastern North America. It was known to be in the northeastern United States in 1898, but it wasn't until 1909 that it was first reported on eastern white pine. It was introduced into Vancouver, British Columbia, in 1921 on a shipment of eastern white pine transplants from France. The fungus is native to Asia and now has invaded the range of five-needle pines in North America and is still spreading. The alternate hosts are *Ribes* spp. (currants and gooseberries). This rust is so devastating because there is little resistance in North American five-needle pine populations. It is widespread in the northeastern United States, but it is not as severe as it is in western North America, perhaps because conditions are not moist enough in late summer and fall in the Northeast to promote basidiospore infection. In western North America *C. ribicola* has spread throughout the North Rocky Mountains, the Cascade Range, and the Sierras. There are isolated populations in New Mexico, Nevada, and Utah and it has recently been reported from Colorado.

TABLE **13.2** | **Common North American Rusts**

Disease	Pathogen	Aecial host	Telial host	Features
White pine blister rust[a]	*Cronartium ribicola*	5-needle pines; *Pinus monticola, P. strobus*	*Ribes* spp.	Blisters on branches and stems
Fusiform rust	*Cronartium quercuum* f. sp. *fusiforme*	*Pinus elliotii, P. taeda,* and other pines	*Quercus* spp.	Spindle-shaped (fusiform) cankers
Stalactiform rust	*Cronartium coleosporioides*	*Pinus contorta,* other 2- and 3-needle pines	Scrophulariacea, e.g., Indian paint-brush (west), cow wheat (east)	Cankers long; length/width = 3 In eastern Canada and in west as far south as southern California
Comandra blister rust	*Cronartium comandrae*	*Pinus ponderosa, P. jeffreyi, P. contorta, P. attenuata*	*Comandra* spp.	In western states; in east length/width = 2–3
Western gall rust	*Endocronartium harknessii*	2- and 3-needles pines; *Pinus radiata, P. contorta, P. ponderosa*	Probably none	Globose galls, yellow/orange
Eastern gall rust, pine-oak gall rust	*Cronartium quercuum* f. sp. *banksiana* and other f. sp.	*Pinus banksiana, P. echinata, P. virginiana,* and other pines	Many oaks (bur, chestnut, pin, red)	Galls like western gall rust
Pine needle rust	*Coleosporium solidaginis*	2- and 3-needle pines (*P. resinosa, P. banksiana, P. sylvestris*)	Goldenrod or aster	Overwinters in needles where it can survive 2–3 years
Cedar-apple rust	*Gymnosprangium juniperi-virgininianae*	Apple	Junipers, especially eastern red cedar	No uredinial host; conifer is telial host
Leaf rust of poplar	*Melampsora medusae*	Conifers	Poplars	Hybrid poplars particularly susceptible; little damage to conifers

[a]Introduced.

Figure 13.9
Aecial blisters on western white pine caused by *Cronartium ribicola* on western white pine.

Figure 13.10
Fusiform rust gall caused by *Cronartium quercuum* f. sp. *fusiforme* on a loblolly pine stem.
(Source: From Skelly et al. 1987.)

In contrast to *Cronartium ribicola, Cronartium quercuum* f. sp. *fusiforme,* the cause of fusiform rust, is a native pathogen, with oak species as alternative hosts (Tainter and Baker 1996). Interestingly, it had little impact before 1930, but is now the most important disease in the southeastern United States. So what happened to make it so important in a region where the fungus had evolved with the hosts? There are a number of reasons related to forest management including (1) fire control—oaks are more sensitive to fire than pine, so fire control has increased the abundance

Figure 13.11
Uredial and telial stages
of *Melampsora medusae*
on poplar leaves.
(Source: From Skelly et al. 1987.)

Figure 13.12
Galls on lodgepole pine
caused by *Endocronar-
tium harknessii,* the cause
of western gall rust.
(Source: From Allen et al. 1996.)

of oaks; (2) off-site planting of slash and loblolly pine in areas where resistant longleaf pine used to grow; (3) thinning and fertilization increased growth of southern pines, but made them more susceptible to rust infection; (4) genetically superior trees were selected only for growth, not for rust resistance; (5) seeds from infected trees were used for early plantings; and (6) infected seedlings from nurseries were widely planted.

Poplar rusts also have increased recently due to plantings of fast-growing susceptible hybrid poplars (*Populus trichocarpa* × *P. deltoides*) in Oregon, Washington, and British Columbia. The most important species involved are *Melampsora medusae* f. sp. *deltoidae,* which is native to eastern North America, and *M. laricipopulina*, the cause of Eurasian poplar rust, which had not been reported in North America before 1991 (Newcombe and Chastagner 1993). Some hybrid clones are very susceptible and whole plantation blocks have been killed by these fungi. *Melampsora medusae* f. sp. *deltoidae* has now spread to southwestern Europe, Australia, New Zealand, Africa, and South America.

Western gall rust, a native disease caused by *Endocronartium harknessii*, has generally not been a problem in the past. However, with recent intensive management of lodgepole pines, where large areas of young trees are regenerating in inland western North America, young trees have become severely infected. Gall formation severely distorts growth (Figure 13.12), kills branches, and can even kill trees. This disease also can be a problem in forest nurseries (Sutherland et al. 1989).

Rust Life Cycles

Rust fungi are obligate parasites (Agrios 1997). They can be long-cycle **heteroecious** rusts (with two hosts) or short-cycle rusts with only one host (**autoecious**). White pine blister rust is a typical heteroecious rust, and the life cycle of the causal fungus is illustrated in Figure 13.13. The spore stages are shown in Box 13.1.

Alternate hosts for white pine blister rust are *Ribes* spp. (currants and gooseberries) (Tainter and Baker 1996). Haploid (1*N*) basidiospores are produced on a septate basidium on the underside of *Ribes* spp. leaves. The basidium arises from dikaryotic (*N*+*N*) teliospores also on the *Ribes* leaves. These teliospores function like the hymenium of the Hymenomycetes and do not disperse. Airborne

Box 13.1

Rust Spores Stages

Spore type	Fruiting structure	Hosts
Spermatia	Spermogonium	Five-needle pines
Aeciospores	Aecium	Five-needle pines
Urediniospores	Uredium	*Ribes* spp.
Teliospores	**Telium**	*Ribes* spp.
Basidiospores	Basidium	*Ribes* spp.

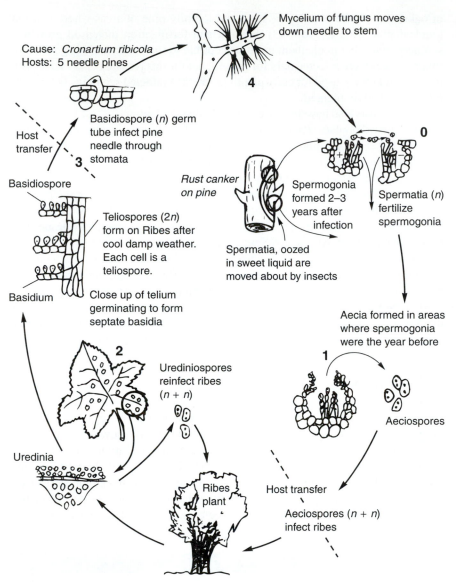

Mycelium of fungus moves
down needle to stem

Cause: *Cronartium ribicola*
Hosts: 5 needle pines

4

Basidiospore (*n*) germ
tube infect pine
needle through
stomata

Host
transfer

3

0

Basidiospore

*Rust canker
on pine*

Spermogonia
formed 2–3
years after
infection

Spermatia (*n*)
fertilize
spermogonia

Teliospores (2*n*)
form on Ribes after
cool damp weather.
Each cell is a
teliospore.

Spermatia, oozed
in sweet liquid are
moved about by insects

Basidium

Close up of telium
germinating to form
septate basidia

Aecia formed in areas
where spermogonia
were the year before

2

Urediniospores
reinfect ribes
(*n* + *n*)

1

Aeciospores

Uredinia

Ribes
plant

Host transfer

Aeciospores (*n* + *n*)
infect ribes

Figure 13.13

Life cycle of *Cronartium ribicola,* the cause of white pine blister rust.

(Source: *Tree Disease Concepts,* 2/e by Manion, Paul D., ©1991. Reprinted by permission of Prentice-Hall, Inc., Upper Saddle River, NJ.).

basidiospores are dispersed at night in late summer or early fall and travel only short distances, usually less than a few kilometers, before they land on their pine hosts. The hyaline basidiospores are very sensitive to UV radiation and do not survive during the day. If basidiospores land on the needles of five-needle pines (e.g., *P. albicaulis, P. flexilis, P. lambertiana, P. monticola,* and *P. strobus*), they germinate if the local environment is favorable (cool temperatures and high needle surface moisture), and penetrate the needle surface through stomates. A small yellow or orange fleck develops at the infection site within a few weeks. The hyphae then move down the needle to the branch or stem by the following spring. In the cambium region, hyphae penetrate between the cells of the host and derive nutrients through haustoria.

Cankers begin to develop on infected branches or stems and in 1 year spermogonia (**pycnia**) are produced. In the **spermogonia** haploid spores known as spermatia are produced in a sweet liquid and moved about by insects. They do not infect plant tissue, but function to exchange genetic material. Insects attracted by the sweet exudate move spermatia from one spermogonium to another. Compatible spermatia fuse with specialized hyphae in the spermogonium and transfer a nucleus so that the fungus **thallus** then has dikaryotic ($N+N$) nuclei. One year later in the spring aeciospores formed in aecia are produced from the dikaryotic hyphae where the spermogonia were. They become airborne and can travel as far as 1,300 km to infect *Ribes* leaves. These spores germinate and then produce urediniospores from uredinia that form on the undersurface of *Ribes* leaves. Urediniospores generally travel only short distances. In the fall telial columns and basidiospores are formed once again. Depending on location and weather conditions, it can take 2 to 3 years to go through the cycle.

Cronartium quercuum f. sp. *fusiforme* on southern pines has a similar life cycle, but the alternate hosts are oaks and the life cycle usually is 1 year or less, because of the warmer environment (Tainter and Baker 1996). Telia are produced on oak leaves in spring not long after infection by aeciospores. They can produce basidiospores and go back to pines before June. Thus uredinia play a smaller role in the life cycle than they do for white pine blister rust, since there is little buildup of inoculum on oak before telia are produced. The life cycle of the poplar rust fungus (Figure 13.14) is interesting because the telial hosts are poplars, and conifers, such as larch and Douglas-fir, are the aecial hosts. A typical short cycle rust is western gall rust caused by *Endocronartium harknessii* (Tainter and Baker 1996). There appears to be no telial host and aeciospores can infect pines directly. It forms globose galls on two- and three-needle pines, especially on lodgepole pine.

The complexity of the rust life cycle must have required a long evolutionary process. Neither the fungus nor the host could gain a strong advantage over the other for any length of time, so that the relationship is for the most part in balance. Why then do rusts cause some of the major forest diseases? The balance is tipped in favor of the fungus when (1) the fungus is introduced into an area where host plants have not evolved a genetic tolerance, (2) cultural practices in forestry and agriculture reduce genetic diversity, and (3) intensive forestry practices increase the incidence of native rust diseases by upsetting ecological balances.

Management of Rust Diseases

A number of strategies can be used to manage rust diseases, including (1) quarantine, (2) genetic resistance, (3) identification of high-hazard areas, (4) correct nursery practices, (5) management of the alternate host, (6) thinning and pruning,

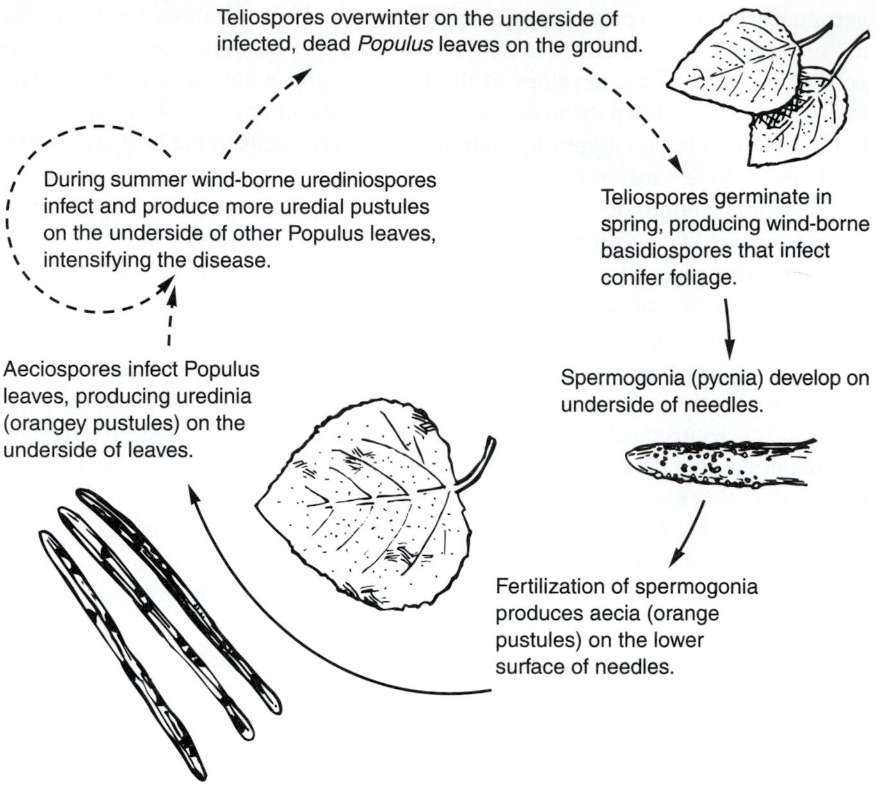

Teliospores overwinter on the underside of
infected, dead *Populus* leaves on the ground.

During summer wind-borne urediniospores
infect and produce more uredial pustules
on the underside of other Populus leaves,
intensifying the disease.

Teliospores germinate in
spring, producing wind-borne
basidiospores that infect
conifer foliage.

Aeciospores infect Populus
leaves, producing uredinia
(orangey pustules) on the
underside of leaves.

Spermogonia (pycnia) develop on
underside of needles.

Fertilization of spermogonia
produces aecia (orange
pustules) on the lower
surface of needles.

Figure 13.14
Life cycle of *Melampsora medusae,* cause of poplar rust.

and (7) fungicides (Tainter and Baker 1996, Maloy 1997). Had quarantine been practiced in the early 1900s, white pine blister rust probably would not have spread so easily. The important quarantine principle now is to ensure that seedlings infected with rusts are not transported to the field. Genetic resistance has shown a lot of promise and has been used successfully for fusiform rust in southern pines, where resistance genes have been incorporated without losing growth gains (Wilcox et al. 1996). Resistant western white pine seedlings now can be purchased, but resistant seedlings of eastern white pine are still not available. The use of resistant clones is particularly important with poplar rusts. However, it must be recognized that breeding programs tend to reduce genetic diversity, a value that is important in modern-day forestry.

Understanding the life cycle and spread mechanisms of *Cornartium ribicola* has been important in implementing silvicultural management strategies. Knowledge of overstory conditions, site characteristics, and topography are particularly important in eastern North America. Early work in the Lake states identified high-hazard areas on the landscape where planting of white pine should be avoided. Airborne basidiospores from *Ribes* plants reach and infect pine needles only near the tops of hills between lakes because of local airflow circulations on cool nights in autumn. Basidiospores released from understory *Ribes* plants are embedded in downslope cold airflows that rise once they reach the lake surface and return to the land at higher elevations where they are impacted on pine needles (VanArsdel 1967).

Infection of seedlings in nurseries also can be reduced by removing the alternate hosts near nurseries to prevent spread of basidiospores. This works well with fusiform rusts and oaks because spread from oak to oak by urediniospores is relatively small. However, it is more difficult to accomplish with white pine blister rust and *Ribes*. The decades spent from the 1930s to the 1960s trying to "grub" out or chemically control *Ribes* in the western states generally failed. However, there is evidence of success of *Ribes* removal in the eastern states. It also is impossible to remove all the larch, Douglas-fir, and hemlock in coastal western North America to protect poplars from rust.

Thinning and pruning can be used effectively for rust control. Thinning removes infected trees and creates conditions that may be less favorable for basidiospore infection. Pruning of infected lower branches to manage white pine blister rust has been effective in western Washington and Idaho. Fungicides are generally not very effective and are usually restricted. However, benlate can be used, particularly in high-value urban areas, and there is potential for fungicide control in nurseries.

REFERENCES

Agrios, 1997. *Plant pathology,* 4th ed. Academic Press, New York.

Allen, E., D. Morrison, and G. W. Wallis. 1996. *Tree diseases of British Columbia.* Forestry Canada, Pacific Forestry Centre, Victoria, B.C., Canada.

Alexopoules, C. J., C. W. Mims, and M. Blackwell. 1996. *Introductory mycology,* 4th ed. John Wiley and Sons, New York.

Daughtrey, M. L., C. R. Hibben, K. O. Britton, M. T. Windham and S. C. Redin. 1996. Dogwood anthracnose. Understanding a new disease in North America. *Plant Disease* 80:349–358.

Funk, A. 1985. *Foliar fungi of western trees.* Pacific Forestry Research Centre, Canadian Forestry Service, BC-X-265, Victoria, B.C., Canada .

Hansen, E. M., and K. J. Lewis (eds.). 1997. *Compendium of conifer diseases.* American Phytopathological Society Press, St. Paul, Minn.

Hughes, S. J. 1976. Sooty molds. *Mycologia* 68:693–820.

Maloy, O. C. 1997. White pine blister rust control in North Amierca. *Annual Review of Phytopathology* 35:87–109.

Manion, P. D. 1991. *Tree disease concepts.* 2nd ed. Prentice-Hall, Englewood Cliffs, N.J.

Newcombe, G., and G. A. Chastagner. 1993. First report of the Eurasian popular leaf rust fungus, *Melampsora larici populina,* in North America. *Plant Disease* 77:532–535.

Scharpf, R. T. 1993. *Diseases of Pacific Coast conifers.* USDA Forest Service, Agricultural Handbook 521.

Sinclair, W. A., H. H. Lyon, and W. T. Johnson. 1987. *Diseases of trees and shrubs.* Cornell University Press, Ithaca and London.

Skelly, J. M., D. D. Davis, W. Merrill, A. E. Cameron, H. D. Brown, D. B. Drummond, and L. S. Dochinger (eds.). 1987. *Diagnosing injury to eastern forest trees: a manual for identifying damage by air pollution, pathogens, insects and abiotic stresses.* National Acid Precipitation Program, Vegetation Survey Cooperative, USDA Forest Service, Forest Pest Management, and Pennsylvania State University, University Park, Penna.

Sutherland, J. R., G. M. Shrimpton, and R. N. Sturrock. 1989. *Diseases and insects in British Columbia forest seedling nurseries.* Forestry Canada and B.C. Ministry of Forests FRDA Report 065, Victoria, B.C., Canada.

Tainter, F. H., and F. A. Baker. 1996. *Principles of forest pathology.* John Wiley and Sons, New York.

Taylor, J. E., and J. W. Schwandt. 1998. *Dothistroma needle blight of limber pine in Montana.* USDA Forest Service, North Region, Forest Health and Protection Report 98-4, Missoula, Mont.

Van Arsdel, E. P. 1967. The nocturnal diffusion and transport of spores. *Phytopathology* 57:1221–1229.

Wilcox, P. L., H. V. Amerson, G. E. Kuhlman, B. H. Liu, D. M. O'Malley, and R. R. Sederoff. 1996. Detection of a major gene for resistance to fusiform rust in loblolly pine by genomic mapping. *Proceedings of the National Academy of Sciences USA* 93:3859–3864.

Ziller, W. G. 1974. *The tree rusts of western Canada.* Canadian Forestry Service, Department of the Environment, Publication 1329, Ottawa, Ontario.

14

STEM AND BRANCH DISEASES

INTRODUCTION

Many trees diseases, such as mistletoes, cankers, galls, decay, vascular wilts, bark diseases, and wood stains, affect stems and branches (Funk 1981, Sinclair et al. 1987, Manion 1991, Allen et al. 1996, Tainter and Baker 1996, Hansen and Lewis 1997). Each of these diseases is considered in this chapter. Rusts also occur on stems and branches, but they are discussed in Chapter 13. Diseases on stems and branches are caused by fungi, bacteria, phytoplasmas, and parasitic plants and cause considerable economic loss (Agrios 1997). They also create habitat for wildlife through production of stem cavities and nesting sites and are of great ecological importance (Bull et al. 1997).

MISTLETOES

There are many species of parasitic plants in a variety of plant families (Kuijt 1969). The majority are dicotyledonous, although there are some gymnosperms. Parasitic plants typically possess haustoria, which function like roots for nutrient and water uptake in host tissue. Excessive hypertrophy and witches'-brooms, chlorosis, and reduced growth and reproductive capacity are common symptoms of infection (Figure 14.1). Portions of infected plants and occasionally the whole plant may die. Many parasitic plants cause problems in agricultural crops, such as witchweed on corn and wheat and dodders on field crops, but these plants are generally not so important in forest trees (Agrios 1997). On the other hand, mistletoes cause a great many problems in forest trees (Sinclair et al. 1987).

There are two broad groups of mistletoes: true or leafy mistletoes in the genera *Phoradendron* in North America (see Figure 10.4) and *Viscum* in Europe and dwarf mistletoes in the genus *Arceuthobium* in the family Visaceae (Tainter and Baker 1996). Mistletoes also occur in the Loranthaceae, and the Eremulepidaceae and the Misodendraceaea, two South American families. *Phoradendron* spp. typically occur on hardwoods, but some species occur on junipers, cedars, and true

Figure 14.1
Excessive brooming on a ponderosa pine tree in eastern Oregon caused by western dwarf mistletoe (*Arceuthobium campylopodum*).

firs. Typically true mistletoes do not occur north of 40 to 45° north latitude (i.e., across the United States from Oregon to New Jersey). They have well-developed stems with leaves 2.5 to 5 cm long. Seeds are disseminated by birds feeding on their berries. Infected branches typically have galls and brooms. True mistletoes can be managed by branch pruning and spraying where the value of trees is very high, such as fruit trees.

Dwarf misletoes are a particular problem in western North America, infecting Douglas-fir, pines, firs, spruces, hemlocks, and western larch (Hawksworth and Wiens 1996). They occur from Mexico northward to Canada and across Canada to the northeastern United States. Species and hosts are listed in Table 14.1. At least 47 species are known from Central and North America, Europe, Africa, Asia, the Azores, and Caribbean Islands. Many are host-specific, like Douglas-fir dwarf mistletoe, but some occur on several hosts. For example, *A. campylopodum* occurs on Jeffrey and ponderosa pines, and *A. tsugense* subsp. *tsugense* occurs on western hemlock, true firs, and lodgepole pine.

Arceuthobium spp. are **dioecious** plants without true leaves. A typical life cycle is shown in Figure 14.2. Seed dispersal normally occurs in the later summer or fall. Local dispersal occurs when seeds are shot from mistletoe plants at velocities as high as 100 kph. Seed dispersal range is usually 6 to 11 m, but some seeds may travel as far as 15 m. Birds or rodents are capable of carrying seeds long distances to initiate new infection centers. Seeds have a viscous coat, enabling them to stick on the host. They commonly overwinter on needles and germinate in the spring. The radicle produced by the seed grows until it reaches a suitable bark crevice, bud, or needle base. A holdfast forms that completes the infection process by wedging into the host tissues and producing an extensive haustorial absorptive system. These "sinkers" penetrate the xylem of the host and are subsequently embedded by additional growth of xylem tracheids to which they are connected. Infected branches are typically swollen and distorted and are commonly called *witches'-brooms*. Infections may be localized or systemic.

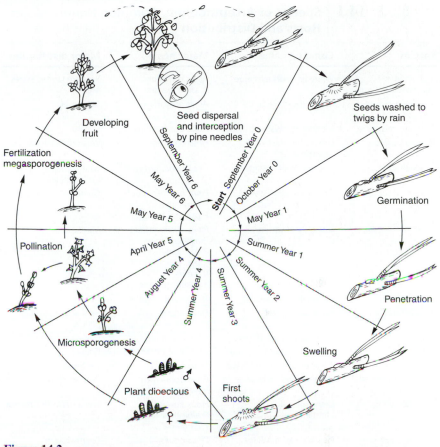

Figure 14.2
Life cycle of a typical dwarf mistletoe.
(Source: From Hawksworth and Wiens 1996.)

Dwarf mistletoes take up nutrients and water from their hosts as well as most of their carbohydrates, although they do photosynthesize (Hawksworth and Wiens 1996). The expanding network of sinkers eventually breaks through the surface to form new aerial shoots, buds, flowers, and seeds. Aerial shoots typically range in height from 2 to 10 cm, and the branching pattern may be whorled or fanlike. Aerial shoots (Figures 14.3 and 10.3) do not emerge until the second or third year (Figure 14.2). Male and female flowers are produced on separate plants in the second year after emergence of the shoots. Male flowers have three or four "petals" that open to expose the pollen sacs, but female flowers are inconspicuous and remain closed. Most flowers are pollinated by insects. Fourteen to 18 months after pollination, the fruit, or fleshy berry, matures, with each berry containing a single seed. From infection to seed production takes 4 to 5 years. Spread is from tree to tree, and infection centers in young stands typically develop around larger and older infected residual trees.

T A B L E **14.1** | **Species of Arceuthobium, Common Names, Major Hosts and Distribution**

Species	Common name	Major hosts	Major distribution
A. abietinum f. sp. concoloris	White fir dwarf mistletoe	Abies concolor A. durangensis	Western United States Mexico
A. abietinum f. sp. magnificae	Red fir dwarf mistletoe	Abies magnifica	California
A. abietis-religiosae	Mexican fir dwarf mistletoe	Abies durangensis	Mexico
A. americanum	Lodgepole pine dwarf mistletoe	Pinus contorta	Western North America
A. apachecum	Apache dwarf mistletoe	Pinus strobiformis	Southwestern United States
A. aureum subsp. aureum	Golden dwarf mistleltoe	Pinus montezumae P. oaxacana P. pseudostrobus	Guatemala
A. aureum subsp. petersonii	Durangan dwarf mistletoe	Pinus michoacana P. montezumae P. oaxacana P. oocarpa P. patula P. pseudostrobus	Mexico
A. azoricum	Azores dwarf mistletoe	Juniperus brevifolia	Azores
A. bicarinatum	Hispanolian dwarf mistletoe	Pinus occidentalis	Haiti and Dominican Repulic
A. blumeri	Blumer's dwarf mistletoe	Pinus ayacahuite	Mexico
A. chinense	Keteleeria dwarf mistletoe	Keteleeria evelyniana	China
A. californicum	Sugar pine dwarf mistletoe	Pinus lambertiana	California
A. campylopodum	Western dwarf mistletoe	Pinus ponderosa P. jeffreyi	Western North America
A. cyanocarpon	Limber pine dwarf mistletoe	Pinus flexilis P. albicaulis P. longaeva	Western North America
A. divaricatum	Pinyon dwarf mistletoe	Pinus edulis P. monophylla	Southwestern United States
A. douglasii	Douglas-fir dwarf mistletoe	Pseudotsuga menziesii	Western North America Texas, Mexico
A. durangense	Durangan dwarf mistletoe	Pinus douglasiana P. michoacana P. pseudostrobus	Mexico
A. gillii	Chihuahua pine dwarf mistletoe	Pinus herrerai P. lumholtzii	New Mexico Mexico
A. globosum	Rounded dwarf misteltoe	Pinus maximinoi P. patula P. teocote	Mexico
A. guatemalense	Guatemalan dwarf mistletoe	Pinus ayacahuite	Guatemala, Mexico
A. hawksworthii	Hawksworth dawrf mistletoe	Pinus caribaea	Belize
A. hondurense	Honduran dwarf mistletoe	Pinus maximinoi	Honduras, Mexico
A. juniperi-procerae	East Farican dwarf mistletoe	Juniperus procera	Kenya, Eritrea, Ethiopa
A. laricis	Larch dwarf mistletoe	Larix occidentalis Tsuga mertensiana	Oregon, Washington, Idaho, Montana, British Columbia
A. littorum	Coastal dwarf mistletoe	Pinus muricata P. radiata	California

(continued)

T A B L E **14.1** | (*concluded*)

Species	Common name	Major hosts	Major distribution
A. microcarpum	Western spruce dwarf mistletoe	*Picea engelmannii* *Picea pungens* *Pinus aristata*	Arizona, New Mexico
A. minutissimum	Himalayan dwarf mistletoe	*Pinus densata*	Bhutan
A. monticola	Western white pine dwarf mistletoe	*Pinus monticola*	California, Oregon
A. nigrum	Black dwarf mistletoe	*Pinus lawsonii* *P. oaxacana* *P. patula*	Mexico, Guatemala
A. oaxacamum	Oaxacan dwarf mistletoe	*Pinus oaxacana* *P. lawsonii* *P. pseudostrobus*	Mexico
A. pendens	Pendent dwarf mistletoe	*Pinus discolor* *P. orizabensis*	Mexico
A. pini	Alpine dwarf mistletoe	*Pinus yunnanensis*	Bhutan
A. pusillum	Eastern dwarf mistletoe	*Picea mariana* *P. glauca* *P. rubens*	Mid and eastern North America
A. rubrum	Ruby dwarf mistletoe	*Pinus cooperi* *P. durangensis* *P. engelmannii* *P. herrerai* *P. teocote*	Mexico
A. sichuanense	Sichuan dwarf mistletoe	*Picea balfouriana* *P. spinulosa*	Bhutan China
A. siskiyouense	Knobcone pine dwarf mistletoe	*Pinus attenuata*	California, Oregon
A. strictum	Unbranched dwarf mistletoe	*Pinus leiophylla*	Mexico
A. tibetense	Tibetan dwarf mistletoe	*Abies forrestii*	China
A. tsugense subsp. tsugense Western hemlock race Shore pine race	Hemlock dwarf mistletoe	*Tusga heterophylla* *Abies amabilis* *A. lasiocarpa* var. *lasiocarpa* *Pinus contorta* var. *contorta*	California, Oregon, Washington, British Columbia, Alaska
A. tsugense subsp. mertensiana	Mountain hemlock dwarf mistletoe	*Tsuga mertensiana* *Abies amabilis* *A. lasiocarpa* *A. procera* *Pinus albicaulis*	California, Oregon, Washington, British Columbia
A. vaginatum subsp. vaginatum		*Pinus patula* + 12 other pines	Mexico
A. vaginatum subsp. cryptopodum	Southwestern dwarf mistletoe	*Pinus durangensis*	Mexico
subsp. vaginatum		*Pinus ponderosa* var. *scopulorum*	Southwestern United States
A. verticilliflorum	Big-fruited dwarf mistletoe	*Pinus arizonica* *P. durangensis*	Mexico
A. yecorense	Yecoran dwarf mistletoe	*Pinus durangensis* *P. herrerai* *P. leiophylla* *P. lumholtzii*	Mexico

Source: Compiled from Hawksworth and Wiens (1996).

Figure 14.3
Aerial shoots of western hemlock dwarf mistletoe (*A. tsugense* subsp. *tsugense*).

Dwarf mistletoes cause growth reduction, loss of host vigor, poor wood quality, branch and top kill, and even mortality (Tainter and Baker 1996). Poor-quality wood results because branch swelling and stem infections cause abnormal grain, resin impregnation, and spongy woody texture. Branch proliferation and swellings also increase the size and number of knots. Squirrels also may feed on infected branches, especially of lodgepole pine, western larch, and Douglas-fir and have caused "red flagging" in young stands in western Canada.

Dwarf Mistletoe Infection Rating Scheme

A 6-class dwarf mistletoe rating scheme was developed in the western United States to help foresters and researchers judge the degree of tree infection (Hawksworth 1977). This scheme, also known as the Hawksworth scheme, is illustrated in Figure 14.4. The first step is to divide the crown of individual trees into thirds (top, middle, and bottom). Then each third is rated as 0, 1, or 2 based on the degree of infection. For example, a rating of 2 is a heavy infection where more than half of the total number of branches in the third is infected. The final step is to add the ratings of the three thirds to get a rating for the total tree. The maximum rating is 6 for a very heavily infected tree.

Ecological Relationships

The distribution of dwarf mistletoe species is strongly influenced by topography, climate, fire, insect outbreaks, and forest succession. For example, Douglas-fir dwarf mistletoe does not occur in western Oregon and Washington. The combined effects of physiography, fire, and forest succession have apparently prevented the immigration and establishment of *A. douglasii* in this region. Climatic warming the last few thousand years has greatly fragmented the populations of dwarf mistletoes in the southwestern United States (Hawksworth and Wiens 1996).

Wildfires seem to be the most important single factor governing the abundance and distribution of dwarf mistletoes (Hawksworth and Wiens 1996). Although high-intensity stand-replacing fires reduce dwarf mistletoe populations, other types of fires can increase populations. For example, spotty fires leave scattered, live,

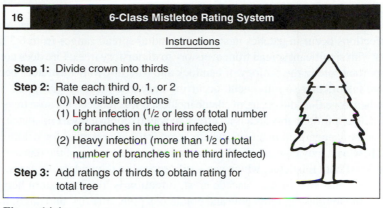

16	6-Class Mistletoe Rating System

Instructions

Step 1: Divide crown into thirds

Step 2: Rate each third 0, 1, or 2
(0) No visible infections
(1) Light infection ($1/2$ or less of total number of branches in the third infected)
(2) Heavy infection (more than $1/2$ of total number of branches in the third infected)

Step 3: Add ratings of thirds to obtain rating for total tree

Figure 14.4
The 6-class dwarf mistletoe rating scheme developed in the western United States. This scheme is known as the Hawksworth scheme.
(Source: Hawksworth 1977.)

infected trees that reinfect regenerating stands. In the Rocky Mountains wildfires tend to reestablish seral lodgepole pine stands, which are highly susceptible to *A. americanum* and limit development of climax spruce and fir species. Heavily infected lodgepole pine stands are very susceptible to fire because of dead wood and brooming.

Dwarf mistletoes also can influence wildlife populations. Mistletoe brooms provide good structures for establishment of birds' nests, and the number of spotted owls nests on the east slopes of the Cascades in Oregon and Washington has increased as a result of an increase in mistletoe brooms due to active fire suppression.

Management

Because of extensive timber losses from dwarf mistletoes, considerable effort has been expended to manage them (Hawksworth and Wiens 1996). There are basically four ways to manage dwarf mistletoes: (1) silvicultural, (2) chemical, (3) biological, and (4) genetic resistance. Dwarf mistletoes are relatively easy to manage silviculturally because of their slow rate of spread, host restriction, and dependence on living trees for survival. Success can be nearly 100% if done correctly. Clearcutting is the most successful method, but infected residual trees must be removed. Ten well-dispersed infected trees per acre (25 trees per hectare) can potentially result in infection of all new and existing trees. Thinning may increase the incidence of dwarf mistletoe by allowing more light in a stand and stimulating the growth of existing mistletoe plants. Outbreaks of mountain pine beetle may increase the proportion of infected trees by killing uninfected trees and increasing the rate of spread of mistletoes. Thus management objectives for both beetles and dwarf mistletoes need to be taken into account.

It is interesting to contrast silvicultural management on the east and west sides of the Cascade crest in Oregon and Washington. In old-growth stands on the west side of the Cascades, infection is high in western hemlock and true firs. As previously discusssed, very little mistletoe occurs on Douglas-fir in this zone. Mistletoe infections occur in patches in stands and radial spread ranges from 0.3 to 1.5 m/year with significant spread from overstory to understory trees. Fire does not play an important role in the ecology of hemlock and true fir dwarf mistletoes because fires are historically so infrequent, occurring only every 200 to 400 years. However, when large fires did occur, mistletoe incidence was reduced because trees were killed. Clearcutting in this area is an effective way to control dwarf mistletoe. New ecosystem management practices, where green trees are retained for wildlife purposes, however, may increase mistletoe infection because many of the retained trees are infected with mistletoe, which can easily spread to the understory trees.

On the east side of the Cascade crest, infection is low in western hemlock, but high in Douglas-fir, lodgepole and ponderosa pine, true firs, and larch. Radial spread is only 0.5 to 0.7 m/year, but there is significant spread from overstory to understory. In this region fire plays an important role in the development of dwarf mistletoes, especially in pine forests. The historically frequent low-intensity fires here killed small infected trees and reduced the rate of spread. Fire suppression has caused increased infection in these stands, and the incidence of infection is now much higher than natural levels. Clearcutting usually is not a viable management option here. Removal of infected trees will reduce infection in un–even-aged stands, but these stands must be revisited after a number of years to remove residual infected trees. Ponderosa pine dwarf mistletoe also may be managed in thinned or unthinned stands by prescribed understory burns, but scorch heights of 30 to 60% of the crown length are required to significantly reduce dwarf mistletoe infections. Use of resistant species also is a viable management option.

Chemical control consists of herbicides and growth regulators. Herbicides have generally proven to be unsuccessful and growth regulators have shown the most promise. The ethylene-releasing growth regulator Florel, whose active ingredient is ethephon (2-chloroethyl phosphonic acid) is the only chemical approved by the EPA for use in the United States. Use of this chemical can delay seed production by the dwarf mistletoe by 2 to 4 years, reduce the rate of mistletoe spread, and protect understory trees. However, it cannot cure infected trees since the endophytic system remains active. Ground spraying of high-value trees in recreational and residential areas has proven effective, and shoot abscission rates of 90 to 100% are possible when coverage is thorough. Helicopter spraying for extensive aerial coverage, however, has proven unsuccessful.

Biological control has promise because many fungi are pathogenic on dwarf mistletoes and some insects, particularly lepidopteran larvae, feed on dwarf mistletoes. However, in practice the use of biological control seems remote. Genetic control has greater potential. In central Oregon small *Pinus ponderosa* seedlings produced by grafting scions from *Arceuthobium campylopodum*–resistant trees to root stocks were planted in a heavily infected stand. After 20 years high levels of resistance were found in grafts of several selections. Evidence for host resistance to dwarf mistletoes has been noted in several host-parasite combinations.

CANKERS

Cankers result from the death of definite and relatively localized areas of bark and/or cambium on branches and stems of trees (Hansen and Lewis 1997). The most serious cankers are caused by fungi, but they also can be caused by non-infectious agents such as frost and sunscald or mechanical injury. Typical canker causing fungi are Ascomycetes, but Basidiomycetes and bacteria also can cause cankers (Tainter and Baker 1996). Cankers caused by rust fungi are covered in Chapter 13 (Foliage Diseases and Rusts) along with cankers that occur only on twigs and shoots, such as anthracnoses. Cankers may also be associated with mistletoes, involving fungi such as *Cytospora abietis*.

Wounding is generally involved in canker formation and repeated callusing usually is necessary before a lesion is termed a *canker,* but not always, as with annual cankers. Death of the bark and cambium is followed by death of the under-lying wood and sooner or later the bark sloughs off, exposing the wood. Canker fungi may or may not penetrate the wood, but wood decay fungi usually will infect the exposed wood surface.

Cankers may be annual or perennial. With annual cankers the agent causing the disease operates for only one season and the damaged tissues are sloughed off or grown over. In the case of perennial cankers the agent, usually a fungus, causes damage year after year. Typically the pathogen enters through a wound and invades and kills the healthy bark, usually during a dormant period. The host then attempts to limit pathogen invasion by forming a layer of callus over the edge of the infected tissue. During the next dormant period the fungus invades the callus, which is followed by new callus formation. This process can be repeated year after year. Eventually the stem may be girdled and the tree may die or it may snap in the wind at the point of the canker where the stem is weakened. Perennial cankers often result in sunken, unhealed wounds on branches and trunks and are usually resinous in conifers.

Cankers vary considerably in size and shape and are classified as target cankers, diffuse cankers, or canker blights. Some fungi cause canker rots. These usually are decay fungi that attack heartwood as well as the living sapwood and cambium. Table 14.2 lists common canker diseases and the type of canker formed. Target cankers are roughly circular in shape and contain abundant callus throughout the canker face and margin. Pathogen spread is slow. Cankers that are elongated ovals and contain little callus at the margins are diffuse cankers. Diffuse cankers grow quickly forming little callus, and often girdle the tree because they can enlarge faster than the radial growth of the tree. With canker blights the cankers are circular to elliptical, have little or no callus, and increase rapidly during a single season.

Cankers on Hardwoods

Cankers tend to be more important on hardwoods than on conifers (Sinclair et al. 1987, Tainter and Baker 1996). The most important canker diseases of hardwoods are chestnut blight (Figure 14.5), Hypoxylon canker (Figure 14.6), beech bark

T A B L E **14.2** | **Common Canker Fungi and Hosts in North America and Europe**

Common name	Causal fungus	Principal hosts
Hardwoods		
Chestnut blight[a]	*Cryphonectria parasitica*	American and European chestnuts, Allegheny chinkapin
Hypoxylon canker	*Hypoxylon mammatum*	Trembling, bigtooth, and European aspen, balsam poplar
Beech bark disease[a,b]	*Nectria coccinea* var. *faginata*	American and European beech
Butternut canker	*Sirococcus clavigignenti-juglandacearum*	Butternut (black walnut rarely)
Fusarium canker	*Fusarium solani* and *Fusarium* spp.	Yellow-poplar and sugar maple
Ceratocystis canker[b]	*Ceratocystis fimbriata*	Poplars, sycamore, rubber and fruit trees
Septoria canker	*Mycosphaerella populorum*	Poplars
Nectria canker	*Nectria galligena*	Sugar and other maples
Strumella canker	*Urnula caterium*	White and red oaks (beech, hickory, red maple, ironwood, basswood, black gum, chestnut occasionally)
Madrone canker	*Nattrassia mangiferea*	Pacific madrone
Dothichiza canker	*Dothichiza populea*	Lombardy poplar
Black knot	*Dibotryon morbosum*	Cherry and plum
Fire blight	*Erwinia amylovora*[c]	Apples, pears, Sorbus and many other species
Thyronectria canker	*Thyronectria austroamericana*	Honeylocust, mimosa
Tubercularia canker	*Tubercularia ulmea*	Siberian elm, Russian olive
Conifers		
Atropellis canker	*Atropellis pinicola*	Sugar, western white, ponderosa, lodgepole pines
	Atropellis piniphila	Lodgepole pine
	Atropellis tingens	Slash, shortleaf pine, and other hard pines
Botryosphaeria canker	*Botryosphaeria ribis*	Giant sequoia and incense-cedar
Cenangium canker	*Cenangium ferruginosum*	Jeffrey pine, other pines and true firs
Cytospora canker	*Cytospora abietis*	True firs and Douglas-fir
Dasyscyphus canker	*Dasyscyphus* spp.	Douglas-fir, pines and true firs
Dermea canker	*Dermea pseudotsugae*	Douglas-fir and grand fir
Phomopsis canker	*Diaporthe lokoyae*	Douglas-fir
Pitch canker	*Fusarium subglutinans* (now *F. circinatum*)	Monterey, Bishop, Alepp, and Italian stone pines
Diplodia canker	*Sphaeropsis sapinea* (*Diplodia pinea*)	Pines (a particular problem in Monterey pine in southern hemisphere plantations)

(*continued*)

TABLE **14.2** | **Common Canker Fungi and Hosts in North America and Europe (*concluded*)**

Common name	Causal fungus	Principal hosts
Conifers (continued)		
Nectria canker	*Nectria fuckeliana*	White fir
Redwood canker	*Seiridium* spp.	Coast redwood
Larch canker	*Lachnellula* and *Trichoscyphella* spp.	European larch and eastern larch
Scleroderris canker	*Ascocalyx abietina*	European pine and spruce; all eastern conifers except balsam fir and northern white cedar
Basal canker of white pine[b]	*Pragmopara pithya*	Eastern white pine
Sirococcus shoot blight	*Sirococcus conigenus*	Red pine; western hemlock; Jeffrey, Coulter, ponderosa, lodgepole, and jack pines; Douglas-fir; white fir; and spruces

[a]Introduced.
[b]Insect/disease complex.
[c]Bacteria.

Figure 14.5
Chestnut blight canker.

Figure 14.6
Hypoxylon canker on trembling aspen.

disease (Figure 14.7), and butternut canker. Cankers caused by *Nectria, Thyronectria, Eutypella, Fusarium, Ceratocystis,* and *Septoria* are of less importance. Fire blight, a bacterial disease caused by *Erwinia amylovora,* severely affects pear and apple orchards in both the northern and southern hemisphere, but is not a problem with forest trees.

The chestnut blight fungus (*Cryphonectria parasitica,* Ascomycete), which is native to Asia, was introduced into the United States about 1900 on imported Asiatic chestnut seedlings (Tainter and Baker 1996). It also was introduced to Europe. By 1940 it had spread across the natural range of American chestnut, nearly eliminating the species, at least in tree form. Stumps may remain alive, but only sprouts form. The chestnut blight fungus spread very rapidly (up to 40 km/yr), because (1) there was virtually no resistance in the trees, (2) the fungus produces copious amounts of easily dispersed spores, and (3) it can survive as a saprophyte on dead tissues. American and European chestnuts and Allegheny chinkapin are very susceptible, but Chinese and Japanese chestnuts are resistant. The fungus also can infect oaks, maples, and other hardwood species, but it does not cause the same injury.

The life cycle of *C. parasitica* is as follows. Ascospores and conidia infect fresh wounds in the living bark. Conidia are produced in a sticky, yellow-orange tendrils and are carried in rain splash or by insects and birds such as woodpeckers. Ascospores are airborne and are also carried by birds and insects. The fungus

Figure 14.7
Beech bark disease on American beech.

grows in the inner bark and cambium and kills cells by enyzmatic action. Infected stems or branches are rapidly girdled and killed and diffuse cankers are formed (Figure 14.5), eventually producing cankers and spores. Although infected trees may produce adventitious shoots, they usually are rapidly killed. The fungus does not invade the roots.

Attempts to control this disease have not been very successful in the United States. No satisfactory resistant strains of American chestnut have been found, but hybrids of American and Asiatic chestnut show some promise. However, chestnut blight is controlled naturally in Italy, and by artificial inoculation of cankers in France with hypovirulent strains of *C. parasitica*. These strains were first noticed in Italy in the early 1950s in areas where old infections were healing. Hypovirulent strains carry viruslike double-stranded RNAs (dsRNAs) that limit the pathogenicity of virulent strains. The dsRNAs appear to pass from hypovirulent to virulent strains through mycelial anastomoses rendering the latter hypovirulent and slowing down or stopping canker development. Hypovirulent strains are being tested in the United States, but the treatment has not been as successful as it is in Europe. Virulent strains in the United States are more variable than European ones requiring multiple hypovirulence factors, and the interaction is more strain-specific.

Hypoxylon canker is extremely devastating to aspen in North America, causing poor growth, mortality, and inadequate stocking (Manion and Griffin 1986). The major host is trembling aspen, but other poplar and aspen species are infected as well as some species of willows. It is caused by the Ascomycete *Hypoxylon mammatum*, spreads by ascospores and conidia, and produces diffuse cankers (Figure 14.6). The infection process is not entirely understood, but insects and wounding are involved. *Hypoxylon mammatum* produces a toxin and also can decay wood. Commonly trees are girdled and killed and snap off at the canker. No practical control measures are known. It is not yet known in Alaska and quarantine is the best method to keep it out.

Beech bark disease (Figure 14.7) is caused by the scale insect *Cryptococcus fagisugal Nectria* complex (Tainter and Baker 1996). The Ascomycete *Nectria coccinea* is the causal fungus in Europe; *N. coccinea* var. *faginata* and sometimes *N. galligena* and *N. ochroleuca* are involved in North America. The conidial stage is *Cylindrocarpon* spp. It has been known since the 1840s in Europe and was first noted in North America in 1920 in Nova Scotia. All varieties of American and European beeches are susceptible to some degree. The first sign is the woolly-appearing, white waxy material secreted by the scale insects on twigs and bark. The scale insect creates a small feeding wound by penetrating the bark, and this wound is opened further by drying and shrinking of the bark, providing access to the windborne ascospores. The greater the intensity of the scale infestation, the greater the tree injury and girdling and tree death can result.

Epidemic development of the fungus occurred between 1920 and 1950 in the northeastern United States, and the complex has continued to move to the South and West since its introduction in Canada, and it is now in Ohio and Tennessee (Houston 1994). Three stages of the disease have been noted: (1) the advancing front, where the scale is present and the fungus is rare; (2) the killing front, where both agents are present; and (3) the final stage, where there are small trees of sprout origin, evidence of former mortality, and few older trees.

Beech stands can be managed by selecting and maintaining the most resistant individuals and removing susceptible ones. Herbicides may be needed to control sprouting from roots and stumps of susceptible individuals as well as advance reproduction. Nursery inspection and prohibition of movement of infected individuals will slow the spread to new areas. An insect predator (twice-stabbed ladybird beetle) and a mycoparasite also may contain the disease.

Butternut canker is a particular problem in Wisconsin, North Carolina, and Virginia (Tainter and Baker 1996). As well as butternut, black walnut is occasionally cankered by the causal fungus *Siroccocus clavigignenti-juglandacearum*. It enters hosts through leaf scars, buds, and bark wounds including insect wounds and bark splits. Butternut cankers are perennial and are elliptical to fusiform-shaped on branches, stems, and even roots. A black exudate oozes from fissures in cankers in the spring. The relatively high incidence and mortality has raised concern about the future of this species. Efforts have been made for at least two decades to search for resistant material, and trees have been planted in hopes of maintaining some part of the gene pool.

Madrone canker has developed dramatically in coastal areas of the Pacific Northwest United States in recent years, particularly in the Puget Sound region. The causal fungus is *Nattrassia mangiferae*. Branch dieback and cankering of the main stem are common symptoms and considerable mortality has occurred. No treatment has yet been developed.

Cankers on Conifers

A variety of cankers also are present on conifers (Hansen and Lewis 1997) (Table 14.2), but generally are not as damaging as those on hardwoods. Pines are particularly susceptible to canker fungi, but cankers also occur on Douglas-fir, true firs, spruces, cedars, hemlocks, coast redwood, and giant sequoia. Diplodia canker has become a particular problem in radiata pine plantations in New Zealand, generally associated with pruning wounds. Pitch canker disease of pines has received considerable recent attention (Storer et al. 1997). Copious resin flows from infected areas (branches, shoots, cones, exposed roots, and the bole) and resin cankers are formed. It was first described in 1946 in Florida and now infects many southern pine species, radiata pine and Douglas-fir in California. It also has been reported from Mexico, Japan, and South Africa. The causal fungus is *Fusarium subglutinans* f. sp. *pini*. A variety of beetles, including *Ips* spp. are involved in the transmission of the fungus. There is great concern that this fungus will spread to other conifer species in the western United States and to radiata pine plantations in the southern hemisphere, where it could be particularly devastating.

GALLS AND OTHER MALFORMATIONS

Galls, witches'-brooms, and fasciations are the most common malformations on trees (Sinclair et al. 1987, Tainter and Baker 1996). Witches'-brooms are most commonly caused by mistletoes and rust fungi, and these have been discussed earlier. Fasciation or broadening and flattening of the stem has been attributed to over-nutrition and bacteria, but in most cases the cause is unknown. It is assumed to be hereditary in some plants. It occurs on both conifers and hardwoods, but is unimportant on forest trees.

Swellings or hypertrophies on the stem or branches are called *galls* and occur commonly on hardwoods and softwoods. They may be caused by rust fungi or dwarf mistletoes (discussed earlier), other fungi, bacteria, insects, or noninfectious agents. If the bark in the gall is similar to normal bark, then it is likely caused by a noninfectious agent. In contrast galls caused by pathogens normally change the appearance of the bark in the gall.

Large noninfectious galls or burls occur commonly on hardwoods such as black walnut and black cherry and are valued for furniture. They may be the result of injury. Their appearance is somewhat similar to bird's-eye maple, but bird's-eye is not associated with galls and occurs in trees with no external indication of abnormal wood. Large galls also occur on the trunks of conifers such as redwood, Sitka, and white and other spruces, but their cause is still unknown. Douglas-fir frequently has small noninfectious burls on the main stem occurring at wounds. Noninfectious galls also have been noted on other conifers.

Infectious galls are caused by bacteria and to a lesser extent by fungi. Bacterial galls are globose to fusiform in shape and occur on hardwoods and conifers. Crown gall is common on some hardwoods, but rarely on conifers. It is called *crown gall* because it occurs near the root crown or root collar of the tree and is caused by the bacteria *Agrobacterium tumefaciens*. Bacteria also cause galls in Europe—for example, *Bacterium savatanoi* var. *fraxini* on ash and *Pseudomonas pini* on conifers. Young Douglas-fir trees have been noted to have small globular galls on the main stem and branches caused by *Erwinia pseudotsugae* (Figure 14.8) where it occasionally causes mortality or death of the leader.

Figure 14.8
Bacterial gall on Douglas-fir.
(Source: From Scharpf 1993.)

Several species of fungi also cause galls. For example, *Phomopsis* spp. causes globose galls on oaks, hickories, maples, and American elm that are similar to crown gall. *Macrophoma tumefaciens* causes almost spherical galls on poplars in the Rocky Mountains and Lake states.

DECAY

Wood decay is the decomposition of wood by the enzymatic activities of microbes, primarily fungi (Manion 1991). Bacteria can decompose wood, but are inefficient compared to fungi. Insects, marine organisms, and UV light also can deteriorate wood.

Knowledge of wood chemistry is extremely important in understanding the decay process. Wood consists primarily of cellulose, hemicellulose, lignin, and extractives. Cellulose, hemicellulose, and lignin are the major constituents of cell walls, whereas extractives are not in the cell wall and typically confer decay resistance. Extractives are primarily phenolic chemicals. Species, such as western redcedar, redwood, and Douglas-fir, with colored heartwood typically have a high extractive content and have higher decay resistance than species with white heartwood such as hemlocks, spruces, and true firs.

Types of Decay

There are three basic types of decay: white, brown, and soft rot (Boyce 1961) as shown in Box 14.1.

White rots are fibrous because some cellulose remains intact until late in the decay process. It is whitish because of the loss of lignin, which is brown, and because the remaining cellulose tends to be white. Cellulose is degraded by a complex of enzymes known as cellulases, which typically break the cellulose into shorter chains and eventually into glucose, the building block of cellulose. White

14.1

Basic Types of Decay

Decay type	Fungus	Color	Texture	Chemistry
White	Basidiomycota	Bleached/white	Fibrous	All components removed, including cellulose and lignin
Brown	Basidiomycota	Brown	Cubical rot, cross-checking	Mostly lignin remains; cellulose and other components removed
Soft	Ascomycota/ Deuteromycete	Bleached or brown	Surface decay	Carbohydrates removed, and some lignin removed

rots are described as stringy, spongy, laminated (separated at the annual rings), mottled, or white pocket. Black zone lines are often present. In white pocket rot and mottled rots, lignin and hemicelluloses are removed in the early stages of decay, leaving areas rich in cellulose. This is called *selective delignification* and is of great interest to the pulping industry since lignin can be removed by a process known as biopulping.

The brown color of **brown rots** results from the selective removal of cellulose and hemicellulose, leaving brown, oxidized lignin. Brown rots are classified as cubical or pocket rots. The wood is not fibrous because of the lack of cellulose. It shrinks on drying and has a cubical appearance because of cross-checking. There are only a handful of "brown pocket rots," and they only occur in living trees. The initial stage of decay of brown rot decay is nonenzymatic, and the fungi involved produce peroxide that breaks the chains of cellulose and hemicellulose into smaller molecules. Enzymes like cellulase can then break down these molecules, releasing sugars that are absorbed by the fungi. Soft rot fungi have limited ability to break down cellulose and lignin and commonly cause cavities or pits in the secondary wall of cells.

Decay Terminology

Wood decay can be described not only by the type of decay, but also by where it occurs in the tree and if it is incipient or advanced. Decay near the top of the tree is called *top rot* and usually results from top breakage or kill. Lower in the bole decay is called *stem* or *trunk rot* and near the tree base it is called *butt rot* (Hansen and Lewis 1997). If the decay occurs in the center of the living trees it is called *heartrot*. If it occurs in the sapwood it is called *saprot*. Saprot most commonly occurs in dead trees or logs. Some fungi cause canker rots, which decay the sapwood. Many of these are Ascomycetes, whereas most of the important decay fungi are Basidiomycetes. *Slash rot* is a term used to describe the decay of logging slash—that is, branches, tops, and stumps.

Two other terms are used to describe decay: *incipient* and *advanced* decay. In the early stages of decay, incipient decay stain typically develops, which usually is brown or reddish-brown. In this case the wood is still firm to touch but the fungus causing the decay is well established. As the decay progresses the wood loses it structural integrity, is soft to touch, and can be easily penetrated with a sharp tool or even a finger. This is advanced decay.

Important Decay Fungi

Table 14.3, on pages 352–353, shows some important Basidiomycete decay fungi, common names, the type of decay they cause, and their hosts. There are so many decay fungi that all of them cannot be listed or discussed here. Many butt rot fungi, such as *Heterobasidion annosum, Armillaria* spp., *Phellinus weirii,* and *Phaeolus schweinitzii* also cause root rots and they are discussed in detail in Chapter 12. The fruiting bodies or conks of several common decay fungi, including *Phellinus pini, Ganoderma applanatum, Phellinus igniarius,* and *Laetiporus sulphureus* are shown in Figures 14.9 to 14.12, respectively. Excellent illustrations of the conks of decay fungi are shown in Sinclair et al. (1987), Allen et al. (1996), and Hansen and Lewis (1997).

Spread of Decay Fungi

The life cycle of a typical Basidiomycete decay fungus is shown in Figure 10.10. Decay fungi produce either annual fruiting bodies or perennial conks. Most are spread by airborne spores that enter through infection courts. A few produce conidia. Spores usually enter through fire scars, wounds, branch stubs, cankers, mistletoe infections, and galls. Wounds result from broken tops, treefall scars, bark beetle attack (e.g., fir engravers), animal damage (caused by deer, elk, cattle, bears, and porcupines), logging damage, pruning, and vandalism (carving and hatchet marks). Once the spores have germinated and invaded wounds, the fungus spreads vegetatively by mycelium. The spread usually is slow, less than 10 cm/year,

Figure 14.9
Fruiting body of *Phellinus pini.*

Figure 14.10
Fruiting bodies (conks) of
Ganoderma applanatum.

Figure 14.11
Fruiting bodies (conks) of
Phellinus igniarius.

T A B L E 14.3 | Important Basidiomycete Wood Decay Fungi, Hosts, and Type of Decay They Cause

Fungus	Common name	Common hosts	Decay type	Comments
Decay in Sapwood or Heartwood of Living Trees				
Echinodontium tinctorium	Indian paint fungus Brown stringy trunk rot	Non-pine conifers	White	Enters through branchlet stubs
Fomes fomentarius	Tinder fungus	Birches, beeches	White	Infects through branch stubs, wounds
Fomitopsis officinalis	Quinine conk	Conifers	Brown	Produces huge white conks, mycelium in thick mats
Ganoderma applanatum	Shelf fungus or artist's conk	Conifers and hardwoods	White	Occurs mostly on dead trees, but does occur on living trees
Inonotus glomeratus	Canker rot	Maples, beech	White	Produces small clinker "conks" in stubs that won't heal
Inonotus obliquus	Clinker or cinder conk	Birches and other hardwoods	White	Produces large "sterile conks"
Laetiporus sulphureus	Sulfur fungus	Conifers and hardwoods	Brown	Fruiting bodies are common on dead trees, logs, and stumps
Phellinus igniarius	False tinder fungus; white trunk rot	Hardwoods	White	Is very common and important
Phellinus pini	Red ring rot	Conifer	White pocket	Produces conks in branch stubs
Phellinus robustus	White heart rot	Conifers and hardwoods	White	
Polyporus squamosus	Scaly or saddleback fungus	Conifers and hardwoods	White	Produces large fan-shaped fruiting bodies
Stereum sanguinolentum	Red heart rot	Conifers	Brown	Causes slash decay; "bleeding" fungus
Root and Butt Rots				
Armillaria spp.	Armillaria root disease	Conifers and hardwoods	White	Produces mushroooms and rhizomorphs
Heterobasidion annosum	Annosus root and butt rot	Conifers and hardwoods	White	Produces perennial conks on roots, stumps, and downed logs

Fungus	Common name	Common hosts	Decay type	Comments
Root and Butt Rots (continued)				
Inonotus tomentosus	False velvet top fungus	Conifers	White	Produces annual fruiting body resembling that of *Phaeolus schweinitzii*
Perenniporia subacida	Stringy butt rot	Conifers and hardwoods	White	Is common on dead coniferous and hardwood trees and slash
Phaeolus schweinitzii	Velvet top fungus Schweinitzii butt rot	Conifers and hardwoods	Brown	Occurs usually on older trees
Phellinus weirii	Laminated root rot	Conifers	White	Produces yellow laminated advanced deacy
Decay of Standing Dead/Logs/Slash				
Bjerkandera adusta	Scorched conk	Hardwoods—poplars, red gum	White	Produces thin, fleshy annual conks
Cryptoporus volvatus	Gray-brown sap rot	Conifers	White	Produces "pouchlike" fruting body, common in bark beetle— and fire-killed trees.
Daedalea quercina	Oak maze-gill	Hardwoods	Brown	Is important in decay of poles and posts
Fomitopsis pinicola	Red belt fungus	Conifers and hardwoods	Brown	Is important log decomposer
Ganoderma oregonense	Varnish conk	Western conifers	White	Is common on stumps and logs
Ganoderma tsugae	Varnish conk	Western conifers	White	Is common on stumps and logs
Gloephyllum sepiarium	Slash conk	Conifers and hardwoods	Brown	Produces shelflike fruiting boides with thick gills
Schizophyllum commune	Sapwood rot	Hardwoods	White	Produces small lobed fan-shaped conks
Trametes versicolor	Rainbow conk, turkey tail	Hardwoods	White	Produce sapwood decay

Figure 14.12
Fleshy fruiting body of
Laetiporus sulphureus.

although some fungi can spread as much as 60 cm/year. It takes a long time for decay to become well established in conifers, and thus decay is uncommon in young trees. It is much more common in old-growth trees where it is has had a long time to develop.

Decay is generally much faster in hardwoods than conifers. Hardwoods are more susceptible to wounding and branch breakage and generally have fewer extractive chemicals in the heartwood to protect against decay. Some hardwoods, however, particularly tropical species and some species of eucalypts in Australia, such as jarrah (*Eucalyptus marginata*) are very decay resistant. Species that stump sprout are susceptible to further infection since the sprouts can be infected via decayed heartwood in the stump.

The Concept of Compartmentalization of Decay in Trees (CODIT)

This concept was developed by Dr. Alex Shigo (USDA Forest Service, NE Forest Experiment Station), who dissected thousands of eastern deciduous hardwood trees and conifers and examined the patterns of decay (Shigo and Marx 1977, Shigo 1984). Unlike animals, trees do not have the ability to regenerate damaged tissues. They can, however, confine damage due to stem decay fungi to certain areas and impede the rate of decay development. CODIT is illustrated in Figure 14.13. Wounded trees possess four walls that compartmentalize decay fungi. Wall 4, or the barrier wall, is formed at the cambium at the time of wounding and is the

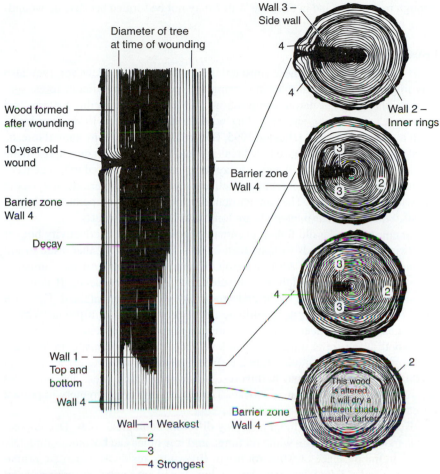

Figure 14.13
The concept of compartmentalization in trees.
(Source: From Shigo and Marx 1977.)

strongest wall. Here phenolic chemicals, gums, and tyloses are laid down in the damaged cells and the response is transmitted around the annual ring. Without further wounding, the fungus will not spread outside that ring and decay can develop only in a column inside the ring. Decay resulting from a wound in a small tree thus is not as serious as decay resulting from a wound in a large tree in terms of the volume of wood available for decay. Wall 3 consists of the ray cells that inhibit radial spread. Wall 2 is the summer wood of the annual rings, and Wall 1, the weakest wall, consists of the ends of the vertically oriented vessels and tracheids that make up wood. Thus fungi can spread most rapidly up and down the tree. There is some argument about whether compartmentalization of decay occurs in all species. There also are cases where decay fungi may enter stems through root grafts, and in this case Wall 4 may not be formed because no wounding is involved

Ecological Role of Decays

Heart and butt rot fungi cause substantial economic losses. However, they also play important ecological roles in altering the speed and direction of forest succession, changing vegetation composition, altering nutrient cycling patterns and long-term productivity, and providing wildlife habitat for cavity-nesting birds, small mammals, and bats (Hennon 1995, Bull et al. 1997). In many natural forests, heart, root, and butt rots cause bole breakage and uprooting or induce trees to die standing, thus creating small gaps. Decay fungi can be considered to be agents of disturbance like fires and insects, particularly between long periods of episodic large-scale disturbances. They are particularly important in temperate and boreal forests, where fire return intervals are long, and in tropical forests.

In coastal Alaska old forests eventually convert to low-productivity bogs if they exist for long periods without disturbances, like uprooting, that result in soil mixing (Hennon 1995). Accumulation of organic matter restricts moisture drainage in this high-rainfall area, resulting in gradual tree death. If trees are uprooted, soil mixing can occur and productivity can be maintained. Heart rot fungi on the other hand may contribute negatively to long-term productivity in these stands because heart-rotted trees are not generally uprooted.

Wood decay fungi initiate a complex succession of organisms in logs and standing boles involving other fungi, N-fixing bacteria, invading plants and mycorrhizal fungi, and animal herbivores, fungivores, omnivores, and predators including salamanders, bats, and bears. Heart rot fungi are extremely important for cavity-nesting birds.

Brown and white rot fungi also play different ecological roles. Decomposition is more complete with white rot fungi, and tree boles can become completely hollow in the presence of white rot fungi. Brown rot residues, which are primarily composed of lignin, last considerably longer and are important sources of soil organic matter. Nitrogen fixation in brown rotted decay is an important source of N to forest ecosystems. Brown rotted logs also can provide wildlife habitat for long periods of time; hundreds of years in large logs. By increasing the variability of structure in forests, heart rot fungi indirectly enhance habitat for many wildlife species. For example, gaps created in forests in southeastern Alaska allow more light to reach the forest floor, increasing understory vegetation browse for black-tailed deer (Hennon 1995).

Interactions among fire, decay fungi, and mountain pine beetles strongly control the vegetation dynamics in the lodgepole pine stands that form an edaphic climax over large areas of the infertile "pumice plateau" in south-central Oregon (Gara et al. 1985). Periodic fires create fungal infection courts in damaged roots; with time, advanced decay caused by Basidiomycete decay fungi develop in the butts and stems of these trees. The mountain pine beetle (*Dendroctonus ponderosae*) preferentially selects and kills these trees during the flight season. As

Box 14.2

Percent Volume of Cull in Douglas-Fir Trees Due to Decay Fungi in Relation to Tree Age

Tree age (years)	Volume of cull (%)
<40	0
40–80	2
80–120	4
120–160	9
160–200	34
>200	85

these outbreaks develop, additional uninfected trees are attacked. As needles drop, snags fall, and logs decay the stage is set for subsequent fires. Ignited logs tend to smolder because the interior of the logs has been decayed, which causes fire scars where logs contact remaining trees. In addition, heat from the burning logs damages or kills tree roots, resulting in tree stress and subsequent bark beetle attack.

Management of Decay Fungi

If the major management objective is timber production, decayed trees should be removed since valuable wood is being decomposed. On the other hand, if the emphasis is on ecosystem management, decay fungi may actually be introduced into living trees to create habitat for cavity nesting birds (see Chapter 16). Decayed trees also may be a danger to humans in high-hazard areas in forests—for example, near campgrounds or along heavily traveled roads (Wallis et al. 1992). Trees with stem decay can be particularly hazardous in urban environments where they can cause considerable property damage, human injury, and even death. Indicators of decay, including the presence of conks, can be used to predict its extent. Tree coring, a Shigometer, or sonic techniques can be used to determine the extent of decay and how much sound rind exists. This information can then be used to assist in decisions for removal of hazard trees.

Clearcutting, even aged management, and short rotations reduce losses due to decay. For example, decay losses in western hemlock in western Oregon and Washington average less than 1% in stands less than 80 years, but increase to 20% in stands that are 120 years old. The amount of cull in Douglas-fir due to fungi like *Phellinus pini*, *Fomitopsis officinalis*, and *Phaeolus schweinitzii* increases exponentially with age as shown in Box 14.2 (Tainter and Baker 1996).

Most yield tables are based on gross yield, but net yield must be considered when heartwood decay is present since growth gains can be offset by decay losses. In light of this relationship, the concept of the **pathological rotation** needs to be considered. The pathological rotation is the age when decay losses exceed annual increment. In most intensively managed stands, the usual rotation is shorter than the pathological rotation. Pathological rotations have been calculated to be 40 to 50 years for aspen in Minnesota, but longer in Utah (80–90 years). For yellow birch it ranges from 70 years in New York to 120 years in Nova Scotia. In conifers the pathological rotation is usually longer—for example, 225 to 275 years for western hemlock in western British Columbia and 100 to 120 years in western Oregon and Washington. Prevention of wounding during thinning also will inhibit decay. If pruning is practiced, it should be done early when branch sizes are small, and done properly to avoid creating internal defects and decay (see section on compartmentalization of decay).

VASCULAR WILTS

Vascular wilt diseases impede transpirational water flow, resulting in wilt symptoms (Agrios 1997). They are among the most devastating of plant diseases because they can kill trees swiftly. The features of the major vascular wilt diseases are shown in Table 14.4. The most important vascular wilt diseases are Dutch elm disease and oak wilt, but there are a several others including elm yellows, Verticillium wilt, persimmon wilt, and on conifers, conifer wilt and black-stain root disease (Tainter and Baker 1996). Black stain root disease is discussed in more detail in Chapter 12 (Root Diseases).

Dutch Elm Disease

Dutch elm disease has been responsible for killing the majority of elms in Europe and North America within a few decades after its introduction, probably initially from Asia (Stipes and Campana 1981, Manion 1991, Tainter and Baker 1996). It was particularly devastating to street and shade trees, and all species of native elms in North America and most in Europe are susceptible. Siberian elms and other Asian elms are resistant, but not immune.

It is named Dutch elm disease because it was first reported in the Netherlands in 1919, although dying elms were noted in France before that. It then spread rapidly through Europe and reached eastern North America in 1930. It has since spread throughout the United States and Canada, reaching Oregon in 1974 and the Puget Sound area in the 1990s. A second more virulent form was introduced into England in 1967, causing a second epidemic in Europe.

Dutch elm disease was originally caused by the Ascomycete fungus *Ophiostoma ulmi,* which is spread by insects and through root grafts to adjacent trees. In the last 20 years in Europe, North America, and Asia, two very aggressive races (Eurasian and North American) and a new species of the fungus (*O. novo-ulmi*) have emerged, which are displacing the less aggressive *O. ulmi* (Tainter and Baker 1996).

The life cycle of the fungus is illustrated in Figure 14.14. The smaller European elm bark beetle (*Scolytus multistriatus*) is the major vector in North America, but it also is spread by the native elm bark beetle (*Hylurgopinus rufipes*). The fungus was introduced into the United States from Europe in logs that contained both the fungus and *Scolytus multistriatus*. The beetles escaped at different

T A B L E 14.4 | Common Vascular Wilt Diseases, Causal Organisms, Principal Hosts, and Mechanisms of Spread

Common name	Organism name	Type of organism	Principal hosts	Mechanisms of spread
Hardwoods				
Dutch elm disease[a]	*Ophiostoma ulmi*	Fungus (Ascomycete)	North American and European elms	Spread by bark beetles and through root grafts
Oak wilt	*Ceratocystis fagacearum*	Fungus (Ascomycete)	Oaks, *Castanea, Castanopsis, Lithocarpus*	Spread by bark beetles and through root grafts
Verticillium wilt	*Verticillium albo-atrum*	Fungus (Ascomycete)	Sugar, Norway, red maples; many hosts in Europe and North America	Usually enters through root wounds; insects not involved
Elm yellows	Not yet named	Phytoplasma	American, red, winged elm	Spread by white-banded leafhopper
Mimosa wilt	*Fusarium oxysporum*	Fungus (Ascomycete)	Mimosa	Usually enters through roots
Persimmon wilt	*Acromonium diospyri*	Fungus (Ascomycete)	Persimmon	Enters through branch or stem wounds caused by animals or insects
Conifers				
Black-stain root disease	*Leptographium wageneri*	Fungus (Ascomycete)	Pines, Douglas-fir	Spread by beetles and weevils
Conifer wilt	*Ophiostoma piliferum* (*Ceratocystis* spp.)	Fungus (Ascomycete)	Pines and other conifers throughout the world	Spread by bark beetles (*Dendroctonus* and *Ips* spp.)—a complex of diseases

[a]introduced disease.

359

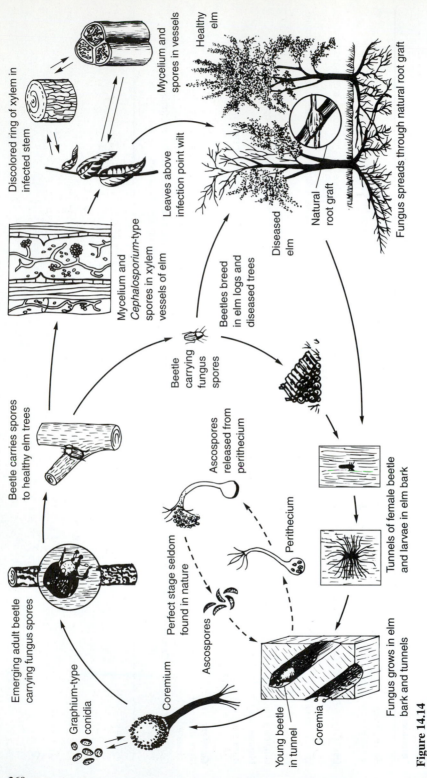

Figure 14.14
Life cycle of *Ophiostoma ulmi*, the cause of Dutch elm disease.

Healthy elm

Mycelium and spores in vessels

Discolored ring of xylem in infected stem

Leaves above infection point wilt

Fungus spreads through natural root graft

Natural root graft

Diseased elm

Mycelium and *Cephalosporium*-type spores in xylem vessels of elm

Beetles breed in elm logs and diseased trees

Beetle carrying fungus spores

Beetle carries spores to healthy elm trees

Ascospores released from perithecium

Tunnels of female beetle and larvae in elm bark

Perithecium

Emerging adult beetle carrying fungus spores

Perfect stage seldom found in nature

Ascospores

Fungus grows in elm bark and tunnels

Graphium-type conidia

Coremium

Young beetle in tunnel

Coremia

places between New York and the veneer factories in Ohio and Indiana. *H. rufipes* occurs throughout the entire range of elm, but *S. multistriatus* is less able to survive cold winters and its northern spread is limited. In Europe the disease is spread by the larger European bark beetle (*Scolytus scolytus*).

The fungus enters the tree through spores carried into the phloem by beetles feeding on branch crotches for sexual maturation. In most areas there are at least two and sometimes three broods of smaller European elm bark beetles per year. Broods that overwinter emerge in spring in May and June and are thought to be responsible for most of the spread, however. Native bark beetle produce one-and-a-half generations per year, emerge in April and feed on larger living branches.

Once inside the host the fungus grows through the cells of the outer sapwood. The tree reacts to the presence of the fungus by producing gums and tyloses, which plug the vascular system, resulting in wilting. Usually an entire branch or section of the tree wilts, followed by wilting of the rest of the tree. A brown ring forms in the outer sapwood that may be continuous or disrupted. A toxin (cerato-ulmin) is also produced by the fungus and may play a role in the early stages of pathogenesis. The fungus produces small conidia or a yeastlike stage (*Sporothrix*), which can be carried rapidly through the vascular system. Adult beetles lay their eggs in the bark of attacked trees. At this stage the fungus is well established in the tree and insect galleries become infected. Special fungal structures called *synnema* or *coremia* line the gallery walls. They consist of balls of conidia in a gelatinous mass on top of black structures about 1 mm in length; this is called the *Pesotum ulmi* stage. Ascospores of the *Ophiostoma ulmi* stage of the fungus also are produced in perithecia in insect galleries as well as the spores of the other two stages, and all three spore types are capable of being spread on the body of the insects to new hosts as they emerge. During breeding, beetles also may introduce the fungus into elm material that was not infected. Elms stressed by pruning or drought can be attacked.

Management strategies involve sanitation, prevention of root grafts, pruning, systemic fungicides, insecticides, resistant species, biological control, and an effective urban forestry program (Tainter and Baker 1996). Sanitation to reduce the spread of the bark beetles is accomplished by destroying any dead or dying elms in which beetles can breed. Debarking also is effective since the beetles can breed in either the bark or exposed wood. Insecticides such as Dursban can be used until the elm wood dries out. Pheromones and sticky traps also have been used to control insects. Severing root systems chemically or mechanically prevents spread through root grafts. Trees more than approximately 9 to 11 m from infected trees have a low chance of infection. A vibratory plow or a mechanical trenchers to a depth of about 1 m disrupts most root grafts. Fumigants such as Vapam also can be used. Pruning to remove the infected part of the tree can be effective in the early stages of disease development, but sometimes pruning will attract beetles. Systemic injection of fungicides have shown some promise for large trees, but small trees may be damaged by the treatment.

Use of resistant trees appears to have promise (Smalley and Guries 1993). Siberian elms are more resistant than native North American elms, but do not have the form of American elms and are susceptible to winter injury. Hybrids between American and Siberian elms incorporating the resistance of Siberian elms have been produced. Biological control has some possibility, including parasitic wasps on bark beetles and parasitic fungi that lower bark beetle reproduction.

Oak Wilt

Oak wilt is extremely important in the north-central United States where oaks are predominant, but it also occurs in the southern states and across to Texas (Tainter and Baker 1996). It is of great concern internationally (Nair 1996). Most oak species and varieties are susceptible, along with three species of *Castanea,* and a species of *Castanospis* and *Lithocarpus.* Red oaks are very susceptible, white oaks range from susceptible to highly resistant (e.g., *Quercus alba*).

The fungus causing oak wilt is *Ceratocystis fagacearum.* Like Dutch elm disease it is transmitted by insects and through root grafts. Ascospores and conidia are spread by sap-feeding and bark beetles. Spores are apparently introduced to healthy trees through wounds. The fungus spreads rapidly through the vascular system of the tree, degrading the cell wall of infected sapwood cells. Tyloses plug the water-conducting tissues of the tree, resulting in wilt symptoms. Brown to black discoloration occurs in the sapwood, usually in the most recent ring. After the tree dies the fungus produces spore mats between the bark and wood. The fermenting odor attracts insects.

The fungus spreads rapidly through root grafts, but grafting occurs only between trees of the same species. The best method of management is the disruption of root grafts either chemically or mechanically. Complete bark removal of infected trees prevents spore mat formation and spread by insects. Systemic fungicides also have been used successfully.

STAINS

Stains produced by fungi are common in many conifers, especially pines and Douglas-fir, and usually are brown, black, or blue in color (Boyce 1961). This topic is covered in more detail in Chapter 16. The fungi involved are mostly Ascomycetes or Deuteromycetes. Blue-stain fungi usually are wedge-shaped because the fungus moves preferentially through the tangential ray cells (Hansen and Lewis 1997) (see Figure 16.5). Stains and wetwood also are common in whitewood species such as spruces, firs, and hemlocks. Wetwood is wood that is high in moisture content.

REFERENCES

Agrios, G. N. 1997. *Plant pathology,* 4th ed. Academic Press, New York.

Allen, E., D. Morrison, and G. W. Wallis. 1996. *Common tree diseases of British Columbia.* Pacific Forestry Research Centre, Canadian Forestry Service, Victoria, B.C., Canada.

Boyce, J. S. 1961. *Forest pathology,* 3rd ed. McGraw-Hill, New York.

Bull, E. L. C. G. Parks, and T. R. Torgensen. 1997. *Trees and logs important to wildlife in the interior Columbia River Basin.* USDA Forest Service, Pacific Northwest Station, General Technical Report PNW-GTR-391, Portland, Oreg.

Funk, A. 1981. *Parasitic microfungi of western trees.* Pacific Forestry Research Centre, Canadian Forestry Service, BC-X-22, Victoria, B.C., Canada.

Gara, R. I., W. R. Littke, J. K.Agee, D. R. Geizler, J. D. Stuart, and C. H. Driver. 1985. Influence of fires, fungi and mountain pine beetles on development of lodgepole pine forests in south-central Oregon, pp. 153–162. In D. M. Baumgartner, R. G. Krebill, J. T. Arnott, and G. F. Weetman (eds.). *Lodgepole pine: the species and its management.* Cooperative Extension Service, Washington State University, Pullman, Wash.

Hansen, E. M., and K. J. Lewis (eds.). 1997. *Compendium of conifer diseases.* American Phytopathological Society Press, St. Paul, Minn.

Hawksworth, F. G. 1977. *The 6-class dwarf mistletoe rating system.* USDA Forest Service, Rocky Mountain Forest and Range Experiment Station, General Technical Report RM-48, Fort Collins, Colo.

Hawksworth, F. G., and D. Wiens. 1996. *Dwarf mistletoes: biology, pathology, and systematics.* USDA Forest Service Agricultural Handbook 709, Washington, D.C.

Hennon, P. E. 1995. Are heart rot fungi major factors of disturbance in gap-dynamic forests? *Northwest Science* 69:284–293.

Houston, D. R. 1994. Major new disease epidemics: beech bark disease. *Annual Review of Phytopathology* 32: 75–87.

Kuijt, J. 1969. *The biology of parasitic flowering plants.* University of California Press, Berkeley and Los Angeles.

Manion, P. D. 1991. *Tree disease concepts,* 2nd ed. Prentice-Hall, Englewood Cliffs, N.J.

Manion, P. D., and D. H. Griffin. 1986. Sixty-five years of research on Hypoxylon canker of aspen. *Plant Disease* 70:803–808.

Nair, V. M. G. 1996. Oak wilt: an internationally dangerous tree disease, pp. 181–198. In S. P. Raychauduri, and K. Maramorosch (eds.). *Forest trees and palms: diseases and control.* Science Publishers, Lebanon, N.H.

Scharpf, R. F. 1993. *Diseases of Pacific Coast conifers.* USDA Forest Service Agricultural Handbook 521, Washington, D.C.

Shigo, A. L., and H. G. Marx. 1977. *Compartmentalization of decay in trees.* USDA Forest Service Agricultural Information Bulletin 405, Washington, D.C.

Shigo, A. L. 1984. Compartmentalization: a conceptual framework for understanding how trees grow and defend themselves. *Annual Review of Phytopathology* 22:189–214.

Sinclair, W. A., H. H. Lyon, and W. T. Johnson. 1987. *Diseases of trees and shrubs.* Cornell University Press, Ithaca, N.Y.

Smallety, E. B., and R. P. Guries. 1993. Breeding elms for resistance to Dutch elm disease. *Annual Review of Phytopathology* 31:325–352.

Stipes, R. J., and R. J. Campana (eds.). 1981. *Compendium of elm diseases.* American Phytopathological Society, St. Paul, Minn.

Storer, A. J., T. R. Gordon, D. L. Wood, and P. Bonello. 1997. Pitch canker disease of pines. Current and future impacts. *Journal of Forestry* 95(12):21–26.

Tainter, F. H., and F. A. Baker. 1996. *Principles of forest pathology.* John Wiley and Sons, New York.

Wallis, G. W., O. J. Morrison, and D. W. Ross. 1992. *Tree hazards in recreation sites in British Columbia.* British Columbia Ministry of the Environment/Canadian Forestry Service, Joint Report 13, Victoria, B.C., Canada.

FOREST DECLINES

INTRODUCTION

Tree mortality occurs naturally in forests (Ciesla and Donaubauer 1994). Suppression competition, insects, diseases, fire, wind, erosion and landslides, volcanic eruptions, adverse weather (especially flooding and drought), and climate change are the major causes of tree death. In old-growth forests of the Pacific Northwest United States approximately, 1 to 2% of the canopy trees die naturally each year (Franklin et al. 1987). Humans also have directly caused tree mortality through indigenous and modern cutting and burning and other activities such as the production of air pollutants and the introduction of exotic pathogens and insects. In most cases scientists have been able to provide an explanation for tree death, including the large-scale mortality resulting from the introduction of pathogens causing chestnut blight, Dutch elm disease, and white pine blister rust in the early 1900s in North America (Manion and Lachance 1992).

There have been many situations, however, where episodes of tree mortality over wide areas have been difficult to explain. These typically have been called *diebacks, declines,* and sometimes *blights.* In most cases, at least in the initial phases of these episodes, the use of these terms indicates that the cause or causes of mortality are unknown. A list of forest declines throughout the world is shown in Table 15.1 on pages 366–367. Note they have occurred on all continents and even on Pacific islands such as the Hawaiian and Galápagos Islands. The declines listed in Table 15.1 have occurred in native species, but declines have also been noted in exotic plantation species, such as *Eucalyptus* spp. in Brazil (Ciesla and Donaubauer 1994). Decline and dieback are used to describe a complex of symptoms involving crown dieback, leaf discoloration, leaf loss, reductions in leaf size and stem radial growth, and twig and branch death that may lead to the death of the whole tree (Manion 1991, Manion and Lachance 1992). Symptoms of decline may be subtle, but are persistent and progressive and usually occur over periods ranging from 3 to 30 years. Periods of decline may be followed by periods of recovery, which may be temporary or complete. Forest declines usually occur over large regions and involve single tree species. Thus the term *species decline* might be a better term than forest decline (Skelly 1989). Decline is a stand-level or population-level phenomenon, where many trees of the same species show symptoms. Dieback usually represents a further progression of the decline, whereby premature defoliation exposes barren branches and dead tops and most cases ends in the death of the tree. The terms are so close in meaning, however, that they often are used interchangeably.

Diebacks and declines were noted as early as the 1700s in Europe. For example, oak decline was recorded in northern Germany from 1739 to 1748, and dieback of European silver fir or *Tannensterben* in central Europe was first reported in 1810 (Manion and Lachance 1992). Tannensterben involved increment decrease, needle loss, dying of branches from the bottom to the top of the crown, and formation of a "stork's nest" on the top of the tree. By 1928 Tannensterben had been reported from almost all the central European countries. The most recent occurrence of this dieback began in the 1950s and peaked in the 1970s.

In the 20th century a number of diebacks, declines, and blights were noted in North American hardwood forests, including ash, birch, and maple diebacks, oak and maple declines, and maple blight. Conifer forests were similarly afflicted with Alaska yellow-cedar decline, pole blight of western white pine, littleleaf disease of loblolly and shortleaf pines in the Piedmont region of the southeastern United States, and ponderosa pine decline in the San Bernadino and San Gabriel Mountains near Los Angeles, California (Table 15.1). No principal pathogen capable of causing diseases in healthy trees was identified in the majority of these cases. Rather they seemed to be caused by a complex of factors. Changing climate or unusual weather conditions have commonly been blamed.

Tree species declines, however, have not been confined to the 19th and 20th centuries. Some interesting and dramatic changes in tree species populations have been identified in the last 5,000 years based on paleoecological studies involving pollen analysis of lake sediments. Margaret Davis of the University of Minnesota observed a dramatic drop in eastern hemlock pollen in eastern North America in an area from New Hampshire to southern Connecticut and across to Upper Michigan about 4,500 radiocarbon (^{14}C) years ago (Davis 1981). Hemlock regained its previous abundance in some areas 2,000 years later, but in other areas it never recovered to previous levels.

A number of causes for the hemlock decline were postulated: climatic change, fire, windstorms, the impact of prehistoric humans, and pathogens or insects. Climatic change was dismissed because a climatic event of this magnitude would surely have affected a number of species, not just hemlock, although this is not necessarily true as you will see from later discussion. Fire was dismissed because there were no charcoal fragments in the sediments at the level of the hemlock decline. Also fire would have damaged beech and maple, which are fire-sensitive and increased following the fall in hemlock.

Wind does not seem to be a factor. Although hurricanes near the coast are known to have blown-down forests over wide areas in the northeastern United States—for example, the 1938 hurricane—this would not explain the simultaneous decline of hemlock in Upper Michigan. The probability that prehistoric man was

T A B L E **15.1** | Important Forest Declines or Diebacks in Europe, North America, the Pacific, Australia and New Zealand, South America, Asia, and Africa

Decline or dieback	Hosts	Region	First reported
Europe			
Oak decline	*Quercus* spp.	Northern Germany	1739
Tannensterben (European silver fir decline)	*Abies alba*	Central Europe	1810
Waldsterben (European forest decline)	*Picea abies* and hardwoods	Germany and other European countries	1979
North America			
Alaska yellow-cedar decline	*Chamaecyparis nootkatensis*	Southeast Alaska	1900
Ash dieback	*Fraxinus* spp.	Eastern United States and Canada	1925
Pole blight	*Pinus monticola*	Western United States	1927
Littleleaf disease	*Pinus elliottii, P. taeda*	Piedmont region of southeastern United States	1929
Birch dieback	*Betula* spp.	Eastern United States and Canada	1932–1935
Balsam fir decline (Fir waves)	*Abies balsamae*	Eastern United States	1950
Oak decline	*Quercus* spp.	Eastern United States and Canada	1951
Sweet gum blight	*Liquidamber*	Eastern United States	1951
Ponderosa pine decline	*Pinus ponderosa*	Southern California	1950s
Sugar maple decline	*Acer saccharinum*	Eastern United States and Canada	1957
Red spruce decline	*Picea rubens*	Eastern United States	1980s
Fraser fir decline	*Abies fraseri*	Eastern United States	1980s

Decline or dieback	Hosts	Region	First reported
The Pacific, Australia and New Zealand			
New England dieback	*Eucalyptus* spp.	New South Wales	1850s
Jarrah decline	*Eucalyptus marginata*	Western Australia	1921
Ohia decline	*Metrosideros polymorpha*	Hawaii	1970s
Nothofagus dieback	*Nothofagus* spp.	New Zealand	1980s
South America			
Canopy dieback	*Scalesia pedunculata*	Galápagos Islands	1935
Cypress decline	*Austrocedrus chilensis*	Southwestern Argentina	1948
Nothofagus dieback	*Nothofagus antarctica*	Argentina (Patagonia)	1988
Africa			
Dieback of *Terminalia ivorensis*	*Terminalia ivorensis*	Ivory Coast and Ghana	1970s
Rain forest dieback	*Newtonia buchananii, Louoa swynnertonii* and other species	Uganda	1984
Neem decline	*Azadirachta indica*	West African Sahel	1990
Stinkwood decline	*Ocotea bullata*	South Africa	1990
Asia			
Top dying of sundri	*Heritiera fomes*	Bangladesh	1915
Cryptomeria japonica decline	*Cryptomeria japonica*	Japan	1970s
Shimagare decline	*Abies veitchii, A. mariesii*	Japan	1976
Canopy dieback of montane forests	*Calophyllum* and *Syzygium* spp.	Sri Lanka	1978

Source: From Innes (1993) and Ciesla and Donanbauer (1994).

responsible for the decline also seems remote. Neolithic agriculture in Europe certainly affected the forests, but the population of humans in eastern North America 5,000 years ago was sparse and little agriculture was practiced. The most likely explanation for hemlock decline is that a pathogen or insect was responsible, although direct evidence is lacking (Taber et al. 1986). No fossilized insects or fungi or hemlock needles showing evidence of predation have been observed. Hemlock looper is one possibility, however. In the Pacific Northwest United States, western hemlock looper has frequently defoliated old-growth coastal forests dominated by western hemlock and heavily defoliated trees may die. With the recent reduction in old-growth forests because of harvesting, however, there has been a reduction in looper outbreaks. The eastern hemlock looper has a similar life cycle to the western hemlock looper, and it has been postulated that a sudden outbreak of hemlock loopers or similar pests caused the hemlock decline 4,500 years ago.

In the early 1980s concern for forest health increased dramatically, largely in response to the potential impact of air pollution and acid precipitation on the forests of Europe, especially Germany, and the eastern United States (Innes 1993). Interest also developed in declines in eucalyptus in Australia, Ohia in Hawaii, and Alaska yellow-cedar in southeastern Alaska, although some of these, especially Alaska-yellow-cedar decline, had a long history, being first reported in 1900 (Hennon and Shaw 1997) (Table 15.1). Declines also have been reported from South America, Africa, and Asia (Ciesla and Donaubauer 1994). In this chapter we discuss theoretical aspects of forest decline and analyze some specific case examples.

THE CONCEPT OF DECLINE DISEASES

Since we don't fully understand forest declines, decline theories are still evolving. However, all the theories agree that a complex of factors is involved, although they differ markedly on which key factors are most important (Shigo 1985, Manion and Lachance 1992). In the 1960s forest pathologists in the eastern United States including David Houston, Wayne Sinclair, and Paul Manion provided the beginnings of a conceptual framework for understanding and explaining decline diseases (Manion and Lachance 1992). At least five theories or concepts have emerged: (1) the environmental stress and secondary organisms theory; (2) the concept of predisposing, inciting, and contributing factors; (3) the climate change or climate perturbation theory; (4) the air pollution theory; and (5) the ecological theory (involving forest disturbances, population and community structures, and forest succession). These are examined below.

The Environmental Stress and Secondary Organisms Theory

This concept was developed by David Houston (Houston 1981, 1992). In his model (Figure 15.1) healthy trees are affected by environmental stress and secondary organisms attack the weakened tissues. As the disease condition develops, tissues and trees dieback, and the trees may ultimately die. Houston emphasized the central importance of organisms of secondary action in the demise of stressed trees, and his model was based on the premises in Box 15.1

Box **15.1**

The Environmental Stress and Secondary Organisms Theory

1. Healthy trees + stress → Altered trees (tissue) (the beginning of dieback).
2. Altered trees + more stress → Trees (tissues) altered further (continued dieback).
3. Severely altered trees + more stress → Trees (tissues) altered further (continued dieback).

   •

   •

   •

n. Severely altered trees (tissues) + secondary organisms = Trees (tissues) invaded. Trees lose ability to respond if conditions improve, decline, and perhaps die.

Figure 15.1 is a graphic conceptualization of this theory.

With removal of stress in the early stages of decline, and if secondary saprogens or insects are not involved, dieback often ceases and trees recover. Stress alone, if prolonged like drought, can cause dieback and death even in the absence of secondary organisms. As the process continues even the removal of stress will not allow trees to recover since they usually have been attacked by insects and disease organisms. Houston recognized at least three triggering mechanisms: insect defoliation, sucking insects, and drought. This is discussed in more detail in the section on case examples later on.

The Concept of Predisposing, Inciting, and Contributing Factors

Following on from Houston's ideas, Wayne Sinclair developed the well-known three-step chain reaction theory of causal factors (Sinclair and Hudler 1988)—that is, predisposing, inciting, and contributing factors—and Paul Manion incorporated this concept into a model known as the disease decline spiral model (Manion 1991) (Figure 15.2). Manion defines *decline* as "an interaction of interchangeable, specifically ordered abiotic and biotic factors to produce a gradual general deterioration, often ending in the death of trees." Predisposing factors include tree age and genetic potential, soil conditions, planting species outside their natural ranges, and air pollution. Inciting factors include defoliating insects; drought; human activities such as site manipulations and application of road salt; and air pollution. Contributing factors, which usually are the cause of mortality, include Armillaria root rot, canker fungi, wood- and bark-boring insects, nematodes, and perhaps viruses and phytoplasmas.

The Climate Change or Climate Perturbation Theory

Another group of scientists from the eastern United States and Canada, led by Allan Auclair, have focused largely on climate change and climatic perturbations as being the dominant mechanism of forest dieback (Auclair et al. 1992). They suggest that there is strong evidence that the progressive dieback and mortality of yellow birch, white ash, and red spruce in northeastern North America, and silver fir and Norway

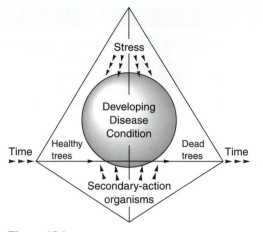

Figure 15.1

A diagrammatic representation of the host-stress-saprogen model. Healthy trees are weakened by environmental stress and attacked in time by saprogens. As the disease develops, trees dieback, decline, and eventually die.

(Source: From Houston 1992.)

spruce in Europe, is strongly linked to climate change. These declines occurred long before regional air pollution was considered to be a problem. Furthermore, large-scale declines have occurred in areas remote from air pollution in the Pacific Rim, in southeastern Alaska (with Alaska yellow-cedar) (Hennon and Shaw 1997), and in black spruce forests in northern Quebec.

Using persistent regionwide dieback as the major symptom, Auclair et al. (1992) noted that many of the diebacks in North America consistently had the following symptoms on individual trees: early leaf coloration and premature leaf fall, leaf chlorosis, small leaf size, tufting of leaves, and stem dieback from branch tips inward toward the bole. Although radial growth reduction was a common symptom, they chose not to use this in their analysis because it is complicated by tree growth and competition between trees. Analysis of crown diebacks of red, black, and white ashes, yellow and white birch, and sugar maples in eastern Canada from 1920 to 1987 revealed they were highly episodic (Figure 15.3) and possessed the following features:

1. Onset years were abrupt and regionwide, but site influences caused local variability.
2. Each major dieback episode included a "lead" species and 5 to 30 associated tree species.
3. Recovery from crown dieback also depended on local factors and was equally as abrupt as the onset.
4. Once dieback was initiated a background or chronic level of dieback persisted on the lead and most associated dieback species.

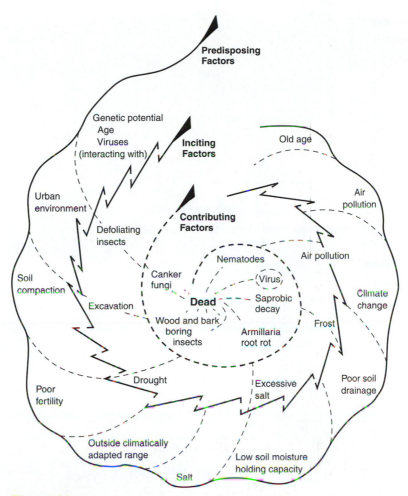

Figure 15.2

Manion's decline disease spiral model, showing predisposing, inciting, and contributing factors.

(Source: *Tree Disease Concepts*, 2/e by Manion, Paul D., ©1991. Reprinted by permission of Prentice-Hall, Inc., Upper Saddle River, NJ.)

5. Successive major dieback episodes this century appear to have become more severe in terms of species, areal extent, and mortality.

Their theory is illustrated in Figure 15.4. With global warming it is assumed that temperature variability is increased and thaw-freeze events may be more common in winter. The universal mechanism to explain regionwide diebacks is xylem cavitation injury caused by extreme freezing and/or moisture fluctuations at local and regional levels. The progression of crown dieback is directly proportional to the degree of irreversible cavitation injury, resulting in water stress in the growing seasons following injury.

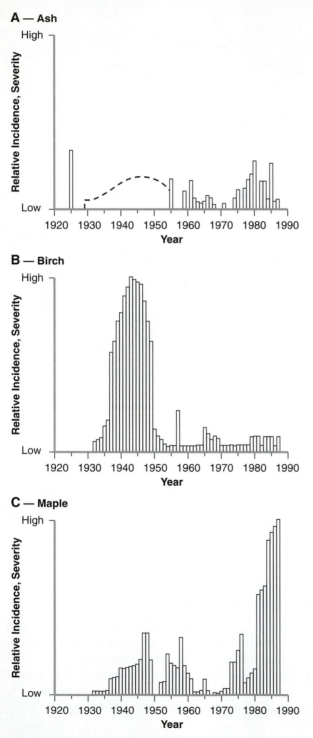

Figure 15.3
A, The relative incidence/severity of crown dieback on black, white, and red ash; **B,** yellow and white birch; and **C,** sugar maple hardwoods of eastern Canada reconstructed between 1920 and 1987.
(Source: From Auclair et al. 1992.)

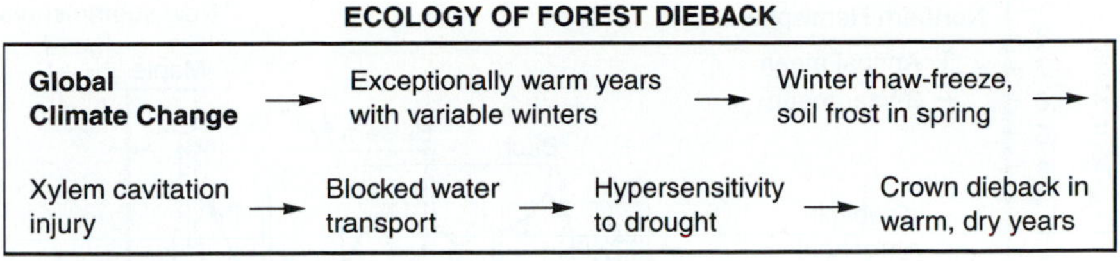

ECOLOGY OF FOREST DIEBACK

Global Climate Change →	Exceptionally warm years with variable winters →	Winter thaw-freeze, soil frost in spring →
Xylem cavitation injury →	Blocked water transport →	Hypersensitivity to drought → Crown dieback in warm, dry years

Figure 15.4

Conceptual model linking global change to forest dieback in the north temperate and boreal zones.

(Source: From Manion and Lachance 1992.)

How does xylem injury due to cavitation occur? Warm "false spring" conditions in the winter, followed by an extremely cold period at the end of winter can result in cavitation of both vessels and tracheids in the sapwood. During the warm period snow melts, soil warms, and the sap begins to flow. Sudden drops in temperature following the false spring then result in deep soil freezing, because of the loss of insulating snow, and the cessation of sap flow. Sudden freezing of sap can result in the formation of air bubbles in sapwood as dissolved gases come out of solution. Root frost damage further exacerbates the conditions, and warm conditions in spring when the soil is still frozen then damage tissues by dessication and prevent the normal reversal of cavitation.

Examination of the weather and stream discharge data in Quebec for 1925, 1937, and 1981—years of onset of sugar maple decline—revealed unusual weather conditions. All years had at least one pronounced thaw-freeze event. Also the stream discharge in the Chaudiere River watershed in Quebec in the area of severe dieback exceeded 325% of the monthly normals, indicating winter thaw events. The evidence for irreversible cavitation, however, is circumstantial since no measurements were taken in the field, but from laboratory experiments it known that birch is more susceptible to cavitation than associated species such as trembling aspen.

Auclair et al. (1992) argued that there is strong relationship between climate warming in the northern hemisphere and dieback of ash, birch, maple, and balsam fir in eastern Canada (Figure 15.5). Note the "dieback threshold," which is assumed to be the annual hemispheric temperature deviation from the long-term mean in 1925, which was the year of the first major regionwide dieback. The consistency of the global change/forest dieback link was illustrated by the following: (1) each onset year of a major dieback coincided with a period (3–9 years) of rapid hemispheric warming—each successive dieback episode occurred at a point the mean annual temperature exceeded the previous record annual mean temperature, and involved a new lead species; (2) recovery from dieback symptoms occurred in years below the dieback threshold; and (3) severe dieback on balsam fir occurred in 1954 in areas within the northernmost limits of northern hardwoods. The mean annual hemispheric temperature deviation in 1954 exceeded that in the previous year, and there was a marked thaw-freeze event in the winter of 1954.

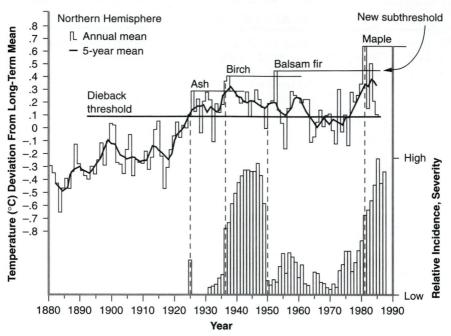

Figure 15.5
Relative incidence of forest dieback on ash, birch, balsam fir, and maple of eastern
Canada in relation to annual and 5-year mean temperatures in the northern hemisphere.
Dashed lines identify onset years of ash, birch, balsam fir, and maple dieback episodes
and recovery of birch. The dieback threshold is the mean annual hemispheric tempera-
ture of the first episode on ash. Subthresholds are the record mean annual hemispheric
temperatures within each dieback episode.
(Source: From Manion and Lachance 1992.)

How universal is the global change theory? Can the onset of pole blight of
western white pine in the Pacific Northwest in 1936, the onset of decline in Norway spruce in Europe from 1979 to 1980, and the sporadic diebacks of jarrah in
western Australia from 1921 to 1982 be explained by this hypothesis? There is evidence from weather data that climate and thaw-freeze events played a role in the
pole blight and Norway spruce declines such that freezing, stem cavitation, root
frost damage, and mortality typical of northern hardwood diebacks very likely
occurred. *Abies* species in particular are very sensitive to cavitation dysfunction.
The situation with jarrah dieback is different. The dieback years were associated
with exceptionally heavy rainfall, leading to the suggestion that Phytophthora root
rot was the major cause of the decline since *Phytophthora* incidence increases in
wet soils. However, root xylem vessels in waterlogged soils embolize, and though
they may be quickly sealed off with tyloses, transpiration continues and roots cav-
itate. The prolonged wet, cloudy weather during the dieback periods was followed
by clear, bright, hot weather systems, leading to plant water stress. Thus cavitation

could play a role in jarrah dieback as it does in the dieback of conifers and hardwoods in the northern hemisphere. Wet soil conditions are suspected to play a role in the dieback of Alaska yellow-cedar in southeast Alaska and cypress decline in southwest Argentina. Aging also may play an important role in diebacks. As trees age there is a loss of energy due to gradually increasing respiratory wood volume/leaf area ratio.

The Air Pollution Theory

In the early 1980s scientists in central Europe, particularly in Germany and Switzerland, claimed that a large-scale forest decline was occurring largely due to air pollution and acid precipitation (Figure 15.6). Following this lead similar claims were made about forests in the eastern United States, particularly in high elevation forests. Specific declines were noted at Mount Mitchell in North Carolina and Camels Hump in Vermont. In Germany the decline was called *Waldsterben* (forest death), and it was maintained that this decline was quite different from those observed previously or in other locations. The features of Waldsterben (Schutt and Cowling 1985) are shown in Box 15.2.

Excess N inputs due to air pollution also are thought to negatively influence forests (Aber et al. 1998) (see Chapter 9). Typical N inputs in unpolluted forested areas are 1 to 5 kg ha^{-1} yr^{-1}. In some areas of Europe, particularly the Netherlands, inputs can be as high as 70 kg ha^{-1} yr^{-1}. This excess N may fertilize forests initially leading to growth responses, but it also may extend the growing season, allowing trees to be more susceptible to frost damage. The process of nitrification in the soil also is enhanced, and this can lead to soil acidification and leaching of cations such as Mg enhancing the acid rain effects noted in Box 15.2.

A decade of research on the problem in Germany, however, suggested that the decline attributed to air pollution and acid precipitation was not novel and that needle loss in conifers, acute yellowing of Norway spruce, and loss and discoloration of foliage in hardwood species did not occur concurrently. Retrospective studies, including comparisons of historic and modern photographs of German forests, showed that similar levels of crown transparency in Norway spruce to those observed in the early 1980s had occurred in the past, as well as recurrent decline episodes in the main tree species (Kandler 1990). In the late 1980s many of the German forests appeared to have recovered, although an extensive liming program was initiated to increase soil pH. Some forests, however, still have poor growth and mortality; these sites are generally restricted to higher altitudes with poor nutrient supply. **Endemic diseases** probably contributed to the decline, but there was little mention of them.

During the same period in the 1980s, reports of "unprecedented" declines in high-elevation spruce-fir and sugar maple in the eastern United States dominated the popular press and television. This was dramatized through photographs of dead and dying trees blamed on acid rain and air pollution (see Figure 1.11), but few reports were substantiated. Air pollutants were basically the only reported causes of injury, yet we know that the drought and sometimes severe rime ice events

Figure 15.6
Dead and dying trees in high-elevation forests in Switzerland initially attributed to acid rain.

during that period had a considerable influence on forest health. Perhaps the greatest omission was the impact of the introduced balsam woolly adelgid, which was responsible for most of the mortality in spruce-fir, particularly on Mount Mitchell in North Carolina on Fraser fir (see Chapters 1 and 9).

This is not to say that air pollution and acid precipitation have not influenced forest ecosystems. Acid precipitation has certainly impacted streams and lakes, and we know that stratospheric ozone has had considerable influence on forests in southern California, the eastern United States, and Europe.

The Ecological Theory

Our analysis of forest decline so far has been pathological and physiological rather than ecological in nature. As mentioned in the introduction, tree death is a normal process in natural forests, but large numbers of dying trees are generally not considered to be natural or normal unless a forest has been burned by a large fire or attacked by insects. Let us examine dieback from an ecological perspective by examining species populations and community dynamics relative to forest disturbances caused by fire, wind, weather, and insects. Is forest decline a disease or a natural process of forest aging?

Natural fire is thought to keep a forest healthy (see Chapter 4). There are two extreme fire regimes: (1) frequent and low intensity and (2) infrequent and moderate to high intensity. In areas of high fire frequency, such as the east slopes of the Cascade Range in North America, natural fires burn every 10 to 15 years. Stands usually are uneven-aged and are dominated by early-successional species such as ponderosa pine. In these stands dieback phenomena are rare. However, in areas where fire protection has been actively practiced, dieback becomes common, usually as result of late-successional species, such as Douglas-fir and true firs, being attacked by bark beetles and defoliators such as the Douglas-fir tussock moth and spruce budworm (Franklin et al. 1987). Even the remaining early-successional pine

Box 15.2

Features of Waldsterben-European Forest Decline

1. All the important tree species were involved, including both conifers and hardwoods, and showing similar symptoms, not recognized before.
2. The syndromes stood apart from ordinary tree diseases.
3. The main symptoms were decrease in increment, loss of foliage and feeder roots, transparent crowns, yellowing of foliage, premature senescence, altered branching habit, and abnormally high seed production.
4. Rapid development—Norway spruce and pines were expected to deteriorate so rapidly that they would have to be cut within 3 years of showing the first symptoms, and silver fir and Norway spruce in the Black Forest were expected to die within 10 years. A visit to the German forests in the mid 1980s revealed that many stands appeared relatively healthy because dead and dying trees had been removed.
5. The most likely cause was a complex disease triggered by cumulative stress from increasing air pollution and deposition of acid precipitation, sulfur dioxide, nitrogen oxide, and ozone.
6. Acid deposition caused increased soil acidity resulting in increased soil Al levels that are toxic to fine (feeder) roots and mycorrhizae, and also depleted the soil of cations, particularly Mg, resulting in yellowing foliage. Ozone further damaged the foliage, resulting in leaching and further loss of Mg.

species are susceptible to attack because of the increase in stem density and increased stress. Root diseases and dwarf mistletoes also tend to increase, adding to the stress and mortality and increasing the tendency for a large stand-replacement fire. These forests are unhealthy.

In areas of moderate fire frequency (every 150–200 years), such as the lodgepole pine forests of the Yellowstone plateau, stands are typically composed of mosaics of spatially separated even-aged cohorts because fires jump around the landscape leaving unburned patches (i.e., they do not burn uniformly). As these forests age, stress increases and periodic large-scale bark beetle outbreaks occur, resulting in large-scale mortality. This increases the fuel loading, making larger-scale fires more likely. Fires, when they occur, start the cycle again. A similar situation occurs in the boreal forests of Canada and Alaska with white spruce and bark beetles. Fire suppression in these forests changes stand dynamics, but does not influence successional patterns and forest health to the same degree that fire suppression does in areas with high fire frequency.

Where fire frequency is low, such as the western slopes of the Cascade Range and adjacent coastal forests, fire does not play such an important role in maintaining forest health. Fire suppression has little influence on species succession. Here the major disturbances are wind and diseases, and large-scale forest decline is rare. However, as we previously mentioned in this chapter, large-scale decline and mortality in western hemlock caused by the western hemlock looper does occur. Manion's (1991) forest decline spiral model has been adapted to examine tree mortality relative to succession in native western forests (Franklin et al. 1987) (Table 15.2 and Figure 15.7).

In the northeastern United States fire also plays a smaller role in maintaining forest health. Natural disturbances in the deciduous forests of eastern North America are small compared to natural landscape units and usually involve windthrow of individual trees or small groups, although hurricanes can create larger-scale disturbances. Another example of natural disturbance and decline is the wave-regenerated balsam fir forests in the mountains of the northeastern United States, where mortality waves move through the forest every 50 to 70 years (Sprugel and Bormann 1981). The waves move at a relatively uniform pace and the forest always contains dead and dying trees (Figure 15.8). Since consecutive waves are only about 100 m apart, relatively small areas can contain all phases of disturbance. Fir waves are thought to result from buildup of rime ice during winter on taller trees, resulting in stem breakage, and waves move in the direction of the wind. A similar phenomenon occurs in Japan.

Forest declines and diebacks are natural components of forests that have "autosuccession"—that is, the fire-killed stand is replaced by the same species. In tropical forests, at least in multispecies lowland forests, large-scale forest declines are rare and single treefall gaps are the rule rather than large-scale fire disturbances. This results in "chronosequential gap rotation" or small-patch mosaic systems involving many ecologically different tree species as gap fillers. Large-scale disturbances are mostly caused by strong winds, such as hurricanes or cyclones. As we have seen this also is somewhat true for temperate rainforests, but of course fewer tree species are involved.

Gap rotation, however, appears not to apply to tropical forests on islands and in mountains. For example, in Hawaiian Ohia forests three types of dieback/recovery patterns occur: "replacement dieback," "displacement dieback," and "stand-reduction dieback." Displacement dieback occurs on nutritionally rich soils where tree ferns are abundant and Ohia is replaced by other canopy dominants. Stand-reduction dieback occurs when high-stature forests give way to Ohia scrub due to site deterioration. Here older soils are depleted in cations and phosphorus because of leaching due to high rainfall. There are other examples of stand-reduction dieback or successional decline due to nutrients in sub-tropical areas—for example, on sand dunes in Queensland in Australia where ecosystem development or natural succession proceeds to high-stature rainforest and then to Banksia scrub.

Based on this evidence Dieter Mueller-Dombois of the University of Hawaii (Mueller-Dombois et al. 1983, Mueller-Dombois 1992) proposed a natural dieback theory based on four principal generic causal factors:

- s (simplified forest structure)
- e (edaphically extreme sites)
- p (periodically recurring perturbations)
- b (biotic agents)

| TABLE 15.2 | Changes in Causes and Rates of Tree Mortality During Forest Successional Stages in the Douglas-Fir Region of the Pacific Northwest |

	Successional stage				
	Prevegetative closure	Full vegetative cover	Closed tree canopy	Mature forest	Old forest
Approximate period (years)	0–5	5–20	20–100	100–200	>200
Mortality rate	Very high	High	High to medium	Medium to low	Medium to low
Typical causes of mortality	Environmental stress, predation, pathogens	Interspecific competition, environmental stress, pathogens, predation	Intraspecific competition, pathogens, wind	Pathogens, wind, competition	Wind, pathogens, physiological disorders
Predominant diseases	Root diseases	Root diseases, foliage diseases	Root diseases, butt and stem decay	Butt and stem decay, dwarf mistletoes, root diseases	Butt and stem decay, dwarf mistletoes, root diseases

Source: Adapted from Franklin et al. (1987).

Simplification of forest structure—that is, few canopy species, can degrade the soil by soil acidification and nutrient extraction. This has been recognized in agriculture and led to the practice of crop rotation. Edaphically extreme sites may support dieback-susceptible species that are not perfectly adapted to those site—for example, on serpentine soils in Oregon. Unfavorable conditions on these sites also may enhance dieback. Periodically recurring perturbations, such as drought, may temporarily stress a forest stand far beyond normal seasonality. Biotic factors typically provide the coup de grâce and hasten dieback.

The predisposing, inciting, and contributing factors of Sinclair and Hudler (1988) and Manion (1991), can be related to those of Mueller-Dombois in the following way:

Predisposing factors

s = simplified cohort structure and senescence.

e = extreme edaphic and evolutionary stresses.

p = pulse perturbations—that is, periodic physiological shocks from extreme weather.

Inciting factors

p = pulse perturbations that trigger canopy breakdown in demographically weakened stands.

Figure 15.7

Adaption of Manion's Decline Spiral Model for a
Douglas-fir tree in western forests. In this example,
a healthy tree is suppressed by larger trees. If not
released from competition, the tree is predisposed
to attack by defoliators. Once partially defoliated,
the weakened tree is attractive to bark beetles and is
unable to resist the beetles, which carry the blue-stain
fungus. The fungus blocks the transpiration stream in
the tree and causes dessication of the leaves. As the
tree progressses along the spiral, its opportunities to
escape death become more limited.

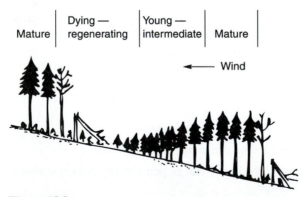

Figure 15.8

Cross section of a balsam fir wave in the northeastern
United States.

(Source: Reprinted with permission from D.G. Sprugel and F.H. Bormann, "Natural
disturbance and the steady state in high altitude balsam fir forests," *Science*
211:390–93. Copyright 1981, American Association for the Advancement of Science.)

Contributing factors

b = biotic factors, such as pathogenic fungi and insect pests, which may overpower a stand weakened by the three preceding causes: *s, e,* and *p.*

Forest declines will continue to occur in the future; some are natural, and some are caused by human activity. Natural and human caused factors interact. Controls of emissions of air pollution gases SO_2, NO_x, and CO_2 (and other greenhouse gases like CH_4) can only help improve forest health. Global change could seriously increase forest declines.

References

Aber, J., W. McDowell, K. Nadelhoffer, A. Magill, G. Bernston, M. Kamakea, S. McNulty, W. Currie, L. Rustad, and I. Fernandez. 1998. Nitrogen saturation in temperate forest ecosystems—hypotheses revisited. *BioScience* 48:921–934.

Auclair, A. N. D, R. C. Worrest, D. Lachance, and H. C. Martin. 1992. Climatic perturbation as a general mechanism of forest dieback, pp. 38–58. In P. D. Manion, and D. Lachance (eds.). *Forest decline concepts.* American Phytopathological Society Press, St. Paul, Minn.

Ciesla, W. M., and E. Donaubauer. 1994. *Decline and dieback of trees and forests. A global overview.* Food and Agricultural Organization of the United Nations, FAO Forestry Paper 120, Rome.

Davis, M. B. 1981. Outbreaks of forest pathogens in forest history, pp. 216–227. In *Proceedings of the IV international palynological conference (1976–77),* vol. 3. Birbal Sahni Institute of Paleobotany, Lucknow, India.

Franklin, J. F., H. H. Shugart, and M. E. Harmon. 1987. Tree death as an ecological process. *BioScience* 37:550–556

Hennon, P. E., and C. G. Shaw III. 1997. The enigma of yellow-cedar decline. What is killing these long-lived defensive trees? *Journal of Forestry* 95(12):4–10.

Houston, D. R. 1981. *Stress triggered tree diseases: the diebacks and declines.* USDA Forest Service Resource Paper NE-INF-41-81, U.S. Government Printing Office: 1981-705-796, Washington, D.C.

Houston, D. R. 1992. A host—stress—saprogen model for forest dieback-decline diseases, pp. 3–25. In P. D. Manion, and D. Lachance (eds.). *Forest decline concepts.* American Phytopathological Society Press, St. Paul, Minn.

Innes, J. L. 1993. *Forest health: its assessment and status.* CAB International, Wallingford, United Kingdom.

Kandler, O. 1990. Epidemiological evaluation of the development of Waldsterben in Germany. *Plant Disease* 74:4–12.

Manion, P. D. 1991. *Tree disease concepts,* 2nd ed. Prentice Hall, Englewood Cliffs, N.J.

Manion, P. D., and D. Lachance (eds.). 1992. *Forest Decline Concepts.* American Phytopathological Society Press, St. Paul, Minn.

Mueller-Dombois, D. 1992. A natural dieback theory, cohort senescence as an alternative to the decline theory, pp. 26–37. In P. D. Manion, and D. Lachance (eds.). *Forest Decline Concepts.* American Phytopathological Society Press, St. Paul, Minn.

Mueller-Dombois, D., J. E. Canfield, R. A. Holt, and G. P. Buelow. 1983. Tree-group death in North American and Hawaiian forests: a pathological problem or a new problem for vegetation ecology. *Phytocoenologia* 11:117–137.

Schutt, P., and E. B. Cowling. 1985. Waldsterben, a general decline of forests in Central Europe: symptoms, development and possible causes. *Plant Disease* 69:548–549.

Shigo, A. L. 1985. Wounded forests, starving trees: the root of dieback and declines. *Journal of Forestry* 83:668–673.

Skelly, J. M. 1989. Forest decline versus tree decline—the pathological consideration. *Environmental Monitoring and Assessment* 12:23–27.

Sinclair, W. A., and G. W. Hudler. 1988. Tree declines: four concepts of causality. *Journal of Arboriculture* 14:29–35.

Sprugel, D. G., and F. H. Bormann. 1981. Natural disturbance and the steady-state in high-altitude balsam fir forests. *Science* 211:390–393.

Taber, D. A., R. F. Moeller, and M. B. Davis. 1986. Pollen in laminated sediments provides evidence for a mid-Holocene forest pathogen outbreak. *Ecology* 67:1101–1105.

16

MANAGEMENT OF FOREST DISEASES AND THE DETERIORATION OF WOOD PRODUCTS

INTRODUCTION

This chapter summarizes the current strategies used for forest disease management. It includes a guide for disease detection and appraisal, modeling forest diseases, major management strategies, and a short section on wood products deterioration. Many of the techniques were discussed in previous chapters with respect to specific diseases. Modern disease management strategies also are discussed in Tainter and Baker (1996) and Hansen and Lewis (1997).

FOREST DISEASE MANAGEMENT

Disease management strategies are different in forests than in agricultural crops because of the relatively low value of forest products. Cost plays an important role in decisions about how to manage forest diseases. Nursery crops, for example, have a higher value than forest crops, resulting in more options.

Disease Detection, Recognition, and Appraisal

The first step in disease management is disease detection, recognition, and appraisal. A guide for disease detection and recognition is shown in Box 16.1 and this information can be used to make a damage appraisal. The presence of a pathogen in a stand does not necessarily mean that it needs to be managed. Further questions need to be asked such as:

1. How fast is the disease spreading and how much mortality is going to occur?
2. How much wood volume will be lost in the future if the disease is not managed?
3. Is the stand better managed for wildlife or for timber purposes?
4. How much can one afford to spend on disease management?

Box 16.1

Guide for Disease Detection and Recognition

1. Determine part of plant affected: foliage, branch, stem, roots, cones, buds, and so on, and symptoms (e.g., resin flow, foliage damage, decay, stain, excess cone crops, wilting).
2. Determine tree species, age, and condition of adjacent plants.
3. Look for patterns of occurrence at the leaf, branch, tree, stand, and landscape level—for example, random versus systematic patterns, rapid versus slow death. Determine if the disease was caused by a biological agent or is a nonbiological injury.
4. Note recent forest management practices and disturbances (fire, wind, etc.) in the area.
5. Look for nonbiological causes such as frost damage, sunscald, spraying, and so on.
6. If a biological agent is suspected, look for aboveground signs—for example, fruiting bodies of fungi, mycelium, rhizomorphs, and insects.
7. If nothing is found examine roots for signs—for example, fruiting bodies of fungi, mycelium, rhizomorphs, and insects.
8. If biotic disease agent is suspected, determine the specific causal organism (e.g., *Phellinus weirii*).

Seven major disease management strategies are discussed below relative to these questions. The questions can also be explored using models as we will discuss later in the chapter.

Major Management Strategies

There are seven major strategies that can be used for disease management: (1) silvicultural, (2) biological, (3) chemical, (4) genetic resistance, (5) quarantine, (6) doing nothing—passive management, and (7) managing to increase decay. The latter might be an option if land is to be managed primarily for wildlife and biodiversity and habitat is needed for cavity-nesting birds, for example (Bull et al. 1997). More than one of these strategies might be employed at a time, and successful strategies often employ multiple strategies. This is known, particularly in the entomological world, as integrated pest management or IPM. Integrated management also involves management of diseases, insects, and fire together. This information easily can be contained on a single CD to be used on a computer.

Silvicultural Management

Silvicultural management techniques are designed to maintain or increase the overall health and vigor of trees and include the use of clearcutting, thinning, pruning, inoculum reduction or removal, species selection, fertilization, prescribed fire, and avoidance of high-hazard areas or times of infection. The philosophy here is to prevent diseases from occurring or to slow the rate of spread of

T A B L E **16.1**	**Summary of Uses of Silvicultural Techniques for Managing Forest Diseases**
Technique	**Examples of uses**
Clearcutting	Management of dwarf misletoes
Thinning	Removal of dwarf mistletoe infected trees; thinning around laminated root rot pockets to prevent disease spread by vegetative spread
Pruning	Removal of infected branches associated with dwarf mistletoe, white pine blister rust, and Dutch elm disease
Sanitation (inoculum removal)	Removal of stumps and roots to manage root diseases (laminated root rot, Armillaria root disease, Annosus root and butt rot), particularly after clearcutting; removal of infected boles with Dutch elm disease
Species selection	Laminated root rot—western redcedar, western white pine Dutch elm disease—siberian elms Port-Orford-cedar root rot—western redcedar
Fertilization	N and K fertilization improves resistance to Armillaria root disease in inland western United States
Prescribed fire	Control of brown spot needle disease in longleaf pine
Avoidance of high-hazard times or areas	Avoidance of thinning in autumn, winter, and spring to prevent stump infection by *H. annosum* in southern pines in the southeast United States; avoidance of thinning in summer, autumn, and spring in Scandinavia to prevent stump in infection of Norway spruce and Scots pine; avoidance of sandy soils to prevent infection by *H. annosum* in southern pines; avoidance of planting susceptible white pines in higher elevations between lakes in the Great Lakes region of the United States

established diseases. Silvicultural techniques used to manage diseases are summarized in Table 16.1.

Clearcutting will reduce the incidence of obligate parasites such as dwarf mistletoes, particularly if all infected trees are removed, including nonmerchantable ones (Hawksworth and Wiens 1996). This practice has reduced the incidence of western hemlock dwarf mistletoe in coastal Oregon and Washington because the spread of dwarf mistletoes is mostly local and slow. Clearcutting and short rotations can be used to manage decay and canker fungi that take considerable time to develop. On the other hand, rust diseases, which also are caused by obligate parasites, are not readily managed by clearcutting. In this case, spores of rust fungi are capable of spreading long distances, and in the case of white pine blister rust and other long-cycle rusts, populations of the alternate hosts are not reduced by clearcutting. Clearcutting alone will also not reduce the incidence of root diseases since the fungi involved, such as *Phellinus weirii, Armillaria ostoyae,* and *Heterobasidion annosum,* can survive as saprophytes in stumps and roots and have the capacity for attacking planted or naturally regenerated trees on the site as they develop (Hadfield et al. 1986). *Phytophthora* spp. can survive as resting spores in the soil after clearcutting.

Thinning also can be used to increase the health of a stand in both uneven- and even-aged stands. Suppressed trees and trees that are heavily infected with dwarf mistletoes in uneven-aged stands in eastern Oregon and Washington can be removed, leaving the stand in a healthier condition. However, unless the stand is rethinned after the initial thinning to remove mistletoe-infected trees that were not noticed at the time of the first thinning, the value of thinning may be lost. The practice of thinning a number of times is known as sanitation and is discussed below. Thinning may reduce the incidence of foliage diseases in conifer plantations, especially in Christmas tree plantations, because it increases airflow and reduces the humid conditions that favor disease development. It also has been suggested that thinning nonsymptomatic trees around a laminated root rot pocket may reduce the spread of the disease because the fungus does not spread as rapidly along or through dead root systems as living root systems. On the other hand, thinning may increase the incidence of other root diseases. *Heterobasidion annosum,* which spreads by airborne spores, may infect trees remaining after thinning by infecting stumps and wounds created during thinning operations. Armillaria root disease also may increase in infected stands since the fungus can build up its inoculum potential by occupying stumps and roots of thinned trees, particularly if thinning stumps are large. Thinning also may increase the incidence of black-stain root disease because the insects that spread the disease may be attracted to slash and fresh stumps (Hansen and Lewis 1997).

Pruning can be used to reduce the incidence of dwarf mistletoes, branch cankers, rust diseases, and vascular wilts such as Dutch elm disease, but other diseases may increase as a result of improper pruning. For example, decay fungi can enter through the pruning wounds, especially on hardwoods, and in radiata pine in New Zealand cankers caused by *Diplodia pinea* have developed (Hansen and Lewis 1997). Pruning of branches of young white pine trees infected by *Cronartium ribicola* in western Washington and Idaho has been successful in preventing the development of white pine blister rust, but requires frequent monitoring.

Sanitation or **Inoculum Reduction** can be utilized for structural root diseases such as Armillaria root disease, laminated root rot, and Annosus root and butt rot (Morrison et al. 1991, Thies and Sturrock 1995). Here infected stumps and roots are removed from the soil during clearcutting operations to reduce the inoculum. This is done with tractors and excavators after harvesting, or by tree pushing at the time of harvesting. However, it is all but impossible to completely remove all the infected root tissues and some usually remain. Disease incidence, however, is dramatically reduced using this technique. As mentioned previously inoculum reduction or sanitation also is conducted when uneven-aged stands are rethinned to remove dwarf mistletoe–infected trees not noticed during initial thinning in the inland western United States. Perhaps the most cited example of sanitation is the removal and burning of insect brood material to prevent the spread of Dutch elm disease. Sanitation also is practiced in the management of Phytophthora root disease. Because this disease spreads by spores in the soil, it is important that inoculum is not spread in soil particles on boots, equipment and machinery. A simple washing or treatment in Clorox will reduce or eliminate the potential for spread. In areas with susceptible ornamental trees, surface water flows should be managed to prevent spore spread and the creation of wet soil conditions.

Tree Species Selection has great promise for disease management. Most disease organisms have specific hosts or a range of susceptible hosts so species selection is important (Hansen and Lewis 1997). For example, in areas with laminated root rot, red alder (which is immune), western white pine, or western redcedar (resistant to the fir form of *P. weirii*) could be planted as alternatives to Douglas-fir, recognizing that western white pine is susceptible to white pine blister rust and western redcedar is prone to deer browse in some areas. Western redcedar or Douglas-fir could be planted as an alternative species to Port-Orford-cedar in areas with Port-Orford-cedar root disease.

Prescribed Fire now is accepted as a management tool and has been used to control brown spot needle disease on longleaf pine in the southeastern United States and can be used to reduce dwarf mistletoe inoculum by killing infected suppressed trees in eastern Washington and Oregon. It is of little use, however, in controlling structural root diseases like laminated root rot because temperatures in roots usually do not get high enough to kill the fungus involved.

Fertilizers used in forestry usually are of three types: nitrogen fertilizers (urea or ammonium nitrate), phosphorus fertilizers (super phosphate or rock phosphate), and trace elements (Cu, B, and Zn). Potassium fertilization also has been used to a limited extent. Phosphorus and trace elements usually are deficient in areas where soils are old and weathered (e.g., southeast United States and Australia), or in very sandy soils (e.g., New Zealand). Nitrogen fertilizers typically are used on younger soils, particularly those formed after recent deglaciation (e.g., in the Pacific Northwest United States, Canada, and Scandinavia). Fertilization generally improves tree health and may reduce losses due to root diseases. For example, Douglas-fir stands in Idaho showed less loss to Armillaria root disease when fertilized with urea and potassium (Mandzak and Moore 1994). Urea fertilization may increase the impact of foliage diseases and influence decay fungi by increasing diameter growth, thereby decreasing the percentage loss to decay when fungi are compartmentalized.

Disease Avoidance may be as simple as thinning at certain times of the year to avoid infection or planting in areas where the disease is not present or the environment does not favor disease development. In slash and loblolly pine stands in the southeastern United States, thinning during the summer, when stump surfaces are hot, reduces infection by *Heterobasidion annosum* and promotes growth of competitive fungi such as *Trichoderma viride*. In Scandinavia thinning can safely be conducted in winter when spores are not present. High- and low-hazard areas for white blister rust have been designated in the Lake states. Areas close to lake shores have low hazard because airflows carrying basidiospores from *Ribes* plants do not reach the foliage of white pine trees. Areas on ridges between lakes, on the other hand, are designated as high-hazard areas where resistant white pines should be planted. Soil characteristics have been used to estimate hazard to Annosus root disease and littleleaf disease. In the southeastern United States incidence of *H. annosum* is highest in deep sandy soils with low organic matter. Littleleaf disease caused by *Phytophthora cinnamomi* is worse on eroded soils with hardpans and poor drainage (Hansen and Lewis 1997). Phytophthora root disease also can be avoided in forest nurseries by locating them on well-drained soils.

Biological Control There are several good examples of biological control of forest diseases (Tainter and Baker 1996). *Phanerochaete gigantea* is a saprophytic decay fungus that is capable of outcompeting *Heterobasidion annosum* on stump surfaces. In Scandinavia and England this fungus is applied to stump surfaces at the time of thinning, effectively controlling *H. annosum* (Hansen and Lewis 1997). *Trichoderma viride* and *Phanerochaete gigantea* also act as natural biological control agents on southern pine stumps during the summer months. *Trichoderma viride* outcompetes *H. annosum* when temperatures are above 25°C. Ectomycorrhizal fungi on conifer roots also act as biological control agents, especially in forest nurseries. The fungal mantle acts as a physical barrier to infection by root diseases.

Chemical Control Chemical control is not as widely practiced in forestry as it is in agriculture, because of costs and environmental concerns. Soil fumigation with methyl bromide/chloropicrin is widely used in forest nurseries to reduce fungal pathogen populations. Use of methyl bromide, however, is soon to be restricted. Other chemicals used include borax stump treatment to prevent infection by *H. annosum,* herbicide or growth regulator treatment for dwarf mistletoe, and seed coat and seed fumigation treatments. Injecting elms with fungicides to eliminate Dutch elm disease is another example of chemical control. Foliar sprays for controlling foliage diseases on ornamental and Christmas trees and in radiata pine plantations in New Zealand also have been used successfully (Hansen and Lewis 1997).

Genetic Resistance Genetic resistance is an important tool for disease management (Hansen and Lewis 1997). Resistance to rust diseases occurs in both agricultural crops and forest trees and breeding programs have been developed successfully for white pine blister and fusiform rusts. Resistant white pine and loblolly pine seedlings are now planted. Clones of poplars resistant to poplar rust and Port-Orford-cedar trees resistant to Port-Orford-cedar root disease have been developed. Trees resistant to fungi causing structural root diseases, such as *Phellinus weirii* and *Heterobasidion annosum* have been noted, but to date genetic resistance has not been used to manage structural root diseases. Eastern white and ponderosa pines are very susceptible to ozone damage. Resistant varieties, however, have survived ozone exposure and the rate of mortality, particularly for ponderosa pine in southern California, has declined in recent years.

Quarantine Oceans and mountain ranges are effective natural barriers to the long-distance spread of pathogens. Humans, however, have broken these natural barriers by transporting seedlings, seeds, logs, and other plant parts around the world and accidentally introducing devastating pathogens to Europe and North America, causing diseases such as Dutch elm disease, white pine blister rust, and chestnut blight in the late 1800s and early 1900s. In 1991 the Eurasian poplar rust fungus was first observed in the western United States. More than 20 exotic fungal pathogens and 360 exotic insects, including the European gypsy moth and most recently the Asian gypsy moth in 1992, now attack woody trees and shrubs in North America. Had effective quarantine procedures been in place, introduction of many of these devastating fungi and insects may not have occurred. Quarantines are regulations that restrain the transport of materials that may harbor pests. The word *quarantine* is derived from the French word *quarante,* or forty, representing 40 days of isolation. Shipments of plants into the United States often are held in warehouses for 7 to 28 days and examined for pests by U.S. Customs Agricultural Pest Inspectors before release or being destroyed if exotic pests are found. This is a huge job, considering the volume of material entering the United States and moving around the world.

A recent development has drawn attention to the possibility of once again introducing devastating pathogens to North American forests. The United States traditionally has imported only small quantities of tropical and temperate logs and lumber, and most of the demand for wood products was from internal harvest. This situation now has changed and forest harvest in the United States cannot supply the demand. Large quantities of conifer logs and lumber now are imported from Canada, but this is not of great concern with respect to the spread of exotic diseases (Filip and Morell 1996). However, in 1990 the Animal and Plant Health Inspection Service (APHIS) of the U.S. Department of Agriculture received the first requests to import logs from Siberia, including Siberian larch and pine. APHIS is charged with protecting the country's agricultural and forest resources from plant pests, but at the time did not have regulations in place for logs, lumber, and other unmanufactured wood articles. Relying solely on inspection at the time of importation was not acceptable for addressing new plant pest risks. In 1992 requests were made to import *Pinus radiata* and *Pseudotsuga menziesii* logs from New Zealand and *P. radiata, Nothofagus dombeyi,* and *Laurelia philippiana* logs from Chile. It was deemed that there was great potential for importation of deleterious diseases or pathogenic strains of diseases or insects. This topic is discussed in more detail in Chapter 25 (Forest Insect Quarantine).

A detailed pest risk assessment was then requested by APHIS, and proposed regulations were developed through an environmental impact statement that addressed potential impacts of the use of methyl bromide, as well as the impacts on biodiversity and endangered and threatened species. Methyl bromide treatment was addressed specifically because methyl bromide is classed as an ozone depletor in the atmosphere. This topic is addressed in detail in Chapter 11 (Nursery Diseases and Mycorrhizae). The proposed requirements for universal importation are shown in Table 25.2. Specific importation requirements may be needed for radiata pine logs and/or lumber from New Zealand or Chile, or Douglas-fir logs and/or lumber from New Zealand.

Doing Nothing—Passive Management   In some circumstances it may appropriate to passively manage diseases if the impact is anticipated to be low or if creation of wildlife habitat is a favored management objective. For example, dead and dying trees in root disease pockets caused by *Phellinus weirii* may provide suitable habitats for cavity-nesting birds (e.g., woodpeckers), mammals, and even reptiles. Thus a land owner may decide to manage for biodiversity in this manner.

Increasing Decay   Management guidelines for federal lands in the western United States now dictate that a certain number of standing dead and green trees

be left for wildlife after logging. Some of the remaining green trees are being inoculated with decay fungi to enhance habitat for primary cavity nesters, like woodpeckers, that typically excavate cavities in dead trees in areas softened by decay.
Because they remain alive for a long time, inoculated green trees are safer during
logging operations than standing dead trees. Also because they remain alive, they
provide wildlife habitat for longer time periods than dead trees that soon fall. The
decay fungi that have been suggested for inoculation are *Phellinus pini, Fomitopsis
cajanderi,* and *Fomitopsis pinicola* for all conifers, *Hericium abietus* for western
hemlock, *Fomes hartigii* for western hemlock and Douglas-fir and *Wolfiporia
cocus* for larch (personal communication, Catherine Parks, USDA Forest Service
La Grande, Oregon).

Integration of Disease, Insect, and Fire Management

As we manage forests more as ecosystems we need to integrate disease, insect,
and fire management. An example of this is the recent recommendation from the
USDA Forest Service for managing federal forests in Oregon and Washington
(shown in Box 16.2) (Filip and Schmidt 1990, Campbell and Liegel 1996). The
recommendations for areas east of the Cascade crest and in southwest Oregon
differ from those in areas west of the Cascade crest.

MODELING FOREST DISEASES

Modeling can be used to predict the rate of spread of diseases, the interactions of
insects, diseases, and fire and to determine management strategies for management
(Tainter and Baker 1996) There are three basic approaches to modeling diseases:
(1) differential equations; (2) regression equations; and (3) simulation models that
take into account site, environmental, and cultural factors.

Differential equations can be used to examine the rate of disease spread. For
example, the equation

$$dx/dt = rx$$

is commonly used. Here the change in the amount of disease (dx) with the change
in time (dt) is proportional to the amount of disease (x) at that time multiplied by
r (the infection rate). This is an exponential curve (Figure 16.1) and the solution
to the equation is $x = e^{rt}$. Disease at some future time can be calculated by knowing the present amount of disease and the infection rate. The amount of disease
can be determined using disease surveys and a number of survey techniques have
been developed, particularly for root diseases.

Although diseases may spread exponentially in the early stages, they cannot
do this forever, because the amount of susceptible tissue decreases with time. Thus
a better model for disease spread with time is

$$dx/dt = rx(1 - x)$$

In this equation $1 - x$ is the amount of healthy susceptible tissue or the "cor-
rection factor." As x increases, $1 - x$ decreases and the epidemic slows. The solution to this equation result is a logistical curve (Figure 16.2).

16.2

Integrated Management Recommendations for Maintaining Healthy Forests in Oregon and Washington

East of the Cascade Crest and in Southwest Oregon

- Thin stands to reduce competition, stress, and bark beetle mortality.
- Harvest certain species like lodgepole pine to create a mosaic of age classes across the landscape to prevent bark beetle outbreaks.
- Keep in mind the effects that certain activities (such as thinning, harvest, or replanting with certain species) will have on root diseases and dwarf mistletoes.
- Reduce forest floor fuel loads to prevent destructive stand-replacement fires.
- Introduce prescribed fire to lighten fuel loads and remove Douglas-fir and true firs. Use fire to regenerate larch or quaking aspen that depend on fire for seedbeds or to stimulate root sprouting.
- Design site-specific regeneration (natural or planted) to promote desired species composition and structure.

West of the Cascade Crest

- Shift from single species to multiple species to reduce insect outbreaks and proliferation of diseases. This involves tradeoffs between maximizing timber production and minimizing insect and disease damage.
- Replant harvested areas with seedlings grown from local seed sources or use natural regeneration. The severity of diseases such as Swiss needle cast is much less when trees are adapted to the site.
- Maintain a mix of species and age classes, preventing the landscape from being dominated by uniform, highly susceptible stands.

Differential equations have not been extensively used in plant pathology because of their relative simplicity and the fact that many variables control the infection rate. On the other hand, regression equations have been widely used to predict disease spread, the amount of decay in trees, and the spread of dwarf mistletoes. Multiple regressions have been used because many independent variables can be taken into account. An example of a multiple regression equation is shown below for the spread of dwarf mistletoe (*Arceuthobium campylopodum*) in ponderosa pine in Arizona and California (Dixon and Hawksworth 1979).

For uneven-age stands the following equations is used:

$$y = 11.696x_1 - 0.018x_2 + 0.778x_3$$

where:

y = estimated distance of spread (in feet)
x_1 = 6-class dwarf mistletoe rating of infected trees (see Figure 14.4)
x_2 = density of stand (trees/acre)
x_3 = infection age (years)

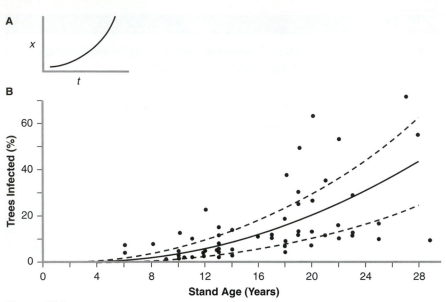

Figure 16.1

A, Exponential curve of disease progression. **B,** An example of this type of disease progression for the increase of dwarf mistletoe in ponderosa pine trees (age 5–28 years) within 9 m of infected stands. Dotted lines indicate 95% confidence intervals for regression line.

(Source: From Tainter, F.H. and F.A. Baker, *Principles of Forest Pathology*, ©1996 by John Wiley & Sons, Inc. Reprinted by permission of John Wiley & Sons, Inc.)

The disadvantage of regression models is that they can be used only in the area where the data were collected. Simulation models can include many more variables, are more universally applicable, and have more utility for ecosystem management. An example of a simulation model is the Forest Vegetation Simulator for the western United States (formerly known as Prognosis), which contains submodels dealing with root diseases and bark beetles known as the Western Root Disease Model (see Chapter 12). This model is designed to be run on a personal computer. Input to the model is stand-level data. The model is run under a number of management scenarios that allow managers to make decisions on root disease treatment. Another example of a simulation model is shown in Figure 16.3, which gives the structure of a model for Douglas-fir seedling root growth, damping-off, and Fusarium root rot development in British Columbia nurseries (Bloomberg 1976) .

In addition to these modeling techniques, economic models also can be used for decision making. For example, after conducting a survey of *Phellinus weirii* in a Douglas-fir stand (Bloomberg et al. 1980), an economic analysis can be done on a personal computer using present net worth analysis for predicting return on investment (Russell et al. 1986). Costs of site preparation and stump removal, planting, thinning, vegetation control, and fertilization can be included

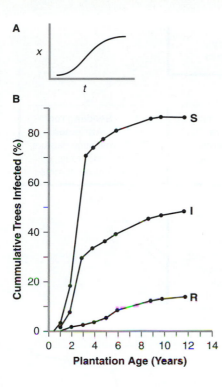

Figure 16.2

A, Logistic curve of disease progression. **B,** An example of this type of disease progression for the increase of fusiform rust on resistant (R), intermediately resistant (I), and susceptible slash pine families (S).

(Source: From Tainter, F.H. and F.A. Baker, *Principles of Forest Pathology,* ©1996 by John Wiley & Sons, Inc. Reprinted by permission of John Wiley & Sons, Inc.)

and volume/acre can be predicted with and without laminated root rot. New computer techniques also can be used for examining the consequences of various land management alternatives. Data from air photos, stand inventories and geographic information systems (GIS) can utilized. An example of this is the Landscape Visualization System (LVS) of Oliver and McCarter (1995) that is currently linked to the Western Root Disease Model (Frankel 1998). Outputs include graphs, charts, spreadsheets and landscape visualizations showing changes in trees and root disease through time.

WOOD PRODUCTS DETERIORATION

Wood products commonly decay in place when exposed to moist conditions. The decay fungi usually involved are Basidiomycetes that occur naturally in forests, but are particularly well adapted for decaying wood in buildings, especially in damp areas like bathrooms and basements, decks, railroad ties, utility poles, wooden bridges, and landscape timbers. Stored pulp also is susceptible to decay.

The most important fungi causing decay in buildings and lumber are *Serpula lacrimans* and *Gloeophyllum sepiarium,* which cause brown cubical decay and are fairly common on ties, poles, and posts. *Phanerochaete gigantea* is a major cause of decay on stored pulpwood in the southern United States.

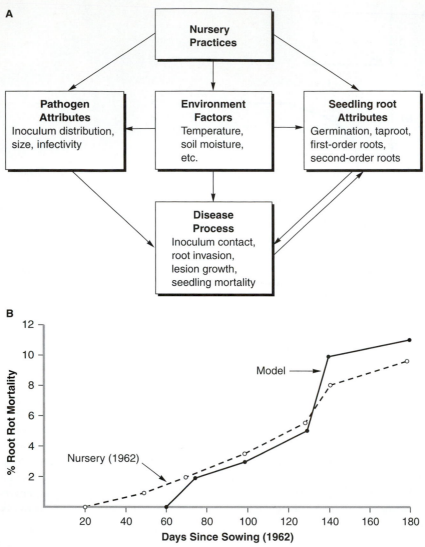

Figure 16.3
A, Example of the structure of a simulation model for Douglas-fir seedling root growth, damping-off, and Fusarium root disease. **B,** Comparison of root rot mortality in Douglas-fir seedlings in a nursery in British Columbia and that predicted by the model. Air and soil temperatures were computed from meteorological records. Inoculum concentration and infectivity were estimated from experimental data.

Wood moisture content is the most important factor in controlling decay. Moisture added to dry wood is absorbed in the cell walls resulting in swelling. Decay fungi require free water so wood must be above the **fiber saturation point** to decay. The fiber saturation point is about 28% moisture content on a dry weight basis.

Figure 16.4
Blue-stain fungi in a ponderosa pine log in eastern Washington.

The best way to control wood decay in products is to keep the wood dry ($< 20\%$ moisture is a good rule of thumb). Building designs should allow for plenty of air movement, especially in crawl spaces under houses and in attics where condensation on the inside of roofs and walls can easily occur. Another good suggestion is to use durable wood; heartwood is better than sapwood. Preservative chemicals extend the life of wood, especially in decks and in landscape timbers. The most commonly used preservative is copper chrome arsenate (CCA), which is applied under pressure to get deep penetration of the fungicide in the wood. Pressure-impregnation is more effective in controlling decay than surface painting or dipping. One of the first wood preservatives to be used was creosote and it is still used, although it is rather messy. Pentachlorphenol, an extremely effective fungicide, was extensively used, but it is no longer available because of its carcinogenic properties. It is particularly damaging to fish if it leaches into water.

A variety of stains occur in wood. The most common and well-known form is blue stain in conifers, especially in pines (Figure 16.4). Blue-stain fungi in the genus *Ophiostoma* are spread by bark beetles, which introduce the fungui when they attack trees. Relationships between bark beetles and fungi are discussed in detail in Chapter 21. The fungus helps the insect by killing the ray parenchyma and resin canal cells in the sapwood, reducing the ability of the host to respond to the attack. Logs stored in the field also are susceptible to beetle attack and the introduction of blue-stain fungi. The rays of stained wood are heavily colonized by the fungus, which does not decay the wood, but may reduce its strength. Blue-stained wood used to be considered unmerchantable, but clever marketing has convinced customers that blue-stained boards can be attractive and they are sold as special decorative wood or "blue pine." Airborne mold fungi also can cause stains on freshly cut stored lumber that is not kiln dried. Sapwood is especially susceptible (Figure 16.5), and with younger, smaller-diameter conifers now being cut into dimension lumber, the proportion of sapwood in boards has increased as has the sapwood stain problem. Kiln drying and surface treatment/dipping with chemicals can prevent this problem.

Figure 16.5
Mold fungi growing on the ends of freshly cut Douglas-fir boards at a sawmill in Washington. Note the high incidence of mold growth in the sapwood.

REFERENCES

Bloomberg, W. J. 1976. *Simulation of Douglas-fir seedling growth, damping-off and root rot.* Canadian Forestry Service, Pacific Forest Research Centre, BC-X-127, Victoria, B.C., Canada.

Bloomberg, W. J., P. M. Cumberbirch, and G. W. Wallis. 1980. *A ground survey method for estimating loss caused by* Phellinus weirii *root rot. II. Survey procedures and data analysis.* Canadian Forestry Service, Pacific Forest Research Centre, BC-RO-4, Victoria, B.C., Canada.

Bull, E. L., C. G. Parks, and T. R. Torgensen. 1997. *Trees and logs important to wildlife in the interior Columbia River Basin.* USDA Forest Service, Pacific Northwest Station, General Technical Report PNW-GTR-391, Portland, Oreg.

Campbell, S., and L. Liegel (tech. coords.). 1996. *Disturbance and forest health in Oregon and Washington.* USDA Forest Service, Pacific Northwest Research Station, General Technical Report PNW-GTR-381, Portland, Oreg.

Dixon, G. E., and F. G. Hawksworth. 1979. A spread intensification model for southwestern dwarf mistletoe in ponderosa pine. *Forest Science* 25:43–52.

Filip, G. M., and J. J. Morrell. 1996. Importing Pacific Rim wood: pest risks to domestic resources. *Journal of Forestry* 94(10):22–26.

Filip, G. M, and C. L. Schmitt. 1990. *Rx for* Abies: *silvicultural options for diseased firs in Oregon and Washington.* USDA Forest Service, Pacific Northwest Research Station, General Technical Report PNW-GTR-252, Portland, Oreg.

Frankel, S. J. (tech. coord.). 1998. *User's guide to the western root disease model,* version 3.0. USDA Forest Service, Pacific Southwest Research Station, General Technical Report PSW-GTR-165, Berkeley, Calif.

Hadfield, F. S., D. J. Goheen, G. M. Filip, C. L. Schmitt, and R. D. Harvey. 1986. *Root diseases in Oregon and Washington conifers.* USDA Forest Service, State and Private Forestry, Forest Pest Management, R6-FPM-250-86, Portland, Oreg,

Hansen, E. M., and K. J. Lewis (eds.). 1997. *Compendium of conifer diseases.* American Phytopathological Society Press, St. Paul, Minn.

Hawksworth, F. G., and D. Wiens. 1996. *Dwarf mistletoes: biology, pathology, and systematics.* USDA Forest Service Agricultural Handbook 709, Washington, D.C.

Mandzak, J. M., and J. A. Moore. 1994. The role of nutrition in the health of inland western forests, pp. 191–210. In R. N. Sampson, and D. L. Adams (eds.). *Assessing forest ecosystem health in the inland West.* Haworth Press, Binghamton, N.Y.

Morrison, D. J., H. Merler, and D. Norris. 1991. *Detection, recognition and management of Armillaria and Phellinus root diseases in the southern interior of British Columbia.* Joint Publication of Forestry Canada and the British Columbia Ministry of Forests, Forestry Canada, FRDA Report 179, Victoria, B.C., Canada.

Oliver, C. D., and J. B. McCarter. 1995. Developments in decision support for landscape management. In *Proceedings of the ninth symposium on geographic information systems,* March 27–30, 1995. Vancouver, B.C., Canada.

Russell, K., R. Johnsey, and R. L. Edmonds. 1986. Disease and insect management, pp. 189–207 In C. D. Oliver, D. P. Hanley, and J. A. Johnson (eds.). *Douglas-fir. Stand management for the future.* Institute of Forest Products, College of Forest Resources, University of Washington, Contribution 55, Seattle, Wash.

Tainter, F. H., and F. E. Baker. 1996. *Principles of forest pathology.* John Wiley and Sons, New York.

Thies, W. G., and R. N. Sturrock, 1995. *Laminated root rot in western North America.* USDA Forest Service Resource Bulletin PNW-GTR-349, Portland, Oreg.

17

INTRODUCTION TO FOREST ENTOMOLOGY

INTRODUCTION

In 1905 Ephraim P. Felt, state entomologist for New York, described the immense impact of forest insects on North American forests. He noted the 1818 spruce beetle [*Dendroctonus (piceaperda) rufipennis*] outbreak in Maine that killed every spruce tree west of Penobscot, and he discussed A. S. Packard's 1890 report that analyzed the wholesale destruction of northeastern North American spruce forests by this insect. Earlier a correspondent for the periodical *Nation* mentioned that not one spruce in 20 was found alive in Essex County, New York.

In 1891 A. D. Hopkins observed outbreaks of the southern pine beetle (*D. frontalis*) covering an area of over 50,000 square miles in West Virginia, Maryland, Pennsylvania, and New Jersey (Hopkins 1899). Hopkins also described the annual financial losses caused by forest insects at the end of the 19th century (in Felt 1905, Table 17.1). He further stated, "To this should be added losses in working up and disposing of defective wood, losses caused by use of the same, all of which, if it could be estimated in dollars and cents, might be placed at $75,000,000. . . ." (Felt 1905). Altogether, Hopkins calculated that about $100 million was annually lost to insect damage. Losses to the forests caused by insects at the beginning of the 1900s were not so serious from an economic point of view as later on. By the mid-20th century timber became more valuable and losses due to insect damage grew accordingly. F. C. Craighead in 1950 reported that annual losses caused by bark beetles were $20 million; defoliators, $20 million; forest products insects, $60 million; and insects affecting shade and ornamental trees, $100 million. In fact, every year in North America, forest insects destroy millions of hectares of forests. The southern pine beetle in 1996 killed over 3 million ha of southern pines; the western and eastern spruce budworms (Choristoneura occidentalis and C. fumiferana, respectively) annually defoliate hundreds of square kilometers of Douglas-firs, firs, and spruces across North America (Figure 17.1). These same **defoliators** slow the growth and kill millions of trees along the length and breadth of Canada (Figure 17.2).

TABLE **17.1** | **An Estimate of Annual Losses Caused by Selected Forest Insects in the United States During the Late 1890s as Reported by A. D. Hopkins**

Insects	Damage	Losses/year ($)
Southern pine and spruce beetles	Cause conifer mortality	5,000,000
Timber beetles and bark borers	Cause defective wood in fallen timber	1,000,000
Insects associated with fire damage	Bore into dead and moribund timbers	1,000,000
Columbian timber beetle, (*Corthylus columbianus*)	Affect the timber quality of hardwood trees in eastern North America	1,000,000
Timber and carpenter worms in oak— *Arrhenodes minutus* and *Prionoxystus robiniae*, respectively	Bore into sound sapwood of standing trees	2,000,000
Longhorned and shorthorned borers and other borers of standing trees	Bore into sapwood of dead and dying trees	4,000,000
Foliage-feeding insects of forest and shade trees	Cause defoliation, resulting in loss of growth, mortality, and aesthetic damage	3,000,000
White pine weevil (*Pissodes strobi*), plant lice, scale insects, etc.	Affect young growth	1,000,000
Insects of dry wood (e.g., powder post beetles)	Bore into dry wood, reducing it to powder	1,000,000

Source: Modified from E. P. Felt (1905).

National Forests in the United States are not exclusively viewed as a source of timber; extraction of commodities is but one resource of the forests. After 1960 Forest Service lands were to be managed according to the Multiple Use and Sustained Yield Act, which stated that the land must be used so that all resources are harmoniously managed and coordinated, ". . . each with the other, without impairment of the productivity of the land, with consideration being given to the relative values of the various resources. . . ." This concept is still changing. Today silvicultural guidelines observe multiple use concepts and, at the same time, maintain those forest structures that serve to protect and stabilize the ecosystem across the landscape. Ecosystem management will require information on the role of insects in maintaining healthy and diverse ecosystems; protecting timber values will remain as only one aspect of forest entomology.

On commercial forest lands, managing the health of forests remains a traditional part of silviculture. Forest health management protects stands from the time

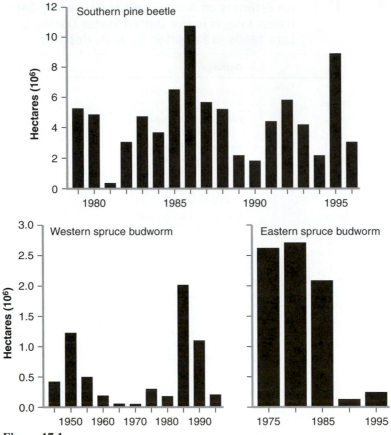

Figure 17.1

Area of southern pines killed by the southern pine beetle (*Dendroctonus frontalis*); area of Douglas-fir and true firs affected or killed by the western spruce budworm (*Choristoneura occidentalis*); and area of true firs and spruces impacted by the eastern spruce budworm (*C. fumiferana*).

(Source: From USDA Forest Service 1997.)

of regeneration through the time of costly intermediate operations to the time of harvesting (Figure 17.3). Forest entomology also is an essential part of urban forestry; even in 1905, E. P. Felt wrote:

> The welfare of the human race is closely connected with that of our trees, and any work looking to their better protection makes for the advancement of mankind.

He went on to write:

> The value of our street and park trees is much greater than the cost of their production, and a city or village blessed with such has treasure which should be most jealously guarded, since these magnificent growths have an important

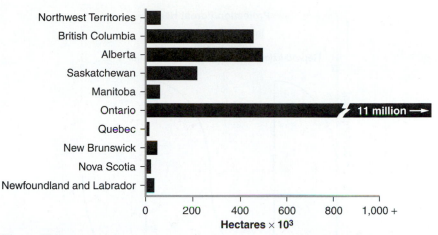

Figure 17.2

The extent of damage by the western and eastern spruce budworms (*Choristoneura occidentalis* and *C. fumiferana*), respectively, to Canadian forests from 1996 to 1997.
(Source: From Natural Resources Canada 1997.)

influence in modifying climatic conditions, besides adding materially to the beauty of the surroundings. This is not only true in cities and villages but also in the country at large. . . .

In 1972 the importance of urban forestry was recognized by the Society of American Foresters as expressed by part of the society's policy statement (Grey and Denke 1978):

> In its broadest sense, urban forestry embraces a multi-managerial system that includes municipal watersheds, wildlife habitats, outdoor recreation opportunities, landscape design, recycling of municipal wastes, tree care in general, and the future production of wood fiber as raw material.

Caring for urban forests will be of major concern in the 21st century as human populations increasingly concentrate in cities and as metropolitan areas begin to spread across vast areas (Figure 17.4).

HISTORICAL PERSPECTIVE

Schwerdtfeger (1973) provided an excellent review of the history of forest entomology. As German foresters developed some of the earliest sustained yield concepts, the necessity of protecting trees from destruction by forest insects became important. In fact, the first publications described immense outbreaks: "Report on a Caterpillar which did Great Damage in Some Places in Saxony Several Years Ago," a translation from a 1752 article by a Regensburg theologian, J. C. Schäffer. Typical of the time, and into the next century, zoologists, physicians,

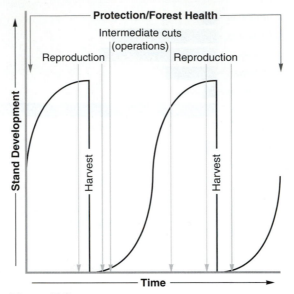

Figure 17.3

Silivicultural events associated with growth of an even aged forest: (1) reproduction occurs a few years before harvest and is encouraged until the stand is successfully established (a few years after harvest); (2) intermediate operations such as thinning, pruning, release, and fertilization occur from stand establishment until the stand is again prepared for harvesting and reproduction; and (3) the health of the forest is protected during the entire reproduction-harvest cycle.

(Source: From Smith, D., *The Practice of Silviculture*, ©1986, John Wiley & Sons, Inc. Reprinted with permission.)

and preachers wrote about insect outbreaks, described the insects and their enemies, and proposed theories on the initiation and collapse of outbreaks. Schwerdtfeger wrote about Schäfer's contributions:

> On a journey he happened to go to a region where caterpillars were doing enormous damage in forests and gardens. Trying to get more information from the residents he was confronted with utter ignorance and thus was stimulated to go into the phenomenon more deeply. Cause of the damage was the gypsy moth (*Porthetria* [*Lymantria*] *dispar*). Shäffer described correctly and in a detailed manner the exterior of the caterpillar, its genesis from the egg, its development, feeding activities, pupation, and the metamorphosis of the pupa into a moth. Shäffer's views as to the causes of the outbreak seem to be almost modern: he mentions favorable weather conditions, failure of natural enemies, and, especially suitable food.

Development of forest entomology in Europe and North America went through at least four stages: (1) a taxonomic and natural history phase; (2) a divergent phase where investigators felt that either biotic or abiotic factors independently were most important in controlling insect populations, whereas others felt

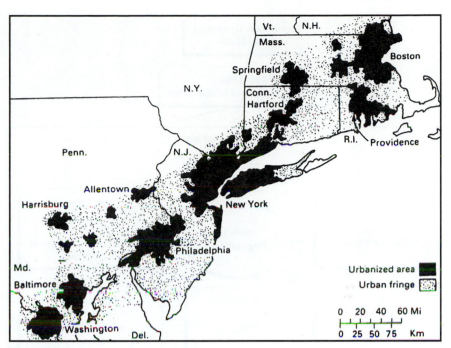

Figure 17.4
The growth of an urban corridor extending from Boston to Washington, D.C.: the formation of a megalopolis.

populations primarily were regulated by density-dependent processes; (3) an ecosystem analysis phase, where forest insect populations were but a reactive portion of a web of relationships inherent to forests; and (4) a population modeling phase, where information on insects, their hosts, and the environment could be organized into mathematical models whose outputs could be used to predict outbreaks—prognosis models (Figure 17.5). Whereas in Europe forest entomology as a science and practice began early in the 18th century, the profession began much later in North America.

Europe

By the end of the 17th century, central Europe feared a wood famine, a fear that was validated during the next century. To relieve this shortage, scientific forestry developed as a theoretical and applied discipline; forest entomology followed. In 1763 the oldest forestry journal, *Allgemeine Oeconomische Forst-Magazin* published a review of forest insects. During the latter two-thirds of the 19th century, literature on forest insects burst forth with the monumental works of J. T. C. Ratzeburg, who published tomes on the life histories of forest insects. His most famous, three-volume monograph was *Die Forstinsekten* (1837–1844), followed by shorter guides to forest insects, then another three volumes on ichneumonid parasites of forest insects. Ratzeburg was born in 1801, died in 1871, and is considered as the

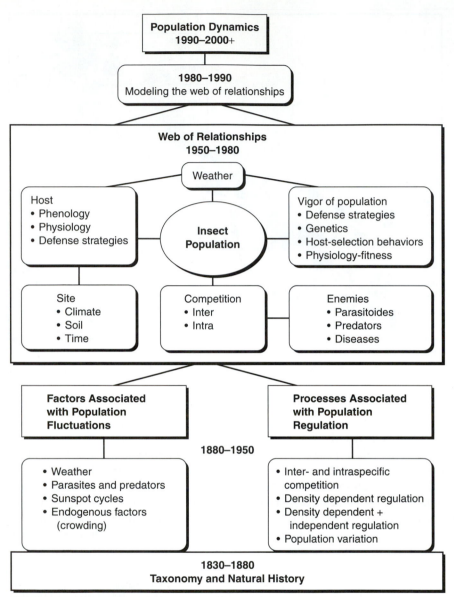

Figure 17.5

Evolution of forest entomology in Europe and North America. From 1830 to 1880, forest insects were cataloged and described; later, certain abiotic factors, biotic factors, or interactive processes were thought to be responsible for the rise and collapse of insect populations. During the last half of the 20th century, fluctuations in forest insect populations were considered to result from changes in the interactive links of the ecosystem; ecosystem studies led to modeling of ecological relationships, all of which in the next century should lead to effective prognosis modeling: a time when the population dynamics of forest insects can be predicted.

Figure 17.6
J. T. C. Ratzeburg (1801–1871).

"father of forest entomology" (Figure 17.6). After his death *Die Forstinsekten* continued to be updated; Karl Escherich, between 1914 and 1942, published a four-volume set titled, *Die Forstinsekten Mitteleuropas,* which contained new material and modernized the older studies. Studies in forest entomology began to proliferate in the scientific literature under the guidance of Escherich when he was editor of *Zeitschrift für angewandte Entomologie* (*Journal of Applied Entomology*), which is still a major voice of this discipline.

During the 20th century research focused on the ecology of insects and how environmental interactions were associated with the buildup and decline of forest insect populations. For example, the periodicity of defoliators was linked with the physical environment, parasites and predators, sun spots, climatic cycles, and so on. Research that concentrated on biological factors or combinations of physical factors that led to population fluctuations resulted in an overvaluation of insect enemies alone, or the role of weather alone, as being the key factors responsible for population dynamics. Over decades, however, forest entomology evolved to where population trends were associated with the multidimensional webs of relationships found in ecosystems. There is great interest in understanding how insect populations react to changes in the environment. Empirical modeling of these changes someday will lead to predictions of population trends (see Figure 17.5) through design of prognostic models. This evolution of the profession in Europe served as a guide to growth of the profession in North America and other parts of the world.

North America

Forest entomology in North America began its taxonomic–natural history period during the 1880s. The profession lagged behind that of Europe because wood supplies in North America, during the 19th century, seemed to be inexhaustible—the public felt this vast timber supply could never be diminished by insects, diseases, or fire (Knight and Heikkenen 1980).

During the mid-19th century, native and introduced forest insects were iden-tified and cataloged and injuries done to individual trees received considerable attention. For instance, Asa Fitch (botanist for New York State) wrote a long series of reports entitled "Noxious, Beneficial and Other Insects of the State of New York" from 1856 to 1870. The first compendium on the natural history of forest insects was *Insects Injurious to Forest and Shade Trees* by A. S. Packard in 1881 and 1890. In 1905 a two-volume monograph by Felt, *Insects Affecting Park and Woodland Trees,* proved to be an excellent reference on the identification and life histories of important forest insects of northeastern North America.

The first professional forest entomologist in the United States was Andrew D. Hopkins, who Escherich called "the American Ratzeburg." He was head of the Division of Forest Insect Investigation, of the U.S. Department of Agriculture (USDA), Bureau of Entomology and Plant Quarantine, established in 1902 in West Virginia; later field stations were established, as the one in Ashland, Oregon (Fig-ure 17.7). In particular, Hopkins studied the natural history and taxonomy of bark beetles. In many bulletins and extensive monographs that were published between 1893 and 1921, he described, named, and worked out the biology of the most destructive North American scolytids, in particular, the genus *Dendroctonus* (Hopkins 1909). He applied the scientific method to his studies, and he illustrated his publications with meticulously detailed pen-and-ink drawings. His work on bark beetles was continued and supplemented by M. W. Blackman in the United States and J. M. Swaine in Canada.

During the mid-20th century, most of the entomological studies were done by the USDA Forest Service at regional experiment stations (Furniss and Wick-man 1998). Besides developing knowledge on the ecology, host selection behav-iors, pheromone interactions, and pest management strategies of important forest insects, the Forest Service (1) published major compendia on U.S. forest insects; (2) began expanded research programs on prominent forest insect pests, which, in turn, produced important scientific contributions; and (3) established major cooperative research programs and international symposia with Canada and Mexico—for example, see Cibrián et al. (1995) and USDA (1994) Forest Service. Examples of compendia are *Insect Enemies of Eastern Forests* by F. C. Craighead (1950), *Insect Enemies of Western Forests* by F. P. Keen (1952), *Eastern Forest Insects* by W. L. Baker (1972), *Western Forest Insects* by R. L. Furniss and V. M. Carolin (1977), *Guide to Insect Borers in North American Broadleaf Trees and Shrubs* by J. D. Solomon (1995), and others. Expanded research programs and their final publications were as follows: *The Southern Pine Beetle* (USDA 1977), *The Douglas-fir Tussock Moth: A Synthesis* (USDA 1978), and *The Gypsy Moth: Research Toward Integrated Pest Management* (USDA 1981). Each of these pub-lications cataloged a myriad of individual research efforts that ranged from life history studies through understanding ecological links controlling populations to population models.

The Forest Service also funds cooperative research with universities. These cooperative programs will be even more important in the 21st century, as the For-est Service is committed to the understanding and management of ecosystems. For

Figure 17.7

A. D. Hopkins (far left) visiting the Ashland, Oregon,
field station of the U.S. Forest Service on an April 1915
inspection trip. Station personnel were (left to right):
W. E. Glendenning, entomological ranger; J. M. Miller,
entomological assistant in charge; and John Patterson,
J. D. Riggs, Philip Sergent, and F. P. Keen (entomological
rangers).

(Photo supplied by B. E. Wickman, USDA Forest Service.)

example, cooperative research with the University of Washington served to
unravel interactions between the mountain pine beetle (*D. ponderosae*), fire, and
fungi in pure lodgepole stands of south-central Oregon (Figure 17.8).

Federally funded cooperative research through the 1887 Hatch Act and the
1962 McIntire-Stennis Cooperative Research Act stimulated significant contributions in forest entomology from the land grant colleges—for example, research
at Washington State University developed population dynamics models of the
Douglas-fir tussock moth (*Orgyia pseudotsugata*) as well of bark beetles; Texas
A&M University provided population models of the southern pine beetle (*D.
frontalis*); and research at Oregon State University, in cooperation with the Boyce
Thompson Institute for Plant Research, Inc., delineated host–bark beetle interaction theory. In addition, the University of California-Berkeley studied bark
beetle–ponderosa pine interactions, urban pollution and associated insect problems; researchers at the University of California-Riverside currently are developing integrated pest management approaches against urban forest pests; Mississippi
State University has provided significant new information on the management of
southern forest insects; and other land-grant institutions are correspondingly
active in providing knowledge on forest insect ecology, ecological theory, host
selection behaviors of forest insects, and pest management strategies.

Canada's forests have always been a major source of the nation's wealth, but
they also have been subject to destructive insect outbreaks (Department of the
Environment Canada 1975). Professional concerns over such destruction led in
1912 to establishment of the Division of Forest Insect Investigations. Under J. M.
Swaine's leadership, the division grew from a few officers based in Ottawa to

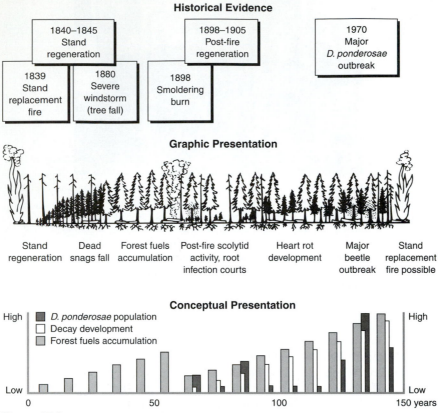

Figure 17.8

Illustration of a 20-year study to understand interaction between fires, fungi, and the mountain pine beetle (*Dendroctonus ponderosae*) in a climax lodgepole pine (*Pinus contorta*) forest of southcentral Oregon. During 1839 a stand replacement fire occurred. This fire eliminated root competition between trees and the stand regenerated. As the stand grew, there was strong competition for light and moisture, resulting in mortality of suppressed trees. Over the ensuing 60 years, these small dead trees provided low-decay-type fuels that laid on top of more decayed logs that were remnants of the 1839 fire. In 1898 the fuel bed was optimal for spread of a smoldering fire. Besides further thinning the stand and establishing reproduction patches, the fire wounded major lateral roots of many residual trees, establishing fungal infection courts. In time a succession of fungi invade the wounded roots and later the boles, causing butt rots. Trees with butt rots eventually became focus trees for another mountain pine beetle buildup, which over several years of the subsequent outbreak supplied fuels for another stand replacement fire (see Gara et al. 1985).

regional laboratories responsible for field investigations and liaison with the forest industry and provincial forestry departments. Later J. J. De Gryse molded the division into an internationally recognized institution with establishment of scientifically based forest insect research, survey, and pest management programs.

De Gryse's forest protection policies were so logical and easily accepted that New Zealand (De Gryse 1955) and later Chile patterned their forest pest management programs after his organizational plans.

Basic and applied forest entomological research doubled in the decade after World War II. The expansion occurred because of catastrophic forest insect outbreaks that awakened public interest to the impact of forest pests to the national economy. To better coordinate pest management research across the nation, forest entomologists and pathologists were placed under the Forest Biology Division in the Science Service of the Department of Agriculture in 1951, and Malcolm L. Prebble became its first chief. This arrangement now has changed considerably, but the contibutions of Canadian forest entomologists to the discipline are world renowned. Among the many distinguished scientists who contributed much to basic knowledge and management of Canadian forest insects is William G. Wellington (Figure 17.9). His studies at the Sault St. Marie, Ontario, laboratory linked synoptic weather patterns with outbreaks of the eastern budworm back to 1900. Later studies led to investigations of relationships between population quality, dispersal, and survival associated with changing climates. He also formulated provocative ideas, such as the notion that the population dynamics of insect species should be based on how individuals behave, not exclusively on natality and mortality data.

With an increasing demand for wood, the demand for forest entomologists has increased. Several industrial organizations employ forest entomologists and pathologists—for example, Weyerhaeuser Company, which is involved with intensive forest management across North America, studies and solves insect problems associated with seed orchards, nurseries, plantations, and forest products. Union Camp Corporation of Savannah, Georgia, is another example. State forestry organizations employ forest entomologists to resolve insect problems on state and private lands; state entomologists also work closely with their U.S. Forest Service counterparts. In particular, annual aerial insect surveys of federal, tribal, and state forests are mostly funded by the federal government and utilize forest entomologists from both jurisdictions.

THE FUTURE OF FOREST ENTOMOLOGY

Advances in communication and travel are placing local, regional, national, and worldwide entomological problems at the doorstep of the discipline. These issues are intensified by growing human populations in developing countries, and a growing demand for their natural resources. This is in contrast to North America where there is need to increase fiber production during a time of changing land-stewardship ethics. Accordingly, pest management options that fulfill the needs of maturing societies while maintaining the integrity of ecosystems will be a challenge to the profession.

Forest entomology often is seen at the industrial level as a way to ensure that tree-growing investments are protected. Pest management that depended on the use of insecticides now is costly and questionable, in terms of (1) pesticide

Figure 17.9
William G. Wellington, 1987, Professor
Emeritus of Plant Science and Resource
Ecology, University of British Columbia,
Vancouver.
(Photo courtesy W. G. Wellington.)

monetary and social costs; (2) unforeseen ecosystem disturbances—for example, use of microbial insecticides, such as *Bacillus thuringiensis,* means that nontarget moths and butterflies also will be killed; and (3) public resistance to traditional pest control in general. Forest entomology will have to face these problems at all political and geographical levels, and pest management will have to be carried out in a sustainable and biorational manner.

Entomology—based on the fundamental, basic, and applied sciences and concepts on how to apply these disciplines in management of forest insects—is changing. New pest management standards, developed from emerging ideas in such areas as population ecology, landscape ecology, urban and wetland ecology, pollution chemistry, meteorology, ethnobotany, and so on will come into play. Most importantly evolving pest management insights will strengthen the support of research on insect population modeling, which, in turn, is based on autecological studies (cause-and-effect interactions) of ecosystems. Once principal ecological links that regulate insect numbers are understood, it will be possible to model the population dynamics of forest pests and their natural controls. During the 21st century, when insect population fluctuations can be predicted, effective pest management, for the first time in the history of the profession, will be possible (Figure 17.10).

REFERENCES

Baker, Whiteford, L. 1972. *Eastern forest insects.* USDA Forest Service. Miscellaneous Publication 1175.

Cibrián, D. C., J. T. Méndez, M. Rodolfo, B. Campos, H. O. Yates III, and J. Flores Lara. 1995. *Forest insects of Mexico/Insectos forestales de México.* D. R. Universidad Autónoma Chapingo, Chapingo, Estado de México, Mexico.

Craighead, F. C. 1950. *Insect enemies of eastern forests.* USDA Forest Service Miscellaneous Publication 657.

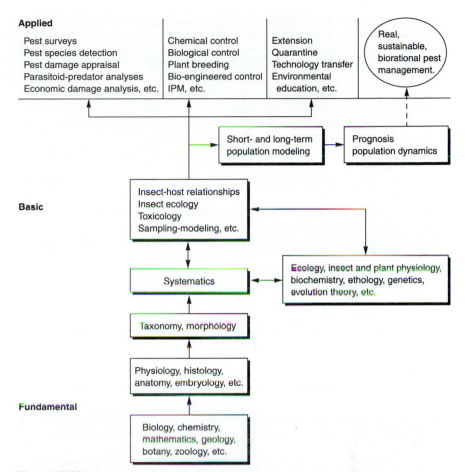

Applied

Pest surveys	Chemical control	Extension	Real, sustainable, biorational pest management.
Pest species detection	Biological control	Quarantine	
Pest damage appraisal	Plant breeding	Technology transfer	
Parasitoid-predator analyses	Bio-engineered control	Environmental	
Economic damage analysis, etc.	IPM, etc.	education, etc.	

Short- and long-term population modeling

Prognosis population dynamics

Basic

Insect-host relationships
Insect ecology
Toxicology
Sampling-modeling, etc.

Systematics

Ecology, insect and plant physiology, biochemistry, ethology, genetics, evolution theory, etc.

Taxonomy, morphology

Physiology, histology, anatomy, embryology, etc.

Fundamental

Biology, chemistry, mathematics, geology, botany, zoology, etc.

Figure 17.10

A simplified chart of forest entomology, a profession based on the fundamental and basic sciences that support forest pest management, an applied science. Forest pest management in the 21st century will depend not only on refinement and integration of traditional pest management practices but also on being able to predict fluctuations in pest populations: the science of population dynamics.

De Gryse, J. J. 1955. *Forest pathology in New New Zealand.* New Zealand Forest Service Bulletin I.

Department of the Environment Canada. 1975. *Aerial control of forest insects in Canada,* M. L. Prebble (ed.). Department of the Environment, Ottawa, Canada, Ottawa K1A OS9.

Felt, E. P. 1905. *Insects affecting park and woodland trees,* vols. 1 and 2 (1906), Memoir No. 8, New York State Museum, New York State Education Department, Albany, New York.

Furniss, R. L., and V. M. Carolin. 1977. *Western forest insects.* USDA Forest Service Miscellaneous Publication 1339.

Furniss, M. M., and B. E. Wickman. 1998. Photographic images and history of forest insect investigation on the Pacific Slope. *American Entomologist* 44:206–216.

Gara, R. I., W. R. Littke, J. K. Agee, D. R. Geiszler, D. R. Stuart, and C. H. Driver. 1985. Influence of fires, fungi, and mountain pine beetles on development of lodgepole pine forests in south-central Oregon, pp. 153–162. In D.M. Baumgartner et al. (eds.). *Lodgepole pine: the species and its management.* Washington State University Cooperative Extension Service, Pullman, Wash.

Grey, G. W., and I. J. Denke. 1978. *Urban forestry.* John Wiley and Sons, New York.

Hopkins, A. D. 1899. *Report on investigations to determine the cause of unhealthy conditions of the spruce and pine from 1880–1893.* West Virginia Agricultural Experiment Station Bulletin 56.

Hopkins, A. D. 1909. *Practical information on the scolytid beetles of North American forests. I. Bark beetles of the genus* Dendroctonus. USDA Bureau of Entomology Bulletin 83.

Keen, F. P. 1952. *Insect enemies of western forests.* U.S. Department of Agriculture Miscellaneous Publication 273.

Knight, F. B., and H. J. Heikkenen. 1980. *Principles of forest entomology.* McGraw-Hill Book Co., New York.

Natural Resources Canada, Canadian Forest Service. 1997. *The state of Canada's forests: learning from history.* Canadian Forest Service, Ottawa, Ontario.

Packard, A. S. 1890. *Insects injurious to forest and shade trees.* U.S. Entomological Commission, 5th. Report, Washington, D.C.

Robinson, W. H. 1996. *Urban entomology.* Chapman and Hall, London.

Schwerdtfeger, F. 1973. Forest entomology, pp. 361–386. In R. F. Smith, T. E. Mittler, and C. N. Smith (eds.). *History of entomology.* Annual Reviews, Palo Alto, Calif.

Smith, D. M. 1986. *The practice of silviculture.* John Wiley and Sons, New York.

Solomon, J. D. 1995. *Guide to insect borers in North American broadleaf trees and shrubs.* USDA Forest Service, Southern Forest Experiment Station, Southern Hardwoods Laboratory, Agricultural Handbook AH-706, Stoneville, Miss.

USDA. 1977. *The southern pine beetle,* Robert C. Thatcher, Janet L. Searcy, J. E. Coster, and G. D. Hertel (eds.). USDA Forest Service Science and Education Administration Technical Bulletin 1631.

USDA. 1978. *The Douglas-fir tussock moth: a synthesis,* Martha A. Brookes, R. W. Stark, and R. W. Campbell (eds.). USDA Forest Service Science and Education Agency Technical Bulletin 1585.

USDA. 1981. *The gypsy moth: research toward integrated pest management,* C. C. Doane, and Michael L. McManus (eds.). Expanded Gypsy Moth Program. USDA Forest Service Science and Education Agency Technical Bulletin 1584.

USDA. 1987. *Western spruce budworm,* Martha H. Brookes, R. W. Campbell, J. J. Colbert, R. G. Mitchell, and R. W. Stark (coordinators). USDA Forest Service Technical Bulletin 1694.

USDA. 1994. *Dynamics of forest herbivory: quest for pattern and principle,* W. J. Mattson, Pekka Niemela, and Matti Rousi (eds.). USDA Forest Service General Technical Report NC-183.

USDA. 1997. *Forest insect and disease conditions in the United States 1996.* USDA Forest Service, Forest Health Protection, Washington D.C.

18

BASIC ENTOMOLOGY

INTRODUCTION

Biology includes the study of plants and animals (botany and zoology). The animals, in turn, are separated into many large groups (each a phylum), and each group is recognized by its discipline: ornithology (study of birds), mammology (study of mammals), ichthyology (study of fishes), invertebrate zoology (study of animals without backbones), protozoology (study of one-celled animals), and others. Invertebrate zoology includes the study of mites and spiders and their relatives (acarology), and segmented worms (nematology). There are several other groups of invertebrates, including the insects (entomology). Entomology seems buried in this hierarchical classification of disciplines and organisms. The insects, however, make up close to 90% of all the animal species on earth; there are more than a million insect species! The main reasons for this incredible diversity is that insect **phylogeny** (i.e., the ancestry of the insects) can be traced back for over 300 million years, and insects reproduce quickly. This means that genetic recombinations and adaptive mutations have occurred for hundreds of millions of years, which has resulted in the immense number of species we know today. Every year hundreds of new insect species are discovered and classified by entomologists. It is small wonder that insects often frustrate forest management objectives; there are insect groups that attack flowers and seeds of trees, even before the seeds can be gathered. Insects invade seeds where they are stored and after they are planted in nurseries. Other insects infest roots, stems, and foliage of seedlings. In natural forests or plantations, insects defoliate, bore, and live in the bark or wood; others live in roots and some transmit pathogens. In old forest stands, insects remove **senescing** older trees, stands, or entire forests. In this sense they carry out the demands of a complex, interactive ecosystem, but often at great financial costs to the fiber-producing economy. To fully appreciate insect diversity and to work in the applied area of protecting forest investments and ecosystems, it is important to understand basic entomology and **systematics** (i.e., classification and evolution) at least at the level of the insect orders. The reference list at the end of this chapter represents an excellent resource for further readings and in-depth knowledge in entomology, especially: Atkins (1978), Chapman (1979), Borror et al. (1981), Evans (1984), Pedigo (1999), and Rowoser and Stoffolano (1994).

THE ARTHROPODA

Insects belong to the phylum Arthropoda which, in part, has these characteristics:

1. An **exoskeleton** which comes in linear segments, giving the arthropods a **bilateral symmetry.**
2. In primitive groups each segment bears an **appendage.** In more evolutionarily advanced taxa, the segmented appendages **coalesce** into other structures (called **homologous** structures) or the appendages are **vestigial** or lost entirely.
3. A tubular, centrally located digestive tract.
4. A dorsal, tubular circulatory organ and a **ventral** nervous system.

The **arthropods** are divided into classes in accordance with their mouthparts and other specialized structures. There is fossil evidence that the arthropods arose from primitive segmented worms, then the primitive arthropods divided into two groups known as the **Chelicerata** and the **Mandibulata** (Figure 18.1).

The Chelicerata

The Chelicerata lack antennae and do not have mouthparts within the mouth. Rather they have external mouthparts that serve as structures to capture prey, suck blood, rasp vegetation, or perform some other function in obtaining food. Many chelicerates also use these appendages to cut up food and place the morsels within the mouth. For example, spiders and scorpions, as well as all other chelicerates, have two pair of bucal (pertaining to the mouth) structures, the **pedipalps** (e.g., the pincerlike claws of scorpions), and the **chelicerae,** which are the structures next to the mouth. The fangs of spiders, which are used to kill or stun prey as well as to feed themselves, are chelicerae. Some common classes of the Chelicerata are the sea spiders (Pycnogonida), the horseshoe crabs (Merostomata), the scorpions (Scorpionida), the pseudoscorpions (Pseudoscorpionida), and the spiders, ticks, and mites (Arachnida) (Figure 18.2).

The **Arachnida** are especially important as they contain the spiders, ticks, and mites. The arachnids are characterized by having a body composed of two regions: the cephalothorax and abdomen. The adults have six pairs of appendages: the chelicerae, pedipalps, and four pairs of walking legs. Spiders belong to the order Araneida, and all are predators and serve as important regulators of insect populations. They are found feeding on insects in all ecosystems and they hunt their prey on the ground, in understory vegetation, and in all parts of trees. Spiders fit into three ecological groups: the hunting spiders, crab spiders, and orb weavers. The hunters (wolf and jumping spiders) are long legged, agile, and either run down their prey or lay in ambush before leaping out to capture their quarry. Crab spiders are delicate and have unusually long legs. These arachnids are well

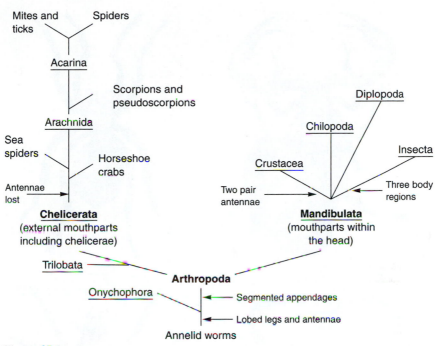

Figure 18.1

The phylum Arthropoda and its two main groups, the Chelicerata and the Mandibulata. The principal feature that separates the two groups is the arrangement of the mouthparts; the Chelicerata have chelicerae and pedipalps held external to the mouth, whereas the Mandibulata have their mouthparts within the bucal (mouth) area. For instance, members of the class Insecta have a labrum, mandibles, maxillae, and maxillary palps, a labium, and labial palps within the mouth.

camouflaged within flowers as they wait for a pollinator to land on the flower. Then with hair-trigger quickness these spiders encircle their prey with their specialized legs and instantly stun the victim with poison-laden chelicerae. Orb weavers spin webs to entangle their prey. Upon sensing a struggling victim, these spiders rapidly move along the web, wrap the hapless offering in webbing, and stun it with their chelicerae.

Although all spiders have venom glands associated with their chelicerae, they rarely bite humans and concentrate on other arthropods, especially insects. There are, however, a few species that can give humans extremely painful bites: the black widow (*Latrodectus mactans*), and the brown recluse (*Loxosceles reclusua*), are two examples. The black widow occurs from southern Canada and south through most of South America. These spiders are found in sheds, dark corners of garages, under houses, rocks, or boards. Their bite cause muscle spasms and if untreated can kill. The brown recluse is found in undisturbed, protected places that offer other arthropods as food—for instance, the back of closets or dark corners in basements; their bites produce a great deal of pain and often lesions that are difficult to heal.

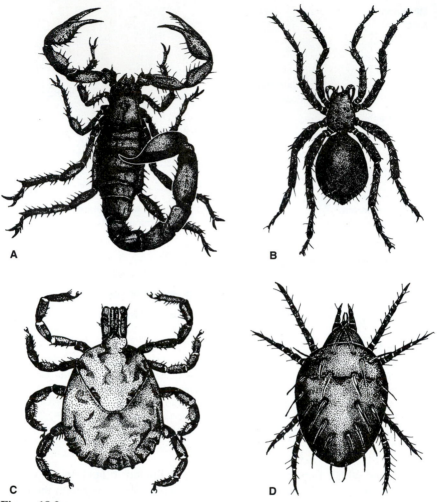

Figure 18.2
Four common arachnids: a scorpion showing its extended abdomen tipped with a stinger (**A**); a spider (**B**); a tick—note the inner chelicerae and outer pedipalps (**C**); and a mite—note the externally placed chelicerae and pedipalps (**D**).
(Source: Drawings by Eduardo Castro, Forestry Faculty, University of Andes, Merida, Venezuela.)

Ticks and mites belong to the order **Acarina.** These arachnids differ from others of their class by having their mouthparts projecting forward on a kind of rostrum and the joint between the **cephalothorax** and **abdomen** is not evident; they seem to be of one continuous body.

North American ticks belong to two families; the Ixodidae or hard ticks have a hard **dorsal** plate called a *scutum;* the soft-bodied Argasidae lack a scutum. The American dog tick (*Dermacentor variabilis*) is an argasid that is a particular nuisance to pet owners, especially those who work hunting dogs. All ticks are parasitic on mammals, birds, and reptiles.

Several species of ticks are annoying and some are important vectors of the following diseases: Lyme disease (a debilitating disease of humans that was first located in New England, but now is widespread), Rocky Mountain spotted fever, relapsing fever, tularemia, and Texas cattle fever. Ticks have a complex life cycle that includes an egg stage, a six-legged larval tick, an eight-legged nymphal tick, and the sexually mature adult. All active stages are blood feeders and stay on their host for several days before they drop off to molt. After **molting** they seek a second or third host; the adult lays eggs on underbrush and low-lying vegetation where an unsuspecting host will brush against clinging strings of larval ticks. Ticks feed by inserting their chelicerae, which have recurved projections, into their hosts. Once these chelicerate mouthparts are firmly inserted, it is difficult to remove the feeding tick, an insurance that the arachnid is satiated before dropping off.

Mites are a ubiquitous group with a multitude of species. Mites also belong to the Acarina, but are extremely small, mostly under a millimeter. They play many ecological roles because the group includes **parasites, predators, herbivores, omnivores,** and **scavengers;** certain groups are aquatic in both marine and fresh water ecosystems. They are particularly abundant in the soil, forest litter and associated organic material where they are important members of the **decomposer trophic** level. In fact, soil mites outnumber all other arthropods and their activity as well as their by-product are important to soil fertility.

Unfortunately many **phytophagous** mites severely damage vegetation by their feeding activity or because they vector virus diseases. Mites are especially damaging to the aesthetic quality of ornamental plants and street trees—a problem in urban forestry. Gall mites (family Eriophyidae), for example, feed on a wide variety of plants and trees and disfigure them by forming bladderlike or pouchlike galls on leaves and branches, but most species of eriophyids cause leaves to brown and fall off prematurely; other species cause distorted or stunted growth. Spider mites (family Tentranychidae) are serious problems of urban plants and trees. They reproduce quickly and infested leaves are soon covered with a silken matrix that contains the mite's various life stages. Heavily infested foliage becomes disfigured and dies. The spruce spider mite, *Oligonychus ununguis,* occurs on several North American conifers (especially spruces) and are important pests of urban tree nurseries and recently transplanted stock.

Other mite species such as the scab or mange mites (family Sarcoptidae) burrow into the skin of humans and animals and produce a searing itch, which when scratched often leads to severe secondary infections. Probably the best-known pest of humans in the eastern, southeastern, and southern states are the chiggers or red bugs (family Trombiculidae) which are ectoparasites (live on another animal) during their larval stage. When attached to humans, the chelicerae contain saliva that dissolves dermal tissues, causing a severe and persistent itch. The other life stages of the trombiculids occur on other organisms.

The Mandibulata

During embryonic development of the Mandibulata, certain body segments and their appendages move around the head and are incorporated within the mouth of the embryo—for example, appendages of the first **postoral segment** form the mandibles. In the Crustacea, however, appendages of the first preoral embryonic segment form a second pair of antennae. All other members of the Mandibulata have only one pair of antennae. Some classes of the Mandibulata are the crabs, lobsters, crayfish, sow bugs (Crustacea), the millipedes (Diplopoda), the centipedes (Chilopoda), and the insects (Insecta).

The crustaceans are mostly aquatic organisms and by and large marine. Most crustaceans have a **cephalothorax,** which often is covered by a shield called a **carapace.** They have two pairs of antennae, the number of legs varies, but all have appendages on each segment of the cephalothorax. A marine crustacean, the gribble (*Limnoria* spp.), attacks wooden piers and pilings and with wave action reduces these structures to a spongelike consistency, especially along a longitudinal area of the piling between low and high tide . Limnorians and shipworms (phylum Mollusca) not only damage pilings but they also invade log storage sites and infest logs awaiting transshipment to other countries. The shipworm (*Bankia setacea*) is an especially severe problem to logs stored in marine waters throughout the Pacific Coast of North America (Gara and Greulich 1995).

The millipedes are elongated tubular creatures that have 30 or more pairs of legs. Bodies of millipedes are cylindrical in cross section and are composed of a head with a pair of short antennae and the rest of the body. Almost all of the body segments appear to have two legs per segment, as in the evolution of the diplopods—every other segment is fused together (**tagmosis**). Millipedes are found in forest litter and in other damp secluded sites, such as under stones and large woody debris. These organisms are mostly **saprophytes** and are involved in the **detritus** food chain. Some species have glands along the side of their bodies that exude an almond-smelling substance, hydrogen cyanide.

The centipedes are segmented, elongated, flattened organisms with more than 15 pairs of legs, one pair per segment. The antennae are long with 14 or more segments and the last two pairs of legs are directed backward. This body plan give chilopods great sensory acuity and mobility. Moreover, the first pair of legs after the head is specialized into a poison claw, which they use to stun their prey; all centipedes are predaceous (Figure 18.3).

THE INSECTS

External Morphology

The insects (class Insecta) are the most specialized Mandibulata as they can be characterized by these features: (1) their bodies are divided into the **head, thorax,** and **abdomen;** (2) most adults have wings; (3) they have one pair of **antennae;** and (4) they have three pairs of legs. The head is the most important area for sensory perception as it includes the eyes (large compound eyes located on the dorsolateral part of the head; most also have simple eyes, **ocelli,** above and between the **compound eyes**); the antennae receive **olfactory, auditory,** and **thigmotactic** stimuli. In certain wood wasps, antennae even receive infrared frequencies from charred trees after a fire sweep through the forest. The head also contains the

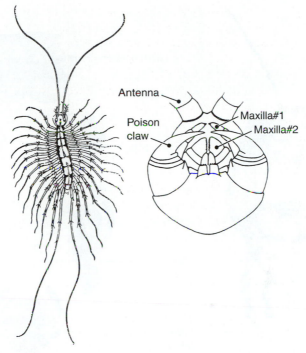

Figure 18.3

The class Chilopoda (centipedes), an arthropod that maintains a pair of appendages per body segment (left), and the specialized appendage that serves a poison claw associated with the mouth as well as two maxillae (right).

mouthparts. The mouthparts of chewing insects, such as locusts, walking sticks, beetles and others, consists of a labrum (an upper lip) and underneath the **labrum** are a pair of articulated biting-chewing structures, the **mandibles.** Under the mandibles are paired maxillae with their **maxillary palps.** The maxillae contain sensory hairs that provide insects with a sense of taste, **gustatory** stimuli, and the fingerlike maxillary palps are analogous to fingers and are used to poke food into the mouth. Then closing the mouth from underneath are a pair of fused segments that form the **labium** (lower lip), the labium subtends paired **labial palps,** which help the maxillary palps shove food in the mouth (Figure 18.4). These basic mouthparts have evolved into specialized feeding organs in the orders that pierce tissues and suck sap or blood, such as in the Hemiptera (e.g., the bugs, aphids, scales, and cicadas) or in certain Diptera (e.g., mosquitoes, blackflies and deer flies). In the Hemiptera the labrum is a vestigial flap, the mandibles and maxillae form nested tubes, **stylets,** and the labium is a folding sheath that houses the stylets. For example, plant-feeding bugs first insert the tip of one threadlike mandible through the leaf cuticle, then the other mandible breaks through, then the first is inserted further and, in this manner, the leaf is punctured. The maxillae follow the path pioneered by the mandibles. The two maxillae are held

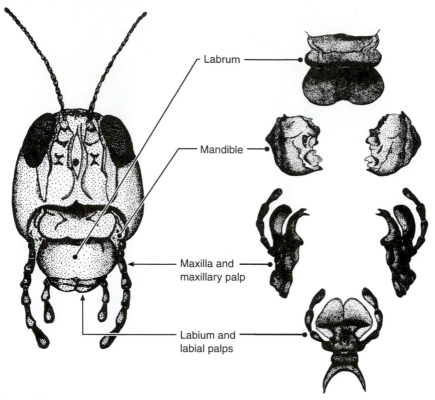

Figure 18.4
The mouthparts of a grasshopper.
(Source: Drawings by Eduardo Castro, Forestry Faculty, University of Andes, Merida, Venezuela.)

together, face-to-face, and the two grooves etched in each maxilla then form two canals; one carries saliva into the puncture and other is a food canal, allowing the plant sap to be pumped into the bug. The saliva contains chemicals that thwart wound protection processes in plants and blood clotting in mammalian hosts. The basic mouthparts vary immensely in other orders with regard to feeding habits. In adult moths and butterflies (Lepidoptera), for example, a substructure of the maxillae, the **galea,** forms a tubular proboscis that lepidopterans use to siphon nectar from flowers.

The three parts of the thorax (**prothorax, mesothorax,** and **metathorax**) contain the legs and wings of insects, the locomotion center. The thorax also contains a pair of **spiracles,** insectan breathing ports. The dorsal meso- and metathorax surface of most adults expands into one or two pair of wings. These are attached dorsolaterally and when extended are beaten, twisted, and warped to give the surfaces aerodynamic lift and allow certain insectan orders the ability to hover or dart or perform intricate aerial maneuvers, especially the flies (Diptera). The ability to fly led to (1) exploitation of new **niches** across the face of the earth as well as new ways to search for food and mates across the landscape; (2) efficient ways to maintain heterogeneous gene pools (dispersal flights over large areas of several populations of the same species); (3) development of intricate host selection behaviors; and (4) development of rapid escape tactics from predators and parasites. Insect wings are double walled and strengthened within by veins. The position and patterns of the wing **veins** are constant within groups and used to identify the various taxa (e.g., families, genera, and species).

Insect legs are specialized for terrestrial or aquatic locomotion. Each segment of the thorax subtends a leg. Legs consist of the following parts: the **coxa,** a ball-like joint that attaches the leg to the thorax; the **trochanter,** a straplike segment that provides added flexibility to the leg; the **femur,** the elongated segment that follows the trochanter; the **tibia,** the second elongated segment; the **tarsus,** a small segment that holds the final (most distal) leg part, the claw. Leg morphology differs between insect groups and is indicative of the mode of life carried out by the insect. Predaceous insects, such as the praying mantis, may have **prehensile** legs for grasping their prey. Fast-running insects, such as cockroaches or tiger beetles, have long spindly legs. Digging insects, such as dung beetles or mole crickets, have **fossorial** (adapted for moving soil) legs, and certain swimming insects, such as water boatmen and diving beetles, have streamlined, hairy legs for propelling themselves rapidly through the water. Taxonomists use leg **morphology,** like the number and appearance of tarsal segments, as a way to identify groups.

The insect abdomen characteristically has 11 segments, and each segment consists of a dorsal tergum and a ventral **sternum.** Each segment has a pair of laterally positioned spiracles. Although the insect abdomen primitively has 11 segments, this number is variable as segments coalesce or are incorporated into other structures used in mating and oviposition. In many male insects, for example, parts of segment 10 are incorporated into complex **copulatory** mechanisms. It is thought that the **oviposition** mechanism of females are derived from appendages of segments 8 and 9.

Internal Morphology

Digestion   The digestive tract is essentially a tube that is divided in three parts (**foregut, midgut,** and **hindgut**) and each part has a distinctive function (Figure 18.5). Food is processed by the mouthparts and carried by **peristaltic** action along a narrow tube of the foregut called the **esophagus** and transported to an enlarged portion called the *crop*. There the food is further ground up in insects that have **proventricular teeth** (e.g., the grasshopper) and then metered by a valve into the midgut. Most of the digestion occurs in the midgut. The midgut is lined with living epithelial cells that secrete digestive **enzymes.** These enzymes extract the vital amino acids, fatty acids, sugars, and other essential nutrients from the food bolus. Nutrients then pass through the permeable wall of the midgut where they are transported by blood to the metabolizing tissues. Undigested food waste is metered by way of a valve into the hind gut where by peristalsis it is conveyed to the rectum. Many insects have **rectal papillae** that extract moisture from the feces

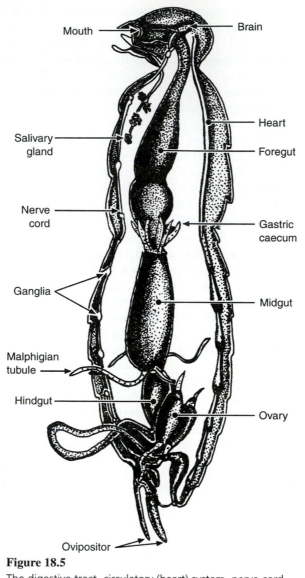

Figure 18.5

The digestive tract, circulatory (heart) system, nerve cord and brain, excretory structures (Malphigian tubules), and reproductive system of a generalized insect.

(Source: Drawings by Eduardo Castro, Forestry Faculty, University of Andes, Merida, Venezuela.)

before it is excreted from the anus as dry **fecal pellets.** Insects that feed on sap, such as true bugs, aphids, and other Hemiptera, are literally plugged into the sap stream of their hosts. These insects have a **filter chamber** that shunts the majority of the incoming sap into the hind gut where it is continually extruded as honeydew.

Circulation   The body of insects is a hermetically closed cavity (**hemocoel**). A tubelike **dorsal vessel** (heart) extends above the alimentary canal from the head to the abdomen. The posterior part of the heart contains several slitlike openings (**ostia**); when the heart is pumped by peristaltic compression of wing-shaped muscles (alary muscles), from the posterior to the anterior, blood enters the ostia, moves forward, and empties into the head above the brain (Figure 18.5). In this manner circulating blood bathes all the hemocoele, carrying dissolved nutrients to all organs and living cells. Insectan blood contains a clear liquid phase (**hemolymph**) as well as nutrient-carrying hemocyte cells and other cells that **encapsulate** foreign bodies and plug wounds. Insect blood is greenish or yellowish. Only in certain insect species, such as midge larvae, which live in waters where oxygen levels are low, is the blood reddish because it contains a **hemoglobin** pigment.

Excretion   As cells metabolize nutrients, wastes are produced, primarily sodium, potassium salts, and nitrogenous by-products (chiefly uric acid). These wastes are released into the blood where they are selectively absorbed by tiny blind-ended tubes, the **Malpighian tubules.** These tubes arise from the anterior part of the hindgut and generally spread throughout the insect body. Thus as the wastes are picked up, they empty directly into the hindgut and are flushed out the **anus** together with the fecal pellets (Figure 18.5).

Respiration   Terrestrial insects directly use atmospheric oxygen to breathe. Air enters through the two thoracic spiracles and the numerous abdominal spiracles. These openings lead to lateral conduits (lateral **tracheae**), which run just inside the body wall. These lateral tracheae ramify into smaller and smaller tracheae called **tracheoles** whose ends ultimately are incorporated into metabolizing tissues. As cells consume oxygen a diffusion pressure deficit is established that draws air into the respiratory system. Some fast-flying insects require bursts of oxygen to bathe their flight muscles; in these cases expansions in lateral tracheae form bladders that force air on the active flight muscles—in effect a supercharger. In many aquatic insects or endoparasites (larvae living inside another animal), parts of the body wall are thin and evaginated (tracheal **gills**) and ramified with tracheae. In aquatic forms oxygen diffuses from the water and through the thinned body wall and into the tracheae.

Reproduction   The gonads of insects are located in the abdomen. Ducts from the gonads (**ovaries** in females and **testes** in males) lead outside near the tip of the abdomen (Figure 18.5). Eggs develop in the ovaries, and as they mature they move down the oviduct toward the **genital** opening (vagina). As they travel near a sperm-storage sac (**spermatheca**), sperm are released and the eggs are fertilized. Accessory glands near the spermatheca may add materials to cover the eggs (e.g., a varnish on overwintering eggs of tent caterpillars) or an adhesive to glue eggs to a substrate. Then as the eggs leave the genital opening, they are placed on or into the required substrate with the external **ovipositor.** In males paired testes produce the sperm, which move down the **seminal vesicle** and as they pass accessory glands they may be provided with a fluid that serves as a carrier. In many insects this fluid hardens and forms a capsule around the sperm, called a **spermatophore.** Then the sperm in its carrier or the spermatophore travel down the ejaculatory duct to the external copulatory organ called the **aedeagus.**

Most insects reproduce through sexual reproduction. But many groups exhibit **parthenogenesis** where offspring are produced without mating. Parthenogenesis may be constant (males are never present), or cyclic (males are present after several generations occur without mating), or **facultative** where males may or may not be present (e.g., among the Hymenoptera, if eggs are not fertilized only males are produced). If eggs hatch after they are deposited , reproduction is **oviparous;** if eggs hatch internally and immatures are deposited, reproduction is **ovovivipa- rous;** and if eggs hatch internally and the female nourishes the young through **pseudoplacenta,** it is called **viviparous.**

Coordination   All active and passive activity of the insects is coordinated by their central nervous system. The system consists of a three-part brain in the head located above the esophagus and near the eyes and antennae. A pair of large con- necting nerves (**circumesophageal connectives**) go around the esophagus and connect with a large ganglion under the esophagus (**subesophageal ganglion**). This major part of the nervous system enervates the eyes, antennae, mouthparts, and the foregut. Paired nerves lead from the subesophageal ganglion to the ven- tral nerve cord that consists of a ganglion at each segment connected by paired nerve cords. However, in many groups the number of ventral ganglia is reduced as individual ganglia coalesce. Small nerves radiating from each ganglion ener- vate the corresponding organs, appendages, and sensory and effector cells asso- ciated with each segment or groups of segments.

Insect Development

When the insect egg is fertilized a **zygote** is formed and after several waves of divi- sion, an **embryo.** As the embryo develops and matures, it consumes the **yolk** for nourishment. Eventually the definitive immature insect escapes from the egg by chewing the eggshell (**chorion**), exerting hydrostatic pressure, or by means of spe- cial structures (e.g., the egg burster of some maggots). At this point the juvenile is dedicated to feeding and growth. However, insects have exoskeletons and thus growth becomes impaired by this relatively inelastic body wall. When further growth reaches this impasse, the insect molts. Molting refers to the ability of insects to shed their exoskeleton and replace it with a larger one. Then after a period of feeding the insect would molt again and so forth until it reaches the adult stage. Insects do not molt after they become adults. The number of times insects molt varies from 2 to 20, depending on the species. The insect itself between molts is called an **instar.** Thus a species that molts six times would have seven instars. The time between molts is a **stadium** and the number of times an insect popula- tion completes a life cycle (goes from egg to adult) is called a **generation.**

Most insects change in form as they go through their different instars. This change is called **metamorphosis** (meta = change; moph = form). There are four types of metamorphosis (Figure 18.6): **no metamorphosis, incomplete meta- morphosis, gradual,** and **complete metamorphosis.** Extremely primitive insects

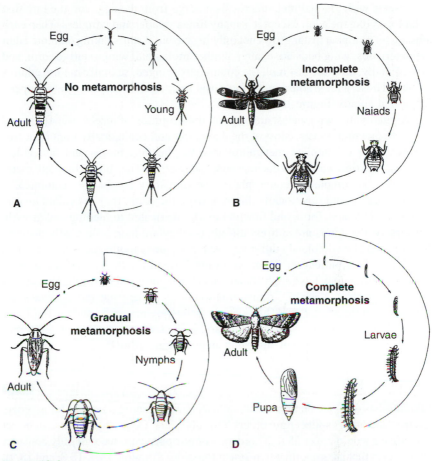

Figure 18.6

The four kinds of metamorphosis found in insects.

(Source: Reproduced with permission from *Trumans Scientific Guide to Pest Control,* figure 2-A, pgs. 24–27. Copyright © by Advanstar Communications, Inc. Advanstar Communications, Inc., retains all rights to these figures.)

that are primordially wingless, such as springtails (order Collembola) and silverfish (order Thysanura) have no metamorphosis (Figure 18.6A). Each instar looks exactly like the previous one except that subsequent instars are larger. The adult looks just like the immatures (nymphs) except that it is sexually mature.

The mayflies (Ephemeroptera), dragonflies (Odonata), and stone flies (Plecoptera) have aquatic nymphs as well as having incomplete metamorphosis (Figure 18.6B). Each instar is larger than the previous, and each subsequent instar starts to take on adult characteristics—for example, wings start to form. The last nymphal instar crawls out of the water, and a completely different-looking insect emerges after this last molt: the adult. In mayflies there is a strange, quasi-adult (**subimago** stage) that emerges; then this adult-looking insect molts again to become the definitive adult.

Insects with gradual metamorphosis emerge from the egg, and the tiny first instar looks like the adult except it is many times smaller and wingless. Then each subsequent nymphal instar grows not only in size but it starts taking on more adult characteristics, wing buds arise; after another molt small wings start to form, and so on. Finally, after the last molt the completely winged, sexually mature adult is formed. Those with gradual metamorphosis include the termites, grasshoppers, thrips, and aphids (Figure 18.6C).

Arguably the two greatest advances in the evolution of insects was flight and complete metamorphosis. Most of the damaging and ecologically important forest insects have complete metamorphosis; the beetles, moths and butterflies, sawflies, parasitic wasps, predacious neuropterans and flies have complete metamorphosis. In complete metamorphosis the immature stages are anatomically, physiologically, and functionally different from the adult; they are called *larvae*. Larvae are in shape, form, and function totally dedicated to feeding and growth. Conversely, the adults are entirely different—they are morphologically distinct; they may feed on completely different hosts (e.g., many moths feed on nectar, and their larvae feed on leaves). Their wings are formed internally, and when sexually mature they may have sophisticated mate selection and host selection strategies. The **pupa** is a resting, transformation stage where the larval tissues are reformed into the adult. Pupae, however, are relatively immobile and the most vulnerable stage of the life cycle. In summary, insects with complete metamorphosis have an egg, larval, pupal, and adult stage (Figure 18.6D).

INSECT ORDERS

Taxonomists disagree on the number of insect orders; some scientists lump orders together and others split certain orders into suborders. Generally a course in insect systematics would cover 25 to 27 orders. For our purposes, we will only consider a few orders that are important to forest entomology (Figures 18.7, 18.8, and 18.9).

The Collembola (Springtails)

Springtails are common in the forest floor where they feed on microorganisms and fungal hyphae associated with detrital breakdown, and as such are key links in the soil mineralization.

These primitively wingless, tiny insects, often less than 4 mm, are great jumpers. They in fact get their common name from a jumping appendage (**furcula**) that arises from the ventral part of the fourth abdominal segment. When not in use, the furcula is bent forward under the third segment of the abdomen, where it is held down under tension by a hook (**tenaculum**). When the collembolan is disturbed, it unclasps the furcula. Tension on the furcula is released, which drives this spring downward and backward and the insect jumps out of danger—it seems to disappear instantly! They have no metamorphosis and chewing mouthparts are within a pouch.

Order	Common Name	Example	Front Wings	Hind Wings
Ephemeroptera	Mayflies		Triangular, membranous; many veins and cross-veins	Smaller, rounded (may be absent)
Odonata	Dragonflies, damselflies		Long, slender, membranous; many veins and cross-veins	Similar to front wings
Plecoptera	Stone flies		Slender, membranous, with numerous veins	Usually wider than front wings; vannus present
Embioptera	Webspinners		Slender, membranous but often smoky; few veins present	Very similar to front wings
Dictyoptera	Cockroaches, mantids, termites		Elongate, often thickened; usually with many veins	Wider than front wings and with vannus (except in most termites)
Orthoptera	Grasshoppers, crickets, katydids, walking sticks		Long and slender, thickened, with many veins (may be absent)	Wider than front wings, membranous, with vannus (may be absent)
Dermaptera	Earwigs		Very short, padlike, leathery (may be absent)	Large, membranous, folding fanlike under front wings (may be absent)
Psocoptera	Barklice, booklice		Membranous, with few veins (may be absent)	Similar to but somewhat smaller than front wings (may be absent)
Hemiptera	True bugs, cicadas, leafhoppers, aphids, etc.		Membranous or thickened, with few veins (may be absent)	Membranous, shorter but often somewhat wider than front wings (may be absent)
Thysanoptera	Thrips		Very slender, with a wide fringe of hairs, few veins or none (may be absent)	Same as front wings
Phthiraptera	Lice		None	None

Figure 18.7

Eleven orders of insects, from the mayflies to the lice.

(Source: From *Insect Biology: A Textbook of Entomology,* by Howard E. Evans; Copyright © 1984 by Addison-Wesley Publishing Company. Reprinted by permission.)

Order	Common Name	Example	Front Wings	Hind Wings
Megaloptera	Dobsonflies, snakeflies		Elongate, membranous, with numerous veins and cross-veins	Similar to or somewhat wider than front wings
Neuroptera	Lacewings, ant lions		Similar to Megaloptera but often with much branching near margin	Similar to front wings but sometimes slightly smaller
Coleoptera	Beetles		Hardened, protective "elytra," which meet in a straight line on back	Membranous, fold complexly beneath front wings (may be absent)
Mecoptera	Scorpionflies		Membranous, slender (especially basally), with numerous cross-veins	Similar to front wings (both pairs may be absent or reduced and modified)
Trichoptera	Caddis flies		Elongate, with few cross-veins, covered with hairs	Similar to front wings but broader and slightly shorter
Lepidoptera	Moths, butterflies		Slender to rather broad, clothed with scales; relatively few cross-veins	Similar to front wings but usually shorter, broader, and more rounded; often attached to front wings
Diptera	Flies, gnats, midges		Membranous with relatively few veins and cross-veins (may be absent)	Absent as functional wings; forming small, knobbed "halteres"
Siphonaptera	Fleas		Absent	Absent
Hymenoptera	Sawflies, wasps, ants, bees		Membranous, with relatively few veins but usually several cross-veins (may be absent)	Smaller than front wings and capable of attachment to them by a series of hooklets

Figure 18.8

Nine orders of insects, from the megalopterans to the wasps, bees, ants, and sawflies: the Hymentoptera.

(Source: From *Insect Biology: A Textbook of Entomology,* by Howard E. Evans; Copyright © 1984 by Addison-Wesley Publishing Company. Reprinted by permission.)

Order	Suborder	Example	Distinguishing Features
Dictyoptera	Blattaria (cockroaches)		Flattened, with a shieldlike pronotum; 2 ocelli; legs fitted for running; cerci with several segments
	Mantodea (mantids)		Elongate; front legs modified for grasping prey; 3 ocelli; cerci with several segments
	Isoptera (termites)	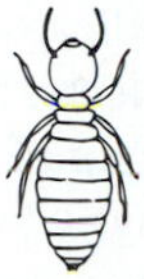	Front and hind wings similar, with a basal fracture line; 2 ocelli or none; cerci very short
Orthoptera	Ensifera (crickets, long-horned grasshoppers)		Jumping hind legs; very long antennae; tympanic organs on front tibiae; long ovipositor
	Caelifera (short-horned grasshoppers)		Jumping hind legs; shorter antennae; tympanic organs on abdomen; short, horny ovipositor
	Phasmida (walking sticks)		Legs slender, fitted for walking; no sound production or tympanic organs; ovipositor small, mostly concealed
Hemiptera	Heteroptera (true bugs)	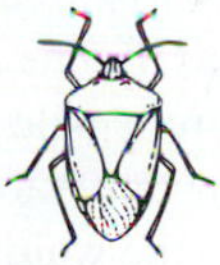	Wings folding flat on abdomen, fore wings thickened basally and membranous apically; beak arises toward front of head
	Homoptera (leafhoppers, cicadas, aphids, etc.)	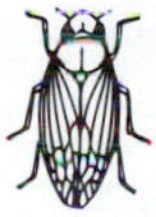	Wings usually sloping over sides of body, fore wings of uniform thickness; base of beak close to front coxae
Phthiraptera	Mallophaga (biting lice)		Biting mouthparts; tarsi with 1 or 2 segments, terminating in 1 or 2 claws
	Anoplura (sucking lice)		Piercing-sucking mouthparts that can be withdrawn into head when not in use; tarsi with 1 segment and 1 large claw

Figure 18.9

Four combined orders that include several suborders.

(Source: From *Insect Biology: A Textbook of Entomology,* by Howard E. Evans; Copyright © 1984 by Addison-Wesley Publishing Company. Reprinted by permission.)

The Ephemeroptera (Mayflies)

Mayfly nymphs are aquatic. They are cylindrical, flattened, or elongated in shape; they mostly have abdominal gills and most families have three tails (**caudal** filaments). These immature stages live under stones and gravels, under decaying organic debris, or cling to rocks of fast-moving streams. They feed on microorganisms that develop on dead leaves and other debris as well as on algae and diatoms. Mayflies are the main **primary consumers** of freshwater ecosystems, they also are important fish food; accordingly, they are a vital part of aquatic food chains. Many ephemeropteran species are intolerant to changing stream conditions and thus serve as stream-health indicators.

Adults are medium sized, delicate, fragile insects with soft bodies and three threadlike tails that arise from the tip of their abdomen. Mayflies have two pairs of many-veined, membranous wings which they hold over their bodies; they cannot fold their wings. The front wings are triangular and larger than the rounded hind wings. Adults have vestigial mouthparts; their antennae are small bristles.

The Odonata (Dragonflies and Damselflies)

The dragonflies and damselflies also are important members of aquatic ecosystems. Both the relatively large, often beautifully colored and patterned adults, and their aquatic nymphs are voracious predators. Adults have two pairs of long, multiveined wings; both pairs are nearly the same length. The dragonflies hold their wings horizontally at rest; damselflies hold them along the abdomen or tilted upward; neither can actually fold their wings flat along their back. Dragonflies tend to be large and robust (25–75 mm long); damselflies are smaller and more delicate. Dragonfly nymphs breathe with internal anal gills while damselfly gills are three frondlike structures arising from the tip of the abdomen. A nymph stalks its prey at the bottom of lakes, ponds and littoral edges of streams. As it nears its prey, it slowly unhinges its prehensile labrum, which is tipped with a pair of movable claws, and at the right instant, rapidly extends the modified labrum to impale the prey and whisk it back to the waiting mandibles. Mature nymphs crawl out of the water and transform into the adult.

The Plecoptera (Stone flies)

Stone flies are the last member of the group of aquatic insects with incomplete metamorphosis. They are small to medium-sized insects with soft flattened bodies. Most species have two pairs of membranous wings with many cross veins. The fore pair is narrower and more elongated than the hind pair. The hind pair usually is folded in pleats when at rest. As a matter of fact, the stone flies are the first insect group that have the necessary hinges (axillary sclerites) to fold their wings flat along their back. They are poor fliers and are found flitting along the banks of streams or rocky lake shores. Their antennae are long with many segments, and they generally have two filaments (elongated **cerci**) at the end of their abdomen. They have chewing mouthparts. Stone fly nymphs have elongated, soft bodies with fingerlike gills on the thorax and at the bases of their legs. Nymphs are found under stones of rapidly moving streams or along rocky lake shores;

some species are algal feeders, others are predaceous and others are omnivores. Aquatic entomologists also use different assemblages of stone fly nymphs to assess stream quality.

The Dictyoptera (Suborder Isoptera) (Termites)

Termites are small to medium-sized, flattened, soft-bodied insects with **castes,** a well-organized social system, and that live in colonies. Each caste performs tasks that are essential for maintenance of the colony. All termites north of Mexico have chewing mouthparts, gradual metamorphosis, and when winged, they have two pairs of membranous, equal-length wings. Their thorax is broadly joined to the abdomen, and they have long beadlike antennae, characteristics that rapidly distinguish them from carpenter ants. The wings are long and when folded along the back they extend beyond the end of the abdomen. Termites feed on cellulose and digest it via the cellulase enzyme produced by protozoans housed within their digestive tract. The reproductive castes are sexually mature males and females, they are fully winged and have compound eyes; their bodies also are darker than the other forms. The reproductives generally have yearly swarms where they mate, chew off their wings along a line of weakness, and begin a new colony. Colonies begin in buried wood, damp wood, old stumps, and large woody debris. The worker caste are nymphs or sterile adults; they are pale colored, lack compound eyes, and have small mandibles. These carry on most of the work of the colony: extending tunnels in wood, building external mud tunnels to traverse exposed areas (e.g., subterranean termites), feeding the queen and brood, and so forth. The soldier caste consists of sterile adults that have enlarged heads to house their exceedingly large mandibles and muscles. When a colony is attacked or a tunnel is broached, the soldiers attack the intruders and often use their massive heads to plug holes punctured in their tunnels or galleries. Termites are largely beneficial as they play an important role in mineral cycling, but when they infest houses and other human structures, they can be extremely destructive. In the United States and Canada, hundreds of millions of dollars are spent annually in structural repairs and termite control.

The Orthoptera and Other Dictyoptera (Crickets, Grasshoppers, Walking Sticks, Mantids, Roaches, etc.)

The winged orthopterans (some are wingless) have two pairs of wings; the front pair often is narrow and parchmentlike, the hind wings are broad and membranous. This diverse group has gradual metamorphosis, chewing mouthparts, and the females of many species have well-developed ovipositors. Crickets (family Gryllidae) are medium-sized orthopterans with long, threadlike antennae; the hind legs are specialized for jumping, and the females have spear-shaped ovipositors. Some crickets are pests of pine nurseries in the southern pine region of the United States. The familiar shorthorned grasshoppers (family Acrididae) are not destructive to trees, but during outbreaks, they not only damage crops but they also defoliate range plants. Walking sticks (family Phasmidae) are long, thin cylindrical insects with very long legs and antennae. They are mostly wingless and at rest they are motionless and closely resemble twigs. The mantids (family Mantidae) are elongated, often large, slow-moving predators. They have large eyes and a large triangular head. The front femora and tibiae are studded with spines and form a set of prehensile pincers. They wait for their insectan prey, and when one comes within reach, the prehensile pincers (armed femora and tibiae) reach out with lightning speed and grasp the hapless insect. Mantids are important elements of biological control of damaging insects. Roaches (family Blattidae) have oval bodies, long hairy legs, a horizontally positioned head, and the females lack a prominent ovipositor. Besides being domestic pests, roaches are found scavenging dead organic matter under the bark of trees.

The Thysanoptera (Thrips)

The thrips are tiny, slender-bodied insects, about 0.5 to 5.0 mm long. Adults are either wingless or have four long, narrow wings with few or no veins and fringed with long hairs. Thrips have rasping-sucking mouthparts and gradual metamorphosis. The upper surface of their mouthparts are etched and form a file. The insect feeds by rasping this file across the cuticle of tender plant tissues, the rasping ruptures the plant cuticle, then the thrips sucks the exposed sap. They feed on many plants, attacking flowers, leaves, fruits, and buds; some even are virus vectors. Certain species, however, are predaceous. Thrips occasionally are serious nursery pests as they damage young tender leaves and shoots—for example, the onion thrips (*Thrips tabaci*) attacks new leaves of conifer seedlings.

The Hemiptera (Suborder Heteroptera) (True Bugs)

The true bugs are sap suckers as they have piercing sucking mouthparts; some species are predaceous. They have gradual metamorphosis, and two pairs of wings that at rest are held flat along the back. Each of the front wings is divided: The upper part is hardened and pigmented; the lower part is membranous (hemi = half, ptera = wing). The hind wing is entirely membranous. Some bugs are damaging to seeds; they stand outside a cone and thrust their stylets through the cone and suck the contents of developing seeds.

The Hemiptera (Suborder Homoptera) (Aphids, Cicadas, Leafhoppers and Planthoppers, and Scales)

The homopterans are all sap suckers. They have gradual metamorphosis and when winged, two pairs of wings. The homopterans are similar to the bugs but are distinguished by their uniformly textured wings, which they hold tentlike over their backs. The stylets of hemipterans appear to arise from the front part of the head; in the Homoptera they arise from the underside of the head. Certain homopterans protect themselves by living under secretions of froth (spittlebugs) or wax (woolly aphids), and others cover themselves with either soft or hard waxy shells (soft and hard scales). Homopterans continuously exude honeydew. Some homopterans have complex life cycles with alternate hosts, bisexual generations alternating with parthenogenetic generations, winged and wingless generations as well as having several forms of the same species. There are many species that are harmful to trees, especially ornamental and street trees.

The Coleoptera (Beetles)

The beetles embrace the largest insect order with perhaps a half million species. Beetles have complete metamorphosis, chewing mouthparts, and two pairs of wings. The front wings, however, easily distinguish the beetles from any other order. The forewings are hardened and act as wing covers that protect the membranous wings, which are folded under the forewings; these hard-front wings are called *elytra*. Beetles associated with trees are a diverse group that includes predators, which as both larvae and adults, feed on other insects; root feeders; foliage feeders, some as both adults and larvae; some feed on foliage only to mature their gonads; flower, cone, and seed feeders; stem feeders, some groups feed externally and others enter stems and trunks and feed in the **subcortical** region; **xylem** feeders of both living and dead trees, and other groups feed in decaying large woody debris. Losses caused by beetles feeding in nurseries, young forests, mature forests, and timber products exceed losses caused by any other destructive agent. Conversely, beetles are linked to both the causes and results of forest ecosystem changes and energy fluxes—for example, when hundreds of square miles of conifer forests are killed by bark beetles (family Scolytidae).

The Trichoptera (Caddis Flies)

Caddis fly larvae live in ponds, lakes, streams, and rivers. Many families are case makers, which they construct from pebbles, twigs, leaves, and other materials; others build nets between stones of fast-moving streams, and others are free-living. The case makers graze on algae and other material, and net builders feed on organic material that drifts downstream. Most of the free-living species are predaceous. Trichopterans are sensitive to changes in aquatic ecosystems and the richness of their taxons are used as biological indicators.

Trichopterans have two pairs of membranous wings, which are held tentlike over their abdomen at rest. The wings are covered with hairs. They have complete metamorphosis and mandibulate mouthparts. Their mandibles are reduced, their palps elongated, and they mostly feed on nectar. Adults are also easily distinguished by their long, filamentous antennae.

The Lepidoptera (Moths and Butterflies)

The lepidopterans are the second most abundant order, and many species cause severe problems to forest management goals. Larvae of some species are **defoliators** of forest and ornamental trees. Certain forest defoliators respond to ecosystem changes with huge population increases, which, in turn, accelerate **successional** changes in the forest; two notorious examples are the spruce budworm (*Choristoneura* spp.), in coniferous forests and the gypsy moth (*Lymantria dispar*) in hardwood forests.

Moths and butterflies have complete metamorphosis; their mouthparts are in the form of a coiled tube, which they use to siphon nectar; and they have two pairs of wings. The wings are covered with small scales that rub off like dust when handled; the wings are triangular in form, and the forewings are larger than the hind ones. Butterflies at rest hold their wings vertically over their bodies; moths hold them horizontally. Other differences are that butterfly antennae are somewhat clubbed, moth antennae are variable; butterflies tend to be day fliers and moths corpuscular or night fliers; and butterfly bodies are slender and delicate, moths are more corpulent. Lepidopteran larvae (caterpillars) are distinctive. They have three thoracic segments and 10 abdominal segments. Each thoracic segment has a pair of jointed legs; typically each abdominal segment, of segments three to six, have a pair of fleshy prolegs (**pseudopods**) as does the last segment. In some groups some or all of the prolegs are absent. The larvae have chewing mouthparts, silk glands associated with the mouth, and several simple eyes. Many larvae use silk to protect their pupae in cocoons, to travel through the air on silken threads and to travel about their hosts in seeking new feeding or pupation sites.

The Hymenoptera (Ants, Parasitic Wasps, Bees, Hornets, Sawflies, and Others)

Hymenopterans are important members of forest ecosystems. Although some—the sawflies, wood wasps, and carpenter bees, feed on foliage, xylem tissues, or perforate wood, others—the parasitic wasps in particular—regulate populations of damaging forest insects. Similarly ants and omnivorous wasp species serve as predators. Bees are important pollinators of meadow and forest ecosystems. Some species are wingless. The winged hymenopterans have two pairs of membranous wings; the hind pair is smaller than the front pair. A set of tiny hooks on the anterior margin of the hindwings serve to couple the fore- and hindwings together when the insects fly. They have complete metamorphosis. Most hymenopterans have chewing mouthparts, but the bees, which ingest liquid food, have a set of lapping mouthparts.

The Hymenoptera is divided into two suborders, the Apocrita and the Symphyta. The latter includes the sawflies and woodwasps; the Apocrita are the wasps, ants, bees and the rest of the order. Differences between the two suborders are as follows:

1. The thorax of the Symphyta is broadly joined to the abdomen. In the Apocrita the thorax is joined to the abdomen by a constricted basal segment of the abdomen called the *propodeum.*
2. Almost all Symphyta larvae are phytophagous; many of the Apocrita are parasitic.
3. The ovipositors of sawflies are sawlike for cutting slits in foliage or awl-like in woodwasps for inserting eggs into woody tissues. Ovipositors of the Apocrita range from long and flexible for inserting eggs into host insects (often first having to penetrate bark and other plant tissues) or modified into offensive or defensive stingers.
4. The larvae of sawflies are caterpillarlike (eruciform); those of the Apocrita are grublike or maggotlike. The larvae of sawflies can be distinguished from those of moths and butterflies by having more than five pairs of prolegs; prolegs are without **crochets,** and they have one pair of ocelli.

The Neuroptera (Lacewings, Antlions, Alderflies, and Dobsonflies)

This order contains a variety of groups; some have aquatic larvae—for example, the alderflies (family Sialidae) and dobsonflies (family Corydalidae), these are important predators of aquatic insects. In fact all neuropterans are predators and efficient biological control agents. The lacewings (family Chrysopidae), for example, are one of the main natural population regulators of aphids and are commonly called *aphidlions*. Adult neuropterans have two pairs of large, membranous wings that they hold tentlike over the abdomen while at rest. They have complete metamorphosis and mandibulate mouthparts. The predaceous larvae have long and slim, needlelike mandibles by which they impale their prey and suck out the liquids from their victims.

The Diptera (Flies)

Flies are the most evolved insects; they fill many niches and do so by being specialists at what they do. Many species are destructive agricultural pests and others are forest pests of nurseries, seed orchards, and plantations; they feed within tissues and are called leaf, bud, and seed miners, gall makers, or twig, branch, and root borers. Blood-sucking forms are not only nuisances to humans and their animals (e.g., mosquitoes, blackflies, no-see-ums, deer flies, and others) but, in tropical areas, many of these are vectors of debilitating diseases such as malaria, yellow fever, dengue, sleeping sickness, filariasis, typhoid fever, dysentery, and others. Other flies are scavengers of dead plants and animals, and some feed on dung. Others are predators and parasites of destructive pests, and some species of these have been introduced as biological control agents.

Winged dipterans have one pair of wings; the second pair (the hind pair) are reduced to small knobbed structures called **halteres**. Halteres are balancing organs and used in intricate flight maneuvers, such as flying in place like a helicopter. Flies have complete metamorphosis. Their mouthparts range from piercing (as in the horseflies and mosquitoes) to sponging (as in the houseflies) to nonfunctional as found in botflies. Dipteran larvae are legless, wormlike, and feed by use of mouth hooks; they are called *maggots*. The exoskeleton of the last instar serves as a pupal case and is called a *puparium*.

REFERENCES

Atkins, M. D. 1978. *Insects in perspective.* Macmillan Publishing Co., New York.

Borror, D. J., D. M. DeLong, and C. A. Triplehorn. 1981. *An introduction to the study of insects.* Saunders College Publishing, New York, 5th ed.

Chapman, R. F. 1979. *The insects: structure and function.* Elsevier North Holland, New York.

Evans, H. E. 1984. *Insect biology: a textbook of entomology.* Addison-Wesley Publishing, Menlo Park, Calif.

Gara, R. I. and F. E. Greulich. 1995. Shipworm (*Bankia setacae*) activity within the Port of Everett and the Snohomish River estuary: defining the problem. *The Forestry Chronicle* 71:186–191.

Pedigo, L. P. 1999. *Entomology and pest management.* Prentice-Hall, Upper Saddle River, N.J.

Romoser, W. S., and J. G. Stoffolano, Jr. 1994. *The science of entomology.* Wm. C. Brown Publishers, Dubuque, Iowa.

19

PRINCIPLES OF FOREST INSECT MANAGEMENT

INTRODUCTION

Public land policy in the United States encourages biodiversity. This policy protects threatened and endangered species and safeguards air and water quality. The central aim is to preserve biodiversity across landscapes: the intent of ecosystem management.

In the evolving concepts of ecosystem management, timber production is no longer emphasized, it is just one output of a managed ecosystem. Because of this uncertain timber supply, Pacific Coast timber companies have been enlarging their southern operations, considering the importation of raw material from other countries (Orr 1994), and practicing intensive forestry on their own land base. Intensive forestry means managing forests with short **rotations** (e.g., 45–60 years) in conjunction with tree improvement programs, optimized nursery production, **vegetation management,** programmed **thinning** and **pruning** schedules, forest fertilization, and so on. This is capital intensive forestry.

In accordance with a focus on ecosystem stewardship, the practice of forest entomology on public lands will be to protect the mandates of federal policies with the comprehensive mission of maintaining healthy forest ecosystems. In the context of private forestry, the purpose of forest entomology is to protect capital investments necessary to produce fast-growing, high-quality trees. These two missions, however, overlap as timber production from federal lands always will be substantial and responsible timber companies will be obliged to maintain the productivity of their lands (Figure 19.1).

PUBLIC FORESTRY AND FOREST HEALTH

Ecosystem management depends on knowing the key interactions that support a particular forest ecosystem. As seen in Figure 19.2, a forest site in combination with the dynamics of competition, and understory development, is responsible for the structure and diversity of the forest at any given time. As succession proceeds, the

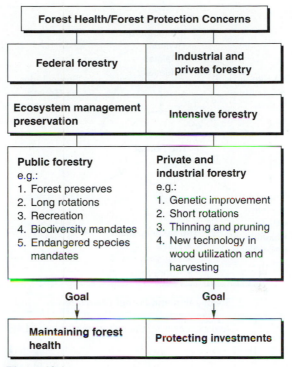

Figure 19.1
Roles played by the disciplines of forest protection or forest health within the context of public and private forestry. Public land managers practice ecosystem management and deal with maintaining the biodiverse structures and the natural integrity of the landscape. The professions of forest entomology, pathology, and fire ecology serve to maintain the tenets of ecosystem management by understanding and manipulating disturbance events. Private and industrial forest entomologists, pathologists, and fire managers protect the financial investments associated with intensive forest practices.

health or vigor of the forest declines in response to factors of competition (e.g., among trees for light, water, and minerals) and weather (e.g., a continuing drought). Later, combinations of weather and stands of low vigor would prepare the forest for major disease and insect outbreaks as well as provide fuel accumulations for wildfires. Stand mortality would favor the **decomposer** chain, and in time the forest would regenerate—a new stand replacing the dead and dying forest. Forest insects, pathogens, and wildfires kill stands, which in turn affect natural succession and plant diversity. These disturbances create habitats for a variety of wildlife species. Dead trees are linked tightly to nutrient cycling, which contributes to the

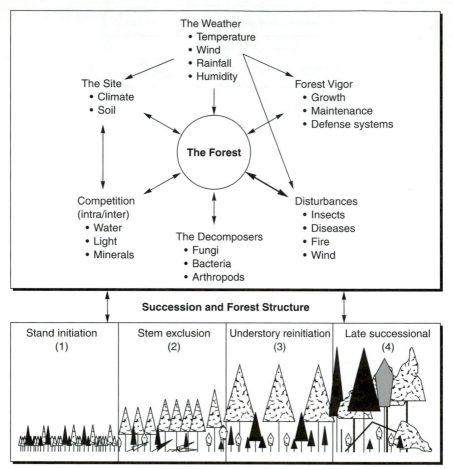

Figure 19.2

A simplified portrayal of a forest ecosystem and forest succession. The type of forest present in an area is defined by the site (topography, climate, and soil). As the forest develops there is strong competition for light, water, and minerals. Competition leads to change of forest composition, structure, and vigor. As the vigor of the forest declines and, under suitable weather conditions, the factors of disturbance (e.g., insect outbreaks, incidence of diseases, and fires) serve to create openings across the landscape. Forest mortality caused by disturbances energizes the decomposer chain, which reintegrates nutrients back into the forest ecosystem. After disturbances create openings across the landscape, pioneering species invade the sites (1), develop quickly and compete intensively for water, light, and minerals and the reproducing stand thins. In time (2) there is severe competition for light as tree crowns close and suppressed trees die. Later, dominant and codominant trees dominate, and (3) understory, shade-tolerant species continue to grow and understory structure develops from advanced reproduction. As openings in the overstory are created, the shade-tolerant species grow into these spaces, and the multilayered, species-diverse old-growth structure develops (4).

long-term productivity of forest ecosystems. Ecosystem management is based on understanding these key processes. In this context managing forests to optimize forest vigor and reduce stand mortality would be a goal of ecosystem management.

TIMBER MANAGEMENT

Forest insects affect management goals by causing catastrophic losses and by lowering growth rates of managed forests. Catastrophic losses are unpredictable and characterized by great severity with major timber losses across thousands of acres. These losses dislocate long-term forest management and utilization plans of the affected areas. Although less dramatic, growth losses and regeneration delays also can frustrate management goals by lengthening forest rotations.

All forest insects play important ecological roles, either as pollinators, herbivores, decomposers, or **parasitoids** and predators of other insects; a few weaken, kill, or degrade trees. The ones that kill or degrade are the pests, which often must be managed. Some can be periodic pests that occasionally cause **economic damage;** others never do and are called *latent pests.* A few cause yearly damage and are called *chronic pests* (Figure 19.3).

Economic damage can be correlated with the level of an insect pest population that kills trees or affects tree growth to the point where the forest is producing less than the predicted annual **increment.** Economic damage also can be based on the value of a forest commodity, such as seeds, nursery stock, or ultimately the lumber. The Douglas-fir tussock moth (*Orgyia pseudotsugata*) has a 9-year outbreak cycle (Clendenen et al. 1978). These outbreaks cause economic damage to Douglas-fir and true fir stands when high populations defoliate forests to the extent that annual growth is significantly reduced (Wickman 1979). As annual increment decreases, the length of the desired rotation increases. Cone and seed insects that damage seeds in a **seed orchard** often are chronic pests. In this case the genetically selected seeds are expensive and have had a high developmental cost. Accordingly even a small amount of annual cone and seed insect damage would cause economic damage (see Box 19.1).

The control of insect populations is based on a fundamental relationship associated with the reproductive potential of an insect species. This relationship is represented by

$$Mq = \frac{(\text{Fecundity})(\text{Sex ratio}) - 1}{(\text{Fecundity})(\text{Sex ratio})}$$

Mq = mortality quotient
Fecundity = average number of eggs a species lays
Sex ratio = number of females in a sample
of the population divided by the number
of males and females in the sample

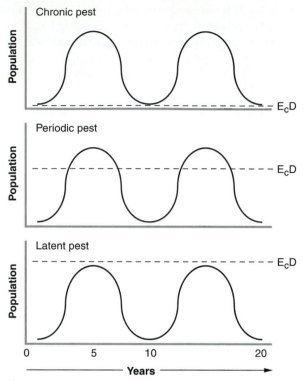

Figure 19.3
Relationship between the level of economic damage (E_cD) and whether an insect is considered as a chronic, periodic, or latent pest. A chronic pest is always causing economic damage, a periodic pest causes economic damage from time to time, and a latent pest historically never causes economic damage.

An example: The Douglas-fir tussock moth lays about 300 eggs. A pupal sample of 1,000 insects revealed that 500 were males and 500 were females; the sex ratio consequently was 0.5. Therefore 99.3% of the population would have to die to prevent the population from increasing further. In other words, 298 of the offspring would have to die (0.993×300) out of each egg mass in order that the survivors would be a male and a female.

INSECT PEST MANAGEMENT

Forest insect outbreaks are controlled by the following methods: (1) mechanical and physical control, (2) chemical control, (3) biological control, and (4) **integrated pest management (IPM)**.

19.1

A hypothetical illustration that shows how a seed orchard manager would determine the level of economic damage caused by the Douglas-fir cone moth, (*Barbara colfaxiana*).

Production costs at a seed orchard (costs/acre)

    Equipment depreciation
    Field labor
    Cultural operations
        Pollination
        Cone bagging
        Rouging
        Fertilizing
        Irrigation
    Miscellaneous

Subtotal	$ 239.20
Cone moth control	80.00
Marketing	10.00
TOTAL COSTS	**$ 359.20**

Expected seed production 850 lb/acre

To make costs, the selling price would
be $0.423/lb, i.e. (850 lb/acre)($0.423) $ 359.20

**To recover costs and make 100% profit,
the market price will be $42.30 + $0.42** $ 42.72/lb

Because of yearly cone moth problems, the seed company has developed the following table:

Production (lb/acre)	Cost/lb	Market price for 100% profit
850	$0.42	$42.72
800	0.45	45.35
750	0.48	48.38
700	0.51	51.81
650	0.55	55.85
600	0.60	60.50

In 2000 the maximum price that the market would bear was $50.00/lb. In terms of seed orchard production, the *economic damage level* was about 725 lb/acre. If production were predicted to fall below 725 lb/acre the cone moth population would have to be reduced—most likely by chemical control.

Mechanical and Physical Control

These methods destroy the insects themselves or change their habitat. In Vietnam and China, for example, light traps are used for pest monitoring of the pine defoliator, *Dendrolimus punctatus,* and hand-picking is used to lower population numbers (McFadden et al. 1981). During a 1990 outbreak in Vietnam, 1,800 workers collected an estimated 50 tons of *D. punctatus* larvae from young plantations of *Pinus massoniana.* This kind of control is effective where labor is inexpensive. In North America cutting and scattering logging slash to prevent buildup of pine engraver beetles (e.g., *Ips pini*) is useful in preventing outbreaks. In this case, larvae of engraver beetles within the slash cannot survive exposure to high temperatures and desiccation. Positioning water dispersion systems above piles of stored logs is effectual in preventing attacks of wood borers (e.g., beetles of the families Cerambycidae and Buprestidae) and pin hole borers or ambrosia beetles (the family Scolytidae) (Figure 19.4). The wet logs are unattractive to these insects as females perceive that the wood would be inhospitable to larval development. Heat treatment of logs in the holds of ships is considered as a way to prevent the importation of forest pests from other countries. Kiln drying of lumber ensures that insects living within the material are killed.

Chemical Control

The usefulness and great variety of organic insecticides has led to considerable dependence on chemical control in agriculture (Ware 1989, Romoser and Stoffolano 1994, Thompson 1994, Bohmont 1997, Pedigo 1999). In forestry, other than in nursery, seed orchard, and other croplike forestry activities, use of insecticides alone to control insects has had marginal results. Chemical control of pests in agriculture is a two-dimensional operation (length and breadth of the crop), but silviculture deals with four dimensions: the length, breadth, height, and duration of the crop. It has been difficult to reduce pests below the level of economic damage in this multidimensional environment by use of chemicals alone. In combinations with other control tactics (i.e., an IPM approach), however, chemical control is useful.

Insecticides are commonly classed as to how they enter the target pest: by contact (enters the insect through the exoskeleton), by ingestion (stomach poison), or by respiration. A more modern designation is based on how a pesticide kills the organism: the mode of action. Although many insecticides are protoplasmic precipitators (e.g., arsenicals, fluorides, methyl bromide), nerve poisons (e.g., organochlorines, carbamates, and organophosphates), respiratory inhibitors (e.g., refined oils and soaps), the exact ways in which toxicants interfere with physiological processes often are poorly understood.

In the use of pesticides, attention must be given to the hazard and toxicity of the products. Hazard refers to the likelihood that under the prescribed condition of use (as specified on the label) application of a chemical will cause injury. Basically hazard is a function of the **toxicity** of the product, the exposure, and the rate of exposure. Toxicity of a pesticide refers to the innate ability of a chemical to cause injury; the toxicity is a fixed property of the chemical. The toxicity of pesticides to mammals is based on extensive dosage/response laboratory tests with rats and other laboratory animals. One of these test series determines the acute toxicity of

Figure 19.4
Sprinkler system used to protect Douglas-fir logs from wood borer and ambrosia beetle infestations.
(Photo supplied by Canadian Forest Service.)

a product and is expressed as a dose that is expected to kill 50% of a test population of rats (the LD_{50}) and is quantified in milligrams per kilogram of body weight. The LC_{50} is an expression of acute toxicity for fumigants and expressed as milligrams per liter. The EPA uses toxicity values to separate insecticides into four categories on the basis of their acute oral, **dermal,** and respiratory toxicities. These categories alert the user to potential hazards of these pesticides (Table 19.1). Although there is not a direct relationship between toxicity and hazard, it is important to obey the label. What's more, many insecticides are toxic to other organisms (e.g., hymenopterous pollinators), so pesticide labels carry additional detailed and relevant information on the proper use of a specific insecticide. Moreover, labels are legal documents and users are obligated to apply the product in accordance with their instructions. Characteristics of some commonly used insecticides follows.

The Botanicals   These are insecticides and **acaracides** (mitecides) derived from plants and include pyrethrins, rotenone, limonene, sabadilla, ryania, and others. Pyrethrins are extracted from chrysanthemums grown in Africa and South America. They have a rapid knock-down effect (i.e., have an immediate effect on the target's nervous system), an LD_{50} of about 1,500 mg/kg, but they are expensive and easily decompose in sunlight. Rotenone is extracted from the tropical *Derris* plant; it is nonphytotoxic and has an LD_{50} of 350 mg/kg. It is effective for about 3 days, and in forestry it is useful against nursery pests: sucking insects as well as thrips, defoliating caterpillars, and cutworms. Sabadilla has been useful in control of nursery defoliators of tropical hardwoods, but it's ineffective against aphids. Ryania, a pesticide derived from a tropical lily, has a demonstrated specificity against defoliating caterpillars and might be used in an IPM program where broad-scale pesticides would not be useful. Limonene is a botanical extracted from lemon peels. It is a new pesticide and has such a high LD_{50} that it is effectively used as a domestic animal dip. This product would be a useful tool in silvipastoral projects.

The Organochlorines (Chlorinated Hydrocarbons)   These were the first synthetic organic insecticides. They were on the market by the mid-1940s with DDT being the most famous. These are nerve poisons (the exact mode of action is still unclear) and based on chlorination of the hydrocarbons. Most organochlorines are

T A B L E **19.1** | **Toxicity Classification of Insecticides**[a]

Hazard indicators	I (Danger—poison)	Toxicity categories		
		II (Warning)	III (Caution)	IV (Caution)
Oral LD_{50}	Up to and including 50 mg/kg	From 50–500 mg/kg	From 500–5,000 mg/kg	>5,000 mg/kg
Inhalation LC_{50}	Up to and including 0.2 mg/L	From 0.2–2 mg/L	From 2–20 mg/L	>20 mg/L
Dermal LD_{50}	Up to and including 200 mg/kg	From 200–2,000 mg/kg	From 2,000–20,000 mg/kg	>20,000 mg/kg
Eye effects	Corrosive; corneal opacity not reversible within 7 days	Corneal opacity reversible within 7 days; irritation persisting for 7 days	No corneal opacity; irritation reversible within 7 days	No irritation
Skin effects	Corrosive	Severe irritation at 72 hours	Moderate irritation at 72 hours	Mild or slight irritation at 72 hours

[a]Based on acute oral, respiratory and dermal toxicity, as well as effects on eyes and skin. In part, information needed to determine the potential hazard associated with each toxicity category.

Source: From "EPA Pesticide Programs, Registration and Classification Procedures, Part II." Federal Register 40:28279.

not recommended for use in forestry. Although their mammalian toxicity generally is low, they are nonspecific, persistent, and have a high affinity for milk and the fatty tissues of vertebrates—the inherent basis for biomagnification in food chains. Methoxychlor, although a broad-spectrum insecticide, has a low affinity for milk and fatty tissues and has a relatively safe LD_{50} of 6,000 mg/kg. It is useful in controlling important greenhouse pests as well as in foggers to control biting flies. Dicofol (Kelthane) is a compound structurally similar to DDT, but it has low insecticidal properties and serves as an acaracide.

The Organophosphates (OPs)   The mode of action of the OPs is known and related to nerve poisons developed for military use. Organophosphates are based on phosphoric acid; some have sulfur instead of oxygen at one or two of the five phosphorus bonds. The ones having sulfur can be called *organothiophosphates.* All of these insecticides irreversibly inhibit **cholinesterase** (acyetylcholinesterase), thereby interrupting nerve impulses across synapses (Figure 19.5). The organophosphates are in contraposition to the organochlorines; OPs are chemically unstable (nonpersistent); some are selective, and they have an extensive toxicity range. For example, parathion has an LD_{50} of 2 mg/kg; acephate's LD_{50} is 700 mg/kg. Acephate (Orthene) has been used to control forest defoliators; many of the other OPs are registered for the control of forest and ornamental trees. Several OPs are systemic—they are taken up by the sap stream and as such kill sap-sucking and rasping insects (e.g., aphids, bugs, and thrips), while not affecting parasites and predators.

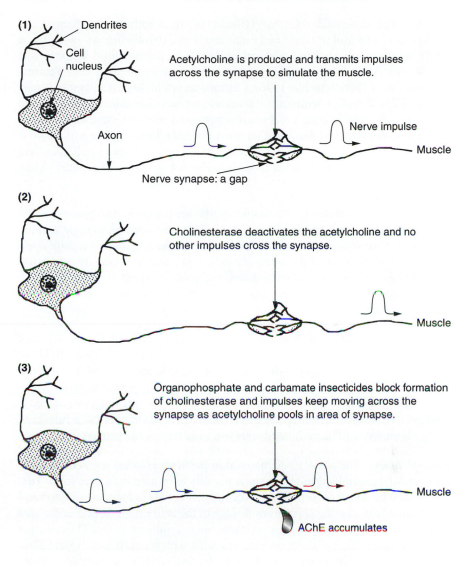

(1) Dendrites

Cell nucleus

Acetylcholine is produced and transmits impulses across the synapse to simulate the muscle.

Axon

Nerve impulse

Muscle

Nerve synapse: a gap

(2)

Cholinesterase deactivates the acetylcholine and no other impulses cross the synapse.

Muscle

(3)

Organophosphate and carbamate insecticides block formation of cholinesterase and impulses keep moving across the synapse as acetylcholine pools in area of synapse.

Muscle

AChE accumulates

(4)   All the insect's nerves are thus continually firing impulses across their synapses and the muscles are continually activated until the organism dies.

Figure 19.5

The mode of action of organophosphate, organothiophosphate, and carbamate insecticides. These insecticides kill animals by blocking formation of cholinesterase.

The Carbamates   These insecticides are derived from carbamic acid and once applied break down readily, but in general not as rapidly as OPs. They also are cholinesterase inhibitors, but as a general rule (with some notorious exceptions such as Temik) less toxic than the OPs. The carbamates are more selective than the other organic pesticides. Carbaryl (Sevin) is a frequently used carbamate and registered as a control of forest and ornamental tree defoliators. As examples, a 2,530 acre western spruce budworm outbreak in north-central Oregon was treated with Sevin-4-oil in 1993 and a similar area was treated near Yakima, Washington, in 1996 (Sheehan 1996). Carbofuran is a highly toxic, systemic carbamate (LD_{50} 8 mg/kg), which also serves as a nematocide and is primarily used to control soil pests. Experimental use in a sustained-release formulation showed that carbofuran could be dissolved within a polymer, buried in the soil where it would deliver systemically minute amounts of the pesticide into small trees, and thus protect them from a tropical shoot borer attack for up to 13 months (Allan et al. 1974).

The Synthetic Pyrethroids   These insecticides are a great improvement over the natural pyrethrins. The pyrethroids are not readily degraded by sunlight and are effective against insects at extremely low doses. Ambush is used to control nursery pests at 0.1 to 0.2 lb of the active ingredient per acre, is relatively safe to use (LD_{50} 247 mg/kg), has a fast knock-down, but is toxic to bees and fish. Research on pyrethroids is intense and new, more selective and efficient products will periodically become available.

The Fumigants   These insecticides are applied in a gaseous form and are used where the vapor can be enclosed as in greenhouses, storage facilities, holds of ships, or under polyethylene tents, or when introduced into the soil. These compounds are usually heavier than air and easily penetrate any porous material. Many of the fumigants are based on halogens (bromine, fluorine, or chlorine) and pose a danger to atmospheric ozone—for this reason, use of the most popular fumigant, methyl bromide, will be prohibited soon after the turn of the century.

Formulations   The actual toxic material in pesticides (called the **active ingredient** or **A.I.**) generally is too toxic and too costly to be used in its pure form. **Toxicants** must be mixed or formulated with other materials to facilitate dilution and an even distribution at the appropriate dosage of the active ingredient over the area treated. In addition, most modern toxicants are insoluble in water. Thus another reason for formulating is to disperse the pesticide with water (it is cheap and plentiful). The insolubility problem is solved by having particles suspended in water or mixed with water in an emulsion. The most common formulations follows.

> **The dusts.** These are formed by mixing the toxicant with finely ground plant materials, such as wheat, soybean, or walnut shells, or more commonly with talc, clays, **diatomaceous earth,** or even sulfur (an insecticide in its own right); these substances are the carriers. The molecules of the toxicant adhere to the carrier particles. The carrier therefore serves as a diluent (toxicant is <10% of the mixture) and as a means to disperse the pesticide on the crop (e.g., nursery plants). Dusts can be applied with simple dusters and give good coverage. Disadvantages include the dangers of inhalation and problems of drift and particulate air pollution. Coarse or granular formulations of dusts have a larger inert

particle size and a lower concentration of the toxicant. Granules are convenient to dispense with a fertilizer dispenser or with granule applicators. Granules often are formulated with systemic insecticides, and when placed on or in the soil gradually provide the active ingredient to plant roots where it is translocated to the rest of the plant.

The wettable powders (WPs). The toxicant adheres to the same kind of carriers as the dusts but at concentrations that range from 25 to 75%. The toxicant-dust mixture plus a wetting agent (a surfactant, a compound that allows the powder to enter the water by breaking the surface tension) is then diluted further by suspension in water. The suspension is applied to the crop. The main problem with WPs is that they must be agitated to prevent settling, so the application equipment is more complex than for dusts or granules.

The emulsifiable concentrates (ECs). These are the most common formulations. Emulsifiable concentrates consist of the toxicant dissolved in a solvent plus an emulsifying agent, essentially a detergent to break the water-oil interface. When the EC is mixed with water, the water turns cloudy as an **emulsion** is formed. Upon application to plants, the solvent volatilizes, leaving a film of the toxicant on the foliage after the water evaporates.

Other formulations. There are several other formulations. Water-soluble powders (SPs) are made from the toxicant adhering to such finely ground clays that a solution is formed when water is added. There are oil **solutions** where the toxicant is dissolved in the oil (frequently diesel oil), and the oil then serves as a diluent and dispensing agent. Ultra-low-volume concentrates (ULVs) are the toxicants themselves in a liquid form or dissolved in a small amount of solvent. They are applied without further dilution by special aerial or ground spray equipment that atomizes the insecticide into micron-sized particles. In this way as little as an ounce of chemical may be spread uniformly over large areas. Since no water or other diluent is needed, airplanes or helicopters can treat a much greater area with a single load of material.

Semiochemical Control

Many forest insects attract mates, locate hosts, colonize susceptible plant material, and determine suitable oviposition sites by producing and/or receiving chemical messages. These materials are known as **semiochemicals** (Figure 19.6). There is a great deal of research on the isolation, identification, and synthesis of semiochemicals as a way to manage pests (Norlund 1981, Roeloff 1981, Skillen et al. 1997, Goyer et al. 1998). Chemicals that are useful in pest management are synthetic sex attractants, **aggregants** as well as repellents, feeding deterrents, and other **kairomones.**

Sex **pheromones** are commonly used to monitor insect activity. A sex lure (synthetic pheromone) is placed in a sticky trap, and the number of insects captured is used to compute the pest population as well as the need for, and timing

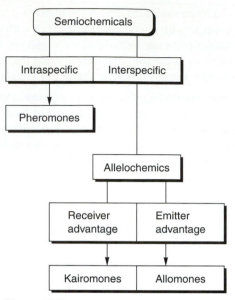

Figure 19.6

A classification system for the semiochemicals.
Pheromones evoke responses among members
of the same species (intraspecific responses).
Allelochemics elicit responses between species
(interspecific responses). Allelochemics which
are produced by one species but the receiver
(another species) has an advantage are
kairomones (e.g., a sex pheromone of one
species also may attract a parasitic species; the
compound is a kairomone to the parasitoid). If
a compound produced by a species protects it
from invasion of an other species, the chemical
is an allomone (e.g., repellents, antifeeding
compounds, poisons, and hormone mimics).

of, suppression programs. Sex attractants are available for many lepidopterans, and
when dispensed in trapping grids they are used for locating the concentration of
pests (e.g., useful in locating isolated populations of gypsy moths). In addition,
they are used in population suppression by trapping out males or in pest control
by having so many individual sources of the sex lure that the males are confused
and can't find the females. An example of this control tactic was the successful
mating disruption of male western pine shoot borers, *Eucosma sonomana*, by the
use of synthetic sex pheromones (Overhulser et al. 1980, Daterman 1990). The
synthetic sex attractant was used to fill thousands of tiny plastic fibers (1.5 cm long
$\times$ 0.3 cm in diameter), which were released over an infested ponderosa pine plantation. After the application there were about 3 fibers m^{-2} of foliage over the treated
areas, each dispensing small amounts of the powerful sex pheromone. Based on

moth responses to pheromone-baited traps, there was 100% disruption of normal female-to-male communication and about an 80% reduction of infested shoots the year following application.

Aggregating pheromones are chemicals used by bark beetles to perpetrate a mass attack on a host tree. Mass attacks serve to overwhelm the defense systems of their hosts and render an attacked tree suitable for oviposition and brood development. Many of the aggregants have been synthesized and are used to lure bark beetle populations away from susceptible stands or lure thousands of beetles to traps or to trap logs. Certain bark beetle species also produce **antiaggregants** (species-specific repellents) as part of their host selection behavior, and synthetic antiaggregants also are used to manage bark beetle populations. An example would be the use of 3-methyl-2-cyclohexen-1-one (MCH) the antiagreggant of the Douglas-fir beetle (*Dendroctonus pseudotsugae*) (Goyer et al. 1998). Synthetic MCH, when dispensed at 50 to 100 mg $hr^{-1}ha^{-1}$, prevents logs from being attacked by beetles that were initially attracted to the hosts. Use of antiaggregating pheromones has the advantage over attractant pheromones of not bringing in thousands of insects to the area being protected.

Biological Control

This type of control is the manipulation of organisms for the purpose of reducing or checking populations of destructive insects (Smith and van den Bosch 1967, Embree 1971, Ryan 1997, Pedigo 1999). **Biological control** does not result in immediate reductions of pest populations, as does the use of insecticides, but over a long time period it may be more effective, economical, and upsets the ecosystem less than chemical control. Biological control includes the use of resistant strains or varieties of plants that have been developed through applied genetics. More traditionally it refers to the suppression of pest populations through the use of predators, parasitoids (called *parasitoids* since "insect parasites" always kill their insectan hosts and only the immature lives in or on the host), and microbial pathogens. The principal distinguishing feature of predators, parasitoids, and pathogens is their potential to respond to increasing pest populations in a density-dependent manner. The concept of density dependency describes the way key environmental factors regulate insect populations. If the intensity by which an environmental factor kills insects is unaffected by the number of individuals present, the factor is said to be density independent. Density-dependent factors limit insect numbers in concert with the numbers present. Examples of density-independent mortality factors would be temperature and humidity extremes, intense rainfall, noncontagious diseases, and insecticides; density-dependent mortality factors would be insect parasitoids, predators, infectious diseases, and food quantity (Table 19.2).

Biological control has been most effective in regulating the numbers of introduced pests. Pests introduced into North America from other countries all too often prosper and dramatically increase their populations over short periods. This is because they are introduced into an ecological system similar to the one left behind, but without their complex of enemies. Conversely, if a key parasitoid or

T A B L E **19.2** | **Difference Between Density-Dependent and Density-Independent Mortality Factors**

	Density dependent	Density independent
Operation	These factors are always killing the host because their existence is dependent on the host.	These factors kill for short periods and independently of the number of host insects present.
Examples	Insect parasitoids, infectious viral, bacterial, or fungal diseases, and food quantity.	Temperature and moisture extremes, windstorms, heavy rain, and insecticides.
Control	Given: 50,000 insect pests, and there is a 90% control, which leaves 5,000 survivors.	Given: 50,000 insect pests, and potential there is a 90% control, which leaves 5,000 survivors.
	The population grows to 500,000 but the kill then increases to 99%, leaving 5,000 survivors.	The population grows to 500,000 but the kill remains 90%, leaving 50,000 survivors.

predator of an introduced pest is successfully brought in from its native land and successfully established, then this density-dependent control agent would concentrate its efforts on its main host, the introduced pest. There are many historical successes in this kind of biological control—a contemporary one is the control of the introduced larch casebearer (*Coleophora laricella*) by ensuring establishment of its key larval parasitoids, *Agathis pumila* and *Chrysocharis laricinellae* (Figure 19.7) (Ryan 1987, Ryan et al. 1987). The larch casebearer first appeared in 1886 in Massachusetts, presumably in ballast piles from European sailing ships. By the turn of the century, the casebearer became a major defoliator of eastern tamarack (*Larix laricina*). After the casebearer was discovered in northern Idaho in 1957, the tiny moth soon became the most serious insect enemy of western larch (*Larix occidentalis*). Between 1979 and 1985, *A. pumila* and *C. laricinellae* were introduced, became established, and by 1980 the parasitoids were regulating the casebearer population. Although the two parasitoids were introduced at about the same time, *A. pumila* proved more successful (Figure 19.8).

The winter moth (*Operophtera brumata*) was introduced to Nova Scotia in about 1930. As the moth caused considerable damage to oak forests, a biological control program was started in 1954 with the introduction of several natural enemies from Europe. By 1970 the winter moth populations were under control by an introduced tachinid fly (*Cyzenis albicans*) and a larval parasite, *Agrypon flaveolatum*. Research has convincingly shown that it was the action of these introduced control agents that stopped further spread of *O. brumata* in the Canadian maritime provinces. In the early 1970s the moth was introduced into Victoria, British Columbia, and into western Oregon. The same parasites that controlled the

Figure 19.7
An introduced parasitoid of the larch casebearer, the eulophid *Chrysocharis laricenellae,* laying an egg in a larch casebearer larva; the eulophid is seen penetrating the casebearer's case with its ovipositor.
(From R.B. Ryan 1987)

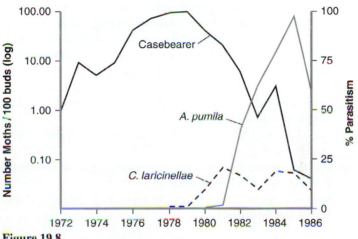

Figure 19.8
The density-dependent relationship and, ultimately, the regulation of larch casebearer populations near Lostine, Oregon, by two species of introduced parasitoids, *Agathis pumila* and *Chrysocharis laricenellae.* In the absence of natural controls, the casebearer population built up between 1972 and 1978 to more than 100 insects per 100 buds. The two parasitoids were released between 1979 and 1985; both species caused a steady decline in casebearer populations.

problem in Nova Scotia were introduced into the affected areas of western North America, and once more winter moth populations are being successfully regulated (Hulme and Green 1984, Roland and Embree 1995).

Introducing and establishing biological control agents to control an exotic pest is called *permanent biological control* and has these advantages:

1. The control is self-perpetuating; once established it's less costly than chemical control, although research costs associated with its establishment are high.
2. Permanent biological control is selective and density dependent, and it produces minimal disturbance to the rest of the ecosystem.
3. Permanent biological control does not create new problems, as chemical control does—that is, there are no chemical residue problems, no water and soil pollution, and no resistance problems.

On the whole permanent biological control has been successful in regulating populations of exotic pests (Table 19.3). There are, however, some disadvantages to the successful establishment of biological control agents, as follows:

1. Permanent biological control by itself is not effective against direct pests. These are insect pests that damage or diminish the value of the products being marketed or directly frustrate management objectives. Insects boring into the sapwood or terminal growth of trees would be direct pests. For example, if capital is invested in plantation thinning and pruning with the goal of producing straight stems with clear wood, then shoot-boring insects would directly defeat this goal. A single Nantucket pine tip moth larva (*Rhyacionia frustrana*), feeding within the leader of a loblolly pine, would cause crooked growth. A single white pine tip weevil larva (*Pissodes strobi*) boring in the terminal stem of a white pine or a Sitka spruce would kill the terminal and cause a crook.
2. For permanent biological control to be effective, some degree of economic damage must be tolerated.
3. Permanent biological control must be practiced over large areas where the use of pesticides is severely restricted.
4. It takes several years of research and development to establish a system of permanent biological control.

Biological control of pests on a temporary basis focuses on the use of bacteria, fungi, viruses, nematodes, and parasitoids that are reared in vast numbers and released to quickly bring a growing pest population under control.

Use of Bacteria   The most common insect pathogens among the spore-forming bacteria include *Bacillus thuringiensis* (Bt) and *B. popilliae* (Angus 1974, Anderson and Kaya 1975, Dubois and Lewis 1981, Entwistle et al. 1993). In 1911, Berliner, a German entomologist from Thuringia, isolated Bt from flour moth larvae. *Bacillus thuringiensis* is one of the rod-shaped bacteria that move through water films by use of flagella, and they produce resting, thick-walled spores. Within the resting spore is a vegetative spore and a curious proteinaceous crystal (Figure 19.9).

| TABLE **19.3** | **Evauation of Biological Control Trials Against Introduced Forest Insect Pests Using Predators and Parasites** |

Pest	Degree of success		
	1910–58	**1958–68**	**1969–80**
Balsam wooly adelgid	Promising	–	–
Larch casebearer	Good	++++	++++
Birch leafminer			+
European spruce sawfly	Good	++++	++++
Satin moth	Good	+++	+++
Gypsy moth			–
European pine sawfly		++	++
Larch sawfly			
Winter moth		++++	++++
Mountain ash sawfly			++
European pine shoot moth		+	+++

Symbols:
-, No control; +, slight pest reduction; ++, local control—distribution restricted or not fully investigated; +++, control widespread but local damage occurs, ++++, control complete.
Source: Modified from Hulme and Green (1984).

The bacteria is a **facultative pathogen** (doesn't need the insect itself for propagation) that infects larvae of lepidopterans, beetles, and flies. This bacteria can be cultured on artificial media and is therefore produced in commercial quantities for agriculture, forestry, and home use. Its best known use is against lepidopterans as Bt has several powerful toxic compounds that kill moth and butterfly larvae. The crystal contains a poison called endotoxin-Δ. Extracts on this toxin kill many species of Lepidoptera and variations of endotoxin-Δ kill larvae of mosquitoes, blackflies, and midges. This material is nontoxic to higher animals. The wall of the vegetative spore contains toxic proteins (α-toxins and β-toxins) that are poisonous to many lepidopteran larvae, but are all nontoxic to mammals. Commercial preparations of Bt (Thuricide, Dipel, Attack, and others) containing both spores and crystals are used to control tree defoliators.

Bt has several desirable characteristics as a means of controlling defoliators. The paralytic action of ingested Δ-toxins results in a rapid cessation of feeding; spores do not persist in the environment, and the material can be used as required once a pest population verges on its economic damage level. The rather specific nature of Bt to kill a limited group of foliage-feeders and not to harm beneficial species is of value in management programs designed to conserve natural enemies.

Fungi   Fungi infect insects under conditions of relative humidity between 75 and 90%. High humidity is necessary for germination of **entomophagous** spores after

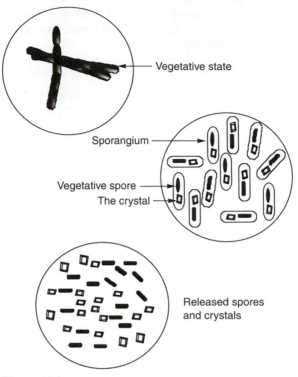

Figure 19.9
The spore-forming, rod-shaped bacteria, *Bacillus thuringiensis*, showing its vegetative state and sporangia. Each sporangium contains a vegetative spore and a proteinaceous crystal (parasporal body). The released spores and crystals are found in a commercial preparation.

they land on the host. Fungi used in controlling pests are mainly the Deuteromycetes (imperfect fungi) and in particular *Beauvaria bassiana* and *Metarrhizium anisopliae* (Hajek and St. Leger 1994, Fuxa et al. 1998). These fungi kill insects in the following manner:

1. The conidium (infective spore) lands on the insect and when environmental conditions are suitable, it germinates.
2. A germ tube forms that penetrates the cuticle by enzymatic action and mechanical force.
3. As the germ tube enters the insect's body (the hemocoel), yeastlike hyphal bodies form, float in the hemolymph, multiply and release toxins that ultimately kill the host.
4. About the time of death, the hyphae enter the insect's fat bodies where they grow rapidly and fill the insect with mycelia. Afterward, the mycelia form conidiophores which erupt through the exoskeleton during suitable conditions and produce spores outside the insect.

In North America and Europe, use of fungi to control forest insects is still being studied, but in Southeast Asia use of *B. bassiana* to control the pine caterpillar, *D. punctatus,* has been an important control method since 1960. In China *B. bassiana* has been shown to be over 80% efficacious if applied when the relative humidity is more than 80% and the temperature above 25°C (Sze-ling 1986). In northern Vietnam and southern China, the fungus is applied by aircraft, backpack dusters, or even by packing a dust formulation of *Beauveria* into firecrackers and exploding them above the canopies of young plantations. The wind then disperses the conidia over large areas.

Viruses   Viral diseases of insects are induced by nucleic acid contained in a virus particle called a *virion*. The virion may be a filament, rod, sphere, or have a complex structure (Ignoffo 1974, Anderson and Kaya 1975, Cunningham and Howse 1984, Fuxa et al. 1998, Pedigo 1999). Virions may be embedded within a proteinacious, polyhedron-shaped matrix or within a simple capsule. The ones that have a polyhedral structure and replicate within the nuclei of insect cells are called the *nuclear polyhedral viruses* (NPVs) and are the most common type (Figure 19.10). Nuclear polyhedroses affect the epidermis, fat body, and blood cells of many lepidopterans and sawflies. The insect host may ingest the virions, or they may enter the insect through the cuticle or be passed from one generation to the next on or within the egg. Once NPVs begin to multiply in the cells of the host, the larvae become sluggish, loose body turgor, and hang limply by their posterior prolegs. At this stage the larval skin ruptures, releasing virus particles into the wind as well as peppering the branches below with particles. Cytoplasmic polyhedroses (CPVs) are similar to NPVs except in their site of action. The CPVs infect the cytoplasm of midgut epithelial cells. The infection debilitates infected larvae, resulting in slow larval growth; infected larvae often become whitish and swollen. The virus is spread when polyhedral bodies are regurgitated or passed out with fecal matter.

The entomopoxviruses (EPVs) are implanted into relatively large (2–8 μm) oval or polygonal-shaped inclusion bodies (Ignoffo 1974). The EPVs replicate in the cytoplasm of adipose tissues. These viruses have been isolated from Lepidoptera, Diptera, and Coleoptera.

The granulosis viruses have a single virion enclosed by a proteinaceous capsule and occur in a few species of Lepidoptera. These viruses replicate in the nuclei or cytoplasm immediately surrounding the nuclei of fatty tissues, or epidermal cells. The disease inevitably kills the host, leaving it hanging as a fragile bag of virus similar to that which results from NPV infection.

Viruses have a promising future in control of insect defoliators. As seen in Table 19.4, an experimental spraying of EPV contaminated with NPV and CPV gave marginal control against the spruce budworm (*Choristoneura fumiferana*). Nevertheless, this degree of control in combination with other control strategies could significantly reduce budworm populations. Viral control also has its own effective mechanisms of spread through a population and persists in the environment from one generation to the next (Table 19.4).

Viruses can be formulated for ultra-low-volume application, and they can be applied with backpack or motorized sprayers where small areas are treated in order

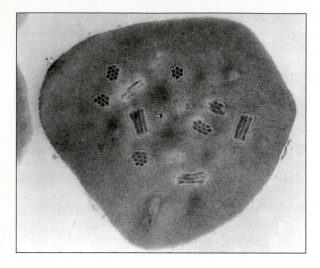

Figure 19.10
A nuclear polyhedral virus, showing the proteinaceous, polyhedron-shaped matrix within which are the embedded virions.
(Photo by USDA Forest Service.)

to establish infection centers. Moreover, viruses are species-specific, they are not harmful to mammals, and they are tolerant to many adverse environmental conditions. Various formulations, registered by the Environmental Protection Agency (EPA), are used to manage defoliators, including the Douglas-fir tussock moth, the gypsy moth (*Lymantria dispar*), the European pine sawfly (*Neodiprion sertifer*), and the redheaded sawfly (*N. lecontei*). However, virus preparations are expensive to produce, and they are susceptible to ultraviolet degradation.

Nematodes and Protozoans   Nematodes are tiny, unsegmented roundworms, some of which parasitize insects (Bedding 1984, Kaya and Gaugler 1993). The infective stage of the nematode penetrates the insect directly or is ingested and reproduces within the insect. Often there is a symbiotic relationship between a nematode and a bacterium—for example, nematodes of the genus *Steinernema* are associated with the *Xenorhabdus* bacteria. These nematodes move from the insect gut into the hemocoele; there they release the bacteria, and the bacteria multiplies and alters the environment of the hemocoele (the inside of an insect). The hemocoele then is ideally suited for massive reproduction of the nematodes. The insect dies as the nematodes exit the moribund host. Parasitic nematodes may not kill the host, but they frequently sterilize their prey. The nematode, *Deladenus siricidcola*, sterilizes the wood wasp, *Sirex noctilio*, and has been used to reduce wood wasp populations in Australian radiata pine plantations.

Protozoans of the order Microsporidia, notably those of the genus *Nosema,* have been used to control insects. The protozoans kill insects slowly; they can provoke sterility or lower the fecundity of infected hosts. Unlike other microbials, protozoans are not species-specific; however, there have been successful epizootics (outbreak of diseases in or on animals) of the microsporidian *Thelohania hyphantriae* when used to control the fall webworm (*Hyphantria cunea*) and the brown tail moth (*Nygmyia phaeorrhoea*) in Europe.

T A B L E **19.4**	**Control of the Spruce Budworm (*Choristoneura fumiferana*) by an Entomopoxvirus Contaminated by Both Nuclear Polyhedrosis and Cytoplasmic Polyhedrosis Viruses**

	Population reduction (%)		Current year's foliage saved (%)	
Year	Balsam fir	White spruce	Balsam fir	White spruce
1972	16	68	44	49
1973	15	83	26	22
1974	0	48	25	40
1975	47	26	1	1

[a]Several 1.07 ha infested stands of white spruce (*Picea glauca*) and balsam fir (*Abies balsamea*) growing in Ontario, were treated by helicopter with the viral combinations in 1971.
Source: Modified from Cunnigham and Howse (1984).

The Parasitoids and Predators   There is much interest in the possibility of releasing thousands of parasitoids or predators to temporarily lower insect populations. The goal is to augment the natural enemies of a pest for purposes of immediate suppression. The most successful approaches have been attempts to lower pests by inundating infested stands with hymenopterous egg parasitoids, in particular the genera *Trichogramma*, *Telenomus*, and *Ooncyrtus* (Smith 1996). The most common egg parasitoids used against forest defoliators are species of *Trichogramma*, but results have been variable (Table 19.5). Inconsistencies in control are due to variability in rearing techniques as well as in release technology, in particular, the timing, frequency, and rate of parasitoid release. There are many factors that influence release, such as the weather, crop, target pest, predation, dispersal, and pesticides. Biological control with *Trichogramma* will improve as new guidlines for standardizing rearing methods, release technology, terminology, and measurements of efficacy are improved.

In stands heavily defoliated by the Douglas-fir tussock moth, islands of green trees surrounded by heavily defoliated trees have been associated with mound-nests of the predaceous *Formica* ants (Campbell and Torgersen 1982, Campbell et al. 1983, Torgerson et al. 1995). In Europe *F. rufa* ants long have been recognized as effective predators of several important lepidopteran and sawfly defoliators. There have been programs on dispersing ant nests in forests to protect stands from outbreaks, but with variable success (Speight and Wainhouse 1989).

Integrated Pest Management (IPM)

Integrated pest management is a system that utilizes all suitable control techniques either to reduce pest populations and maintain them at levels below those causing economic damage or to so manipulate the populations that they are prevented from causing economic damage. As illustrated in Figure 19.11, an IPM system is established by following four guidelines:

TABLE **19.5** | **Examples of Inundative Releases of the Egg Parasite *Trichogramma* spp. Against Various Lepidopteran Defoliators**

Tree hosts	Target lepidoterans	Releases		
		No. releases	Thousands/ha	% Parasitized
Spruce	Spruce budworm	2 at 6 day intervals	600–31,000	80–96
Pine	Pine moth	4–5 at 5–7 day intervals	1,050	45–90
Mixed conifers	Pine moth	2–3 at 5–7 day intervals	200–700	Unknown
Oak	*Lampronadata* spp.	6 at 5 day intervals	30,000	29

Source: (From Smith [1996] and adapted with permission from the *Annual Review of Entomology,* Volume 41, © 1996, by Annual Reviews.)

1. Knowing which are the key pests of a managed forest and determining (a) the economic damage level of each key pest and (b) the life history of each, and the life history of the natural enemies of each. This guideline implies that each management activity may incur potential pests, but only a few (often only a single pest) would cause economic damage.
2. Developing schemes for lowering the average population level (also called the *equilibrium position*) of the key pests. Within this context, IPM efforts manipulate the ecosystem in order to permanently reduce a pest's equilibrium position below the economic damage level. This reduction is accomplished by using one or several fundamental management components—for example (a) encouragement and conservation of natural control agents—it is important to know what these are, how to manipulate them, and how they impact pest populations; (b) utilization and encouragement of pest-resistant or pest-free nursery stock, seeds, or trees— at least, know the mechanisms of resistance and if they can be manipulated; and (c) modification of the pest's environment so as to increase the effectiveness of the pest's natural control agents—it is important to know which silvicultural procedures can be applied to this task.
3. Recognizing that during certain conditions, key pest populations will temporarily exceed the economic damage level. At that time it is essential to have ready remedial measures to suppress a growing pest population—at least, be prepared to use an applied control (e.g., use of *B. thuringiensis*) that is timed to do minimal damage to the natural enemies.
4. Establishing a monitoring system to determine the population level of key pests as well as their natural enemies.

In later chapters, when the various forests pest are described, population management of these will be discussed within an IPM context.

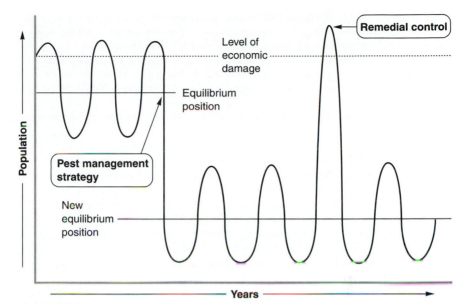

Figure 19.11

Elements of Integrated Pest Management (IPM), showing a forest pest population fluctuating around an equilibrium position (average population density estimated over many years) that periodically causes economic damage; a pest management strategy is established (e.g., a silvicultural treatment) that delineates a new and lower equilibrium position for the population; and application of a short-term remedial control to lower the pest population when needed.

REFERENCES

Allan, G. G., R. M. Wilkins, and R. I. Gara. 1974. Field testing of insecticide-polymer combinations for long term control of the mahogany shoot borer. *International Pest Control* 16:4–11.

Anderson, J. F., and H. K. Kaya. 1975. *Perspectives in forest entomology.* Academic Press, New York.

Angus, T. A. 1974. Microbial control of insects: bacterial pathogens. In F. G. Maxwell and F. A. Harris (eds.). *Proceedings of the summer institute on biological control of plant insect and diseases.* University Press of Mississippi, College Station, Miss.

Bedding, R. A. 1984. Nematode parasites of Hymenoptera. In W. R. Nichol (ed.). *Plant and insect nematodes.* Marcel Dekker, New York.

Bohmont, B. L. 1997. *The standard pesticide users guide.* Prentice-Hall, Upper Saddle River, N.J.

Campbell, R. W., and T. R. Torgersen. 1982. Some effects of predaceous ants on western spruce budworm pupae in north central Washington. *Environmental Entomology* 11:111–114.

Campbell, R. W., T. R. Torgersen, and N. Srivastava. 1983. A suggested role for predaceous ants in the population dynamics of the western spruce budworm. *Forest Science* 29:779–790.

Clendenen, G., V. F. Gallucci, and R. I. Gara. 1978. On the spectral analysis of cyclic tussock moth epidemics and corresponding climatic indices with a critical discussion of the underlying hypothesis, pp. 279–293. In C. Shugart (ed.). *Time series and ecological processes.* Society for Industrial and Applied Mathematics (SIAM), Philadelphia.

Cunningham, J. C., and G. M. Howse. 1984. Viruses: application and assessment. In J. S. Kelleher and M. A. Hulme (eds.). *Biological control programmes against insects and weeds in Canada 1969–1980.* Commonwealth Agricultural Bureaux, Farnham, United Kingdom.

Daterman, G. E. 1990. Pheromones for managing conifer tree pests in the United States, with special reference to the western pine shoot borer. In R. L. Ridgway, R. M. Silverstein, and M. N. Inscoe (eds.). *Behavior-modifying chemicals for insect management.* Marcel Dekker, New York.

Dubois, Normand R., and F. B. Lewis. 1981. What is *Bacillus thuringiensis*? *Journal of Arboriculture* 7:233–240.

Embree, D. G. 1971. The biological control of the wintermoth in eastern Canada by introduced parasites, pp. 217–226. In C. B. Huffaker (ed.) *Biological control.* Plenum Press, New York.

Entwistle, P. F., J. S. Cory, M. J. Bailey, and S. Higgs (eds.). 1993. *Bacillus thuringiensis: an environmental biopesticide.* John Wiley and Sons, Chichester, United Kingdom.

Fuxa, J. R., R. Ayyappath, and R. A. Goyer. 1998. *Forest health technology: pathogens and microbial control of North American forest insect pests.* USDA Forest Service FHTET-97-27.

Goyer, R. A., M. R. Wagner, and T. D. Schowalter. 1998. Current and proposed technologies for bark beetle management. *Journal of Forestry* 96(12):29–33.

Hajek, A. E., and R. J. St. Leger. 1994. Interactions between fungal pathogens and insect hosts. *Annual Review of Entomology* 39:293–322.

Hulme, M. A., and G. W. Green. 1984. Biological control of forest insect pests in Canada 1969–1980: retrospect and prospect. In J. S. Kelleher and M. A. Hulme (eds.). *Biological control programmes against insects and weeds in Canada.* Commonwealth Agricultural Bureaux, London.

Ignoffo, C. M. 1974. Microbial control of insects: viral pathogens. In F. G. Maxwell and F. A. Harris (eds.). *Proceedings of the summer institute on biological control of plant insect and diseases.* University Press of Mississippi, College Station, Miss.

Kaya, H. K., and Randy Gaugler. 1993. Entomopathogenic nematodes. *Annual Review of Entomology* 38:181–206.

McFadden, Max W., D. L. Dahlsten, C. W. Berisford, F. B. Knight, and W. M. Metterhouse. 1981. Integrated pest management in China's forests. *Journal of Forestry* 79:722–725.

Norlund, Donald A. 1981. Semiochemicals: a review of the terminology. In D. A. Norlund, R. L. Jones, and W. Joe Lewis (eds.). *Semiochemicals: their role in pest control.* John Wiley and Sons, New York.

Orr, Richard. 1994. Importation of logs, lumber, and other unmanufactured wood articles. Environmental Impact Statement. USDA Animal and Plant Health Inspection Service, Hyattsville, Md.

Overhulser, D. L., G. E. Daterman, L. L. Sower, C. Sartwell, and T. W. Koerber. 1980. Mating disruption with synthetic sex attractants controls damage by *Eucosma sonomana* (Lepidoptera: Tortricidae, Olethreutinae) in *Pinus ponderosa* plantations II. aerially applied hollow fiber formulation. *Canadian Entomologist* 112:163–165.

Pedigo, L. P. 1999. *Entomology and pest management.* Prentice Hall, Upper Saddle River, N.J.

Roeloff, W. L. 1981. Attractive and aggregating pheromones. In D. A. Norland et al. (eds.). *Semiochemicals.* John Wiley and Sons, New York.

Roland, Jens, and D. G. Embree. 1995. Biological control of the winter moth. *Annual Review of Entomology* 40:475–492.

Romoser, W. S., and J. G. Stoffolano, Jr. 1994. *The science of entomology.* Wm. C. Brown Communications, Dubuque, Iowa.

Ryan, R. B. 1987. Classical biological control. *Journal of Forestry* 85:29–31.

Ryan, R. B. 1997. Before and after evaluation of biological control: the larch casebearer (Lepidoptera: Coleophoridae) in the Blue Mountains of Oregon and Washington. *Environmental Entomology* 26:703–715.

Ryan, R. B., Scott Tunnock, and F. W. Ebel. 1987. The larch casebearer in North America. *Journal of Forestry* 85:33–39.

Sheehan, Katherine A. 1996. *Effects of insecticide treatments on subsequent defoliation by western spruce budworm in Oregon and Washington: 1982–92.* USDA Forest Service. General Technical Report PNW-GTR-367.

Skillen, E. L., C. W. Berisford, M. A. Camann, and R. C. Reardon. 1997. *Forest health technology: semiochemicals of forest and shade tree insects in North America and their management applications.* USDA Forest Service FHTET-96-15.

Smith, R. F., and Robert van den Bosch. 1967. Integrated control. In W. W. Kilgore and R. L. Doutt (eds.). *Pest control: biological, physical, and selected chemical methods.* Academic Press, New York.

Smith, S. M. 1996. Biological control with *Trichogramma*: advances, successes, and potential for their use. *Annual Review of Entomology* 41:375–406.

Speight, M. R. and David Wainhouse. 1989. *Ecology and management of forest insects.* Claredon Press, Oxford, United Kingdom.

Sze-ling, Ying, 1986. A decade of successful control of pine caterpillar, *Dendrolimus punctatus* Walker (Lepidoptera: Lasiocampidae), by microbial agents. *Forest Ecology and Management* 15:69–74.

Thompson, W. T. 1994. *Agricultural chemicals. Book I: Insecticides.* Thompson Publications, Fresno, Calif.

Torgersen, R. F., David C. Powell, K. P Hosman, and F. H. Schmidt. 1995. No long-term impact of carbaryl treatment on western spruce budworm populations and host trees in the Malheur National Forest, Oregon. *Forest Science* 41:851–863.

Ware, G. W. 1989. *The pesticide book.* Thompson Publications, Fresno, Calif.

Wickman, B. E. 1979. A case study of a Douglas-fir tussock moth outbreak and stand conditions 10 years later. USDA Forest Service, Pacific Northwest Forest and Range Experiment Station, Research Paper PNW-224.

C H A P T E R **20**

INSECT DEFOLIATORS

INTRODUCTION

Insect defoliation of the forest can be recognized by the thinning or absence of foliage and the raining of frass (insect fecal matter and bits of foliage). There are several types of **defoliators.** Leaf chewers completely devour leaves or needles; skeletonizers feed only on the softer parts of leaves, leaving a skeletal network of leaf veins. The elm leaf beetle (*Pyrrhalta luteola*), for example, is a skeletonizer of elm foliage and an important urban forestry problem. Leafminers, on the other hand, bore inside and eat the tissues between the upper and lower epidermal walls of the foliage; the lodgepole pine needle miner (*Coleotichnites milleri*) is a notorious case in point.

The most damaging defoliators consume the foliage entirely. Some, like the western spruce budworm (*Choristoneura occidentalis*), specialize in new growth, whereas others, like the Douglas-fir tussock moth (*Orgyia pseudotsugata*), consume both new and old foliage. Some defoliators feed on many hosts (**polyphagous** feeders); others are **oligophagous** and feed only on a few tree species, and a few species feed on a single host. The gypsy moth is a polyphagous feeder that feeds mainly on hardwood foliage but consumes conifers under certain conditions (Table 20.1). The western spruce budworm is an oligophagous pest as it feeds on true firs, Douglas-firs, spruces, and western larch. The larch casebearer is an introduced monophagous pest that feeds only on larch trees, and was on its way toward becoming a serious economic pest before it was successfully controlled by introduction of its principal parasite, the braconid wasp, *Agathis pumila*. In fact, introduced **monophagous** insects are the best subjects for biological control.

Trees die when defoliation is severe, frequent, or persistent over several years. Partial or less frequent defoliation always results in growth loss (Kulman 1971). Growth loss seriously affects management plans by lengthening rotations (Figure 20.1). Defoliators often weaken their hosts and set up the stands for attacks by **secondary insects,** such as bark beetles. In eastern North America the hemlock looper (*Lambdina fiscellaria fiscellaria*) invades overmature and

462

T A B L E **20.1**	Some Host Preferences for the Gypsy Moth in the Northeast[a]	

Favored hosts by all instars	Favored hosts by later instars	Moderately/slightly favored hosts (depending on population level)
Alder	Beech	Birch (black and yellow)
Apple	Cedar (red)	Butternut
Aspen	Hemlock	Cedar (southern white)
Birch (gray, white, and red)	Plum	Cherry
Box elder	Pine	Elm
Gum (red)	Spruce	Gum (black)
Hawthorn		Hickory
Hazelnut		Maple (various spp.)
Larch		
Linden		
Mountain ash		
Oaks		
Lombardy poplar		
Roses		
Sumac		

[a]Conifers become become particularly susceptible when grown in mixtures with the more favored species.

senescing hemlock-balsam fir stands. Looper outbreaks weaken the hemlocks, which brings in the hemlock borer (*Melanophila fulvoguttata*). The hemlock borer normally breeds in logs and windthrown hemlocks, but after a looper outbreak, it dismantles entire hemlock–balsam fir stands.

Defoliator outbreaks may be sporadic. The pest species remains at a **latent population** level, until severe external factors (e.g., droughts, wildfires, root rots, other insects, etc.) adversely affect the condition of trees; lingering droughts typically trigger outbreaks of the western hemlock looper (Figure 20.2). In contrast, periodic outbreaks occur at rather set intervals, as is the case of the Douglas-fir tussock moth (*Orgyia pseudotsugata*). Outbreaks of this defoliator occur every 9 to 10 years (Figure 20.3). Characteristically these outbreaks build up rapidly and are followed by an abrupt collapse of the population.

DETECTION, EVALUATION, AND CONTROL

Defoliation is caused by larvae of Lepidoptera, Hymenoptera (the sawflies), and Coleoptera. Adult beetles also feed on foliage, such as the adult elm leaf beetle (*P. luteola*). As trees are defoliated, discoloration of damaged foliage is visible before the insect itself is noticed. The majority of the defoliators are moths. Although caterpillars are easy to detect and identify, many are difficult to sample

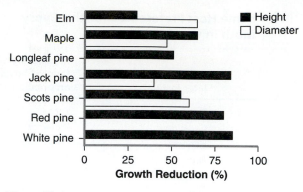

Figure 20.1

Effects of insect defoliation on the height and diameter growth of seven tree species; data on diameter for only elm, maple, jack pine, and Scots pine.

(Source: From Kulman 1971. Adapted with permission, from the *Annual Review of Entomology,* volume 16, © 1971, by Annual Reviews.)

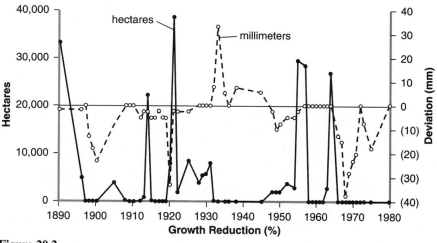

Figure 20.2

Relationship between large outbreaks of the western hemlock looper (*Lambdina fiscellaria lugubrosa*) in western Washington and deviations from the average precipitation.

in a meaningful way: a task necessary for determining if an **economic damage threshold** is being reached. The problem is that larvae may occur on different parts of trees at distinct times, and pupae and eggs may be widely distributed over the host, rather than clustered on the foliage. Accordingly, if populational trends are to be estimated, knowledge of a defoliator's life history must be known. This includes the timing and extent of natural controls, such as occurrence of parasitoids, predators, and infectious diseases. All of this information is needed for selecting a sampling unit and determining when to sample.

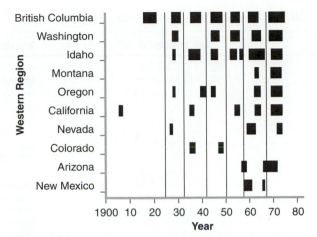

Figure 20.3

Patterns of Douglas-fir tussock moth (*Orgyia pseudotsugata*) outbreaks in western North America.

(Source: From Clendenen et al. 1978.)

Detection

Becoming aware early of when and where stands are being defoliated is essential to an IPM program (Carolin and Coulter 1975). Pest activity is detected through indirect and direct surveys. Indirect approaches are nonsystematic observations and reports done by trained field personnel—observations that often represent an initial awareness that stands are being defoliated. As frequently happens, trail crews, road crews, field foresters, recreation managers, timber cruisers, and so on are the first to notice that stands suddenly have more light, or that **frass** is raining down on the forest floor. These observations should be reported on a "pest alert form" and thus become the first step in a direct, systematic detection plan.

Direct methods include aerial surveys, where reconnaissance flights are made along flight lines that follow topographic contours. Defoliated areas are sketched on maps and estimates are made on the degree of defoliation. Ground surveys then are made in areas that were classed from the air as light, medium, and heavily infested. Defoliation intensity can be classed as no feeding (class 0), less than 50% defoliation (class 1), between 50 and 90% defoliation (class 2), and greater than 90% defoliation class 3 (Dorias and Kettela 1982). More accurate ocular classifications can be made but in all cases, the accuracy of interpreting defoliation damage from the air is refined by subsequent ground surveys.

Aerial surveys are followed by sampling systems (Mason 1969, 1979) that estimate the populational trends of a defoliator and whether there is a need for applied control. Egg surveys provide initial and predictive information on the infestation level of a pest, as with the eastern spruce budworm (*Choristoneura fumiferana*) (Table 20.2). Egg and larval sampling within a sequential sampling scheme is useful in determining populational trends at the beginning of an outbreak when insect numbers are low. The basic sample unit is a collection of

T A B L E **20.2** | **Results of C. *fumiferana* Egg Surveys[a]**

Locality	Egg masses/10 m²	Infestation forcast
Ontario	01–53	Light
	54–215	Moderate
	216+	Severe
Quebec	0	None
	01–100	Light
	101–200	Moderate
	201–600	High
	601+	Very high
Maritime provinces	01–108	Light
Maine	109–260	Moderate
	261–434	High
	435–1,086	Very high
	1,086+	Extreme

[a]The number of budworm egg masses per 10 m² of sampled foliage is used to forcast an infestation level in Ontario, Quebec, the maritime provinces and the state of Maine.
Source: Modified from Dorias and Kettela (1982).

branches (subsamples) taken representatively from a tree. In the case of the Douglas-fir tussock moth (USDA 1978, Mason 1981), twigs are used for larval sampling and whole branches are used for sampling egg masses. Population size is estimated by calculating the density of larvae or eggs per 6,452 cm² (1,000 in.²) of branch area:

$$\text{Larvae or egg mass per 6,452 cm}^2 = \frac{\text{Total larvae or egg mass on subsamples}}{\text{Total branch area (cm}^2) \text{ of subsamples}} \times 6,452$$

Statistically based sets of parallel regression lines are plotted, with intercepts and slopes based on values generated from binomial distributions (or other recognizable distributions). These graphs represent a sequential plan useful in knowing if a Douglas-fir tussock moth population is reaching epidemic proportions where control actions must be taken. As shown in Figure 20.4, if the cumulative number of infested sample units is above the upper decision line, the population size may be heading for an outbreak (heavy population). Values falling between the lines denote intermediate levels of infestation and oblige that sampling of trees should continue until a decision line is crossed. If the lines are not crossed after some discretional limit, the population would be classed as intermediate.

Cocoon sampling also is done to predict larval densities for a subsequent generation (USDA 1978). For example, Douglas-fir tussock moth cocoons are sampled in autumn in order to predict the degree of defoliation expected in spring (Table 20.3).

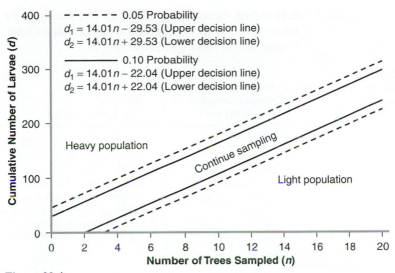

Figure 20.4

Sequential sampling graph of Douglas-fir tussock moth (*Orgyia pseudotsugata*) larvae showing decision lines, at two confidence levels, that divide a potential outbreak into being, light, intermediate, or heavy.

(Source: From Mason 1969, USDA Forest Service.)

Pheromone trapping of adult males is useful for providing early warnings of potential population increases (USDA 1978). Baited traps are useful in surveying remote areas, which make frequent trap checking impractical. Long-lasting weak baits have served this purpose in monitoring for the presence of *O. pseudotsugata* in Oregon and Washington. Pheromone trapping with strong baits, whose synthetic pheromone release rates are standardized, are being developed to evaluate potential larval densities estimated from male catches.

Applied Control

Biological evaluations may indicate that defoliator populations are expanding and that stands will sustain extreme future defoliation and growth loss. For example, third and fourth instar samples of the eastern spruce budworm feeding on balsam fir may indicate over 80% defoliation when 0.18 larvae per 45 cm branch sample (length from tip toward the stem) is estimated (Figure 20.5). If for given stand conditions this damage level is unacceptable, control by aerial spraying may be necessary (Sheehan 1996). If spraying with organic insecticides or microbials is recommended, the applications would be made under the provisions of pertinent federal and state regulations. Because of the necessity of preparing an environmental impact statement, a lead time of 6 to 8 months is needed before final planning of an operation can begin. As an example, in 1990 the Yakama Indian Nation prepared an **EIS** to treat a 24,200 ha western spruce budworm outbreak in south-central Washington; a budworm population that had been gaining destructive

T A B L E **20.3** | **Estimating Cocoon Densities of the Douglas-fir tussock moth to predicted larval densities and status in the next generation**

Cocoon densities sampled in lower crown of hosts (no./6,452 cm²)	Predicted larval midcrown densities (no./6,452 cm²)	Predicted population status
<0.01	<2.0	Low density: no defoliation
0.01–0.30	2.0–20.0	Suboutbreak: little or no visible defoliation
0.31–0.70	21.0–40.0	Moderate outbreak: defoliation visible on most host trees
>0.70	>40.0	Severe outbreak: defoliation intense in upper crowns of host trees with some trees completely defoliated

Source: From Mason et al. (1993).

momentum since 1985. An evaluation of the outbreak by the U.S. Forest Service in 1988 found that significant damage would be done to the tribal forests if the outbreak continued unabated. An EIS was prepared, approved by the Tribal Council and *Bacillus thuringiensis* var. *kurstaki* (Thuricide) was applied aerially on three spray blocks (one of 8,842 ha, another of 8,906 ha, and the third of 11,096 ha). The objective of the treatment was to reduce the budworm population within the treatment areas to levels that would not cause unacceptable tree damage for several years after insecticide application. A specific budworm population reduction target was not established for the project; rather the prevention of damage would be anticipated for at least 3 years, and populational trends were monitored for this period. An incident command system organization (a basic fire suppression organization) was modified to fit the needs of the budworm project (Figure 20.6). Boundaries of the spray blocks were marked in the fall of 1989 for treatment during the summer of 1990.

Prior to spraying, 119 three-tree plots were sampled to determine early larval densities for the three spray blocks. This survey was done to determine if the budworm populations were high enough to meet the treatment criteria of at least six larvae per 45 cm, midcrown branch tip. Western spruce budworm larval development and host tree foliage phenology were monitored to time release of the spray blocks for treatment. The criteria for release was that less than 15% of the larvae were to be in the second and third instars, and at least 95% of the new shoots had to be unfurled. These standards were reached and the three blocks were sprayed between June 20th and 29th.

Postspray population sampling again included the sampling of three-tree plots. Three lower crown branches with new shoots were sampled on each tree. A 5.1 × 10.2 cm white cloth stretched on an aluminum frame was held under the outermost

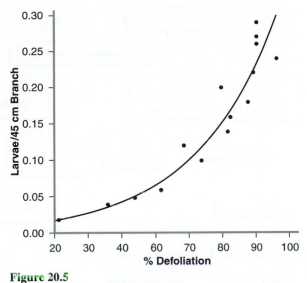

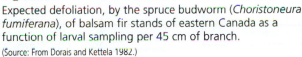

Figure 20.5

Expected defoliation, by the spruce budworm (*Choristoneura fumiferana*), of balsam fir stands of eastern Canada as a function of larval sampling per 45 cm of branch.

(Source: From Dorais and Kettela 1982.)

45 cm of each branch while the branch was struck repeatedly with a stick to dislodge larvae and pupae (Figure 20.7). Results of these entomological analyses showed that the number of larvae per branch was significantly reduced after treatment, but the budworm population was still active a year after the spraying.

THE DEFOLIATORS

There are many species of defoliators that have spread across millions of hectares and killed millions of cubic meters of timber in North America (Craighead 1950, Furniss and Carolin 1977). The principal native defoliator of forests in northeastern North America is the spruce budworm (*C. fumiferana*) (Wellington et al. 1950, Morris 1963, Department of the Environment Canada 1975). Canadian woodsmen, as early as 1770, reported on its ability to kill millions of spruce and balsam firs across the landscape. Suggestions for control of the spruce budworm through stand management have been proposed in the United States since the 1920s. During the 1940s biological and mathematical relationships were developed to rate the susceptibility of stands to budworm outbreaks. Silvicultural prescriptions, based on these relationships, were adopted to lessen the impact of outbreaks and, in addition, thousands of hectares of North American forests have been treated with insecticides. These spruce-fir forests also are sporadically defoliated by the spruce sawfly (*Neodiprion abietis*), the larch sawfly (*Pristiphora erichsonii*), the larch casebearer (*Coleophora laricella*), and other less important insects.

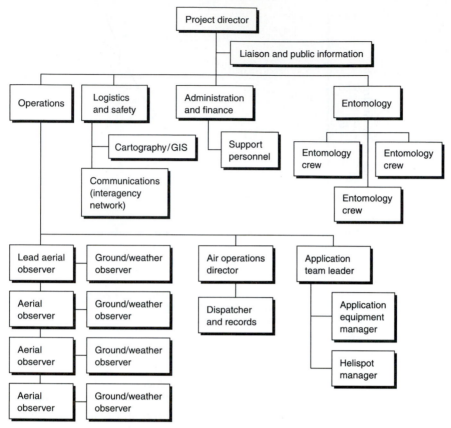

Figure 20.6

An incident command system organized to the operational needs of a 1990 western spruce budworm (*Choristoneura occidentalis*) suppression project on the Yakama Indian Reservation, near Yakima, Washington.

The other catastrophically important defoliator of northeastern forests is the gypsy moth (*Lymantria dispar*) (USDA 1981). Outbreaks of this introduced moth have drastically changed the composition and economic viability of these forests, and valuable oak species are declining as a result of outbreaks of this pest. Moreover, the gypsy moth is spreading across the hardwood forests of United States and Canada and poses unacceptable threats to forest management plans of northern hardwood forests—a valuable resource that will be increasingly important over the decades to come. Another problem that affects intensive management plans of these hardwood stands is tent caterpillar outbreaks—in particular, those of the forest tent caterpillar (*Malacosoma disstria*).

Several sawfly species occasionally cause short-term damage to southern pine. For example, the redheaded sawfly (*Neodiprion lecontei*) defoliates young jack pine and red pine in the northern part of its range as well as all southern pine species in the Southeast and South. The loblolly pine sawfly (*N. taedae taedae*)

Figure 20.7

Sampling for Douglas-fir tussock moth (*Orgyia pseudotsugata*) larvae by holding a hand-held drop cloth under a branch and beating the branch with a stick.

(Source: From Mason 1979, USDA Forest Service.)

is an occasional problem in understocked loblolly pine stands. Probably the most damaging sawfly is the blackheaded sawfly (*N. excitans*) which defoliates commercial pine forests late in the season. Severe outbreaks of this sawfly in Florida and East Texas have weakened trees to the extent that localized *Ips* spp. outbreaks have occurred.

Western North American coniferous forests are defoliated by several important lepidopterous pests; historically the most damaging have been the western spruce budworm (*C. occidentalis*), the Douglas-fir tussock moth (*O. pseudotsugata*), the pine butterfly (*Neophasia menapia*), the western hemlock looper (*Lambdina fiscellaria lugubrosa*), the pandora moth (*Coloradia pandora*), and the western black-headed budworm (*Acleris gloverana*) (Table 20.4).

Eastern and Southern North American Forests

The Eastern Spruce Budworm   In late summer females lay about 200 eggs in a shinglelike manner on needles of their hosts. Balsam fir (*Abies balsamea*) is its main host and the tree species that suffers by far the greatest mortality, especially after the fifth year of an outbreak. White spruce (*Picea glauca*), red spruce (*P. rubens*) and, to a lesser extent, black spruce (*P. mariana*) also are hosts. The first instars emerge after 10 days and one of the two major larval dispersal periods occurs at this time, approximately mid-August. **Dispersal** for many larvae occurs as they drop on threads and are blown to other trees within the stand or beyond. Larvae that remain on, or disperse to, host material do not feed but spin overwintering shelters, called **hibernacula.** Hibernacula are located in bark crevices, under bud and bark scales, under lichens, and in the remains of old **staminate flowers.** Early in May of the subsequent year, second instars emerge from the

T A B L E **20.4** | **Historical Defoliator Outbreaks in Western North America**

Insect (Lepidoptera)	Years	Location	Infested area (thousand ha)
Western blackheaded budworm (*Acleris gloverana*)	1948–55	SE Alaska	11,000
	1952–56	British Columbia	2,000
	1959–65	SE Alaska	30
Western spruce budworm (*Choristoneura occidentalis*)	1949–62	Oregon and Washinton	10,000
	1951–70	Inland states	10,000
	1992–96	Pacific Northwest and inland states	6,346
Douglas-fir tussock moth (*Orgyia pseudotsugata*)	1946–47	Idaho and Oregon	556
	1971–75	Oregon and Washinton	800
Pine butterfly (*Neophasia menapia*)	1893–95	Washington	150
	1952–54	Idaho	255
Pandora moth (*Coloradia pandora*)	1937–39	Colorado	100
Western hemlock looper (*Lambdina fiscellaria lugubrosa*)	1918–22	Oregon	30
	1929–32	Washington	60
	1962–64	Washington	80

Source: Modified from Furniss and Carolin (1977).

hibernacula. The second major larval dispersal occurs at this time when these larvae move to the branch tips and spin down on silken threads. As they spin down, the wind wafts many away. This behavior is termed *ballooning.* Survivors (those that land on suitable host material) soon establish feeding sites. The newly opening staminate flowers are the preferred feeding sites on balsam fir. If flowers are unavailable, larvae either mine old needles or mine into the expanding vegetative buds. By late May to early June third instars are all feeding on the new, unfurling foliage. Development of the fourth to sixth instars lasts into July and, if all the new foliage is consumed, these larvae begin to devour older needles. During outbreaks, destruction of the current shoots causes severe intraspecific competition, and many larvae drop from the defoliated overstory and feed on seedlings and advance regeneration. The sixth instars pupate in late summer (around mid-July) on old needles, twigs, and branchlets. Emerging adults mate as males respond to female sex pheromones. Males as well as females (which have laid some eggs) are active flyers and, especially with the passage of a cold front, may be borne aloft and transported long distances (Wellington 1952, Wellington and Trimble 1984).

Budworm outbreaks start and gain momentum when there is a large proportion of mature and overmature balsam fir in the forests. Large balsam firs also bear abundant staminate flowers, which are highly nutritious to young budworms. Susceptibility of a stand is determined more by the proportion of mature balsam fir than by the characteristics of individual trees. Susceptibility can be lessened by decreasing the proportion of balsam firs, which can be done by favoring establishment of spruce (Figure 20.8). Vulnerability of stands to losses can be reduced

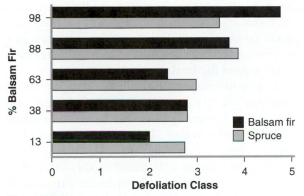

Figure 20.8

Relative spruce budworm (*Choristoneura occidentalis*) defoliation of conifer stands as a function of percent of balsam fir in mixed fir-spruce stands in six counties of Maine. Defoliation classes are as follows: 1 = no feeding apparent, 2 = 1 to 5% of current year foliage missing, 3 = 6 to 20% foliar loss, 4 = 21 to 50% loss, and 5 = over 50% loss.

(Source: From School of Forest Resources, University of Maine, CFRU Bulletin-3, 1981.)

by selectively removing firs and spruces with poor live crown:stem ratios. Other management practices that reduce losses include (1) lowering rotation ages in order to prevent substantial staminate cone crops—this tactic is desirable in areas where a wood chip market is high; (2) planting mixtures that include a high proportion of spruce; and (3) removing mature balsam fir and white spruce overstories while reestablishing mixtures of spruce and white pine.

High rates of parasitism of mature larvae by hymenopterous parasitoids such as *Meteorus trachynotus, Apanteles fumiferanae, Glypta fumiferanae,* and *Phaeogenes hariolus* and the dipteran *Lypha setifacies* are common in the Northeast. Bird predation also is an important element in the natural regulation of budworm populations, especially evening predation by grosbeaks (*Hesperiphona vespertina*). Once a budworm outbreak is under way, it continues until the larvae consume much of the available foliage. As a result, applications of insecticides often are necessary to supplement natural control. Aerial application of the biological insecticide, *B. thuringiensis,* is the most common treatment. Spraying does not kill all the larvae because some are protected by the webbed foliage within which they feed. The goal of these applications, however, is not to kill all larvae but to reduce larval populations sufficiently so that trees retain 35% or more of their foliage.

The Gypsy Moth   *Lymantria dispar* is the worst defoliation threat to deciduous trees in northeastern North America (USDA 1981). The gypsy moth is known for cyclic and devastating population explosions followed by sudden collapses (Campbell 1979). A single instance of complete defoliation can kill conifers, and three consecutive total defoliations can kill most deciduous trees, especially the oaks, the insect's favored food. Gypsy moth caterpillars are a nuisance in cities, towns, and recreational areas where, in addition to disrupting human activities, they injure and kill urban greenspaces and ornamental plantings (Figure 20.9). Larval hairs may also cause allergic reactions to humans in the urban environment. These reactions range from flulike symptoms to rashes.

The species is native to Europe and Asia. In 1869 the pest was introduced accidentally into North America by E. Léopold Trouvelot, an artist, naturalist, mathematician, and astronomer. Apparently he wanted to develop a silk moth strain that was resistant to diseases and thus start a cottage silk industry in Medford, Massachusetts. Regrettably some of gypsy moths escaped. Trouvelot understood the potential magnitude of this accident and notified local entomologists but no action was taken. In 1889 the escaped gypsy moth population exploded. In spite of control efforts, the insect invaded all the New England states and since then has caused major shifts in the species structure of affected forests. Despite all control efforts since its introduction, the moth has rapidly moved north to Canada, west to Wisconsin, and south into the Ozarks and along the Appalachian Mountains to North Carolina. Besides establishment in these areas, incipient populations are found throughout the United States and southern Canada. Gypsy moth larvae defoliate millions of hectares of trees annually.

The first instars emerge from overwintering egg masses in spring whenever there are 2 to 4 days with temperatures of 16.6°C; this is the time that buds are breaking on oaks and other hardwood species. Depending on locality, the hatching period can last from 10 to 30 days. These caterpillars are about 4 mm long and covered with hollow, long, thin, silky hairs, and many use these buoyant hairs for short-distance ballooning. This is the most effective dispersal system used by the gypsy moth. The first three instars feed on foliage day and night. Once they reach the fourth instar, they move off the foliage and hide under bark scales, in secluded areas around the crotches, or come down the tree to seek shelter during the day and feed only at dusk and part of the night. During outbreak conditions, with heavy competition for food, however, larvae also feed during the day. There are five or six instars (males and females, respectively) and their activity lasts 2 to 3 months and extends through May and June, and often into July. During the larval stages a gypsy moth devours an average of 1 m^2 of foliage; the last instar is the most voracious feeder. During this period (usually July) the caterpillars consume 60 to 70% of the foliage they will eat in their lifetime. When the caterpillars are mature they seek a sheltered spot, and construct a loose cocoon of silk and bits of leaves and other debris. After 9 to 17 days the moths begin to emerge, with males emerging a couple of days before the females. The males and females are easily distinguished from one another. Males are brownish-gray, whereas females are white with black markings and larger in size. Unlike the males, the females are incapable of flight.

Upon emergence from the pupae, female moths emit a powerful sex pheromone that attracts males from long distances. Males follow the odor plume with a rather erratic flight while tracking the increasingly strong gradient of the pheromone to its source. Once within a few meters from the source, visual cues are used to locate the female. In particular, males detect the contrast of the light-colored female resting on the dark-colored tree bark. Mated females lay oval-shaped egg masses, which

Figure 20.9
Gypsy moth (*Lymantria dispar*) female.

are covered with hairs from their bodies. These buff-colored egg masses contain 100 to 1,500 eggs and are laid on the underside of tree limbs, bark, rocks, and structures such as buildings, campers, mobile homes, car bodies, utility trailers, boats, and so forth. In fact, long-distance spread of the gypsy moth occurs when egg masses are unknowingly transported from infested areas on vehicles.

Several factors influence forest susceptibility to gypsy moth infestations. The most important is forest composition. Site quality in terms of drainage, soil, and exposure also plays a role. Open stands on dry sites are more susceptible than dense stands on wetter sites. In contrast, resistant stands grow in well-drained soils where trees are vigorous and numerous enough to form a thick canopy. The characteristic species associated with stands that suffer the least damage are sugar and silver maple, white ash, basswood, white elm, water birch, and eastern hemlock. Mixed hardwood, hardwood-hemlock, and maple-beech associations are most resistant to the gypsy moth. Small mammals are more plentiful in these habitats, and they are more successful in slowing buildups of dangerous gypsy moth populations.

A variety of natural controls serve to keep gypsy moth populations in check. These agents include over 20 parasitoid species and many predatory insects. Many of these biological controls were introduced over the last 100 years from Europe and Asia. Small mammals, especially shrews, voles, and mice, appear to be the most important gypsy moth predators, especially when moths are at low population densities. Birds also are important predators at latent population levels. Known bird predators are the blue jay, robin, black-capped chickadee, gray catbird, northern oriole, scarlet tanager, and many others, including a wide variety of migratory birds.

The gypsy moth is killed by infectious diseases (caused by bacteria, viruses, and fungi), which are transmitted by contact from one larva to another. The most effective pathogen is a nucleopolyhedrosis virus, *Borrelinavirus reprimes*, that affects the larvae and pupae. Infected larvae become flaccid and hang from the foliage. Their bodies rupture, releasing huge quantities of virus, which infect other larvae on contact. **Epizootics** of the virus kill over 80% of the population. The disease apparently came from Europe where, because of the larval symptoms, it bears the German name, *Wipfelkrankheit*. Several bacteria—*Streptococcus faecalis, Serratia marcescens, S. liquefaciens, Pseudomonas aeruginosa,* and *Bacillus cereus* — are common natural controls and account for 15% of the natural larval mortality.

The insects also are vulnerable to several fungi including *Beauvaria bassiana.* These are not important controlling agents as they are effective only under particular warm and humid conditions. Nevertheless, all these mortality factors are linked and serve in the ultimate collapse of gypsy moth outbreaks.

Silvicultural approaches to lower the susceptibility of stands to gypsy moth infestations include (1) reducing the proportion of favored hosts while keeping oaks, by removing gray birch, larch, willow, poplar, apple, and others; (2) encouraging white pine over oaks on dry sites; and (3) removing **dominant** and **codominant** mature trees on dry sites—these trees are less capable of mobilizing **allelochemic** defenses. As stand compositions change, however, other problems may arise. For example, increasing proportions of elms in the southern Appalachian Mountains favor outbreaks of the elm spanworm (*Ennomus subsignarius*).

Chemical control is used to control outbreaks in areas with valuable stands of oaks, in heavy-use recreational sites, in urban situations where the presence of countless larvae is unacceptable for aesthetic and health reasons, and in areas where the gypsy moth is a recent colonist. The buildup of populations is detected by use of pheromone traps (Figure 20.10). The traps are set out annually (usually at a density of one trap per 2.6 km^2) at a time before male flight. If certain traps catch more males than others, the area should be searched for egg masses. If a local population is found, the area should be monitored with pheromone traps more intensely, perhaps four traps per 2.6 km^2. It takes 3 to 4 years for a population to build up to epidemic proportions and chemical treatment is applied before stand losses begin. Outbreaks are commonly sprayed with *Bacillus thuringiensis,* when the foliage is about one-third developed; earlier if the population is exceptionally high and the insects are eating leaves as they appear.

Sawflies   These insects are hymenopterans (suborder Symphyta) whose larvae are herbivorous (Wagner and Raffa 1993). Sawfly larvae are superficially similar to caterpillars (Lepidoptera) but differ in the following ways: Sawflies have five or more pairs of prolegs (false legs on abdominal segments), whereas caterpillars have five or less; sawflies do not have hooklike attachments (crotchets) on the bottom of the prolegs, whereas caterpillars do; and sawflies have one pair of simple eyes, and caterpillars have several pairs of simple eyes (Figure 20.11).

There are many species of sawflies that feed on conifers and hardwoods. The life cycles of the various species can be generalized by knowing when during the season the larvae feed (Figure 20.12). For example, species whose larvae feed in spring would include *Neodiprion taedae linearis* (Arkansas pine sawfly); *N. taedae taeda* (loblolly sawfly); *N. nanulus nanulus* (red pine sawfly); *N. pratti pratti* (Virginia pine sawfly); *N. swainei* (Swaines jack pine sawfly); *N. sertifer* (European pine sawfly) (on hard pines in northeastern America, west to Illinois); *N. abietis,* balsam-fir sawfly; *N. tsugae,* hemlock sawfly; *Pristiphora erichsonii,* larch sawfly; *Diprion similis,* European pine sawfly (Figure 20.13) and many others that feed on conifers. Examples of sawflies that feed in summer include *N. lecontei* (redheaded pine sawfly) (on young pines in eastern North America); *N. pinetum* (white

Figure 20.10

Trap baited with the artificial sex pheromone of the gypsy moth (*Lymantria dispar*).

(Photo by the Washington State Department of Agriculture.)

Differences Between Sawflies and Caterpillars

Sawflies: Hymentoptera	Caterpillars: Lepidoptera
1. Sawflies have 1 pair of eyes—2 ocelli.	1. Caterpillars have 6 ocelli grouped in the form of a letter "C."
2. Sawflies have 5 or more prolegs.	2. Caterpillars have 5 or less prolegs.
3. Sawflies prolegs lack crotchets.	3. Caterpillar prolegs have crotchets.

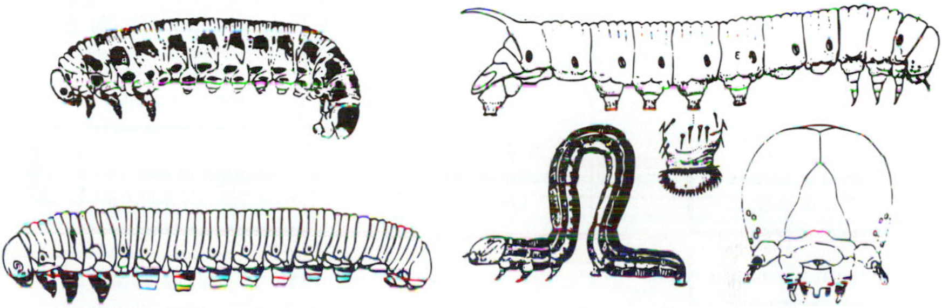

Figure 20.11

The morphological differences between sawfly larvae (Hymenoptera) and caterpillars (Lepidoptera).

(Source: Courtesy USDA.)

pine sawfly); *N. burkei* (lodgepole pine sawfly), *N. excitans* (blackheaded sawfly) (on southern pines) and several others that specialize on conifers; *Hemichroa crocea* (the striped alder sawfly), is a transcontinental sawfly that defoliates alder.

Arguably, the most economically important sawfly is the larch sawfly (*P. erichsonii*) (Wagner et al. 1979, Drooz 1960). Outbreaks completely defoliate thousands of acres of larch or tamarack in the northern hardwood-conifer forests of North America; it also occurs in Europe and Asia, and is probably native to the Old World. In North America its main host is tamarack (*Larix laricina*), but larvae also feed on practically all other larch species across the continent. Historical outbreaks have reduced the growth of tamarack stands and caused a great deal

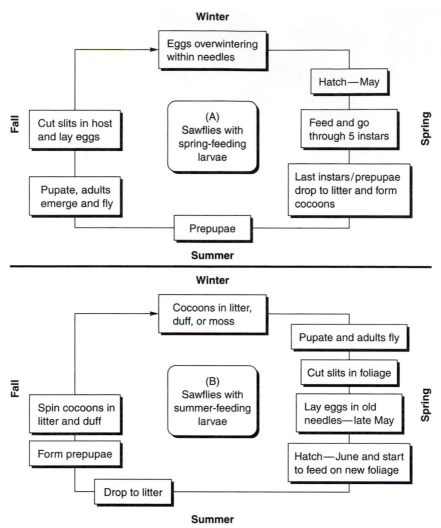

Figure 20.12
Diagram of the two life cycles commonly found in sawflies; one type is typified by larvae that overwinter in foliage and feed in spring, the other type overwinter as prepupae in the litter and larvae feed in summer.

of mortality. For example, between 1910 and 1926 about 4.7 million m³ of tamarack was killed in Minnesota. Recent outbreaks have not caused great mortality, but significant growth loss has frustrated timber management plans in the Lake states as well as in the neighboring Canadian provinces. Outbreaks also are a chronic problem in the boreal forests—a 1996 outbreak defoliated over 243,000 ha in Alaska. An outbreak close to Fairbanks resulted in considerable larch mortality and predisposed huge forested areas to outbreaks of the larch beetle (*Dendroctonus simplex*). The life cycle of *P. erichsonii* is as follows:

Figure 20.13
European pine sawfly (*Diprion similis*) ovipositing in needle of eastern white pine.
(Photo by A. Drooz.)

1. They overwinter as prepupae within compact cocoons scattered in the moss and litter of the forest floor.
2. There is a prolonged adult emergence that occurs between May and the end of July with a certain proportion of the population remaining in diapause (a resting stage) within cocoons and remain there for an additional year or two.
3. Most of the emerging insects are parthenogenic females (<1% are males), and these females fly to new shoots.
4. Eggs are laid chainlike along the twig; each female lays about 100 to 120 eggs.
5. Eggs hatch in about a week, the larvae feed in colonies, and succeeding instars consume all the foliage.
6. When the fifth and final instar have completed feeding, they drop to the ground, crawl into the litter, and spin cocoons.

Smaller but intense outbreaks with well-defined and monitored boundaries are controlled by spraying Malathion, carbaryl, and methoxychlor—pesticides registered for controlling incipient larch sawfly outbreaks in early July. Natural controls have been effective in controlling larch sawfly outbreaks, especially after introduction of the European ichneumonid wasp (*Mesoleius tenthredinis*). Introduction of this natural control agent became a classic success story—the control of an introduced herbivore by introducing and establishing one of its most effective parasitoids. Until the 1950s the parasitoid was effective as over a 60% parasitism was common in the Lake states. Then, **density-independent** mortality **factors** customarily caused outbreaks to collapse. By the mid 1950s the parasitoid no longer was effective. The sawflies were immune to parasitism—the eggs of *M. tenthredinis* were being encapsulated by blood hemocytes of the host. Studies demonstrated that blood cells of *P. erichsonii* flattened out and encapsulated the parasitoid's egg and asphyxiated it. This parasitoid-host relationship is a good example of insects becoming resistant to biological control efforts.

The Larch Casebearer   The larch casebearer (*Coleophora laricella*) belongs to a family of tiny moths, the Coleophoridae. This casebearer appeared in Massachusetts in 1886 when individuals flew from ballast piles left by sailing ships to nearby eastern larch trees. It soon spread to larch forests throughout the northeastern United States and the maritime Canadian provinces. By the early 1900s outbreaks were occurring in tamarack forests of Minnesota and western Ontario. In 1957 an outbreak was discovered in western larch growing in the area of St. Anthony, Idaho. When this outbreak was discovered, other heavily defoliated larch forests were discovered in northern Idaho and eastern Washington; the insect now is found throughout the Pacific Northwest.

Moths of the larch casebearer emerge from pupae during May through June and mate on the hosts that contained their pupae. Each female lays about 50 single eggs on the underside of needles. Approximately 2 weeks later the first instars drill through the bottom of their eggs and into the needles. They feed as leaf miners for 2 months. The third instars hollow out needles, line the inside with silk, and unfasten the needle base from the short-shoot (the rounded projection on the stem that gives rise to the needles). The insects live within these cases and drag them around with them from needle to needle. They feed by fastening their cases firmly to a needle with a pad of silk and then proceed to hollow out the needle. The larch casebearer overwinters as a third instar inside its case. During fall each larva drags its case to a short-shoot and attaches itself to this structure before it begins to hibernate. In spring the larvae resume feeding and by April, the fourth instars are rapidly devouring the new needles. Pupation takes place within the case and new adults begin to emerge by the end of April/early May. Damage results chiefly from the fourth instar feeding activity. In heavy infestations the needles are destroyed as soon as they appear. Fortunately larch can produce two needle crops. A European parasitic wasp, *Agathis pumila,* a parasite of the larch casebearer, was introduced during the 1970s. From a timber management standpoint, the introduction was successful as the parasite is firmly established in the forests and the pest is under natural control (Ryan 1997). In urban situations it may be necessary to spray the emerging moths as control by the parasite alone means that some damage may occur.

Tent Caterpillars   Lasiocampidae is a small family, but contains several destructive North American species (Craighead 1950, Furniss and Carolin 1977). Lasiocampid larvae feed as colonies in silken nests, thus the name, tent caterpillars. The eastern tent caterpillar (*Malacosoma americanum*) feeds on several Rosaceae, chiefly the *Prunus* and *Malus*; the western tent caterpillar (*M. californicum*) feeds on oak, willow, cottonwood, birch, alder, as well as several desert brush species. The most damaging, the forest tent caterpillar (*M. disstria*) feeds on cottonwood, willow, alder, birch, oak, maple, and other deciduous trees of the northern hardwood forests as well as on many important bottomland hardwoods of the Southeast. The forest tent caterpillar is widely distributed over much of North America. Although outbreaks of this tent caterpillar are scattered and unpredictable, they can be spectacularly large, and the growth of stands across millions of hectares is affected. For example, outbreaks in the 1960s in the Lake states and southern Canada covered about 195,000 km^2. Outbreaks normally subside after 3 to 4 years, but not before causing economic damage—for example, to maple syrup production in the Northeast. *Bacillus thuringiensis* sprays have been effective in controlling local outbreaks.

Adult *M. disstria* fly in early or mid summer, and females oviposit in characteristic rings around twigs of underbrush or trees and varnish the egg masses with a waterproof covering. The tiny caterpillars form in late autumn, pass the winter within the eggshells, and emerge when the leaves of their hosts unfold in spring. Young larvae feed on expanding buds, then the foliage, and through the first three instars they live gregariously under silken mats. As they reach maturity they begin to wander individually over the trees and other vegetation in search of pupation sites. They pupate in yellowish silken cocoons anywhere on the tree or on any other convenient object. There is a single generation per year.

Western North American Forests

Western Spruce Budworm   This is the most destructive forest defoliator in Western North America and occurs along the east side of the Cascades, the intermountain West, and into the Southwest (USDA 1987, Powell 1994, Shephard 1994). The main hosts of the western spruce budworm (*C. occidentalis*) include grand fir, white fir, subalpine fir, Douglas-fir, western larch, and the spruces. The life history of *C. occidentalis* is similar to that of the eastern budworm (*C. fumiferana*). Probably the biggest difference is that most western spruce budworm larvae overwinter in hibernacula located under bark scales and fissures of the main stem; few *C. fumiferana* overwinter on trunks of balsam fir.

The main reason why the budworm causes significant losses to western forests is that outbreaks are large and last many years (Figure 20.14). Besides loss of diameter and height growth, which result in longer rotations, these long-lasting outbreaks produce trees with dead tops, double tops, and other defects. If outbreaks coincide with droughts, significant tree mortality occurs across thousands of hectares.

The most susceptible hosts are those that break buds early in spring. This is a period when budbreak is in synchrony with the time budworms disperse and seek feeding sites. Hosts that tend to be in synchrony are the shade-tolerant late-successional species (USDA 1987) (Table 20.5). Pure, multistoried stands of grand fir are more severely impacted by budworm infestations than mixtures of Douglas-fir/true firs. Stands made up of 45% ponderosa pine and 55% Douglas-fir/true firs are only lightly infested with the budworm. Budworm damage increases dramatically as the density of Douglas-fir and/or true firs increases. Douglas-fir growing in mixture with true firs break buds after budworm emergence and are less susceptible, but where Douglas-fir is growing on a dry site or during droughts, then Douglas-fir breaks its buds earlier and becomes susceptible to massive defoliation.

Thick, dense stands of Douglas-fir and true firs have high total **foliar biomass** and provide budworm populations with high-quality food. Moreover, larval dispersal is reduced in dense stands. Dense stands are under competition for moisture, nutrients, and light. Accordingly food reserves become limiting and budworm outbreaks produce more top-kill, growth loss, and **crown class** shifts. Multistoried host stands are better budworm habitat than even-aged single-storied stands. Tall, dominant trees serve as targets for wind-dispersing larvae and the **intermediate** crown layers tend to reduce subsequent larval losses by assuring the immediate

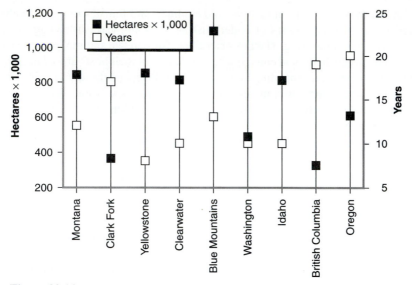

Figure 20.14

Areas encompassed by various western spruce budworm (*Choristoneura occidentalis*) outbreaks across the Pacific Northwest and Inland West, and the number of years each outbreak lasted.

(Source: From USDA 1987.)

availability of a food source. During the progression of a typical budworm outbreak, fifth and sixth instars deplete their food source (i.e., new growth) on dominant trees and at that time the insects spin down on threads in search of new food sources to complete their life cycles. In doing so, the insects readily find new food sources in codominant and intermediate trees. These lower-canopy hosts tend to be true firs, the budworm's favorite hosts.

Fast-growing stands are less susceptible to budworm damage than stagnating, stressed stands, because (1) foliage quality is better for budworm development—that is, less defensive chemicals; (2) older, stagnant stands have more male cones, and this is the food source that results in the most vigorous populations with the highest biotic potential; and (3) ultimately, starch reserves are limited in stressed trees with the accompanying greater impact on growth, defense, and recovery (USDA 1987).

In summary, the most susceptible stands are uneven-aged, multistoried stands with shade-tolerant firs predominating in a dense understory (Figure 20.15)—a stand condition that was brought about by 80 years of wildfire suppression and selective cutting (high-grading) of ponderosa pine. Late-successional species are most susceptible when their buds break in synchrony with budworm emergence. Mature host stands suffer more damage than younger stands because (1) at maturity trees devote less energy to production of defensive foliar chemicals; (2) there are more overwintering sites for construction of hibernacula; (3) there are greater

| T A B L E **20.5** | Susceptibility of Four Host Species to Infestations by the Western Spruce Budworm (*Choristoneura occidentalis*)[a] |

Host species	Susceptibility to attack	Synchrony with budburst	Shade tolerance
Subalpine fir	•••••	•••••	•••••
White fir & grand fir	••••	••••	••••
Engelmann spruce	•••	•••	•••
Douglas-fir[b]	••	••	••

[a]Susceptibility based on synchrony of budburst with emergance of budworm larvae in spring and on shade tolerance of the hosts.
[b]Douglas-fir during droughts or growing on dry sites (south-facing slopes) break buds earlier.

cone crops; and (4) trees tend to be taller and more capable of intercepting wind-blown larvae. Consequently budworm damage is reduced by precommercial and commercial thinnings and application of silvicultural techniques that favor pioneering species and nonhosts (ponderosa pine) and intermediate operations that include prescribed underburnings.

Douglas-Fir Tussock Moth   The Douglas-fir tussock moth (*Orgyia pseudotsugata*) has one generation a year and defoliates stands of Douglas-fir and true fir (Wickman 1963, Wickman et al. 1973, Clendenen, Gallucci, Gara 1978, USDA 1978, Wickman 1979, Mason 1981). Adults appear in late summer. The males are winged, but the female is wingless and does not move far from the cocoon from which she emerged (Figure 20.16). The egg mass, which is formed on the female's cocoon, contains about 300 eggs. These are covered with hairs adhering to a gelatinous substance exuded by the female's accessory gland. The insects overwinter in the egg stage, and in late May or early June, after new foliage has appeared on both Douglas-fir and firs, the eggs hatch. Newly hatched larvae crawl to the edge of the crown, drop on silken threads, and disperse with the wind to new locations. Early instars feed on the new growth but later instars feed on old foliage as well. Pupation occurs in mid summer and lasts 10 to 14 days. Emerging females then call males via a powerful sex pheromone.

In recent decades defoliations by the tussock moth have been of significant economical importance, especially when outbreaks coincide with those of the western spruce budworm (Powell 1994). However, in the more distant past, Douglas-fir tussock moth outbreaks have been rare—for example, the moth wasn't even described by McDunnough until 1921. Outbreaks generally occur in Douglas-fir/fir stands growing on dry sites. Because logging of ponderosa pine and fire suppression continue, the transition of stands from pine toward Douglas-fir and true fir also will continue. As a result of these stand-conversion practices, the likelihood of tussock moth outbreaks also will continue.

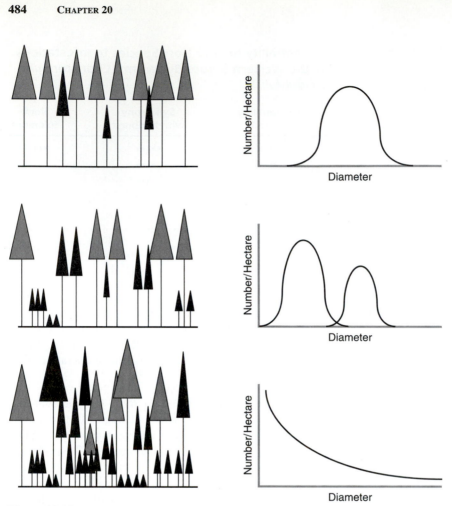

Figure 20.15
Stand structure, in order of susceptibility, from least to most susceptible: even-aged
(top); two-age class, storied (middle); and uneven-aged (bottom).
(Source: From USDA 1987.)

Controlling outbreaks by aerial spraying of pesticides has had some success in reducing mortality and growth loss. Direct control was justified because stand mortality continues for 3 to 4 years after collapse of an outbreak, as a result of the heavy defoliation itself and by bark beetle activity. Moreover, direct control will continue to be useful because the beginning of population increases can be detected by use of pheromone traps. Mating disruption techniques by aerial dispersal of the synthetic pheromone within plastic fibers also is a promising control method; in one application an over 80% reduction in reproduction occurred. This control method implies that if defoliation can be reduced one year, then subsequent defoliation will not cause a major impact on the stand. This assumption is based on the notion that trees must suffer almost total defoliation before trees are killed and surviving trees

Figure 20.16
Douglas-fir tussock moth
(*Orgyia pseudotsugata*)
female just after
emerging from its
cocoon.
(Source: USDA Forest Service
photograph.)

will recover with little or no long-term damage. Prevention of outbreaks requires reestablishment of pine on dry sites while discouraging late-successional species; thinning of dense Douglas-fir and fir stands; and underburning of seral stands to discourage development of multilayered grand and white fir stands.

Pine Butterfly   The pine butterfly (*Neophasia menapia*) is potentially one of the most destructive pests of ponderosa pine in northwestern North America (Furniss and Carolin 1977). Historical outbreaks have decimated thousands of hectares of ponderosa pine. For example, a 1953 outbreak in Idaho killed 25% of the pines in productive areas. A previous outbreak, augmented by western pine beetle activity, on the Yakama Indian Reservation in Washington, killed pines over a 61,000 ha area with a resulting 4.7 million m^3 of timber killed. The affected area on the Reservation is still evident by large even-aged forests that developed after the outbreak. Outbreaks of the pine butterfly primarily occur in mature forests. With the current emphasis on ecosystem management, large areas of old ponderosa pine will once more develop and pine butterfly will play a major role in nutrient cycling.

Recorded outbreaks have lasted only a few years as effective natural controls return pine butterfly populations back to a latent state. An important biological control agent is the ichneumonid wasp, *Theronia atalantae*. In one case over 90% control during the third year of an outbreak in Idaho has been recorded for this parasitoid.

Pandora Moth   The pandora moth (*Coloradia pandora*) is a large moth belonging to the family Saturnidae. During outbreaks they completely defoliate stands
of ponderosa pine. In 1963 a 1,620 ha outbreak in southern Oregon resulted in
noticeable growth loss during the subsequent 6 years, and because of stress some
of the most defoliated stands were killed by bark beetles (Furniss and Carolin
1977). Outbreaks occur at intervals of about 20 to 30 years. During outbreaks,
feeding may be heavy without much loss. This is because the terminal buds are
not eaten, and the insect has a 2-year life cycle. Such a long life cycle means that
feeding occurs on alternate years and trees have a chance to recover between periods of defoliation. Normally outbreaks of the Pandora moth are not controlled by
spraying. Natural controls regulate populations effectively; in particular, a **nucleopolyhedrosis virus** annihilates outbreak populations after the third presence of
mature larvae—that is, 6 years into an outbreak. As the pupae are in the soil,
ground squirrels and chipmunks dig up and destroy large quantities of pupae.

Western Hemlock Looper   Two subspecies of the hemlock looper feed on and
cause extensive tree mortality in mature hemlock forests of North America. These
are the western (*Lambdina fiscellaria lugubrosa*) and eastern hemlock loopers (*L.
fiscellaria fiscellaria*) (Otvos et al. 1973). In both regions the area of mature forest cover will increase due to growing appreciation for habitat provided by these
forests and reduced timber harvests in rural areas and National Forests. The western hemlock looper occurs in coastal and inland forests of the Pacific Northwest
and in the coastal forests of Alaska (Torgerson and Baker 1967). Susceptible
forests have high concentrations of mature and overmature western hemlock. The
susceptibility of these stands may be associated with the inability of old trees to
maintain a diverse energy budget that would include synthesis of expensive chemical defenses typically found in younger more vigorous hemlocks. In terms of
ecosystem dynamics, looper outbreaks in these senescing stands may be essential in recharging the forest system by recycling nutrients, a hypothetical concept
proposed by Mattson and Addy (1975) for the eastern spruce budworm.

Moths emerge in late summer and fall. They mate and each female lays about
100 eggs singly or in small groups on twigs, branches, barks, or hidden under epiphytes growing on the trunk or in the forest litter. The pin-size, bluish eggs overwinter and hatch in May through early June. Larval development is complete by
late August, and pupation takes about 3 weeks. Larvae preferentially feed on new
foliage. As the larvae mature the insects eat old foliage and even small twigs.
Loopers are not efficient feeders, tending to eat only part of a needle, leaving
browning needle remnants, a characteristic sign of an incipient infestation. During severe infestations stands are completely defoliated as well as the understory
species. In August to September the last instars lower themselves to the base of
the tree or litter where they pupate in protected places.

During heavy outbreaks associates species such as Douglas-fir and Pacific silver fir are heavily fed upon. Often surviving trees are killed later by bark beetles or
by windthrow, as the defoliation opens the stands to severe wind action. Looper outbreaks usually last 3 to 4 years when they are brought under control by parasitoids,

predators, and both fungal and viral diseases. The final collapse is associated with an epizootic of a nucleopolyhedrosis virus. Density-independent mortality factors such as heavy rains are important in regulation of looper populations; collapses also are preceded by cooler and wetter than normal conditions.

When large contiguous areas of mature western hemlock were common, major looper outbreaks occurred after droughts (see Figure 20.2). More than a century of logging, however, has reduced the area and continuity of late-successional stands dominated by western hemlocks. Clearcutting western hemlocks over huge areas accelerated after the 1940s and, because of the diminished numbers of susceptible hosts, large looper outbreaks (>2,500 ha) have been uncommon for many decades (Kinghorn 1954, Torgerson and Baker 1967). With the advent of ecosystem management, the return of mature western hemlock stands will occur. Accordingly the best silvicultural control to prevent outbreaks is to break the continuity between older western hemlock stands. Separating preferred species stands by young stands or any age stands of nonhost species would keep outbreaks from developing. Favoring mixed-species stands comprised of less than 50% western hemlock, avoiding cedar-hemlock mixes, and preferring nonhost species also will decrease susceptibility.

Historically spraying infested stands with DDT (last application in the Willapa Bay area, Washington, 1963) was successful in protecting stands from destruction. Aerial spraying worked because western hemlock trees can recover with only a 50 to 75% defoliation. Treatment took place after all the eggs hatched and while larvae were still in the early instars. Consequently, when looper outbreaks once more are common, an IPM management system would be feasible, which might include *B. thuringiensis* applications to reduce defoliation and increase stand survival.

Blackheaded Budworm   The range of the blackheaded budworm (*Acleris gloverana*) is from northern California and areas in the Rocky Mountains northward into Yukon and Northwest Territories in Canada and southeast Alaska (Hard 1974, Holsten et al. 1985). The most serious defoliations, however, occur across vast areas of western hemlock–Sitka spruce forests of southeastern Alaska and coastal British Columbia. They commonly infest second-growth stands, but their appearance in mature British Columbian forests often is in combination with high populations of the western hemlock looper. Outbreaks of the budworm cause reduced tree growth, tree top-kill, and some mortality. Its main hosts are western hemlock buds and the current year's foliage. Older hemlock needles and Sitka spruce foliage are eaten less frequently.

Eggs of the budworm are laid singly in late summer to early fall on the underside of needles, near the branch tips in the upper crown. The eggs hatch in spring and the small larvae feed on the new foliage as buds open. Larvae complete development between mid-July and mid-August. Pupation takes place within webs spun among the foliage and debris on the twigs. Adults emerge late in August and early in September.

Budworm emergence in spring coincides with western hemlock budbreak; Sitka spruces break their buds later. For this reason, Sitka spruce is not a preferred

TABLE **20.6** | **Predicted Defoliation Levels for the Western Blackheaded Budworm Based on Numbers of Eggs Per Branch**

No. of eggs sampled per 45 cm long branch sample	Predicted defoliation
1–5	Trace defoliation (evidence of feeding barely discernible)
6–25	Light defoliation (some branch and upper crown defoliation, barely visible from the air—25% defoliation)
26–59	Moderate defoliation (crown discoloration and noticeably thin foliage; obvious top defoliation—26–60% defoliation)
60+	Severe defoliation (top and many branches completely defoliated—61% defoliation)

Source: From Hard (1974).

host. During large outbreaks in the southern part of its range, other conifers also are fed upon, such as Douglas-fir, *Abies* spp., and Engelmann spruce. Outbreaks in coastal forests last 2 to 3 years before declining to low levels. In Alaska the budworm is at the northern extremity of its range, and unfavorable weather conditions together with increased parasitism decimate high budworm populations. In British Columbia increased parasitism (to an extent not seen in Alaska) and epizootics of a polyhedrosis virus cause outbreaks to crash.

Increasing stand vigor by attention to precommercial and commercial thinning of dense western hemlock stands protects them from severe budworm defoliation. Short-term strategies for managing western blackheaded budworm is to identify high-hazard stands containing the highest budworm population levels (Table 20.6) and treating selected areas with *B. thuringiensis*. Damage thresholds are reached when young stands (<60 years old) have already undergone 1 year of heavy defoliation.

REFERENCES

Campbell, R. W. 1979. *Gypsy moth forest influences*. USDA Forest Service Agriculture Information Bulletin 423.

Carolin, V. M., and W. K. Coulter. 1975. Comparison of western spruce budworm populations and damage on grand fir and Douglas-fir trees. USDA Forest Service, Pacific Northwest Forest and Range Experiment Station, Research Paper PNW-195.

Clendenen, G., V. F. Gallucci, and R. I. Gara. 1978. On the spectral analysis of cyclic tussock moth epidemics and corresponding climatic indices with a critical discussion of the underlying hypothesis, pp. 279–293. In C. Shugart (ed.). *Time series and ecological processes*. Society for Industrial and Applied Mathematics (SIAM), Philadelphia.

Craighead, F. C. 1950. *Insect enemies of eastern forests*. USDA Forest Service Miscellaneous Publication 657.

Department of the Environment Canada. 1975. *Aerial control of forest insects in Canada.* M. L. Prebble (ed.). Department of the Environment, Ottawa, Canada.

Dorais, F., and E. G. Kettela. 1982. *A review of entomological survey and assessment techniques used in regional spruce budworm,* Choristoneura fumiferana *(Clem.), surveys and in the assessment of operational spray programs.* Eastern Spruce Budworm Council, Department of Energy and Resources, Quebec.

Drooz, A. T. 1960. *The larch sawfly: its biology and control.* USDA Technical Bulletin 1212.

Furniss, R. L., and V. M. Carolin. 1977. *Western forest insects.* USDA Forest Service, Miscellaneous Publication 1339.

Hard, John S. 1974. Budworm in coastal Alaska. *Journal of Forestry* 72:26–31

Holsten, E. H., P. H. Hennon, and R. A. Werner. 1985. *Insects and diseases of Alaskan forests.* USDA Forest Service, Alaskan Region, State and Private Forestry, Report 181.

Kinghorn, J. 1954. The influence of stand composition on the mortality of various conifers, caused by defoliation of the western hemlock looper on Vancouver Island, British Columbia. *Forestry Chronicle* 30:380–400.

Kulman, H. M. 1971. Effects of insect defoliation on growth and mortality of trees. *Annual Review of Entomology* 16:289–324.

Mason, R. R. 1969. Sequential sampling of Douglas-fir tussock moth populations. USDA Forest Service Pacific Northwest Forest and Range Experiment Station, Research Note PNW-102.

Mason, R. R. 1979. *How to sample Douglas-fir tussock moth larvae.* USDA Forest Service Agricultural Handbook 547.

Mason, R. R. 1981. A numerical analysis of the causes of population collapse in a severe outbreak of the Douglas-fir tussock moth. *Annals of the Entomological Society of America* 74:51–57.

Mason, R. R., D. W. Scott, and H. G. Paul. 1993. Forecasting outbreaks of the Douglas-fir tussock moth from lower crown cocoon samples. USDA Forest Service, Pacific Northwest Research Station, Research Paper PNW 460.

Mattson, W. J., and N. D. Addy. 1975. Phytophagous insects as regulators of forest primary production. *Science* 190:515–522.

Morris, R. F. 1963. The dynamics of epidemic spruce budworm populations. *Memoirs Entomological Society of Canada* 31:1–332.

Otvos, Imre, D. MacLeod, and D. Tyrell. 1973. Two species of *Entomophthora* pathogenic to eastern hemlock looper (Lepidoptera: Geometridae) in Newfoundland. *Canadian Entomologist* 105:1435–1441.

Powell, David C. 1994. *Effects of the 1980s western spruce budworm outbreak on the Malheur National Forest in northeastern Oregon.* USDA Forest Service, Pacific Northwestern Region, R6-FI&D-TP-12-94.

Ryan, R. B. 1997. Before and after evaluation of biological control of the larch casebearer (Lepidoptera: Coleophoridae) in the Blue Mountains of Oregon and Washington, 1972–1995. *Environmental Entomology* 26:703–715.

Sheehan, Katherine A. 1996. *Effects of insecticide treatments on subsequent defoliation by western spruce budworm in Oregon and Washington: 1982–92.* USDA Forest Service, Pacific Northwest Research Station, PNW-GTR-367.

Shepherd, R. F. 1994. Management for forest insect defoliators in British Columbia. *Forest Ecology and Management* 68:303–324.

Torgerson, T., and B. Baker. 1967. The occurrence of the hemlock looper, *Lambdina fiscellaria* (Guenée) (Lepidoptera: Geometridae) in southeast Alaska, with notes on its biology. USDA Forest Service, Pacific Northwest Forest and Range Experiment Station, Research Note PNW-61.

USDA. 1978. *The Douglas-fir tussock moth: a synthesis,* Martha A. Brookes, R. W. Stark, and R. W. Campbell (eds.). USDA Forest Service Science and Education Agency Technical Bulletin 1585.

USDA. 1981. *The gypsy moth: research toward integrated pest management,* C. C. Doane, Michael L. McManus (eds.). Expanded Gypsy Moth Program, USDA Forest Service Science and Education Agency Technical Bulletin 1584.

USDA. 1987. *Western spruce budworm,* Martha H. Brookes, R. W. Campbell, J. J. Colbert, R. G. Mitchell, and R. W. Stark (coordinators). USDA Forest Service Technical Bulletin 1694.

Wagner, Michael, and K. F. Raffa. 1993. *Sawfly life history adaptations to woody plants.* Academic Press, San Diego, Calif.

Wagner, M. R., T. Ikeda, D. M. Benjamin, and F. Matsumura. 1979. Host derived chemicals: the basis for preferential feeding behavior of the larch sawfly, *Pristophora erichsonii* (Hymenoptera: Tenthredinidae) on tamarack, *Larix laricina. Canadian Entomologist* 111:165–169.

Wellington, W. G. 1952. Air mass climatology of Ontario north of Lake Huron and Lake Superior before outbreaks of the spruce budworm, *Choristoneura fumiferana* (Clem.), and the forest tent caterpillar, *Malacosoma disstria* Hbn. Canadian *Journal of Zoology* 30:114–127.

Wellington, W. G., J. J. Fettes, K. B. Turner, and R. M. Belyea. 1950. Physical and biological indicators of the development of outbreaks of the spruce budworm, *Chorisoneura fumiferana* (Clem.) (Lepidoptera: Tortricidae). *Canadian Journal of Research* 28:308–330.

Wellington, W. G. and R. M. Trimble. 1984. Weather, pp. 399–425. In C. B. Huffaker and R. L. Rabb (eds.). *Ecological Entomology.* Wiley and Sons, New York.

Wickman, B. E. 1979. A case study of a Douglas-fir tussock moth outbreak and stand condition 10 years later. USDA Forest Service, Pacific Northwest Forest and Range Experiment Station, Research Paper PNW-224.

Wickman, Boyd E. 1963. Mortality and growth reduction of white fir following defoliation by the Douglas-fir tussock moth. USDA, Forest Service, Pacific Southwest Forest and Range Experiment Station, Research Paper PSW-7.

Wickman, B. E., R. R. Mason, and C. G. Thompson. 1973. *Major outbreaks of the Douglas-fir tussock moth in Oregon and California.* USDA Forest Service, Pacific Northwest Forest and Range Experiment Station, GTR, PNW-5.

BARK BEETLES AND THEIR MANAGEMENT

INTRODUCTION

Bark beetles belong to the family Scolytidae, a family that contains extremely destructive forest insects. Scolytid outbreaks frequently kill hundreds of square kilometers of coniferous forests. In this context they influence tree age, size, and species distributions across the landscape. The bark beetles, together with pathogens, wind, and fire are a preeminent factor in forest succession.

The Scolytidae form two ecological groups: the bark beetles and ambrosia beetles (latter to be discussed in Chapter 22). Bark beetle broods feed on subcortical tissues (phloem tissues), and ambrosia beetles are wood-boring insects whose larvae feed on fungi (Figure 21.1). Scolytids can be classified on the physiological condition or vigor of their hosts (Rudinsky 1962). Beetles that normally select vigorous hosts are **primary** scolytids, whereas those that normally select weakened and thus susceptible hosts are secondary scolytids. Most of the economically important bark beetles are considered secondary: During latent (endemic or low) population levels, they preferentially attack downed trees, logs, fire-damaged trees, moribund trees, or trees temporally weakened by drought, defoliation, root rots, or other destructive agents (Table 21.1). When populations of secondary bark beetles increase, however, they can become primary and kill healthy trees. For example, the Douglas-fir beetle (*Dendroctonus pseudotsugae*) and *Ips* spp. are secondary scolytids that preferentially attack windthrown trees, logs, and fire-damaged trees. When there is a surplus of brood material, their populations can increase and mass attack living trees.

Associations with Fungi

Scolytids have **symbiotic** relationships with fungi (Berryman and Ashraf 1970, Barras and Perry 1972, Berryman 1972). Fungi associated with aggressive scolytids have an active role in killing trees as developing **mycelia** block vascular tissues. In scolytids invading pines, successful penetration of the **mutualistic** fungi produce a blue stain in the sapwood. The association between bark beetles

491

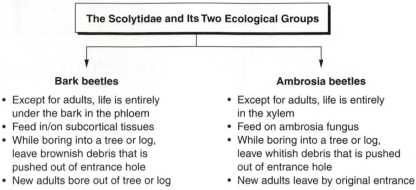

Figure 21.1

The two ecological groups found in the family Scolytidae: the bark beetles and the ambrosia beetles.

and blue-staining fungi arose from the requirement that both the fungi and the beetles need susceptible trees for their development. For instance, when ponderosa pines are unable to muster up their resinous defense systems (e.g. , a reduced **oleoresin exudation pressure**), a condition brought on by water stress, the blue-staining fungi can successfully develop in sapwood of these trees. Spores of these fungi are introduced into the succulent phloem of these trees by invading *Dendroctonus brevicomis, D. ponderosae,* or *Ips* spp. Beetles either transport spores adhering to their bodies, in specialized pits, or within specialized structures, termed **mycangia.** Most staining fungi associated with bark beetles belong to the genus *Ophiostoma*—for example, *O. pini* is carried by *Ips calligraphus, I. grandicollis,* and *I. pini*; on the other hand, the fir engraver beetle (*Scolytus ventralis*) vectors the brown-staining fungus, *Trichosporium symbioticum.*

Fungal interactions with subcortical tissues also serve to enhance the **palatability** and nutrient content of scolytid food. There are scolytid and also some weevil (Curculionidae) species that infest roots of stumps, logging debris, abandoned logs, or the bases of moribund hosts. Some of these **Rhynchophora** (a taxonomic classification that contains the weevils and bark beetles) include the bark beetles, *Hylastes nigrinus, Hylastes ater,* and *Hylurgus lignipurda* (latter two native to Europe); examples of weevils include *Hylobius pales* (pales weevil), *Pissodes fasciatus,* and *Steremnius carinatus.* Brood development in decaying material poses two problems: (1) the fermenting substrate may not provide sufficient energy for reproductive development and (2) larvae developing in this decaying **substrate** may form relationships with pathogens. In the former case, emerging adults feed at the root crown of newly planted seedlings for sexual maturation and kill the seedlings—as is the case with *H. ater, H. pales, P. fasciatus,* and *S. carinatus.* In the latter example, *H. nigrinus, P. fasciatus,* and *S. carinatus* are vectors of the black-stain root disease fungus, *Leptographium wagenerii* (see Chapter 12). When Douglas-fir stands are thinned in southwest Oregon, the fungus is transmitted to new stumps as spore-laden insects move into this new food source

TABLE **21.1** | **Classification of Scolytidae with Regard to Host Selection Preferences**[a]

Beetles invading	Condition of subcortical tissues	Attack behavior of beetles consiered as
Living, healthy trees	Physiologically unchanged, defense systems normal	Primary: *Xyleborus destruens, X. ferrugineus* and other tropical species
Permanently or temporarily weakened trees	Phloem succulent with carbohydrate decreased content and limited defenses	Secondary: *Dendroctonus, Ips, Scolytus, Hylastes, Hylurgus, Piytogenes, Dryocetes, Polygraphus,* and others including ambrosia bettles
Fresh logs windthrows; freshly dead trees; moribund fire-killed trees; suppressed trees; severely defoliated trees, etc.	Phloem moisture decreased but still succulent; phloem may be discolored as it ferments; moisture increasing toward saturation as tissues decompose	
Year-old dead trees; old logs and old windthrows	Subcortical region discolored; fermentation well along; minimal carbohydrate content	Saprophagous: source cambium beetles, etc.

[a]Primary scolytids prefer to attack healthy trees; secondary ones prefer to attack weakened hosts.
Source: Rudinsky 1962; with permission, from the *Annual Review of Entomology,* Volume 7, © 1962, by Annual Reviews.

(Goheen and Hansen 1978). Studies in Oregon demonstrated that 15-year-old dense stands, thinned to 600 stems per hetacre, developed black-stain root disease in 15% of the leave trees during the next 6 years. Plantation mortality by these rhynchophorans is of concern to intensive silvicultural applications around the world and a major issue in implementation of **quarantine** regulations.

In the case of *Scolytus multistriatus*, vector of the Dutch elm disease (caused by the fungus *Ophiostoma ulmi*), association between bark beetle and fungus allows for spread of the fungus and concomitant weakening of elm trees. As infected elms become moribund, they are susceptible hosts for attack of *S. multistriatus* (see Chapter 14). Mortality of elms in European and North American cities is of major economic and aesthetic concern to urban forestry.

Biology of Bark Beetles

Bark beetles have common traits and habits (Wood 1982). They all excavate **egg galleries** in fresh phloem. The larvae feed away, at right angles from the egg gallery, and feed in the succulent phloem tissues. Patterns formed by both the original egg galleries and the **larval mines** are characteristic for each bark beetle species, as shown in Figure 21.2 for five species that infest southern pines.

Bark beetles complete their life cycle under the bark of their hosts, with the exception of a short flight period. A flight period is essential to bark beetles that colonize living trees. By flying, these insects first find, then overcome, the resistance

Four-spined engraver   *Ips   avulsus*   Eichh.

Five-spined engraver   *Ips   grandicollis*   Eichh.

Six-spined engraver   *Ips   calligraphus*   Germ.

Southern pine beetle   *Dendroctonus   frontalis*   Zimm.

Black turpentine beetle   *Dendroctonus   terebrans*   (Oliv.)

Painted by Richard Kilofflu

Figure 21.2
Egg gallery and larval mine patterns produced by five species of scolytids that infest southern pines: (1) the black turpentine beetle (*Dendroctonus terebrans*) infesting the basal portion of its host, creating a cavitylike gallery in the phloem; (2) the southern pine beetle (*Dendroctonus frontalis*) killing the tree and creating netlike galleries; and (3) three engraver beetles (*Ips calligraphus, I. grandicollis,* and *I. avulsus*) that share the phloem region once the tree is moribund.

of these trees through mass attacks (Hodges et al. 1985). Similarly, a dispersal flight (Figure 21.3) is vital to species that depend on finding scattered and ephemeral, temporary habitats—hosts with rapidly degrading phloem, as found in windthrows, moribund, and fire-damaged trees. The flight period also serves to widely scatter the flying population, thereby assuring gene pool heterogeneity.

Figure 21.3
Adult southern pine beetle taking off in a spring dispersal flight.

In many of the highly destructive bark beetles, the colonization process is initiated by beetles that emerge early in the flight season; these are known as pioneer beetles. These pioneers select susceptible trees or downed logs. Stress, caused by drought, disease, competition for light or nutrients, pollution, fires, defoliation, damage during intermediate operations, and other causes are factors predisposing trees to this initial attack. Forest entomologists disagree as to how dispersing bark beetles initially select their hosts (Gara 1967, Heikkenen 1977, Hynum and Berryman 1980, Moeck et al. 1981, Wood 1982, Gara et al. 1984). Some studies show that bark beetle landing rates on weakened and unweakened trees are about the same. These randomly landing beetles, while crawling on the bark, then determine whether the tree is susceptible to attack. Other studies suggest that dispersing beetles are guided by odors (primary attractants) from weakened trees. In any case, after pioneering beetles select their hosts, the insects release **aggregation pheromones,** secondary attractants that attract both males and females. This attraction provokes a mass attack on the selected trees or logs by attracting later-emerging beetles within range of the chemical cue (Figure 21.4).

Once the selected brood material is mass attacked, the insects mate, and each female cuts an egg gallery in the succulent phloem. As egg galleries are constructed, each female chisels egg niches along the sides of the egg galleries and deposits an egg in each niche. Upon hatching, each larva feeds in the phloem perpendicular to the egg gallery. Depending upon the bark beetle species, pupation takes place in a pupal niche at the end of the larval mine or within the outer bark.

MANAGEMENT OF BARK BEETLES

An IPM system for bark beetles includes the following objectives: (1) development of a damage threshold where pest suppression is deemed necessary, (2) establishment of a detection-population monitoring system, and (3) development of silvicultural techniques that lower the population by interfering with the host selection behaviors of dispersing beetles (Hedden 1978, Vité 1990).

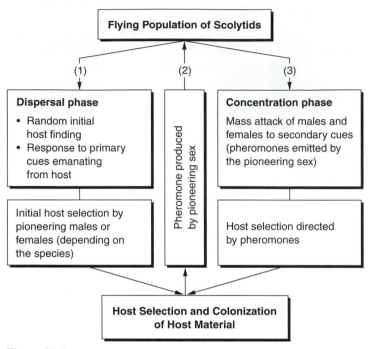

Figure 21.4

Host selection behavior model for bark beetles. Spring and early summer emergence and proper flight temperatures cause beetles to fly: the flying population. These beetles fly with the wind in a random dispersal flight: the dispersal phase. After initial host selection, the pioneering sex (e.g., males in *Ips* spp. and females in *Dendroctonus* spp.) emit a pheromone that is sensed by the flying population. The population, responding to this cue, flies upwind toward the selected host: the concentration phase. In this manner, the flying population is guided to the hosts and colonization of the material is assured.

Damage Thresholds

Catastrophic losses by bark beetles (e.g., the genus *Dendroctonus*) are unpredictable and characterized by great severity with major timber losses across thousands of hectares. These losses dislocate long-term forest management and utilization plans for the affected areas. In intensive forest management operations, damage thresholds can be calculated by knowing the capital inputs and determining the level of losses that can be sustained by bark beetles. In other words, a timber-growing operation operates according to a production model that relates the input of investment in land, labor, and capital in the output (Stern 1973). The effects of insect-caused losses then could be determined on the predicted output. This analysis could be used to approximate an economic damage threshold. The development of techniques to detect and manage western pine beetle populations in the western United States utilizes an economic damage concept based on timber values, beetle establishment dynamics and salvage parameters.

The economic damage level for southern pines being decimated by the southern pine beetle was developed from present net worth models developed by the U.S. Forest Service (Leuschner 1978). In East Texas a spot of dead pine trees killed by the southern pine beetle (*D. frontalis*) of 10 or more is considered to have reached a threshold that would cause economic damage. This low number is because *D. frontalis* outbreaks have the potential of spreading explosively and killing timber on hundreds of hectares.

Detection

Bark beetle detection can be done from ground surveys in well-roaded areas. Over large areas bark beetle surveys are done from the air. Aerial surveys, however, generally evaluate evidence that lags about a year behind beetle attacks. This is because trees killed by bark beetles often do not change color during the season of attack. The time lag between attack and symptoms varies with beetle species, tree species, time of season when attacked, and local environmental factors. Information recorded during aerial surveys includes number and distribution of dead trees per square kilometer and the number of dead trees per group. Four degrees of infestation are recognized according to the number of trees per square kilometer: light, medium, heavy, and extreme. Ground checks subsequently evaluate the degree of damage and determine more accurately the boundaries of infested areas (Figure 21.5).

In southern forests, where there are multiple generations of the southern pine beetle (SPB), estimating whether or not an infestation, or spot, is actively expanding is difficult. Because of rapid rates of beetle development in summer, beetles will have emerged from trees by the time they are detected. In expanding spots there is no problem as SPBs will be found in green-crowned trees at the periphery of the spots. In practice, small spots detected from the air are likely to be inactive when ground checked. Accordingly spots reported in summer will have a 10-tree minimum. Even with a 10-tree threshold, detection crews may have difficulty distinguishing between small spots in the process of expansion and those likely to be inactive when ground checked (Figure 21.6). All data from aerial surveys and ground checks are entered into a database. Data entered into the Texas Forest Service's computerized informational system is summarized in Table 21.2.

Pest Control Strategies

Chemical control of bark beetle infestations has a long history. Reviews of research on the efficacy of this method have shown that expenses have not justified the results—that is, in few instances have outbreaks been suppressed by pesticides alone. Chemical control is predicated on the notion that if most of the infested spots can be located, entered, and all scolytid broods killed with insecticides, the outbreaks will collapse. The approach is flawed because when spots are located from the air, found on the ground and finally treated, most of the beetles have dispersed: Human intervention simply is a step behind the population.

Because of complex host selection strategies and the fact that most of their life cycle is under the bark, scolytid outbreaks are difficult to control. Host availability,

Figure 21.5
Sketch mapping southern pine beetle kills (beetle spots) during an aerial survey of coastal plain pine forests of East Texas (**A**); subsequent ground checking of spots found during the aerial survey (**B**).

(Photos provided by R. F. Billings.)

susceptibility, and suitability are the main determinants of bark beetle population size. Integrated pest management (IPM), based on maintenance of vigorous forests, with minimal amounts of moribund or downed breeding material, is at the core of successful IPM strategies. However, when outbreaks occur, suppression of populations involves not only increasing stand vigor but also exploiting a weak link in the host selection behavior of each bark beetle species, such as the expedient use of synthetic semiochemicals (Goyer et al. 1998).

THE DENDROCTONUS

The genus *Dendroctonus* contains the most damaging bark beetles—the name means "tree killer." They yearly kill hundreds of square kilometers of coniferous timber in North America. All species are not equally damaging black and red turpentine beetles, (*D. terebrans* and *D. valens* respectively) are strictly secondary, but continually kill several pine trees per hectare—a continual drain on the inventory. The most damaging of the *Dendroctonus* are the southern pine beetle (*D. frontalis*), the mountain pine beetle (*D. ponderosae*), the western pine beetle (*D. brevicomis*), the Douglas-fir beetle (*D. pseudotsugae*), and the spruce beetle (*D. rufipennis*) (Table 21.3).

The Southern Pine Beetle (*D. frontalis*)
As shown in Figure 21.4, flying populations of randomly searching *D. frontalis* females locate susceptible hosts, called *focus trees*. Selection of these trees establishes pheromone sources that attract (concentrate) the flying population. Attacks on surrounding trees occur as beetles concentrate on the focus trees, but as the

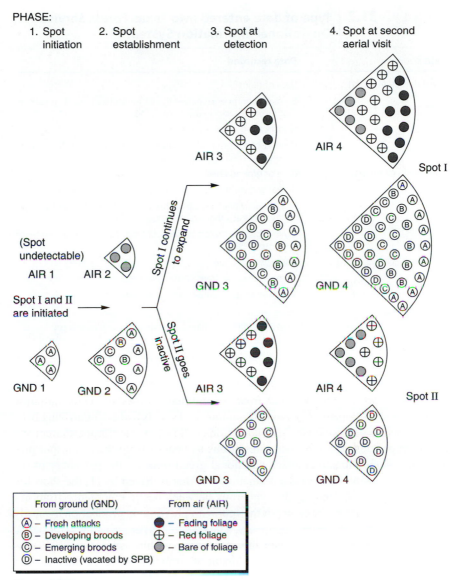

Figure 21.6

Relationship between patterns of infested trees visible from the air and progression of southern pine beetle attacks on the ground in two spots that have different outcomes. Although both spots appear similar at detection, inactivity in Spot II becomes evident at the second aerial visit by the absence of trees with fading, yellowish crowns.

(Source: From Texas Forest Service 1979.)

focus trees become unattractive by interactions between male pheromones and host volatiles, incoming beetles begin to attack neighboring stems—a behavior called the **switching mechanism** (Coster and Gara 1968).

TABLE **21.2** | **Type of data entered into Texas Forest Service Operational Information System[a]**

Data source	Data recorded
Aerial detection report	Date of flight
	Estimated number of active trees (fading trees signify presence of recent attacks in adjacent stand)
	Grid location of mapped spot
	Ground-check priority, based on level of current southern pine beetle activity
Ground-check report	Date of ground check
	Actual grid location of spot (accurate mapped location)
	Relative activity of spot (number and location of recently attacked trees that are still green)
	Control priority (based on level of beetle activity and volume of infested timber)
Spot management report	Date spot controlled
	Number of active trees controlled (these trees are felled)
	Inventory of spot, detailing timber values lost

[a]Data gathered from *Dendroctonus frontalis* detection/evaluation flights, ground check reports and spot-management activities.
Source: From Billings (1974).

The spread and collapse of *D. frontalis* outbreaks consist of (1) the initiation of new infestations (spots) in previously uninfested stands by beetles arriving from distant sources and establishing an attraction center (new pheromone source) and (2) the expansion of established infestations as beetles from the same spot produce new generations that attack additional green trees on the periphery of the spot. Spot growth in spring is a dramatic event that is driven by (1) the short life cycle of *D. frontalis* and (2) the synchrony between beetle emergence from initially attacked trees in the center of the spot with pheromone production by newly attacking females at the spot periphery. Thus a self-perpetuating cycle of beetle development and new brood establishment is set up within a spot (Gara 1967). As seen in Figure 21.7, as long as establishment of attraction centers coincides with beetle emergence, the spot spreads. However, if for any reason (low temperatures, heavy rain, or removal of freshly infested trees) the synchrony is disrupted, emerging beetles will disperse, and the infestation stops growing.

Host selection information has been used in the management of southern pine beetle infestations (Billings 1974, Hedden and Billings 1979, Goyer et al. 1998). In practice this is done by finding the freshly attacked trees at the periphery of a spot; these are the trees that attract flying beetles that emerge within the outbreak. Freshly attacked trees still have green foliage, fresh pitch tubes, and fresh boring dust in bark crevices and at the base of the trees. Beetles can be heard landing, boring, and stridulating (making a chirping sound) in trees undergoing attack.

TABLE **21.3** | **Characteristics of the Genus *Dendroctonus***

Species	Egg gallery pattern in phloem	Larval mines	Pupal cells	Host species
D. frontalis	Winding in S-like fashion	Mostly in outer bark	In outer bark	All southern pines
D. ponderosae	Somewhat linear, going with the grain	Radiate at right angles from egg gallery	In phloem at end of larval mine	All western pines, especially lodgepole, mature western white, and pole-size ponderosa pine
D. brevicomis	Winding and crossing adjacent egg galleries, giving a netlike appearance	Mostly in outer bark	In outer bark	Ponderosa pine
D. rufipennis	Vertical, going with the grain	Grouped on opposite side of egg gallery	In phloem, scoring sapwood	All species of spruce, especially, white, red, Engelmann, and Sitka
D. pseudotsugae	Mostly linear, with the grain	Fan-shaped groups in phloem, on alternate sides of egg gallery	In phloem	Douglas-fir

Source: From Knight and Hiekennen, *Principles of Forest Entomology*, 5e, ©1980 by McGraw-Hill Book Co., reproduced with permission of The McGraw-Hill Companies.

These trees are cut and salvaged together with a 12 to 25 m buffer of green trees (Figure 21.8). In this manner synchrony between beetle emergence, flight, and sources of attractants is interrupted and the treated outbreaks stop expanding as emerging beetles disperse (see Figure 21.7). During the height of an outbreak and when trees cannot be salvaged in a timely manner, the cut-and-leave control method is used. This method involves felling infested trees as well as a buffer of uninfested trees and leaving them in the forest. The treatment disrupts spot growth and emergent beetles are forced into dispersal.

Similarly recognizing that outbreaks expand in a tree-to-tree fashion, dense stands can be thinned to prevent or significantly decrease the spread of spots. Prevention of outbreaks is best accomplished by attention to widespread thinning regimes (Hedden 1978, Nebeker et al. 1985). Thinning southern pines to basal areas of about 9 m^2ha^{-1} minimizes infestations, and has these benefits: (1) there is an increase in host resistance to initial attack, (2) there are changes in stand structures that prevent formation of conditions conducive to outbreaks, and (3) there is an overall revitalization of stands that reduce many linked biological hazards that ultimately support southern pine beetle outbreaks.

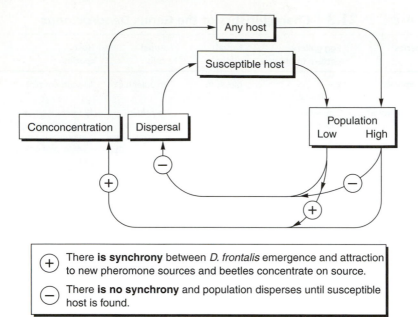

Figure 21.7

An illustration of conditions associated with expansion or collapse of a local southern pine beetle infestation: a spot. During normal host selection behavior, when populations are low, beetles locate susceptible hosts during a dispersal flight. The spot enlarges as beetles responding to new attacks are drawn to recently attacked trees from throughout the forest. However, when populations are high and a spot is enlarging rapidly, adults emerging from within the outbreak are triggered into a concentration flight as their emergence is synchronized with availability of pheromones at the perimeter of the outbreak: a self-generating outbreak. If the synchrony between new pheromone sources and emergence is broken, the emerging beetles disperse and the spot becomes inactive.

The Mountain Pine Beetle (*D. ponderosae*)

Mountain pine beetle infestations of 80- to 90-year-old ponderosa pine and lodgepole pines in the Pacific Northwest are closely associated with overstocked stands and droughts. Mountain pine beetles find and colonize their hosts as shown in Figure 21.4. The first-attacked trees can be termed as *focus trees*. As focus trees attract more beetles (Pitman and Vité 1969), the responding males produce an anti-aggregation pheromone, exo-brevicomin, which masks the female attractant. These trees become unattractive and no longer serve as cues to the concentrating beetle population. Instead, the incoming beetles attack other trees in the vicinity of the focus trees, again the switching mechanism. In lodgepole pine the distance that beetles will switch from a focus tree to an adjacent tree is directly related to the diameters of nearby trees. This means that incoming beetles will invade the

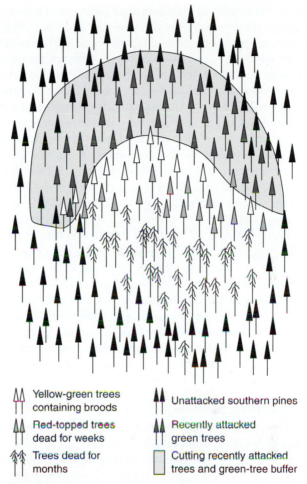

Yellow-green trees containing broods		Unattacked southern pines	
Red-topped trees dead for weeks		Recently attacked green trees	
Trees dead for months		Cutting recently attacked trees and green-tree buffer	

Figure 21.8

A salvage system to manage southern pine beetle outbreaks. Once spots are identified from an aerial survey, the infestation is found and the active trees with the spot are marked (trees undergoing attack that haven't faded). A buffer of unattacked trees around the active head of the outbreak also is marked. The buffer should be as wide as the average height of trees in the spot, about 12 to 25 m. The active trees and the buffer should be cut first, as this would guarantee that synchrony between beetles emerging from the brood trees and sources of attraction will be broken, thus driving emerging beetles into dispersal.

(Source: From Texas Forest Service.)

next largest tree in the vicinity of the most recent focus tree. This relationship can be used in design of thinning prescriptions (Geiszler and Gara 1978, Mitchell et al. 1983) that prevent spread of a mountain pine beetle infestation (Figure 21.9).

In dense, second-growth ponderosa pine stands, however, the spread of an outbreak more closely resembles that of *D. frontalis*. First the focus trees are attacked, then the next nearest neighbors are attacked in a serial fashion.

The frequency and intensity of mountain pine beetle outbreaks can be reduced by thinning (Sartwell and Stevens 1975). Increased host resistance to mountain pine beetle attack is generally regarded as the mechanism that reduces host mortality after thinning. Thinning studies by the U.S. Forest Service show that tree mortality fell immediately after thinning, even though tree vigor indices did not increase until after the next *D. ponderosae* flight season (Schmid et al. 1991). In general, mountain pine beetle–caused mortality is always several times higher in unthinned pine stands (Schmid and Mata 1992). Thinning experiments were done during a mountain pine beetle outbreak in stagnating, 80- to 90-year-old ponderosa pine stands in eastern Washington State. These studies were done to see if three thinning regimes (4, 6, 8 m between trees, each replicated three times by using 0.25 ha plots plus six unthinned control plots) would affect the switching mechanism of attacking beetles—to see if the tree-to-tree spread could be broken. Results showed that even 4 m between trees would stop the spread of a localized outbreak. Accordingly mountain pine beetle management in second-growth ponderosa pine stands can be accomplished by attention to intermediate operations—as also is the case with mountain pine beetle and *Ips* outbreaks in the Rocky Mountain area (Table 21.4).

The Douglas-Fir Beetle (*D. pseudotsugae*)

During years with normal amounts of fallen or severely weakened trees, emerging Douglas-fir beetles fly and disperse with the prevailing wind. In time, pioneering females find scattered windthrown or stressed trees. These females bore into the phloem and release the aggregating pheromone that guides the dispersing population to the newly found food material. The insects will then mate, and each female will cut egg galleries within the phloem and lay about 30 to 50 eggs. Broods develop within the logs and emerge the following spring as new adults.

In a year with large amounts of blowdown material, there will be vastly more food material to colonize than during normal years. With surplus food and minimal host selection flight, the population will magnify dramatically. The year following the blowdown event, and assuming no additional major blowdowns, the increased Douglas-fir beetle population will further multiply, and reattack the first-year's brood material, scattered stems that fell in winter, and possibly living trees.

During the third year, and assuming no major blowdowns to absorb the now immense Douglas-fir beetle population, attacks on living trees is probable. In addition, if stands are under drought stress, a major Douglas-fir beetle outbreak can occur (Table 21.5). This scenario has been the historical pattern as large and unsalvaged-blowdowns were followed by large Douglas-fir beetle outbreaks. An exception to the rule was the famous 1962 Columbus Day storm, when over 20,000 ha of

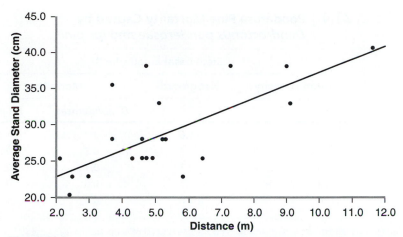

Figure 21.9

Relationship between lodgepole pine stand diameter and distance over which mountain pine beetles will switch attacks from a focus tree (a tree initially selected and attacked) and a neighboring tree, called a recipient tree. For example, if the average stand diameter is 30 cm, then thinning trees to a spacing of about 5.8 m between trees would stop the outbreak from expanding or prevent an outbreak from developing.

western Washington timber was blown down, but timely salvage operations prevented subsequent bark beetle infestations. Clearly maintenance of stand vigor and prompt salvage of downed logs will prevent Douglas-fir beetle outbreaks.

The Spruce Beetle (*D. rufipennis*)

The spruce beetle destroys spruce forests of North America. Over the last two decades, *D. rufipennis* outbreaks have been exceptionally destructive to Lutz spruce (*Picea lutzii*) stands in south-central Alaska, and white spruce (*P. glauca*) stands of interior Alaska (Werner and Holsten 1983, Reynolds and Hard 1991). The bark beetle attacks all spruce species, with historically devastating outbreaks in northeastern North America during the 1920s, an immense outbreak in the Rocky Mountains of Colorado from 1940 to 1952, and a recent outbreak that expanded over hundreds of square kilometers in British Columbia from 1970 to 1980. Although epidemics of the spruce beetle destroy spruce over large areas, latent beetle populations normally colonize windthrown or moribund trees. Outbreaks begin in unmanaged mature forests when beetle populations build up in windthrow patches or logging slash, and invade living trees when their favored downed material is exhausted.

In spring populations of beetles disperse until pioneering females find downed material. Then the pheromone produced by these females stimulates the previously dispersing population to aggregate on the selected hosts. During outbreaks, pioneering females land not only on windthrows but also on certain trees, suggesting a response to olfactory cues emanating from weakened trees. Studies

TABLE **21.4** | Ponderosa Pine Mortality Caused by *Dendroctonus ponderosae* and *Ips pini*[a]

	Stem basal area(m²ha⁻¹)				
	Stand density		Net growth	Mortality	
Spacing (m)	1967	1972		*D. ponderosae*	*Ips pini*
Unthinned	39.2	34.6	−4.6	2.7	1.0
3.7 × 3.7	26.6	25.9	−0.7	0.7	0.5
4.6 × 4.6	19.5	20.2	0.7	0.1	0.0
5.5 × 5.5	13.9	14.6	0.7	0.0	0.0
6.4 × 6.4	8.0	8.6	0.6	0.0	0.0

[a]Occurred 5 years after three thinning regimes in the Black Hills forest of South Dakota. Response to thinning is represented by cumulative net growth.
Source: From Sartwell and Stevens (1975).

show that spruce beetles land on ethanol-treated trees, suggesting the role of this compound as a putative primary attractant. Secondary attraction (response to pheromones) begins once a host tree comes under attack. As the attack rate on the focus trees augments, beetles simultaneously land on neighboring trees (Figure 21.10), in numbers proportional to the rate of attack on the focus tree.

Spruce beetle outbreaks are closely related with dense stands and prolonged drought. Alaskan studies demonstrated a significant decreased radial growth preceded outbreaks by 5 years. Outbreaks occurred in these stagnant, slow-growing stands after strong winds provided breeding material for dispersing beetles. The vastly augmented population then proceeded to kill living trees, with a preference for the largest-diameter trees.

Prevention of *D. rufipennis* outbreaks includes thinning of stands and prompt salvage of windthrown timber. Another outbreak prevention method involves using trap logs or trees, where the synthetic pheromone of *D. rufipennis* attracts beetles to selected log piles or trees. The material utilized can be treated with an insecticide to kill the developing brood, or the material can be salvaged and chipped. In this manner beetle populations are reduced within the treated area, and beetles that fly to the traps from long distances will not attack trees outside the treatment. In operation it is better to salvage the trees rather than use insecticides. Predators, such as the clerid beetle, *Thanasimus undalutus*, and parasitoid wasps, also respond to frontalin (the spruce beetle's pheromone).

The Western Pine Beetle (*D. brevicomis*)
The western pine beetle generally infests overmature ponderosa pine and Coulter pine (Miller and Keen 1960, Furniss and Carolin 1977). Populations of *D. brevicomis* kill senescing, water-stressed, and root rot infected pines that have been

T A B L E **21.5**	**Sequence of Events That Convert the Douglas-Fir Bark Beetle from an Opportunistic, Secondary Insect (Living in the Phloem of Blowdowns) to a Primary Insect (Killing and Colonizing Phloem of Living Trees)[a]**

Susceptible Host material—logs and blown-down trees	Population phase (endemic, extensive, intensive)	Population activity
Yr 1—Extensive blowdowns occur	Endemic	Normal host finding colonizing
Yr 2—More food material than population can utilize	Beginning of extensive phase	Normal host finding and colonizing
Yr 3—Just enough food to maintain attacking population	Extensive population increase	Attack on living trees possible
Yr 4—Less food available to maintain huge population	Intensive phase and population collapses	**Attack of living trees occurs**

[a]When the population is high and the standing Douglas-fir forests are weakened, as after a series of drought years, attacks on living trees (primary attack) occur and outbreaks develop.
Source: Modified from Rudinsky (1962).

growing slowly and losing their foliage for many years. During droughts these susceptible trees are located and infested. When droughts persist, *D. brevicomis* populations increase to epidemic proportions, and beetles kill even vigorous hosts with diameters over 25 cm.

Dispersal flights begin in spring or early summer. There are two generations per year in the southern part of its range and a single generation in the northern part, such as in Oregon and Washington. Initial attacks occur about midstem, and subsequent attacks fill in above and below. Small pitch tubes and red boring dust are indications of successful mass attacks. Egg galleries of individual pairs wind both laterally and longitudinally within the subcortical zone, forming netlike patterns as galleries of adjacent pairs cross one another. Only one instar occurs in the phloem, the rest are in the outer bark. During summer, when most of the larvae and pupae are within the bark, the populations are subjected to intense woodpecker predation. Ponderosa pines killed by *D. brevicomis* are discernible from those killed by the mountain pine beetle, because the bark of the former is often dug out by these avian predators.

Sanitation logging has been the only effective way to prevent western pine beetle infestations (Vité 1971, Fiddler et al. 1989). High-risk trees can be identified

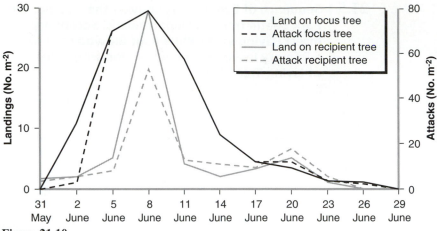

Figure 21.10

Patterns of spruce beetles (*D. rufipennis*) landing and attack on trees baited with the synthetic attractant (focus trees), as well as subsequent patterns of landing and attack on neighboring trees (recipient trees).

and removed from the stand (Miller and Keen 1960). These are trees that have been suffering from water stress and can be easily identified because of thinning crowns and bark characteristics (Figure 21.11).

THE ENGRAVER BEETLES (*IPS* AND *SCOLYTUS*)

The Engraver Beetles (*Ips spp.*)

The *Ips* beetles normally breed in downed pine and spruce material such as logging slash and blowdowns (Speight and Wainhouse 1989, Goyer et al. 1998). But as in other secondary bark beetles, when their populations increase, they also colonize trees and often create outbreaks, especially in combination with drought.

In boreal forests of the north, *I. borealis* and *I. pertubatus* attack host trees in a more opportunistic fashion than congeneric groups of temperate climates. For example, *I. borealis*, in the vicinity of the Eli River of Alaska, respond to terpenes exuding from freshly wounded spruces. If the wounds are not severe, the attacking population is pitched out and killed. This behavior is in contrast to endemic *Ips* populations from temperate regions where *Ips* seldom attack nonsusceptible material. The opportunistic host selection behavior of *I. borealis* is an adaptation for species dealing with a short, unpredictable flight season.

The most widespread pine engraver beetle in North America is *I. pini*. Its hosts consist of *Pinus strobus, P. resinosa, P. ponderosa, P. contorta, P. banksiana, P. jeffreyi,* and *P. virginiana*. Three spruce species also are attacked, although attacks on spruces are infrequent. As with all *Ips* species, males initiate the selection of new brood material by boring into downed logs or branches. They bore into the phloem and construct a nuptial chamber. Following completion of the chamber, they release

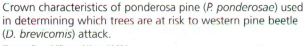

Figure 21.11

Crown characteristics of ponderosa pine (*P. ponderosae*) used in determining which trees are at risk to western pine beetle (*D. brevicomis*) attack.

(Source: From Miller and Keen 1960.)

a pheromone that causes aggregation of males and females toward the recently attacked host material. One to five *I. pini* females join each male in the nuptial chamber. After mating, females construct egg galleries away from the nuptial chamber and proceed either with or across the grain of the wood. Each female cuts 35 to 55 egg niches at regular intervals in the walls of the gallery and deposits an egg in each niche. Males stay within the gallery, removing frass and pushing it out the entrance hole; unlike *Dendroctonus,* pine engraver beetle galleries are clean. After hatching, the larvae feed in the phloem, make pupal cells at the end of the larval mines, and later emerge as new adults. Depending on the climate of a region, there may be one to three generations of *I. pini* per year.

During latent population levels, beetles may kill widely scattered single trees. In outbreak years they may kill groups of 50 to more than 500 trees, especially in unthinned stands. Drought-stressed ponderosa pine in the 60- to 80-year age class, averaging 15 to 20 cm diameter at breast height (dbh), are most susceptible to attack.

Ponderosa pine management policies often encourage *I. pini* infestations. For example, management of stands near Billings, Montana, between 1980 and 1993, included 12-month logging and thinning regimes, coupled with intermittent pre-commercial thinnings and continual fire suppression activities. These policies, together with a drought that covered the same period, provided a continually high *I. pini* population as a steady supply of logging slash and weakened trees became available. These conditions encouraged a chronic 2 to 5% ponderosa pine mortality. Key elements to the management of these *Ips* populations included knowing (1) the flight period of each generation, (2) the period when the three key predator species were active; and (3) when logs dried to the point where they were unsuitable for brood material.

Ips pini have three flight periods in southeastern Montana (Figure 21.12). Supplies of fresh logging slash and thinning residues are fundamental in providing food for the March to June emergence of the overwintered *Ips* population. Studies on the Montana *Ips* problem revealed the following:

1. Three beetle predators (*Enoclerus lecontei, E. sphegeus,* and *Temnochila chlorodia*) partitioned the periodic availability of *I. pini* adults and broods (Figure 21.13).
2. A model, based on the suitability of slash in the process of drying, provided an Index of Population Change. The index predicted whether an *I. pini* population would increase, decrease, or stay the same from one generation to the next. An index value of 1.0 meant that it maintained its size. For example, logging slash created during the winter (e.g., January–February 1992) had an index value of 3.32, which meant that the population increased more than three times its original size. However, logging slash produced in October 1991 had an index value of 0.91, meaning the population was decreasing (Figure 21.14). Slash produced in winter did not dry out and remained succulent until the overwintering *I. pini* generation utilized this material in spring. On the other hand, slash created in October had about 3 months to dry out before it was frozen, so that by the time the spring scolytid flights occurred, the slash was too dry for development of *Ips* broods.

With this information, an effective pest management program that prevented future *I. pini* outbreaks included that following recommendations:

1. The latest time that logging and thinning activities would be allowed would be in late August through December. Adherence to this rule would prevent the second *I. pini* generation from finding overwintering host material (see Figure 21.12).
2. Fresh pine slash would be minimized during January through March. In this manner the overwintering broods would not have fresh host material to absorb their population in spring. Alternatively the same results would occur if slash produced in winter were piled and burned before the spring dispersal flights.
3. Until further study, use of trap logs and pheromones to trap out *I. pini* were not recommended since three predator species have partitioned the three flight

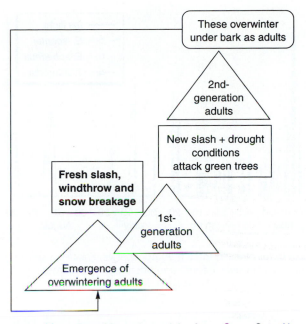

Feb   Mar   Apr   May   Jun   Jul   Aug   Sep   Oct   Nov   Dec   Jan

Figure 21.12

Ips pini activity patterns and type of host material infested within ponderosa pine forests of the Northern Cheyenne Indian Reservation, Lame Deer, Montana

periods of *I. pini.* Studies show that, together with the drying of scolytid host material, the predators effectively regulate low-level *Ips* populations.

In the oak-pine region of North America, *I. grandicollis, I. calligraphus,* and *I. avulsus* can kill southern pines (Werner 1972). Of these, *I. calligraphus* is the most aggressive and small group kills are common. As with all *Ips,* the southern complex prefer felled trees and slash as host material, but when this material is abundant, populations augment extensively. This increased population, coupled with dense unthinned pine stands, which in addition are water stressed, can produce group kills of 5 to 100 trees. This problem is especially serious in housing developments and city parks where individual trees have high aesthetic and monetary values—as, for example, in suburban Houston or Atlanta. Under these urban conditions, the *Ips* hazard can be reduced by adhering to the following rules:

1. Trees left at building and landscaping sites should be selected for maximum vigor and carefully protected during construction. If trees cannot be adequately protected, it is better to remove them before construction begins.
2. Thinning dense stands of pines reduces intertree competition.
3. Deep watering during dry periods and proper fertilization will reduce stress and improve vigor among residual trees.

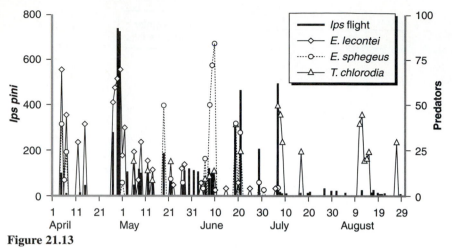

Figure 21.13
Seasonal flight patterns of *Ips pini* and three predator species, *Enoclerus sphegeus,*
E. lecontei, and *Temnochila chlorodia.* Information obtained in ponderosa pine forests
of the Northern Cheyenne Reservation, Lame Deer, Montana.

4. Damaged trees, downed or uprooted trees, those broken by storms should be
   disposed of promptly to avoid a buildup of engraver beetle populations.
5. Finally, especially valuable trees can be given preventive insecticide sprays
   of recommended products.

There are many important species of *Scolytus.* In western North America there
are 18 species that invade many conifers, in particular the genera *Abies, Picea,
Tsuga, Pseudotsuga,* and *Larix.* Other *Scolytus* attack hardwood species, such as
S. rugulosis, which is a major pest of apple, cherry, plum, pear, and other fruit
trees; it also attacks mountain mahogany. By far, the most important is *S. multistriatus,* which invades moribund elms and is the vector of the Dutch elm disease
fungus (Barbosa and Wagner 1989, Schowalter and Filip 1993).

The *Scolytus* Engraver Beetles
The fir engraver beetle, *S. ventralis,* is a killer of true firs and a major problem in
management of grand and white fir. The egg galleries and larval mines are easily
identified. The fir engraver attacks trees from pole size to full maturity. Periodically outbreaks expand along the Cordilleran range from British Columbia to
southern California. Outbreaks are associated with root pathogens (i.e., *Heterobasidion annosum* and *Armillaria ostoyae*) and triggered by prolonged droughts.
Trees severely defoliated by the western spruce budworm and the Douglas-fir tussock moth are also subsequently killed by the fir engraver. After *S. ventralis* outbreaks have drastically thinned fir stands, the residual stands are overall more
healthy and have an enhanced ability to repel future attacks. The results of a more
vigorous stand are seen during the third and fourth year after a defoliator outbreak

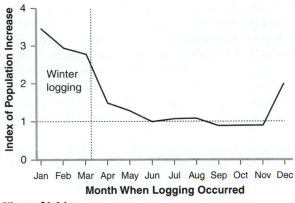

Figure 21.14

Population change of *Ips pini* infesting ponderosa pine slash in southeastern Montana. A population change of 1 means the population is remaining constant from one generation to the next; a population over 1 means the population is increasing; under 1 means the population is decreasing.

subsides. An important result of this recovery is the marked increase in unsuccessful fir engraver beetle attacks that leads to collapse of the beetle population.

Fir engraver attacks also are correlated with logging. This is because leave trees go into shock when stands are opened and beetle populations can breed in slash. When engraver populations are high, harvesting activities should be confined to mid-August to December in order to prevent a buildup in the slash. When poorer sites are harvested, reforestation should be done with pine seedlings. In particular, pine should be encouraged on sites where grand or white fir compose over 70% of the area; stands that have over 70% crown closure (as estimated from aerial photos); and fir stands that receive 89 cm or less of annual precipitation.

The smaller European elm bark beetle (*S. multistriatus*) breeds in weakened and dying elm trees. The dying elms are located by dispersing males and females as they respond to host-produced volatile materials. As unmated females begin their egg galleries, they release a pheromone that concentrates the population to the selected host. When beetles arrive, females initiate more egg galleries while males crawl over the bark, locating and inseminating the tunneling females. Once mated, females stop pheromone production and attraction to the breeding sites declines. In spring, after larvae develop and pupate in the bark of their dead hosts, the new adults are covered with spores of *Ophiostoma ulmi,* the Dutch elm disease pathogen (see Figure 14.4). A certain percentage of this population flies to the tops of healthy elm where females feed on twigs. Twig feeding is a behavior necessary to mature their ovaries, a behavior that injects the healthy host with the Dutch elm pathogen.

Control of *S. multistriatus* has been attempted by prompt removal of dead and dying hosts. However, these efforts have lagged the host selection and attack behavior of the beetles. Mass trapping of beetle populations by means of synthetic pheromone baits has had little effect on suppressing epidemic *S. multistriatus* populations. The maximum effectiveness of this treatment is realized in urban centers where brood trees and competing pheromone sources are low. Accordingly mass trapping may be a management component of an IPM program that includes sanitation and other pest and disease suppression techniques.

REFERENCES

Barbosa, P., and M. R. Wagner. 1989. *Introduction to forest and shade tree insects.* Academic Press, New York.

Barras, S. J., and T. J. Perry. 1972. Fungal symbionts in the prothoracic mycangium of *Dendroctonus frontalis* (Coleoptera: Scolytidae) is synonymous with a secondary female character. *Zeitschrift für angewandte Entomologie* 71:95–104.

Berryman, A. A. 1972. Resistance of conifers to invasion by bark beetle-fungus associations. *BioScience* 21:598–602.

Berryman, A. A., and M. Ashraf. 1970. Effects of *Abies grandis* resin on the attack behavior and brood survival of *Scolytus ventralis* (Coleoptera: Scolytidae). *Canadian Entomologist* 102:1219–1236.

Billings, R. F. 1974. The Texas Forest Service southern pine beetle control program: analysis of survey and control records to provide a basis for making improved operational decisions, pp. 54–57. In T. L. Payne, R. N. Coulson, and R. C. Thatcher (eds.). *Southern pine beetle symposium.* Texas A&M University, College Station, Texas.

Coster, J. E., and R. I. Gara. 1968. Studies in the attack behavior of the southern pine beetle. II. Response to attractive host material. *Contributions from the Boyce Thompson Institute* 24:69–76.

Fiddler, G. O., T. O. Fiddler, D. R. Hart, and P. M. McDonald. 1989. Thinning decreases mortality and increases growth of ponderosa pine in northeastern California. USDA Forest Service Research Paper PSW-194.

Furniss, R. L., and V. M. Carolin. 1977. *Western forest insects.* USDA Forest Service Miscellaneous Publication 1339.

Gara, R. I. 1967. Studies on the attack behavior of the southern pine beetle. I. the spreading and collapse of outbreaks. *Contributions from the Boyce Thompson Institute* 23:349–354.

Gara, R. I., D. R. Geiszler, and W. R. Littke. 1984. Primary attraction of the mountain pine beetle to lodgepole pine in Oregon. *Annals of the Entomological Society of America* 77:333–335.

Geiszler, D. R., and R. I. Gara. 1978. Mountain pine beetle dynamics in lodgepole pine, pp. 182–187. In D. M. Baumgartner (ed.). *Theory and practice of mountain pine beetle management in lodgepole pine forests.* Washington State Cooperative Extension Service, Pullman, Wash.

Goheen, D., and E. M. Hansen. 1978. Incidence of *Verticicladiella wagnerii* and *Phellinus weirii* in Douglas-fir adjacent to and away from roads in western Oregon. *Plant Disease Reporter* 62:495–498.

Goyer, R. A., M. R. Wagner, and T. D. Schowalter. 1998. Current and proposed technologies for bark bettle management. *Journal of Forestry* 96(12):29–33.

Hedden, R. L. 1978. The need for intensive forest management to reduce southern pine beetle activity in East Texas. *Southern Journal of Applied Forestry* 2:19–21.

Hedden, R. J., and R. F. Billings. 1979. Southern pine beetle: factors influencing the growth and decline of summer infestations in East Texas. *Forest Science* 25:547–566.

Heikkenen, H. J. 1977. Southern pine beetle: a hypothesis regarding its primary attractant. *Journal of Forestry* 75:412–413.

Hodges, J. D., T. E. Nebeker, J. D. DeAngelis, B. L. Karr, and C. A. Blanche. 1985. Host resistance and mortality: a hypothesis based on the southern pine beetle—microorganisms—host interactions. *Bulletin of the Entomological Society of America* 4:31–35.

Hynum, B. G., and A. A. Berryman. 1980. *Dendroctonus ponderosae* (Coleoptera: Scolytidae): preaggregation landing and gallery initiation on lodgepole pine. *Canadian Entomologist* 112:185–191.

Knight, F. B. and H. J. Heikkenen. 1980. *Principles of Forest Entomology.* McGraw-Hill, New York.

Leuschner, W. A. 1978. Impacts of the southern pine beetle. In R. C. Thatcher, J. L. Searcy, J. E. Coster, and G. D. Hertel (eds.). *The Southern Pine Beetle.* USDA Forest Service Science and Administration Technical Bulletin 1631.

Miller, J. M., and F. P. Keen. 1960. *Biology and control of the western pine beetle.* USDA Forest Service Miscellaneous Publication 800.

Mitchell, R. G., R. H. Waring, and G. B. Pitman. 1983. Thinning lodgepole pine increases tree vigor and resistance to mountain pine beetle. *Forest Science* 29:204–211.

Moeck, H. A., D. L. Wood, and D. L. Lindahl. 1981. Host selection behavior of bark beetles attacking *Pinus ponderosa* with special emphasis on the western pine beetle, *Dendroctonus brevicomis. Journal of Chemical Ecology* 7:49–83.

Nebeker, T. E., J. D. Hodges, B. L. Karr, and D. M. Moehring. 1985. *Thinning practices in southern pines—with pest management recommendations.* USDA Forest Service Technical Bulletin 1703.

Pitman, G. B., and J. P. Vité. 1969. Aggregation behavior of *Dendroctonus ponderosae* (Coleoptera: Scolytidae) in response to chemical messengers. *Canadian Entomologist* 101:143–149.

Reynolds, K. M., and J. S. Hard. 1991. Risk and hazard of spruce beetle attack in unmanaged stands on the Kenai Peninsula, Alaska, under epidemic conditions. *Forest Ecology and Management* 43:137–151.

Rudinsky, J. A. 1962. Ecology of Scolytidae. *Annual Review of Entomology* 7:327–366.

Sartwell, C., and R. E. Stevens. 1975. Mountain pine beetle in ponderosa pine: prospects for silvicultural control in second-growth stands. *Journal of Forestry* 73:136–140.

Schmid, J. M., and S. A. Mata. 1992. Stand density and mountain pine beetle–caused tree mortality in ponderosa pine stands. USDA Forest Service, Rocky Mountain Forest and Range Experiment Station, RM-515.

Schmid, J. M., S. A. Mata, and C. B. Edminster. 1991. Periodic annual increment in basal area and diameter growth in partial cut stands of ponderosa pine. USDA Forest Service, Rocky Mountain Forest and Range Experiment Station, Research Note RM-509.

Schowalter, T. D., and G. M. Filip (tech. eds.). 1993. Beetle pathogen interactions in conifers. Academic Press, London.

Speight, M. R., and D. Wainhouse. 1989. *Ecology and management of forest insects.* Claredon Press, Oxford, United Kingdom.

Stern, V. M. 1973. Economic thresholds. *Annual Review of Entomology* 18:259–280.

Vité, J. P. 1971. Silvicultural and the management of bark beetle pests. *Proceedings of the Tall Timbers Conference on Ecological Animal Control by Habitat Management* 3:155–168.

Vité, J. P. 1990. Present and future use of semiochemicals in pest management of bark beetles. *Journal of Chemical Ecology* 16:3031–3041.

Werner, R. A. 1972. Aggregation behavior of the beetle *Ips grandicollis* in response to host-produced attractants. *Journal of Insect Physiology* 18:425–427

Werner, R. A., and E. H. Holsten. 1983. Mortality of white spruce during a spruce beetle outbreak on the Kenai Peninsula in Alaska. *Canadian Journal of Research* 13:96–101.

Wood, D. L. 1982. The role of pheromones, kairomones, and allomones in the host selection and colonization behavior of bark beetles. *Annual Review of Entomology* 27:411–446.

22

AMBROSIA BEETLES AND THEIR MANAGEMENT

INTRODUCTION

Logs exported to Asia from the Pacific Northwest represent an income of about $1.5 billion. Even though log export volumes decreased during the 1990s (CINTRAFOR 1994), revenues declined just slightly (Figure 22.1). A major portion of this export was to Asia, and the fact that log revenues remained high implies that premium prices were paid for high-quality raw material. Members of the Scolytidae and Platypodidae, the ambrosia beetles, directly affect these profits. This is because ambrosia beetles invade the sapwood of felled trees or standing trees dying from fire, bark beetles, defoliators, or root diseases. They also infest logs, cants (logs that have been squared), freshly peeled logs, or green lumber. Their damage (holes in the wood) is greatest in the outer 8 to 10 cm of the sapwood, the highest-quality, knot-free portion of large logs (typically logs >80 yrs old) (Figure 22.2). Damage by ambrosia beetles can be so harmful that entire shipments of logs, cants, or lumber are ruined and unacceptable to potential customers. For example, a marine surveying company (Alexander Gow, Inc. of Seattle) reported in 1981 that an infestation of the striped ambrosia beetle (*Trypodendron lineatum*) in Alaska degraded entire shiploads of high-grade Sitka spruce and western hemlock bound for Asian markets. Their report states, "We found literally hundreds of thousands of holes approximately 1 mm in diameter in wood which was in storage at the time of the incident and which had not yet been shipped out." In another part they state, "The entire mill area and vessel were inundated by swarms of flying beetles which descended on the entire storage area and in the vessels cargo holds. . . . Fire hoses from shore and from (the ship) played over the cants in an attempt to flush or wash the beetles off the piles of lumber." Damage by ambrosia beetles is as follows (McClean 1985):

1. **Degrade** As these insects produce a multitude of black-stained holes in the first few centimeters of logs, it is not possible to saw clear, select-grade lumber from the material.

516

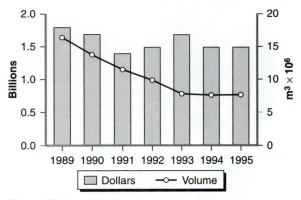

Figure 22.1
Relationship between volume of coniferous log exports from the Pacific Northwest and the income of these shipments to the region.

2. Export Problems   Often log-importing countries demand high-quality logs and lumber without any defect. Home construction in Japan, for instance, is by a traditional post-and-beam, labor-intensive method that requires clear, straight grained material. Holes caused by insects are unacceptable.
3. Remanufacturing and Repacking   When there is a strong demand for high-grade lumber, boards with insect holes are reprocessed. Often bundled lumber must be disassembled and then reassembled after the infested boards are removed. All of these operations are expensive and ultimately reduce profits.
4. Inconvenience   Because of the vulnerability of stored logs or unseasoned lumber in the forest or processing areas, the inventory must constantly be moved. The luxury of retaining stored and stacked materials is dampened.
5. IPM Costs   Managing damage caused by ambrosia beetles is high and adds to production costs.

AMBROSIA BEETLES AND SOFTWOODS

The striped ambrosia beetle (*T. lineatum*) is the most important ambrosia beetle in the West as well as in the mountains of western North Carolina. It is a **holarctic species** that also degrades timber values in Europe (Chapman and Kinghorn 1958, Chapman 1962, Chapman and Dyer 1969, Borden 1988). The striped ambrosia beetle infests practically all conifers and has been reported to attack four angiosperm genera: *Alnus, Betula, Malus,* and *Acer.* Adults emerge in spring from overwintering sites at the base of stumps and debris and begin dispersal flights whenever air temperatures reach 15° to 15.5°C (Figure 22.3). Later they invade logs and blowdowns that were created or fell before the end of January. This host material will have aged 3 to 5 months and has undergone **anaerobic metabolism,**

Figure 22.2
Three species of ambrosia beetles found in the Pacific Northwest, left to right: *Platypus wilsoni*, *Gnathotrichus sulcatus*, and *Trypodendron lineatum* (above). Douglas-fir log infested by *T. lineatum* showing fungus-stained galleries and entrance holes (below).
(Photo by N. E. Johnson.)

which results in production of ethanol. Ethanol serves as a primary attractant or **arrestant** for dispersing beetles. Upon finding a suitable host in this manner, females produce a powerful pheromone after they begin boring into the xylem. The attractive compound is lineatin, which beetles synthesize from ethyl alcohol and α-pinene of their host (Borden 1990). Flying males and females within range of this olfactory cue now direct their flights to these scattered and degrading hosts.

Females construct tunnels in the sapwood. As they bore into the wood, they cut cradles (chambers in which to oviposit) above and below their tunnels (Figure 22.4). These tunneling females also inoculate the walls of these galleries with spores of an ambrosia fungi, principally *Monilia ferruginea*. The spores are carried in specialized structures called *mycangia* found embedded in the prothoracic cuticle. The fungal hyphae penetrate the wood to some extent but mostly proliferate within the tunnels and stain them black. These mycelial gardens provide food for developing larvae. The new generation of beetles emerges together with reemergent parent adults in summer when temperatures persist above 24°C. New adults do not strongly respond to pheromones as their main task is to locate suitable overwintering sites in litter around stumps and other protective large woody debris found on the forest floor.

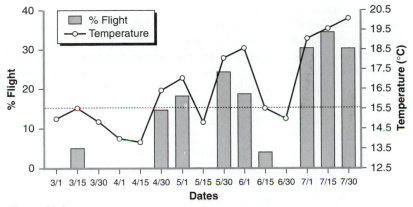

Figure 22.3

Relationship between *Trypodendron lineatum* dispersal flights and ambient temperatures.

(Source: Modified from Chapman and Kinghorn 1958.)

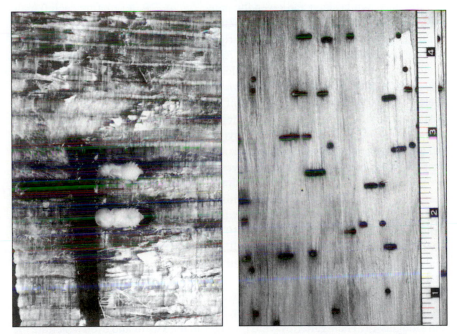

Figure 22.4

Trypodendron lineatum pupae developing in cradles (left) and a damaged board sawn from an infested western hemlock log (right).

(Photo by N. E. Johnson.)

The most viable integrated pest management (IPM) strategy is to limit the amount of available habitat because populations of the striped ambrosia beetles are host dependent. Water misting of log decks is useful in protecting high-value logs from attack, probably by altering the microenvironment on the log surface. Another method of protecting logs involves the use of repellents extracted from pulpmill wastes, such as terpenoids, oleic acid, or pine oil. Attracting beetles during their spring flights away from vulnerable logs (Lindgren et al. 1983) also has been successfully tried in timber-processing areas, where chance of success is greater than in the forest (Table 22.1).

Gnathotrichus sulcatus and *G. retusus* are another pair of western ambrosia beetles that infest coniferous logs, blowdowns, and freshly sawn material in the Pacific Northwest (Furniss and Carolin 1977). Both of these *Gnathotrichus* species infest Douglas-fir, western hemlock and pine. *G. sulcatus,* in addition, is reported from redwood, western redcedar, and other conifers—*G. sulcatus* is the economically most important. Adult *G. sulcatus* emerge from within their hosts in April and reach peak numbers by June; emergence continues throughout the summer. The entrance tunnel and part of the main gallery is initiated by the male beetle after selection of a suitable host. The males emit an aggregating pheromone that has been synthesized and given the name *sulcatol.* After mating each female constructs an egg gallery within the xylem of their host log. The egg galleries lie within the sapwood in a single plane at right angles to the grain, and are concentric with the annual rings (Figure 22.5). As many as 60 eggs are laid in ellipsoidal niches along the galleries. Soon the galleries are stained with the ambrosia fungi *Ambrosiella sulcati* and *Raffaelea sulcati,* which have been introduced by male beetles. The larvae complete development in the wood and overwinter in all stages. Dispersal flights begin when ambient temperatures are between 13° and 26°C. These dispersing beetles respond to ethyl alcohol being emitted from host material after a period of anaerobic metabolism. Moreover, beetles sense this primary host-selection cue within 5 days after western hemlock logs are cut.

Management of *G. sulcatus* populations is similar to that of the lined ambrosia beetle. Operational IPM programs in British Columbia have successfully curtailed major *G. sulcatus* damage within mill yards by establishing beetle-trapping stations around the mill (McClean and Borden 1979). Traps baited with **racemic** sulcatol were placed in key areas around the mill; after the beetle flights began, the percentage of damage expected was reduced by more than 65%.

The platypodid (family Platypodidae) *Platypus wilsoni* is an ambrosia beetle that ranges from British Columbia to California and eastward into Idaho. The beetle invades weakened, injured, and recently dead true firs, western hemlock, and Douglas-fir. Their galleries wind throughout the xylem and even penetrate the heartwood. Side branches are cut at intervals along the main tunnels where the eggs are laid. *P. abietis* is a small beetle that attacks white fir in Utah, Arizona, and New Mexico. *P. pini* attacks ponderosa pine logs, slash, and dying trees in Arizona and New Mexico.

The principal ambrosia beetle that attacks southern pines is *P. flavicornis* and found from Texas east to Florida and New Jersey (Baker 1972). This *Platypus*

TABLE 22.1 | **IPM Approaches and Tactics That Exploit Behavior of *Trypodendron lineatum* Populations**

Observations	IPM approaches	
	Strategy	**Tactics**
Population levels dependent on supply of hosts available for development of brood	Regulate amount of resource: availability of host material	Reduce amount of logging debris and culled logs left in the woods; remove felled and bucked timber from forest before beetles fly in spring, especially that felled before January; and process log inventories rapidly
	Protect hosts from infestations	Kill beetles on hosts with residual insecticide formulations; deter beetles from boring into logs by water misting; and repel beetles from logs with artificial repellents or masking pheromones
Aggregation of population on available hosts caused by response of beetles to semiochemical cues	Modify overwintering habitat around timber-producing areas	Cut back forest margin to increase hazard to flying beetles as they seek overwintering sites
	Increase distance from overwintering habitat in forest to supply of new hosts in spring	Set up logging plans so new cutblocks are as far away as possible from previous year's cutblocks.

Source: From Borden (1988).

infests weakened, dying, or freshly cut pines and unseasoned pine lumber. They bore into the sapwood and heartwood of logs or lumber and make hundreds of pin-size holes, each darkly stained with fungal mycelia. Female *P. flavicornis* lay eggs in small batches in the tunnels, and developing larvae dig small cells extending from the main tunnels parallel with the grain of the sapwood (Figure 22.6). There are several generations per year.

Piles of fluffy, whitish boring powder around the bases of trees indicate that there is a sector of deadwood within a tree being infested by this ambrosia beetle. Timber-marking rules suggest that when boring powder occurs around more than a third of a tree's circumference, the tree must be marked for cutting. These beetles attack standing southern pines in increasing numbers within 5 to 10 days after attack by the southern pine beetle. Prompt utilization of dead and dying trees and rapid drying of lumber reduces losses from *P. flavicornis*. In addition, log decks can be protected from attack by installation of water-misting systems. Kiln schedules commonly used for lumber and veneer kill ambrosia beetles and render the products safe from further attacks.

Figure 22.5
Primary and secondary tunnels cut by *Gnathotrichus sulcatus* in sapwood of a western hemlock log.
(Photo by N. E. Johnson.)

AMBROSIA BEETLES AND HARDWOODS

The Columbian timber beetle (*Corthylus columbianus*) is an exception to the rule that ambrosia beetles infest weakened, dying, or dead hosts—these prefer healthy trees (Nord 1972, Solomon 1995). Fungal stains extend from tunneling activity of this beetle several centimeters around each tunnel, damage that greatly affects the quality of valuable hardwood lumber. Damaged wood cannot be used for **veneer, cooperage,** or furniture. Hosts attacked by *C. columbianus* include maple, oak, yellow poplar, beech, box elder, sycamore, birch, basswood, elm, and several other angiosperms. Problems caused by this scolytid occur from Massachusetts south to Florida and west to Kansas and Michigan.

Adults overwinter in protected sites such as in the litter at the base of trees. In spring adults either reattack previous hosts or find new trees to attack. Males initiate the entrance and, presumably via a pheromone, are soon joined by females and other males. Females cut a primary gallery and several additional tunnels that branch away from the primary gallery. Soon after tunneling begins, two fungi (*Ambrosiella xylebori* and *Pichia* spp.) begin growing within the tunnels with mycelia penetrating the tunnel walls. Eggs are laid singly in brood cells or cradles, and developing larvae feed on the fungi; pupation occurs within these cells

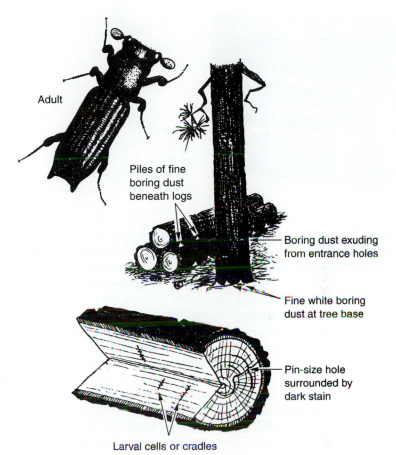

Adult

Piles of fine
boring dust
beneath logs

Boring dust exuding
from entrance holes

Fine white boring
dust at tree base

Pin-size hole
surrounded by
dark stain

Larval cells or cradles

Figure 22.6

The ambrosia beetle, *Platypus flavicornis,* and its damage.
(Source: USDA Forest Service.)

(Figure 22.7). The Columbia timber beetle has two to three generations per year. Silvicultural practices to lessen damage caused by this insect are being developed, but direct control tactics are still unknown.

An ambrosia beetle (Solomon 1995) that attacks weakened as well as dying and dead hardwood hosts is the yellow-banded timber beetle (*Monarthrum fasciatum*). Typical hosts include oak, maple, birch, sweet gum, poplar, hickory, apple, and others. The beetle ranges throughout eastern North America and westward to Mississippi and Minnesota. This beetle also tunnels through the wood of their hosts, producing a great amount of defect. Management of this pest includes removing dead and dying trees quickly, especially after a forest fire. Freshly cut logs must be removed from the forest in a timely fashion, and logs that cannot be utilized promptly should be submerged in log ponds or provided with sprinkler systems.

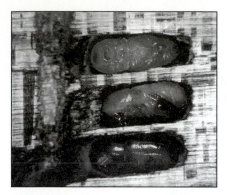

Figure 22.7
Pupation of the Columbian timber beetle (*Corthylus columbianus*) within cradles cut in sapwood of a hardwood tree.
(Photo by J. C. Nord.)

North America also accommodates the extremely large genus of ambrosia beetles, *Xyleborus*. This genus is worldwide, many infesting conifers, but the majority attack hardwood trees. There are over 1,000 species. Some have complex social organizations that include polygamy and parthenogenesis. Although many *Xyleborus* are pests of North American forests, the great majority are tropical and produce enormous economic damage—for example, to coffee plantations. Most of the North American species attack weakened or dead and dying trees. Probably the most important species is the cosmopolitan ambrosia beetle (*X. ferrugineus*). It infests oak, hickory, water tupelo, sweet gum, and many others; close to 200 host species have been recorded worldwide, especially in the tropics. It also attacks conifers, such as pines and bald cypress. The best management of damage is prompt salvage and use of infested trees and logs. Burning infested debris and slabs as hog fuel is recommended. Stored logs should be provided with a sprinkler system.

Two members of the genus were introduced from Europe and have caused substantial damage; they are *X. saxeseni* and *X. dispar.* Both species attack a large number of hardwood species and both have been reported from conifers. *Xyleborus saxeseni* is the more damaging and ranges from southern Canada to Florida along the eastern seaboard and from British Columbia to Baja, California, and Hidalgo, Mexico. As with most ambrosia beetles, *X. saxeseni* infests injured and dying trees and freshly cut logs. Trunks 5 to 50 cm in diameter are most subject to attack, but branches down to 2.5 cm also are susceptible. Infested timber should be harvested and processed rapidly. Along the Gulf Coast, logs cut during warm weather should be removed from the forest quickly and used within 1 to 2 weeks. In addition, two species of predaceous beetles exert a significant natural control: *Colydium lineola* and *Enoclerus sphegeus*.

REFERENCES

Baker, Whiteford L. 1972. *Eastern forest insects.* USDA Forest Service Miscellaneous Publication 1175.

Borden, J. H. 1988. The striped ambrosia beetle, pp. 579–596. In A. H. Berryman (ed.). *Dynamics of forest insect populations.* Plenum Publishing, New York.

Borden, J. H. 1990. Use of semiochemicals to manage coniferous tree pests in western Canada, pp. 281–315. In R. L. Ridgway, R. M. Silverstein, and M. N. Inscoe (eds.). *Behavior-modifying chemicals for insect management.* Marcel Dekker, New York.

Chapman, J. A. 1962. Field studies on attack flight and log selection by the ambrosia beetle *Trypodendron lineatum* (Oliv.) (Coleoptera: Scolytidae). *Canadian Entomologist* 94:74–92.

Chapman, J. A., and E. D. A. Dyer. 1969. Characteristics of Douglas-fir logs in relation to ambrosia beetle attack. *Forest Science* 15:95–101.

Chapman, J. A., and J. M. Kinghorn. 1958. Studies of flight and attack activity of the ambrosia beetle, *Trypodendron lineatum* (Oliv.) and other scolytids. *Canadian Entomologist* 90:362–372.

CINTRAFOR. 1994. *Primary wood product export volumes decrease, but values remain stable.* Center for International Trade in Forest Products, University of Washington, Fact Sheet 14, Seattle, Wash.

Furniss, R. L., and V. M. Carolin. 1977. *Western forest insects.* USDA Forest Service Miscellaneous Publication 1339.

Lindgren, B. S., J. H. Borden, L. Chong, L. M. Friskie, and D. B. Orr. 1983. Factors influencing the efficiency of pheromone-baited traps for three species of ambrosia beetles (Coleoptera: Scolytidae). *Canadian Entomologist* 115:303–313.

McClean, J. A. 1985. Ambrosia beetles: a multimillion dollar degrade problem of sawlogs in coastal British Columbia. *Forest Chronicle* 61:295–298.

McClean, J. A., and J. H. Borden. 1979. An operational pheromone-based suppression program for an ambrosia beetle, *Gnathotrichus sulcatus*, in a commercial sawmill. *Journal of Economic Entomology* 72:165–172.

Nord, John C. 1972. Biology of the Columbian timber beetle, *Corthylus columbianus* (Coleoptera: Scolytidae), in Georgia. *Annals of the Entomological Society of America* 65:350–358.

Solomon, J. D. 1995. *Guide to insect borers in North American broadleaf trees and shrubs.* USDA Forest Service Agriculture Handbook AH-706.

CHAPTER 23

WOOD PRODUCTS INSECTS

INTRODUCTION

Ambrosia beetles reduce wood values by boring in the sapwood of trees and logs and by having fungi stain their tunnels. Other insect groups also degrade wood by boring into sapwood, and their boring activity can be so extensive that their excavations affect the mechanical properties of the wood. As with ambrosia beetles some of these insects attack healthy trees, but most infest weakened hosts, logs, or blowdowns. Others infest dry wood and, as such, are destructive to structures and finished products. Wood borers occur in several orders, from the primitive Isoptera to the highly evolved orders: the Coleoptera, Lepidoptera, Hymenoptera, and Diptera. Detailed information on the biology and damage caused by these groups is found in Solomon (1995), Metcalf and Metcalf (1993), Johnson and Lyon (1988), Furniss and Carolin (1977), Baker (1972), and Hickin (1968).

Borers of Living Trees

Some lepidopteran families are wood products pests as well as ornamental tree pests—especially those whose larvae bore into living hardwood tree species. The clearwing moths, the Sesiidae, are a good example. The western poplar clearwing, (*Paranthrene robiniae*) feeds in the bark phloem and sapwood of black cottonwood, other poplars, willow, and birch. The larvae tunnel at first in phloem tissues; later they move into the wood. Infested trees are deformed and their growth slows down. The large *Synanthedon* spp. complex of clearwing moths damages urban trees, shrubs, and diverse ornamentals; others bore into timber trees, such as *S. culiciformis,* a pest of alder and birch. Other invaders of standing trees are the carpenter worm moths, the Cossidae; broods of this family excavate large galleries in branches, stems, and roots of hardwoods. An important cossid is *Prionoxystus robiniae,* whose host trees include elm, ash, birch, black locust, oak, cottonwood, maple, willow, and many other timber and ornamental species. In fast-growing poplar plantations, the large larva bores 1.3 cm diameter holes in the trunk and main branches and can frustrate biomass production goals. Degrade from its damage can

Figure 23.1
Damage of the carpenter worm (*Prionoxystus robinae*) to the trunk of a hardwood tree (**A**), and carpenter worm-produced defects in the stem of an oak (**B**).
(Source: From University of Georgia Cooperative Extension Service 1995.)

reach 15% of the value of sawn oak lumber (Solomon 1995) (Figure 23.1). Although carpenter worm moths have an effective complex of natural enemies that regulate their populations, intensive forest practices may require application of an integrated pest management (IPM) approach. In this context trunk-applied insecticides timed with the use of synthetic sex pheromones to synchronize with egg laying is an effective tactic in preventing attack.

Possibly the most important boring pest of poplar plantations is a weevil introduced from Europe in 1882, the poplar-and-willow borer (*Cryptorhynchus lapathi*). Preliminary studies in Canada of hybrid poplars revealed infestation rates as high as 50%. Poplars are most heavily attacked, but other hosts include birch, alder, and willow; quaking aspen is rarely attacked. Attacks occur on stems 1.3 to 15.0 cm in diameter but commonly on those 2.5 to 10.0 cm in diameter. Active attack sites are seen as a mixture of sap combined with reddish-brown and white frass at entrances, in bark fissures, and at the base of the tree (Figure 23.2). Irregular splits develop in the bark and eventually roundish exit holes begin to appear. Developing larvae bore in the sapwood and pack their galleries with fibrous residue. Repeatedly attacked stems are honeycombed with tunnels, and infested trees are prone to wind breakage, growth deformities, and mortality. Adults are poor flyers so that cultural and sanitation practices can minimize damage—for example, cutting and removing all infested stems around new plantings delay infestations. With an increase in **hybrid** poplar plantations, the best solution to *C. lepathi* problems is to select and breed resistant hybrids.

There are many longhorned wood borers (Coleoptera: Cerambycidae) that attack living hardwood trees. A notorious example is the poplar borer (*Saperda calcarata*) whose tunneling larvae often kill young poplars and degrade the value of older plantations. An additional problem caused by *S. calcarata* occurs when oviposition punctures serve as infection courts for the hypoxylon canker disease.

One of the principal pests of eucalyptus, the cerambycid, *Phoracantha semipunctata,* was discovered in Orange County, California, in 1984 (Hanks et al. 1993). This beetle is native to Australia and has been reported as a major threat to eucalyptus plantations in Spain, North Africa, New Zealand, and Chile. The insects normally attack water-stressed hosts and, as their population increases, they

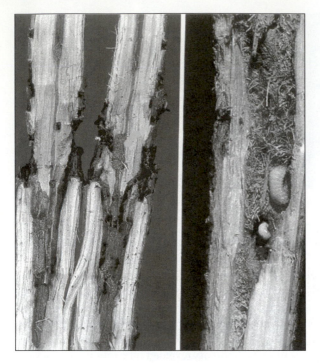

Figure 23.2
The poplar-and-willow borer (*Cryptorhynchus lapathi*) in the stem of a hybrid poplar (**A**), and stem riddled by larvae (**B**).
(Source: From Johnson and Lyon 1988.)

can decimate plantations during droughts years (Figure 23.3). Parasites of *P. semi-punctata* have been introduced into California and some have been established. Because water-stressed trees are most susceptible, timely irrigation or installation of permanent drip irrigation systems will prevent outbreaks of this insect. Prompt removal of beetle-infested trees, logs, and slash as well as use of a trap logs system should be considered as part of an IPM program. Current research at the University of California at Riverside has shown the value of eucalyptus species trials as a way to select species resistant to cerambycid attack (Hanks et al. 1995).

The western cedar borer (*Trachykele blondeli*) (Coleoptera: Buprestidae) infests living western redcedar trees as well as junipers (*Juniperus* spp.), Monterey cypress (*Cupressus macrocarpa*), and possibly incense-cedar (*Libocedrus decurrens*). In spring, adult western cedar borers feed on foliage of their hosts for sexual maturation. Later eggs are laid under bark scales on branches. Larvae enter the branches and bore into the trunk of the tree and extend their mines into the heartwood. Their life cycle may take 2 to 3 or more years. The last instar forms a pupal cell just under the bark in the fall, and new adults cut through the surrounding bark the following spring. Larval mines degrade lumber cut from infested trees and ruin trees selected for poles, shingles, and other products requiring uninfested wood (Figure 23.4).

Two additional buprestids that invade living hosts are the golden buprestid (*Buprestis aurulenta*), and the turpentine borer (*B. apricans*). The golden buprestid, a western insect, lays eggs around old tree wounds, fire scars, or lightning-caused scars of pine, spruce, Douglas-fir, and fir. Larvae mine extensively in the wood

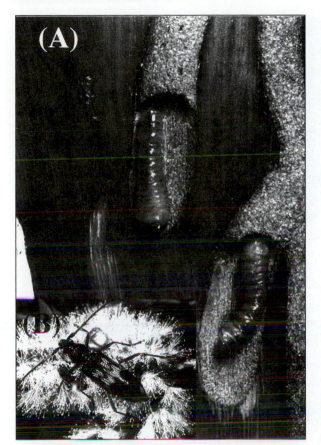

Figure 23.3
Larvae of the eucalyptus longhorned borer (*Phoracantha semipunctata*) developing under bark of host (**A**), and photograph of adult (**B**).
(Photos supplied by T. D. Paine.)

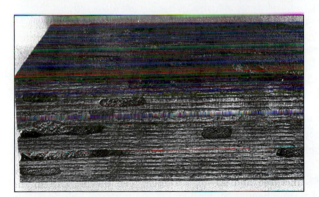

Figure 23.4
Degrade of western redcedar lumber caused by larval galleries of the western cedar borer (*Trachykele blondeli*).
(Source: USDA Forest Service Miscellaneous Publication 1339, 1977.)

and degrade any subsequent product made from their host tree. They are problems in buildings, especially in the Pacific Northwest. They mine timbers, beams, and boards and often emerge through finished surfaces; if the wood is dry, they may have been mining in the material for more than 10 years—there are records of the

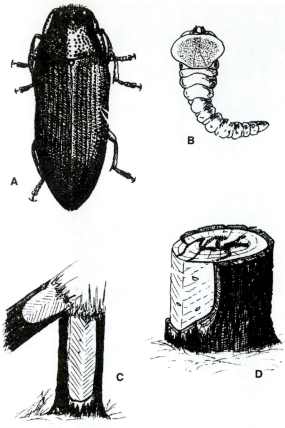

Figure 23.5
The turpentine borer (*Buprestis apricans*) a major problem
in the southern naval stores industry, as beetles are attracted
to the damaged stems; the adult (**A**), the larva (**B**). Larval
feeding may weaken stems and cause wind breakage (**C**).
Larval damage is characterized by tunnels packed with resin
and boring dust as well as the elliptical emergence holes (**D**).
(Source: From W. H. Bennett and H. E. Ostmark, USDA Forest Service 1972.)

golden buprestid mining in timbers for 26 years. The turpentine borer produces
the same kind of damage in southern pines and has been a special problem in the
naval stores industry (Koch 1972) (Figure 23.5).

Borers of Weakened Trees, Windthrows, and Logs
The most important wood-boring beetles of dead and dying material belong to the
Cerambycidae (Linsely 1959) (longhorned or roundheaded borers) or to the
Buprestidae (shorthorned beetles, metallic beetles, or the flatheaded wood borers)
(Figure 23.6). Although the most commercially important cerambycids are wood
borers, there are several that are phloem miners and occupy a similar ecological

Woodborers: Cerambycidae and Buprestidae

Cerambycidae	**Buprestidae**
1. Adults generally have long, flexible antennae; sides of pronotum often serrate or spiked.	1. Adults flattened in appearance, posterior of elytra pointed, often metallic in color.

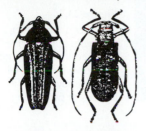 

2. Larvae have rounded, ball-like prothoracic region.

2. Larvae have flattened, anvil-shaped prothoracic region; have hardened plates on both dorsal and ventral surface of prothoracic area.

3. Larvae generally loosely pack their galleries with frass; the frass has a fibrouslike texture (in some species the frass is fine textured, but under a microscope it's seen as elongated fibers).

3. Larvae tightly pack their galleries with frass; the frass is tamped down in a "quarter moon–like" fashion.

4. Adult exit holes are rounded.

4. Adult exit holes are oblong.

Figure 23.6

Differences between the longhorned wood borers (Cerambycidae) and the shorthorned wood borers (Buprestidae).

niche as bark beetles. A common example is the ponderosa pine bark borer (*Acanthocinus princeps*); their larvae are important biological control agents since they compete directly with bark beetle larvae for the same resource. Some cerambycids are associated with forest fires and are good indicators of which charred trees will recover and which will be claimed by cerambycids and bark beetles. The ribbed pine borer (*Rhagium inquisitor*) is a good example, it arrives at fire-damaged firs, spruce, pines, or Douglas-firs within 2 to 3 hours after a fire is out (Figure 23.7). These cut **lenticular** niches in the charred bark and lay two to four eggs in each niche; the larvae then bore through the bark and feed in fresh phloem.

The ponderous borer (*Ergates spiculatus*) is a destructive longhorned beetle of fire-killed coniferous trees, logs, and windthrows. It is the largest cerambycid of western forests; adults are 3.8 to 5.7 cm long and their huge, tunneling larvae quickly riddle their hosts. Eggs are laid in bark **fissures** of dead trees, and the larvae excavate large tunnels and pack them with coarsely chewed wood debris. The first instars feed in the phloem, but as the phloem ferments the larvae bore in the

Figure 23.7
The ribbed pine borer,
(*Rhagium inquisitor*)
removed from a charred
lodgepole pine tree while
the cerambycid was in
process of ovipositing.

sapwood, eventually reaching the heartwood. Fully grown larvae are about 6 cm long, and they bore back toward the surface and pupate a few centimeters below the bark. Attacks by the ponderous borer can be so serious that fire-killed material may be unsalvageable.

Another important group of wood-boring cerambycids are the sawyers, various members of the genus *Monochamus*; they damage wood of recently dead and downed conifers, such as pines, spruces, firs, and Douglas-firs. In general, *Monochamus* eggs are laid in slits cut in the bark of their hosts; larvae feed for a period in the phloem, but then tunnel into the sapwood and heartwood. As they bore through their host, they construct oval tunnels that eventually bend toward the surface. There the last instar pupates, and the adult emerges by gnawing a round hole through the thin layer of wood and bark separating the pupal cell from the surface. The life cycle of the transcontinental white spotted sawyer (*Monochamus scutellatus*) takes a year but can be 2 years in Alaska, Canada, the northern United States, and the high mountains of North Carolina. Larvae of the pine sawyer (*Monochamus titillator*) cut 0.8 to 1.3 cm in diameter tunnels through the sapwood and heartwood of logs, dead and dying, and fire-charred southern pines. Pines felled in summer are immediately attacked by adult pine sawyers; during the hottest parts of the season, unprotected logs are destroyed in 3 weeks. There often are three generations per year of this cerambycid in the South—for example, East Texas and Louisiana. If infested logs are sawn into lumber and the lumber is air dried, the life cycles of the pine sawyer can continue, and 15- to 20-year life cycles in air-dried lumber have been reported. Of course, kiln drying of lumber kills any boring insects.

Another example of wood-boring cerambycids causing structural damage occurred during spring 1999. A yacht club in Seattle, Washington, built roofs over boat slips with 2-in. tongue-and-groove Douglas-fir lumber that was infested with larvae of the newhouse borer (*Arhopalus productus*). Many of these completed their life cycle; the new adults emerged from the lumber, bored out through the roof, and created leaks whenever it rained. Most likely the lumber was sawn from trees killed during a forest fire which, in turn, was infested by *A. productus* (Ebling 1975).

Cerambycid damage can be reduced by (1) rapid salvaging and utilization of dead and dying trees and prompt conversion of logs into lumber, (2) peeling the bark from logs not promptly utilized, and (3) water storage of logs or placement of water-sprinkler systems over stored material.

The flatheaded wood borers, the buprestids, also have members that feed exclusively in the phloem and compete with bark beetles. Most, however, attack weakened, dead, and recently felled trees. Females glue eggs under bark scales, bark fissures, or under bark epiphytes. The larvae then bore through the bark and feed in the succulent phloem. Although the boring debris of cerambycids is fibrous and unaligned, buprestid larval galleries are tightly packed with frass. The beetles emerge in spring and summer through elliptical exit holes in the bark. The California flatheaded borer (*Melanophila californica*) is destructive to water-stressed ponderosa and Jeffrey pines as well as other pines in the Pacific Northwest. The early instars feed in the phloem for several months, even years. They usually score the sapwood without killing the tree. If the tree is killed, the larvae continue their development under the bark and form pupal cells under the bark; new adults emerge the following spring. This association with water-stressed trees serves to mark weakened trees or predispose the hosts for subsequent bark beetle attacks.

The flatheaded fir borer (*M. drummondi*) attacks Douglas-fir, true firs, hemlock, western larch, spruce, and other conifers throughout the West. It preferentially infests dying or recently felled trees; sometimes they kill living, but weakened trees, especially during droughts or after forest fires. One of the most interesting members of this genus is *M. acuminata*, which has infrared receptors in pits below its front legs and embedded in its antennae (Figure 23.8). The insect is able to receive and process infrared signals and, in this manner, cue in on smoldering hosts.

The most economically important buprestids are those that bore into the sapwood. Many species begin feeding in phloem and later bore into the wood. The flattened oval tunnels are tightly packed with boring dust and often wind tortuously through the wood so as to riddle it with holes. Even a few such holes greatly lower the mechanical properties and value of infested lumber. The sculptured pine borer (*Chalcophora angulicollis*) is the largest of the western flatheaded borers. The larvae feed within the wood of dead pines, firs, and Douglas-fir. As with the longhorned beetles, the best way to manage flatheaded wood borers is to utilize fire-killed trees rapidly, store logs under water or under a sprinkler system, and utilize the material as soon as practical. When harvesting extremely valuable trees, such as old-growth western redcedar, it is important to know the main flight season of wood borers. For example, the western cedar borer (*Trachykele blondeli*) has its main flight during the first week in May (Figure 23.9); yarding of material out of the forest has to be accelerated during May, and the material must be utilized immediately.

Horntail wasps (Hymenoptera, family Sericidae) are borers of recently killed trees or fresh logs. There are 12 species distributed among the genera *Xeris, Sirex, Urocerus,* and *Tremex.* Except for *T. columba*, which attacks a variety of hardwoods, all the other species oviposit in conifers (Table 23.1). The adults are large, broad-waisted wasps (Figure 23.10), and the females have long stingerlike ovipositors that

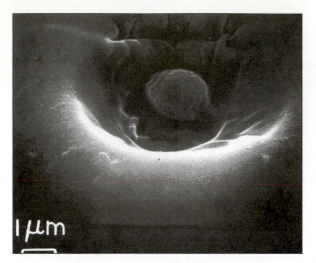

Figure 23.8
Scanning electron micrograph of the sixth antennal segment of *Melonophila acuminata*, showing one of the pits with domelike receptor in the center; structure is thought to be an infrared receptor.
(Source: Scott and Gara 1975.)

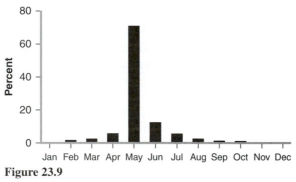

Figure 23.9
Seasonal flight pattern of the western cedar borer (*Trachykele blondeli*); data are expressed as percent of insects caught in flight traps positioned above trap logs.

extend back horizontally from the insect at rest. The biology of the Sericidae of North America is poorly known. Adult horntails fly in spring and early summer. They often are attracted in large numbers to forest fires where they readily attack badly scorched and dying conifers. Undoubtedly females have infrared sensors that vector them to an area of smoldering trees. Fire fighters frequently observe and report seeing female wood wasps stuck in trees as they cannot extract their ovipositors. When a recently killed host is found, females insert their long flexible ovipositors into the wood, often 25 mm or more, and lay their eggs. As the ovipositor extrudes its eggs, it also coats them and the walls of the tunnel with a glandular, mucuslike, phytotoxic secretion that contains spores of the symbiotic fungus *Amylostereum* spp. The fungus, which is carried in mycangia located at the

TABLE 23.1	Principal Western Horntails (Hymenoptera: Sericidae) and Their Known Hosts

Species of horntail	Hosts
Sirex ariolatus	*Abies concolor, Cupressus macrocarpa, Juniperus occidentalis, Libocedrus decurrens, Pinus contorta, P. jeffreyi, P. lambertiana, P. radiata, Pseudotsuga menziesii,* and *Sequoia sempervirens*
S. behrensii	*Cupressus macrocarpa, Pinus jeffreyi, P. lambertiana, P. ponderosa,* and *P. radiata*
S. cyaneus	*Abies concolor, A. lasiocarpa, Picea engelmannii, P. glauca, Pinus ponderosa,* and *Pseudotsuga menziesii*
S. juvencus	*Abies lasiocarpa, Cupressus macrocarpa, Larix occidentalis, Picea sitchensis, Pinus contorta, P. jeffreyi, P. ponderosa, P. radiata,* and *Pseudotsuga menziesii*
S. longicauda	*Abies concolor* and *A. magnifica*
Tremex columba	*Acer, Betula, Platanus, Quercus, Ulmus* and other hardwoods
Urocerus albicornis	*Abies amabilis, A. lasiocarpa, Larix occidentalis, Picea, Pinus, Pseudotsuga menziesii, Thuja plicata* and *Tsuga heterophylla*
U. californicus	*Abies concolor, A. lasiocarpa, A. magnifica, Larix occidentalis, Pinus contorta, P. monticola, P. ponderosa, Pseudotsuga menziesii,* and *Tsuga heterophylla*
U. gigas flavicornis	*Abies, Larix occidentalis, Picea, Pinus,* and *Pseudotsuga menziesii*
Xeris morrisoni	*Abies concolor, A. grandis, A. lasiocarpa, Larix occidentalis, Libocedrus decurrens, Picea engelmannii, P. pungens, P. sitchensis, Pinus contorta, P. ponderosa, Pseudotsuga menziesii,* and *Tsuga heterophylla*
X. spectrum	*Abies concolor, A. grandis, A. lasiocarpa, Larix occidentalis, Picea engelmannii, P. pungens, P. sitchensis, Pinus contorta, P. ponderosa, Pseudotsuga menziesii,* and *Tsuga heterophylla.*
X. tarsalis	*Cupressus macrocarpa, Juniperus occidentalis,* and *Thuja plicata*

Source: From Furniss and Carolin (1977).

Figure 23.10
The horntailed wasp (*Urocerus gigas flavicornis*) probing the bark of a ponderosa pine for an oviposition location.
(Source: USDA 1977.)

base of the ovipositor, prepares the wood for the larvae by reducing its moisture content, causing decay, and it is thought that the larvae feed on the decaying material. The developing larvae bore at right angles to the egg gallery and stay within the sapwood for several weeks, then they turn and tunnel toward the heartwood. The tunnel becomes tightly packed with frass. In time the larvae begin to bore toward the surface of the wood and there they pupate.

Water treatment of stored logs and prompt utilization are the best way to prevent injury by woodwasps. Kiln drying of green lumber is also recommended as larvae in surface-dried lumber will continue to develop and often are reported emerging from the inside dwellings.

Borers of Dry and Seasoned Wood

The most costly form of insect damage to wood is when finished products are infested (Hickin [1968] presents excellent information on these wood products pests). This is because the entire investment in manufacturing and installation is threatened. Chief among the insects that cause such damage is a cerambycid that infests dry wood, a collection of families (Anobiidae, Lyctidae, and Bostrichidae) known as the powderpost beetles, carpenter ants, termites, and the great carpenter bee (*Xylocopa virginica*).

The Old House Borer   A cerambycid known as the old house borer (*Hylotrupes bajulus*) was introduced from Europe around 1900 (Figure 23.11). It is firmly established in eastern North America and will inevitably become a problem in the West. In Europe the old house borer is so firmly established in buildings that Hickin (1968) states, "It is now extremely rare to find this species in the wild— outside buildings, so that it must be assumed that the indigenous race died out somewhere round the turn of the century." Adults lay eggs in cracks and fissures in the wood of pine, spruce, fir, and larch. The larvae bore for years in sapwood before they pupate. One generation requires 3 to 6 years, but extremes of 32 years have been reported. In the southeastern United States, adults disperse in June and July; in Texas adults have been found in early May; in the Northeast adults are active until early fall. In areas where infestations are common, only kiln-dried lumber should be used in construction.

Powderpost Beetles   Powderpost beetles of the family Anobiidae include the furniture, deathwatch, and drugstore beetles. All of these infest dry wood, which they render to powder. A notorious example, which was introduced from Europe at the turn of century, is the furniture beetle (*Anobium punctatum*). In North America this anobiid tunnels in the sapwood of old furniture (a bane to the antique furniture trade) and less commonly, exposed structural timbers. A commercially significant anobiid in the West is the Pacific powderpost beetle (*Hemicoelus gibbicollis*) a species that commonly infests structural timbers, subflooring, and other dry lumber along the coast from Alaska to California (Figure 23.12A). This beetle infests dry, unrotten sapwood of practically any tree species; *H. umbrosus* also is found along the Pacific Coast but, in addition, extends its range eastward into

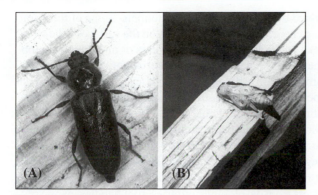

Figure 23.11

The old house borer (*Hylotrupes bajulus*) adult (**A**) and a pupa exposed inside the wall stud of a house (**B**).

(Photograph provided by USDA Forest Service.)

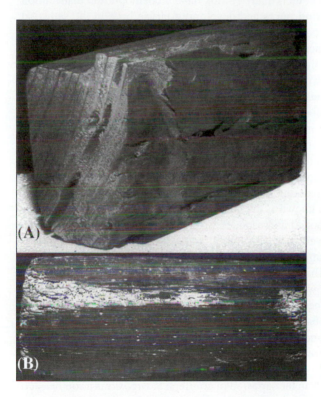

Figure 23.12

A piece of planking from an old, lumber-carrying schooner, the *Wanona* (being restored in Seattle) that is infested with the Pacific powderpost beetle (*Hemicoelus gibbicollis*) (**A**), and a joist from a deck infested by carpenter ants (*Camponotus* spp.) (**B**).

the Canadian maritime provinces. The most destructive anobiid of northeastern, midwestern, and southeastern United States is *Xyletinus peltatus*. A survey of houses in Baton Rouge, Louisiana, found that 75% of the structures were infested by this powderpost beetle.

Female anobiids typically lay eggs in fissures or checks along the length of an exposed beam or joist. Larvae bore into the wood for a few millimeters, then turn and follow the wood grain; they mostly feed in the springwood but leave heartwood intact. Minute pellets excreted by boring larvae fill the tunnels. Some of these tiny pellets are expelled, causing characteristic sawdust piles on the surface of the wood and ground. The life cycle of anobiids is at least 3 to 5 years or more, but the problem is that subsequent generations reinfest the same material until there is substantial structural damage. Damage is difficult to prevent; in houses without basements proper ventilation should be maintained by close adherence to building codes. When severe damage occurs, entire structures are fumigated.

The Lyctidae are referred to as the true powderpost beetles, and they have a preference for hardwoods. These powderpost beetles attack all seasoned wood products and range throughout the world, from the tropics to most temperate regions. Through international exchange of wood products (e.g., rattan furniture, wooden carvings and artifacts from developing countries, tool handles, exotic plywoods, moldings, flooring, etc.) many lyctids have a cosmopolitan distribution. Eggs are inserted into **wood pores** and developing larvae bore in the sapwood, reducing it to flourlike consistency. Near maturity, larvae bore circular holes through a veneer of surface wood they have left and push out their boring powder; often this is the first sign that the material is infested. Adults emerge and normally reattack the same wooden structure.

Measures should be taken to prevent buildup of lyctids in warehouses. Stocks should be inspected regularly, especially in facilities where imported artifacts are held. Storage of flooring and cooperage materials as well as wooden handle stocks are prone to lyctid infestations, especially by the southern lyctus beetle (*Lyctus planicollis*) in the South and the western lyctus beetle (*L. cavicollis*) in the West. Surface treatments of exposed wooden items will prevent attacks—staining handle stock, varnishing, or painting stored materials, and so on.

The Bostrichidae are known as the false powderpost beetles, and they are easily identified by their hooded and rounded pronotum that covers the head. The bostrichids, as with the lyctids, are cosmopolitan pests; the most damaging members are tropical as they attack plantation crops of rubber, coffee, cocoa, and timber species. Unlike the other powderpost beetles, bostrichids in temperate regions invade seasoning lumber while it is at a higher moisture content than that suitable for the other powderpost beetle families. They also attack recently cut trees or logs. Adults bore through the bark of their host and into the sapwood. They then tunnel across the grain just under the sapwood-phloem interface. Larvae feed mostly in the sapwood but may reach the heartwood. Feeding continues until the wood is dry. Most bostrichids bore in hardwoods, but *Stephanopachys substriatus* is a holarctic species that feeds in the sapwood of conifers; *S. rugosus* is a southeastern U.S. species that infests freshly sawn southern pine lumber. An interesting species is the leadcable borer (*Scobicia declivis*) which commonly is found in a variety of western hardwood species, especially important in boring into wine kegs; in earlier times when buried telephone lines were sheathed in lead, this insect burrowed through the lead, short-circuiting communications.

Carpenter Ants Carpenter ants belong to the large genus *Camponotus* with more than 1,000 described species, the largest number of species occurring in the tropics. The most destructive northeastern and southern species is *C. pennsyl-vanicus,* and in the West, *C. herculeanus, C. laevigatus,* and *C. vicinus* are the ones commonly found in houses. They are termed carpenter ants because they construct smooth tunnels in the wood of stumps, logs, snags, dead central portions of living trees, and wooden framing of houses. Wood is not ingested by these ants; rather the tunneling is used as nests in which to rear their broods. They are omnivores, feeding on both plant and animal food. They have a strong preference for caterpillars, and honeydew of aphids; in houses they are effective scavengers, feeding on human detritus—for example, bread crumbs, bacon rinds, and bits of cheese. They normally select moist and partially decayed wood in which to begin construction of their galleries. The queen and her brood are usually located in a humid part of the nest, but satellite colonies often are located in sound wood. As workers enlarge their galleries, they tear off tiny wooden splinters that they expel to the outside; this is the telltale evidence of an infestation. Excavations follow the wood grain and are found mostly in spring wood (Figure 23.12*B*).

Management and prevention of carpenter ant damage include solving moisture-causing problems in houses, then replacing damaged wood; removing discarded wood from underneath houses; halting the custom of piling firewood next to houses; and if need be, calling a pest control operator to find and destroy the colony as well as treating the perimeter of the structure with nontoxic, long-lasting pesticides. Finely ground clays or diatomaceous earth, for example, can be used; ants crawling through the formulation would suffer severe abrasions of their exoskeleton.

Termites   Termites are the most destructive pests of wooden structures and cellulose products—for example, stored rolls of pulp or paper. They are social insects that live in large colonies and utilize cellulose-containing material as food. There are over 2,000 species of termites, most of them tropical, but only 200 species are known as pests and about 40 species cause problems in North America. The economic importance of a few pest species is so dramatic (Truman et al. 1982, Robinson 1996), that the importance of termites in breaking down woody tissues and returning nutrients to the soil is obscured.

Colonies are composed of several castes. The king and queen establish the nest and the king remains with the founding female. The queen lays hundreds of eggs that hatch into nymphs, and nymphs may develop into several forms: (1) workers, which are responsible for tunneling in wood, feeding developing nymphs and soldiers as well as the king and queen; (2) soldiers, which are morphologically and behaviorally specialized for guarding the colony from enemies; (3) supplemental kings and queens, which are produced when colonies enlarge or when the original royal pair are killed; and (4) winged reproductives, which leave the colony in swarms to establish new nests (Figure 23.13).

The Pacific dampwood termite (family Hodotermitidae, *Zootermopsis angusticollis*) severely damages structures and stored paper products from southern British Columbia, down the Pacific Coast and into Mexico. These termites swarm in late summer and fall. The founding pair find a moist, semirotten piece of wood

King and Queen
The male and female established the nest for the colony, the male remains with the female throughout life.

Eggs
The queen is the primary egg producer of the colony; she may lay thousands of eggs over a period of years.

Nymphs
The developing termite passes through several stages: first the egg, then the immature or nymphal forms. The nymph may develop into one of several forms. They also feed on wood.

Workers
These forms are responsible for foraging for food and the construction of tunnels in the soil and infested wood. They feed the nymphs, the soldiers, and the reproductives.

Soldiers
These specialized forms are responsible for protecting the colony against ants and other enemies. They are blind and must be fed by workers.

Supplemental King and Queen
When colonies become large or there are environmental opportunities for expansion, additional males and females (kings and queens) may develop from nymphs. They also act as replacements for injured or dead kings and queens.

Winged Reproductives
These are the most visible termites. They leave the colony in large numbers (swarming) to establish nests in new locations.

Figure 23.13

Growth of a typical subterranean termite colony. Starting with the royal pair (the queen and king), the colony expands as workers, soldiers, secondary reproductives, and new sexually mature winged reproductives are produced.

(Source: From Robinson 1996.)

and there they begin their colony. Unfortunately the moist wood may be found outside or within a building, and a developing colony of dampwood termites can cause serious and expensive structural damage. Keeping untreated wood away from sources of moisture, repairing leaks, replacing leaking gutters and other such prevention of moisture problems, prevents damage by *Z. angusticollis*.

Subterranean termites (family Rhinotermitidae) occur throughout North America, and their colonies are found either entirely or partly in the soil. They, however, often feed in wood located above ground, but they always maintain connection with the ground by means of earthen tunnels. As these termites extend their galleries in wooden structures, they always leave a covering veneer of wood intact. Because of this behavior, there usually is no external evidence of an infestation, even though the interior of the wood may be demolished. The first indication of a problem may be the swarming of winged forms, the presence of earthen tunnels over foundation walls, or sudden sagging of supporting beams or flooring.

There are several subterranean termite species in North America. Two important examples are *Reticulitermes flavipes,* which occurs from Ontario south to the Gulf of Mexico and west to the Great Plains, and the western subterranean termite (*R. hesperus*). The western subterranean species occurs along the Pacific Coast from British Columbia to central California. This species swarms in autumn after heavy rains. After mating the king and queen shed their wings and locate a buried piece of wood. In a typical scenario the buried material may be near or under a house, and the colony then finds a windfall of food in the structural components of the building; the moisture of their environment is maintained as they extend their foraging tunnels over the concrete foundations of the house. Prevention of damage includes (1) removal of wooden scraps that may become buried after a house is built, (2) continuous monitoring for earthen tunnels, (3) treatment of soils before concrete slabs are poured, and (4) continual inspection for potential problems that may require services of pest control professionals.

The drywood termites (family Kalotermitidae) attack and live aboveground in sound dry wood. There are several species of drywood termites in North America, and they all have similar life cycles. After mating in swarms, they attack their hosts at any point where the royal pair can find untreated wood. In comparison to other termite families, colonies tend to be small and consist of reproductives, nymphs, and soldiers; there is no worker cast, nymphs serve as workers. Over the years multiple attacks of exposed wood can occur, and these colonies coalesce and appear as a single large colony. The presence of drywood termites is noted by blistering of the wood surface, the sound of movement in the walls, and the presence of fecal pellets ejected from infested wood. Ultimate proof of their residence occurs when swarms of termites collect at windows or within light fixtures.

Carpenter Bees   Carpenter bees (*Xylocopa* spp.) belong to the same family as bumblebees, the Apidae. Bees excavate perfectly round tunnels in bark and wood to a depth of 5 cm or more; their perforations are so perfectly made that they appear to have been bored by a 9 mm drill bit. They stuff a ball of pollen at the far end of the tunnel, lay an egg on the ball, and seal off the egg and its larval food with a partition made from sawdust and saliva. Then they position another pollen ball, egg, and partition next to the first cluster; in this way, five to seven other segments are produced along the tunnel. Although their damage is customarily minor, they can perforate exposed timbers, tanks, poles, and posts.

REFERENCES

Baker, W. L. 1972. *Eastern forest insects.* USDA Forest Service Miscellaneous Publication 1175.

Ebling, Walter. 1975. *Urban entomology.* University of California, Division of Agricultural Sciences, Berkeley, Calif.

Furniss, R. L., and V. M. Carolin. 1977. *Western forest insects.* USDA Forest Service Miscellaneous Publication 1339.

Hanks, L. M., J. S. McElfresh, Jocelyn G. Millar, and T. M. Paine. 1993. *Phoracantha semipunctata* (Coleoptera: Cerambycidae), a serious pest of *Eucalyptus* in California: biology and laboratory-rearing procedures. *Entomological Society of America* 86:96–102.

Hanks, L. M., T. D. Paine, Jocelyn G. Millar, and John L. Hom. 1995. Variation among *Eucalyptus* species in resistance to eucalyptus longhorned borers in southern California. *Entomologia Experimentalis et Applicata* 74:185–194.

Hickin, Norman E. 1968. *The insect factor in wood decay.* Hutchinson, London.

Johnson, W. T., and Howard H. Lyon. 1988. *Insects that feed on trees and shrubs.* Comstock Publishing Associates, Division of Cornell University Press, Ithaca, N.Y.

Koch, Peter. 1972. *Utilization of the southern pines.* USDA Forest Service Agricultural Handbook 420.

Linsely, Gorton. 1959. Ecology of Cerambycidae. *Annual Review of Entomology* 4:99–138.

Metcalf, R. L. and R. A. Metcalf. 1993. *Destructive and useful insects.* McGraw Hill, New York.

Robinson, W. H. 1996. *Urban entomology.* Chapman and Hall, New York.

Scott, D. W., and R. I. Gara. 1975. Antennal sensory organs of two *Melanophila* species (Coleoptera: Buprestidae). *Annals of the Entomological Society of America* 68:842–846.

Solomon, J. D. 1995. *Guide to insect borers in North American broadleaf trees and shrubs.* USDA Forest Service Agriculture Handbook AH-706.

Truman, L. C, Gary W. Bennett, and W. L. Butts. 1982. *Scientific guide to pest control operations.* Harcourt Brace Jovanovich, Duluth, Minn.

INSECTS OF SEED ORCHARDS, NURSERIES, AND YOUNG PLANTATIONS

INTRODUCTION

Seed orchards and nurseries play important roles in production of fast-growing and high-quality trees. These are expensive and labor-intensive operations—and pest problems, which were formerly overlooked or were considered as minor nuisances, suddenly can become important.

Emphasis on establishing genetically selected seedlings means establishment of trees with restricted genotypes and potential explosion of pests. The problem may be overemphasized, because large clonal plantations have been established for several decades, and these have not experienced serious insect damage (Zobel and Talbert 1984). Besides, genetic studies indicate that even in clonal plantations there may be enough genetic variation to preclude widescale herbivory. Nevertheless, seed orchards, nurseries, and plantations are high-priced operations, which means that acceptable damage caused by insects is low, and applied control operations are similar to those of agricultural pests (as shown in Table 24.1).

Protecting plantations from damage is one of the most important goals of applied forest entomology. Plantation establishment places trees in the natural environment; there, seedlings and young trees are subject to maximum physical destruction, physiological stress, and potential ruin from herbivory. Plantations also represent the raw material for **intermediate operations.** These are the operations that form the forests for particular managerial goals. For example, thinning places maximum growth on the fastest-growing and straightest trees, and pruning ensures maximum commercial value from these stems. Consequently insects that kill newly established stands decrease the chance of selecting the best stems for intermediate operations; insects that kill or deform older trees frustrate the goals of thinning-pruning regimes.

SEED ORCHARDS

Cone and seed insects have a rich food supply, and during good seed years they cause significant damage to seed orchards (Ebel et al. 1980, Ruth 1980, Hedlin et al. 1981). Conditions in **seed orchards** differ from those in the forest. The trees are widely spaced and the cone-bearing branches are near the ground. These differences affect physical factors such as temperature, light, and moisture: conditions that favor the abundance of many cone and seed insects.

Considerable capital and human resources are expended in establishing seed orchards, such as selection of plus trees, grafting, fertilization, and progeny studies, **roguing,** maintenance, and so on. Seeds produced in these orchards therefore are more expensive than ones collected in the forest. Cone and seed insects cause severe economic losses by feeding on pollen, developing ovules, and the cones—or indirectly by injuring foliage, twigs, and branches that bear these structures. This damage greatly increases the operational costs not only by lowering seed production but also by increasing the expenses of cone harvesting and seed extraction.

Several insect orders damage the cones and seeds of conifers; the Lepidoptera are the most damaging. The most damaging insects feed specifically on cones, seeds, and flowers; others feed on foliage and subcortical regions. Cone crops are cyclical, so there are years when food supply for these pests is low and species dependent on flowers, cones, seeds, and foliage have distinct behaviors to cope with reduced food supplies. In temperate regions many insect pests undergo extended **diapause** (period of arrested development) when cone crops are poor. This phenomenon is important to survival of these insects—limited food supply for 1 year, followed by extended diapause, would result in great feeding pressure on a subsequent year, which often coincides with a good seed year. Usually seed production losses in one year are influenced by the size of the previous year's cone crop. Crops produced the year following a heavy cone crop are heavily attacked, whereas those preceded by a light crop will be lightly damaged. Accordingly the best way to alleviate losses is to establish an integrated pest management (IPM) system (Figure 24.1). Successful IPM approaches depend on well designed monitoring protocols that identify or predict (Cameron 1981) (1) seed crops from one year to another, (2) trees with continually poor crops, (3) cone and seed losses as well as identification of associated pests, and (4) analyses of success or failure of control decisions. Seed crops are monitored from initiation of flowers, seed production, and seed viability. Estimates of the number of emerging flowers serve as the baseline for a continuous inventory. In practice, a representative sample of orchard trees is analyzed yearly; a portion of the flowers on sampled trees are tagged and observed periodically until cone harvest. This is done to determine when losses occur as well as to identify insect pests associated with these losses. Samples of mature cones are cut open to quantify the extent and causes of seed losses. This inventory includes seed extraction and germination efficiencies to help identify problems occurring before, during, and after seed extraction. Monitoring losses on a clonal basis isolates genetically controlled traits that may be manipulated to further increase seed yields—for example, identifying genotypes that repel insect attacks.

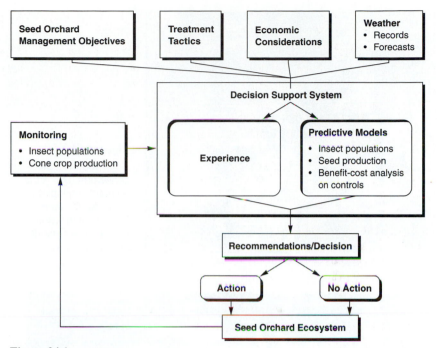

Figure 24.1

Diagrammatic portrayal of a typical seed orchard pest management system.
(Source: Modified from Cameron 1981.)

Due to the many pest species present in seed orchards, identification of key pests for each orchard is important: those present most years that cause reductions of seed crops. Once key pests are known, insect monitoring and control efforts can be concentrated on these few species. For example, in southern pine seed orchards of East Texas, coneworms and seed bugs are key pests. In particular, *Dioryctria amatella, D. clarioralis,* and *D. merkeli* cause yearly damage; the most important coneworm is *D. amatella* (Figure 24.2). Two species of seed bugs, *Leptoglossus corculus* and *Tetyra bipunctata* (Figure 24.3) reduce seed yields throughout the growing season of every year.

Cone and seed insect populations are monitored by systematically checking on cone production. A monitoring system developed for eastern Canadian seed orchards proposes the following approach (Turgeon and DeGroot 1992):

1. Select 2 to 5% of the seed-orchard trees.
2. Count all the receptive seed cones on the sample trees, including those dead or damaged, and use this count to estimate the potential number of cones, seeds, and seedlings that could be produced by the orchard.
3. Select and tag branches and cones, about 30 pine cones and about 25 spruce cones are tagged.

Figure 24.2
Adult southern pine
coneworm (*Dioryctria
amatella*) on loblolly pine
cone (**A**); *D. clarioralis*
larva feeding within
loblolly pine cone (**B**).
(Photo by Scott Cameron.)

Figure 24.3
The shieldbacked pine seed bug (*Tetyra bipunctata*) feeding on seeds of
a loblolly pine (**A**), and *Leptoglossus corculus* whose early instars feed on
conletes and later instars on seeds (**B**).
(Photo by Scott Cameron.)

4. Visit the tagged cones at least two to four times per year, and record the
   number of healthy and damaged cones; schedule the first visit when insects
   begin cone attack, the second when cones are about half their mature length,
   and the third when cones are full length—the frequency and timing of these
   visits is based on the biology of the key pests.

5. After each visit use the information to update and validate population prediction models, predicted crop sizes and, in part, to recommend control tactics.
6. Collect mature cones.
7. Extract and germinate seeds, to calculate both the actual seed potential and efficiencies in seed extraction and germination, and compare these with predictions made at the beginning of the season.

During the carrying out of this monitoring, the damaging insects are identified by extracting the life stages or capturing adults. Many insects directly infesting cones (direct pests) attack by laying eggs when the seed-cones are open to receive pollen. They generally lay eggs between the developing cone scales. The seed chalcids (tiny wasps whose larvae live within developing seeds), however, lay eggs in the seeds when the cones are half their mature length. Detection of direct pests is difficult as their eggs are minute and the larvae develop inside cones or seeds, leaving little evidence of their presence. Indirect pests lay their eggs on or near cones. Attack by indirect pests is initiated by larvae, and their presence is easy to observe; coneworms (e.g., the Douglas-fir coneworm (*D. abietivorella*)), for instance, enter the cone and feed on seeds and cone tissues and expel their frass on the outside of the infested cones. The presence of cone and seed adults also is monitored by light and pheromone traps. Pheromone traps are more sensitive than light traps for detection of low insect populations and mainly catch the insect pest of interest. Light traps are better for monitoring moderate to high populations of adults. These two kinds of adult trapping techniques are useful in control of *Dioryctria* spp. as coneworms pass through long periods of winter dormancy between oviposition and feeding on cones in spring. Population levels may change during this period due to weather factors. Accordingly moderate to large coneworm moths in either kind of traps would signal the need for a larval sample to be taken early the following spring before destructive cone feeding occurs. In the southeastern United States, larvae of *D. amatella* and *D. clarioralis* start to feed on cones soon after they emerge from eggs in spring. Therefore insecticide sprays are applied just after peak moth flight before young larvae tunnel into cones where they are protected from insecticide sprays (Figure 24.4).

There are many insect species that severely damage seed crops; some important genera are as follows:

1. Thysanoptera: the flower thrips (*Frankliniella* spp.), which feeds on developing flower parts.
2. Coleoptera: include the scolytids (*Conopthorus* spp.), tiny beetles that enter cones and lay eggs in subcortical tissues of the cone stem.
3. Lepidoptera: the coneworms (Pyralidae: *Dioryctria* spp.), the most important and cosmopolitan lepidopterous cone pest of North America, as they feed within cones as well as on shoots; olethreutid moths, conemoths and coneborers (Olethreutidae: *Barbara* spp., *Epinotia* spp., and *Eucosma* spp., respectively), *B. colfaxiana* is an important cone moth of Douglas-fir cones (Figure 24.5); the seedworms and shoot moths (Tortricidae:

Figure 24.4
Rotary mist blower applying insecticide in an East Texas seed orchard to control infestations of southern pine coneworms (*Dioryctria amatella*).
(Photo by Scott Cameron.)

seedworms (*Laspeyresia* spp.), shoot moths (*Rhyacionia* spp.), and others), seedworm larvae feed almost entirely on seeds after emerging from eggs laid on cone scales, and shoot moths destroy cones on young trees and on the lower branches of larger trees.

4. Hymenoptera: sawflies and seed chalcids; some sawfly species feed on pollen and others on flower buds (e.g., *Neodiprion pratti pratti,* occasionally a severe pest of jack, loblolly, pitch, and shortleaf pines), but the seed chalcids are major pests of conifer seeds as the wasps oviposit through cone scales and into developing ovules (e.g., the Douglas-fir seed chalcid (*Megastigmus spermotrophus*)) (Figure 24.6).

5. Hemiptera: Coreidae, the leaffooted bugs as well as pentatomid stinkbugs—both these hemipterans poke their piercing sucking mouthparts through cone scales and feed on developing seeds.

6. Diptera: cone maggots and cone midges (*Hylemia* spp. and *Contarinia* spp., respectively)—for example, the Douglas-fir cone gall midge (*C. oregonesis*) lays its eggs on flower scales, maggots begin to feed on the developing seed causing formation of a gall that surrounds the seed and kills it as it fuses to the seed coat.

Since the acceptable economic damage level is low for seed orchards, it usually is necessary to include insecticidal controls as part of the IPM system. Based on the best information available to an IPM system (see Figure 24.1), a spray schedule is developed for particular trees or parts of the seed orchard. For example, in East Texas a spray schedule was developed based on probable oviposition periods for five key pest species (Table 24.1).

NURSERIES

Insects cause significant losses to both **container and bare root nurseries** (Landis et al. 1990). Their herbivory lower nursery yields as well as plant quality, which, in turn, results in economic losses. Nursery IPM programs focus on prevention of pest problems, but when they do develop, eradication of the pest must begin at once to minimize damage to valuable plants. Pest problems are prevented

T A B L E **24.1**	**Insecticide Spray Schedule for a Texas Forest Service Seed Orchard Near Lufkin[a]**

Months	Apr	May	Jun	Jul	Aug	Sep	Oct
Suggested Spray Dates	▲	▲	▲	▲		▲	
Key Pests			*Oviposition Periods*				
Dioryctria amatella							
Dioryctria clarioralis							
Dioryctria merkeli							
Tetyra bipunctata							
Leptoglossus corculus							

[a]Based on Oviposition Periods of Key Pest Species
Source: From Cameron (1981).

by establishing cultural practices that ensure healthy and vigorous seedlings. Plants that are properly spaced, **mycorrhiza** inoculated, weeded, pruned, fertilized, and watered are less susceptible to insect problems than seedlings exposed to undesirable environmental conditions.

There are about 20 arthropods that commonly infest seedlings growing in containers, and nine are important (Figure 24.7); serious infestations are treated with insecticides (Table 24.2). Aphids and cutworms are the most commonly found pests in container nurseries. However, any of the other nine pests can reach outbreak proportions, depending on location and the particular environmental conditions.

It is practically impossible to prevent aphid infestations from seedlings growing in containers, but they can be managed by prompt detection (**chlorotic** spots on foliage and presence of **honeydew**) and timely application of insecticides. It is important to control aphids in the nymphal stage, because they lay overwintering eggs on foliage, and they are tolerant to chemical treatments in this stage, and so infested seedlings often are sent to planting sites. Cutworms are noctuid moth larvae that feed on succulent young shoots, often leaving tiny stumps. A common species (among many) throughout North America is the variegated cutworm (*Peridroma saucia*) and it feeds on most forest tree seedlings as well as on a myriad of agricultural crops. Cutworms feed at night and move from to plant to plant; during the day they can be found under the surface of the growing medium or under the containers. Greenhouses should be screened to exclude noctuid moths, and larval infestations are treated by application of poison baits or containers are treated with appropriate insecticides, mostly through the irrigation system. Root weevils attack practically all plant species grown container nurseries. They are difficult to control as larvae feed on roots and adults gnaw root crown and stem tissues. In the Pacific Northwest the most susceptible hosts are western hemlock and spruce seedlings.

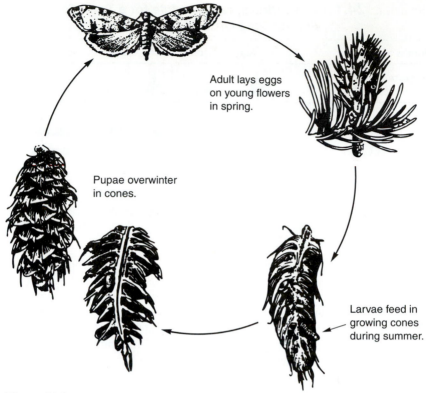

Figure 24.5

Life cycle of the Douglas-fir cone moth (*Barbara colfaxiana*) a pest of Pacific Northwestern seed orchards.

(Source: From Hedlin et al. 1981.)

Many insects cause occasional damage to seedlings of bare root nurseries—for example, grubs of scarab beetles feed on roots of conifer seedlings and adults on foliage in nurseries that were recent pastures. An occasionally important scarab in nurseries of Oregon and Washington is a *Pleocoma* spp. complex whose larvae feed on roots of Douglas-fir seedlings. Other insects, however, cause significant injury to roots, shoots, and foliage of nursery stock (Figure 24.8). In the Pacific Northwest the root weevil complex damage the roots of both conifer and hardwood nurseries; the black vine weevil (*Otiorhynchus sulcatus*) is the most important. Both soil and foliar insecticidal treatments are used to control damage. Foliar treatments seem to be most effective by targeting adults as they emerge from the roots of infested seedlings. Besides these controls, parasitic fungi and nematodes are being used to lower damage in containerized seedlings growing in greenhouses. Because the timing of these applications is critical, monitoring to determine periods of adult activity is essential.

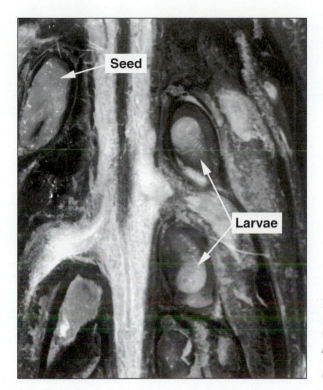

Figure 24.6
Larvae of the seed wasp (*Megastigmus spermotrophus*) developing in a Douglas-fir cone.
(Photo by N. E. Johnson.)

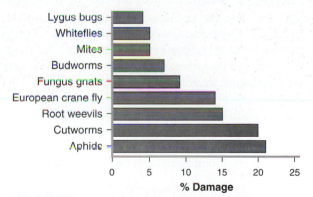

Figure 24.7
The nine most important nursery insect pests in terms of their percent occurrence in nurseries surveyed in the United States.
(Source: From USDA 1989.)

TABLE **24.2** | **Commonly Used Insecticides for Control of Insect Pests of Container Tree Seedlings**

Insect pest	Active ingredient (common name)	Trade name
Aphids	Malathion	Malathion
	Diazinon	Spectracide
	Acephate	Orthene
Cutworms	Carbaryl	Sevin
(*Peridroma* spp.)	Diazinon	Spectracide
	Chlorophyrifos	Lorsban
Root weevil complex	Endosulfan	Thiodan
(*Otiorhynchus* spp.)	Carbaryl	Sevin
	Carbofuran	Furadan
European crane fly	Diazinon	Spectracide
(*Tipula paludosa*)	Acephate	Orthene
Fungus gnats	Diazinon	Spectracide
(*Bradysia* spp.)	Malathion	Malathion
	Dimethoate	Cygon

Source: From USDA Forest Service (1990).

Webworm larvae, of the moth family Pyralidae, strip the outer bark from the upper roots of seedlings (weevil grubs feed on the lower roots) and sometimes kill thousands of seedlings. They are most damaging when they infest transplant beds of 2-year-old seedlings. An important pyralid that infests Douglas-fir roots is the cranberry girdler (*Chrysoteuchia topiaria*). Grass acts as a reservoir for webworms. Within an IPM context, an effective way to reduce the general population of these pyralids is to eliminate grass and herbaceous vegetation from around nurseries by mowing or treating with herbicides. Severe outbreaks are treated with insecticides; often diazinon is applied to the foliage to control the moths, and soil applications have been effective against the larvae. Pheromone traps are used to monitor adult populations to properly time treatments.

Aphid infestations of shoots and foliage are a universal pest of container and bare root nurseries. Aphids have complex life cycles that include both sexual and asexual reproduction. Because aphids produce many generations per year, they can quickly develop large populations that suppress normal development of seedlings across entire nursery beds. Prompt treatment before populations explode may be necessary.

The other pests outlined in Figure 24.8 also can cause unacceptable damage, and their populations must be constantly monitored. Thrips, for example, are tiny and hard to detect. Adult thrips are carried into nurseries by winds moving across agricultural or horticultural areas—for example, the onion thrips (*Thrips tabaci*), which is found on a huge variety of cultivated and wild plants. Thrips attack newly formed foliage and, as they rasp the tender tissues, the foliage twists and curls. To ensure correct diagnosis, affected foliage should be lightly tapped over a piece of paper, then the thrips can easily be identified with a hand lens. Outbreaks, once identified, can be controlled with timely application of insecticides.

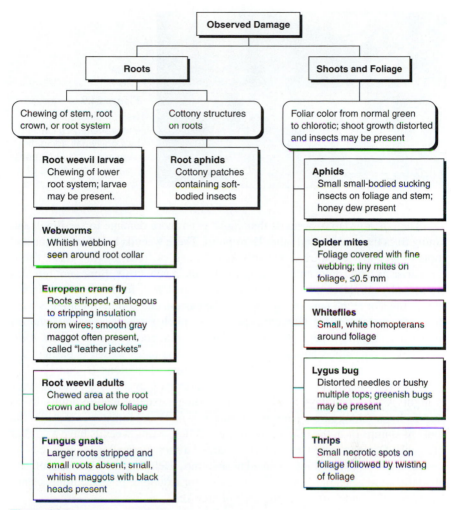

Figure 24.8
Damage key for pests of seedling roots, shoots, and foliage.
(Source: Modified from USDA Agricultural Handbook 674, 1989.)

PLANTATIONS

Newly established plantations are attacked by aphids and phylloxerids. They insert their piercing sucking mouthparts into vascular bundles of young foliage, thereby bypassing defensive chemicals in the **mesophyll** cells—the ecological niche of these hemipterans. For example, the spruce gall adelgid (*Adelges cooleyi*) is a phylloxeran that sucks the sap of young Douglas-fir trees in spring. These are wingless females, easily recognizable by their whitish and fluffy covering; several generations may occur on the foliage and high populations cause needles to become distorted, spotted, and drop prematurely. In later generations, winged individuals fly to spruce where they subsequently damage foliage by producing galls (Figure 24.9).

Figure 24.9
Gall on Sitka spruce
(*Picea sitchensis*) caused
by Cooley gall adelgid
(*Adelges cooleyi*) nymphs
feeding at the bases of
young needles in spring.

There is a group of weevils that cause even more damage to seedlings and young trees than aphids and other Hemiptera. These weevils breed in stumps and roots of recently felled trees, recently killed trees, or fresh logging slash. Problems arise as new adults emerge in spring to select new hosts. Emergent females are sexually immature and, to develop their ovaries, they feed on cortical tissues of seedlings or at the base of saplings. In the eastern and southern United States, there are several of these damaging species: the pitch-eating weevil (*Pachylobius picivorus*); the pales weevil (*Hylobius pales*); the pine root collar weevil (*H. radicis*), and various other damaging *Hylobius* spp. The pales weevil is particularly destructive to intensively managed southern pine plantations (Figure 24.10). Adults emerge in April to June and feed on tender bark of seedlings and saplings at night. After a few days of maturation feeding, sexually mature adults fly to stumps left after logging operations. There they mate and females oviposit in the roots of stumps or in the base of dying trees. Pitch-eating weevils have a similar life cycle. Damage to new plantations, planted a short time after logging, can be severe, especially to seedlings planted near stumps. Delaying establishment of new plantations prevents damage. In East Texas, for example, pine seedlings established 6 to 9 months after logging are not attacked.

In the western United States, adults of the weevil *Pissodes fasciatus* feed on Douglas-fir seedlings and probably other conifers for sexual maturation. A study in western Washington showed that *P. fasciatus* adults emerge from roots of stumps or recently killed trees from mid-May through June. Females feed at the base of seedlings and girdle them. After sexual maturation adults locate moribund trees or stumps; females lay eggs in the root collar and larvae feed upward and downward in the phloem. Several other *Pissodes* spp. attack weakened conifers in eastern forests, one of them, *P. dubius*, breeds in windthrown, dying, or recently killed balsam fir and red spruce. It is one of the most important insects invading trees weakened after spruce budworm (*Choristoneura fumiferana*) defoliation.

Many insects infest the buds and shoots of trees growing in plantations. Destruction of **meristems** may stop further tree growth or cause adventitious buds to take apical dominance and produce deformed growth (Figure 24.11). Repeated

Figure 24.10

Adult of the pales weevil (*Hylobius pales*) (top); newly emerged adults feeding at night on the stem of a seedling (middle); and a chip cocoon formed in the wood, and made from finely shredded sapwood. Here the last instar pupates (bottom).

(Source: From Bennett and Ostmark, 1972, USDA Forest Service.)

Figure 24.11

Deformed growth on pine caused by feeding activity of the European pine shoot moth (*Rhyacionia buoliana*) (**A**); crooked logs resulting from repeated shoot moth infestations (**B**).

attacks by shoot-infesting insects may result in crooked stems, thus defeating managerial goals of producing high-quality timber. Only a few of the more important bud- and shoot-destroying insects will be discussed.

There are four species of North American tipweevils of the genus *Pissodes* that breed in the subcortical region and outer wood of conifers. The most commercially important are the white pine weevil (*P. strobi*) and the lodgepole pine weevil (*P. terminalis*) that infest young pines and spruces. These two tipweevils infest and kill the terminals of their hosts, resulting in growth loss and crooking or branching of the trunk (Figure 24.12). White pine weevil larvae kill the leader of their hosts (three important hosts being eastern white pine, Sitka spruce, and white spruce) by mining and feeding in the cortical zone of these young shoots. Lateral branches from the stem whorl below the damaged leader assume a vertical position and compete for dominance. When stems are attacked over several years, trees are stunted, deformed, and overtopped by competing vegetation to the point where the plantation may be useless.

Sitka spruce plantations established on the west side of Vancouver Island, B.C., and within 1.6 to 3.2 km from the coast of Washington, are relatively resistant to white pine weevil attack (Warkentin et al. 1992). This is because trees growing on sites most influenced by moist and cool maritime air photosynthesize more efficiently and presumably apportion more energy to resin-producing defense mechanisms. Plantations established inland are less influenced by offshore air masses, and these trees are more susceptible to *P. strobi* attack. Accordingly Sitka spruce should not be planted in areas where high weevil attack is evident. Neither systematic clipping of infested leaders (to reduce local populations) or chemical treatments have been successful in controlling tipweevil damage. However, studies currently under way show that even in areas of high weevil attack, some trees show resistance, and a tree-breeding program is under way to produce tipweevil-resistant Sitka spruce (Alfaro et al. 1995, Alfaro 1996). Resistance to tipweevil attack would become one important element in an IPM plan being formulated by the Canadian Forest Service (Figure 24.13).

Pine plantations in both the northern and southern hemisphere are damaged by moths of the family Olethreutidae; two that are economically important to young pine plantations are the European pine shoot moth (*Rhyacionia buouliana*) and the Nantucket pine tip moth (*R. frustrana*). The latter is widely distributed in northeastern North America, and the central and southern states. The pine tip moth infests all pine species within its range, with the exception of longleaf pine and the soft pines. Boring larvae kill terminal and lateral buds of their hosts (primarily loblolly and shortleaf pines), resulting in reduced growth through loss of photosynthetic area as well as tree deformation and occasional tree mortality. Infestations can be reduced by attention to intermediate operations such as judicious use of herbicides to reduce weed competition and by timely fertilization. Tree-breeding programs have developed a model for screening families for tip moth resistance, and pine stock less susceptible to attack are being produced.

Figure 24.12
Studies on the host selection behavior of white pine weevils (*Pissodes strobi*) after males or females are caged on terminals of Sitka spruce; spruce in far left showing signs of a crook caused by weevil infestations below it.

The European pine shoot moth, indigenous to Europe, was introduced to the eastern United States in 1914 and reported in the northwestern states in 1959. This pest also has recently entered the intensively managed radiata pine (*Pinus radiata*) of South America. Extensive damage is controlled by application of contact sprays aimed at the dispersal period of adults in early summer, but on harsh sites where pine stands are particularly susceptible to shoot moth attack, IPM programs are more useful. IPM approaches include monitoring the population by trapping with synthetic pheromones (Kydonieus et al. 1982); lowering the level of damage by selecting tolerant breeding stock; planting on sites where weather conditions permit tree growth but are unfavorable for moth development (i.e., a tactic used in the Midwest); thinning and pruning of injured material; and controlling populations when they reach predetermined thresholds by accurately timing insecticide applications or by suppressing populations by use of pheromones or inhibitors to disrupt mating (Daterman 1974). In areas where *R. buoliana* has recently been introduced, populations are reduced by release of parasitoids—for example, the braconid larval parasite, *Orgilus obscurator*—then other IPM protocols can be used.

The western pine shoot borer (*Eucosma sonomana*) is another olethreutid moth that infests terminal shoots of ponderosa pine and to a lesser extent, Engelmann spruce and lodgepole pine. Larvae mine terminal shoots and reduce vertical growth. Damage is low, about 5 to 12% per year, but chronic and significant damage accrues over the course of a rotation. Economic analyses have shown that treating plantations for 5 years after shoot borer activity begins is financially justified. Mating disruption with aerially applied and ground dispersed synthetic pheromones has been successful (Overhulser et al. 1980). Treatment dosages of

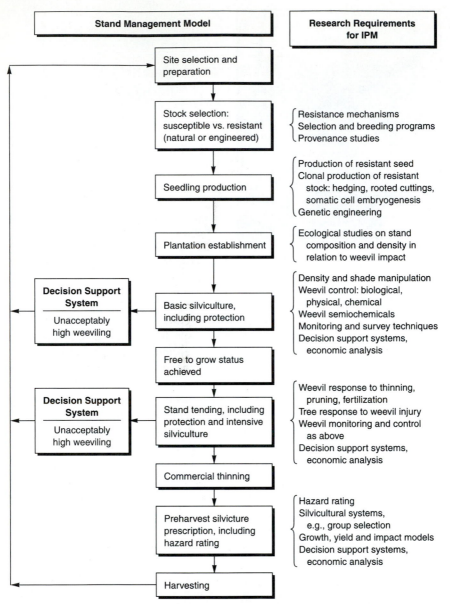

Figure 24.13

Stand management model for Sitka spruce in British Columbia, showing how unacceptably high damage by the white pine weevil, analyzed by a Decision Support System, can recommend abandonment of plantations before maturity (left-hand loop), and on the right research and development program to institute an IPM of *Pissodes strobi*.

(Source: From Alafaro et al. 1995, Canadian Forest Service.)

3 to 20 g/ha of active ingredient have resulted in 70 to 90% fewer terminal shoots infested. Following treatments, damage again increases within 2 years and re-application of treatments may be necessary.

Midges of the dipteran family Cecidomyiidae damage the terminal growth of many hardwoods and conifers; in forests of the high north, for instance, the spruce bud midge (*Rhabdophaga swainei*) is a pest of boreal forests. The gouty pitch midge (*Cecidoymia piniinopsis*) is a pest of ponderosa pine plantations; it causes shoot mortality, deformities of limbs, and stunting. In spring the flies lay eggs on the current year's expanding shoots and later larvae bore into the shoots. Eventually their activity causes galls to form. When these expanded necrotic areas are large, the affected limbs or stem bend and grow in a deformed manner. This pest is managed by clipping deformed branches and in high-value stands by spraying.

Leafcutting ants, *Atta texana*, construct large underground nests in pine plantations of East Texas and western Louisiana, especially in sandy, well-drained soils. They defoliate plantations by their habit of systematically cutting the foliage from young pine trees and carrying the clippings into their nests. There they culture a fungus on the foliage and proceed to feed their broods the fungus. Many control tactics have been used to manage this pest—they include first finding nests before areas are planted, then setting out poison baits (baits with Myrex have been successful, but the product has been banned) or inserting nozzles into the main orifices of the nests and applying aerosols of cyclodienes or chlorinated hydrocarbons. Most of these compounds also have been banned. Current controls still require location of nests and application of registered insecticides under pressure or by thermal fog applications.

The Douglas-fir pitch moth (*Synanthedon novaroensis*) a clearwing moth of the family Sesiidae, is an insect that feeds in the cambium and phloem of Douglas-fir, Sitka spruce, Engelmann spruce, western white pine, and other western conifers. Unlike other insects that threaten recently pruned stands, pitch moths are primary pests and as such attack healthy vigorous trees (Johnson et al. 1995).

Pitch moths also are a major problem in seed orchards after trees are circumferentially wounded to promote cone production. Adult *S. novaroensis* oviposit around the border of fresh wounds, especially pruning wounds; as the larvae feed in the subcortical area, a large amount of resin is produced that forms a pitch mass outside the wound. A pruned Douglas-fir stand had about 6% of the pruned trees infested the year of pruning and, the following year, over 9% of the pruned stems were infested. The percent of infested trees following pruning was even more severe in a second stand: The first year 5% of the pruned stems were infested and over 25% the second year. Larvae develop over a 1- to 2-year period; accordingly additional time is required to occlude both the defect caused by the mining of this insect and the pitch nodule within which the moth pupates. Production of clear wood thus is delayed by attacks of this sesiid. Problems from pitch moth are reduced by thinning and pruning prior to crown closure. This practice perpetuates understory growth, enhances biodiversity, and provides habitat for pitch moth parasitoids.

REFERENCES

Alfaro, R. I. 1996. Role of genetic resistance in managing ecosystems susceptible to white pine weevil. *The Forestry Chronicle* 72:374–380.

Alfaro, R. I., J. H. Borden, R. G. Fraser, and A. Yanchuk. 1995. The white pine weevil in British Columbia: basis for an integrated pest management system. *The Forestry Chronicle* 71:66–73.

Bennett, W. H., and H. E. Ostmark. 1972. *Insects of southern pines.* USDA Forest Service, Sourthen Forest Experiment Station.

Cameron, R. S. 1981. *Toward insect pest management in southern pine seed orchards.* Texas Forest Service, Texas A&M University System, Publication 12, College Station, Texas.

Daterman, G. E. 1974. Synthetic sex pheromones for detection survey of the European pine shoot moth. USDA Forest Service, Pacific Northwest Forest Range Experiment Station, Research Paper PNW-180.

Ebel, B. H., T. H. Flavell, L. L. Drake, H. O. Yates III, and G. L. De Barr. 1980. *Seed and cone insects in southern pines.* USDA Forest Service General Technical Report SE-8.

Hedlin, Alan F., H. O. Yates III, D. Cibrián Tovar, B. H. Ebel, T. W. Koerber, and E. Merkel. 1981. *Cone and seed insects of North American conifers.* Canadian Forest Service, USDA Forest Service Unnumbered Publication.

Johnson, J. M., M. D. Petruncio, and R. I. Gara. 1995. Impact of forest pest problems on intensive management practices such as pruning, pp. 245–251. In D. P. Hanley et al. (eds.). *Forest pruning and wood quality of western North American conifers.* Contribution 77, Institute of Forest Resources, College of Forest Resources, University of Washington, Seattle.

Kydonieus, A. F., Morton Beroza, and Gunter Zweig. 1982. *Insect suppression with controlled release pheromone systems.* VII. CRC Press, Boca Raton, Fla.

Landis, T. D., R. W. Tinus, S. E. McDonald, and J. P. Barnett. 1990. *The container tree nursery manual,* Vol. V. *The biological component, nursery pests and mycorrhizae.* USDA Forest Service Agricultural Handbook 674.

Ruth, D. S. 1980. *A guide to insect pests in Douglas-fir seed orchards.* Forestry Canada, Pacific Forest Research Centre, Unnumbered Publication, Victoria, B.C., Canada.

Overhulser, D. L., G. E. Daterman, L. L. Sower, C. Sartwell, and T. W. Koerber. 1980. Mating disruption with synthetic sex attractants controls damage by *Eucosma sonomana* (Lepidoptera: Tortricidae Olethreutinae) in *Pinus ponderosa* plantations II. Aerially applied hollow fiber formulation. *Canadian Entomologist* 112:163–165.

Turgeon, J. J. and Peter De Groot. 1992. *Management of insect pests of cones in seed orchards in eastern Canada.* Ontario Ministry of Natural Resources and Forestry Canada, Queen's Printer of Ontario, Ontario, Canada.

Warkentin, D. L., D. L. Overhulser, R. I. Gara, and T. M. Hinckley. 1992. Relationships between weather patterns and susceptibility of Sitka spruce, *Picea sitchensis*, to infestations of the tip weevil, *Pissodes strobi. Canadian Journal of Forest Research* 22: 542–549.

Zobel, Bruce, and John Talbert. 1984. *Applied forest tree improvement.* Waveland Press, Prospect Heights, Ill.

FOREST INSECT QUARANTINE

INTRODUCTION

Insects species **coexist** when they **partition resources** and thus avoid severe **inter-specific** competition. Species having many niche **parameters** in common, however, can evolve in separate regions of the world and are called **ecological homologs.** When these kinds of insects are introduced into North America, they often outcompete and displace native insects. Even more serious problems can occur when introduced insects are preadapted to fill an unoccupied niche. Many of these insects reach pest status quickly. This is because they arrive without their natural enemies, and their new host plants lack **coevolved** natural defenses. Examples of introduced insects that have altered forest ecosystems in North America are the gypsy moth, the smaller European elm bark beetle, the balsam woolly adelgid, the hemlock woolly adelgid, and various species of sawflies (Barbosa and Schultz 1987, U.S. Congress, Office of Technology Assessment 1993).

During the 19th century and into the early 20th century, nursery stock and timber products were brought into North America or exported with little concern about the problems of introducing pests. The gypsy moth actually was introduced on purpose (see Chapter 20), but most introductions are accidental. Some of these introductions occurred because immigrants wished to be surrounded by familiar European trees. Because of these practices, exotic forest insects now are established in North America, and many are causing severe damage (Ciesla 1993; Haack and Byler 1993; Heiner 1995; Morrell 1995; USDA Forest Service 1997) (Table 25.1). The current status on some of these insects follows:

1. Balsam woolly adelgid (Phylloxeridae)—*Adelges piceae* in Europe has a complex life cycle that includes **alternate hosts** and males. In North America the insect is entirely parthenogenic and develops its life cycle completely on fir trees. The range of the phylloxerid includes all of the maritime provinces, New England, and down through the Appalachians, and it is found throughout the Pacific Northwest. Currently it devastates

561

TABLE **25.1** | **Forest Insects Introduced into Various Parts of the World Where They Have Become Serious Pests**

Organism	General origin	Year of introduction	Site of introduction
Gypsy moth (European form) (*Lymantria dispar*)	France	1869	Medford, Mass.
Larch sawfly (*Pristophora erichsonii*)	Eurasia	1880	Boston, Mass.
Larch casebearer (*Coleophora laricella*)	Europe	1886	New England
Beech scale (*Cryptococcus fagisuga*)	Europe	1890	Canada
Spruce aphid (*Elatobium abietinum*)	Europe	1900	West Coast
Balsam woolly adelgid (*Adelges piceae*)	Europe	1908	Brunswick, Maine
Lesser European elm bark beetle (*Scolytus multistriatus*)	Europe	1909	New Jersey
European pine shoot moth (*Rhyacionia buoliana*)	Europe	1914	Long Island, N.Y.
Pine sawfly (*Diprion similis*)	Europe	1914	United States
European spruce sawfly (*Gilpinia hercyniae*)	Europe	1922	Canada
Hemlock woolly adelgid (*Adelges tsugae*)	Asia	1924 1950	West Coast Richmond, Va.
European pine sawfly (*Neodiprion sertifer*)	Europe	1925	United States
Winter moth (*Hyphantria cunea*)	Europe	1930	Canada
Eucalyptus longhorned borer (*Phoracantha semipunctata*)	Australia	1984	California
Gypsy moth (Asian form) (*Lymantria dispar*)	Russian Far East Europe/Asia	1991 1993	Oregon and Washington North Carolina
Common European pine shoot beetle (*Tomicus piniperda*)	Europe	1990	Ohio
Asian longhorned beetle (*Anolophora glabripennis*)	Asia	1996	New York

Source: Modified from USDA (1997).

stands of subalpine fir (*Abies lasiocarpa*) (Figure 25.1) and severely affects the growth of silver fir (*A. amabilis*) in many areas. This insect is now an urban pest of ornamental firs and a major Christmas-tree plantation problem, especially with Fraser fir (*A. fraseri*) growing in the Appalachians.

Figure 25.1

Subalpine firs (*Abies lasiocarpa*) in Washington State infested by balsam woolly adelgids (*Adelges piceae*) during 1975 (**A**) and the same trees during 1990 (**B**).

2. Gypsy moth (European)—*Lymantria dispar* has received millions of dollars for research, eradication, and control efforts—more than any other introduced forest insect. **Eradication** has been abandoned in the infested forests of the northeastern United States, and pest management operations focus on reducing foliar losses during outbreaks (Dahlsten and Garcia 1989). States and provinces outside the general infestation area continue to pursue eradication by treating low-level infestations that cause no visible defoliation. These populations are low, and their presence is detected by organized surveys that use sensitive pheromone-baited trapping grids. In spite of these efforts, gypsy moths keep on spreading across the continent and continue to threaten hardwood forests along their path.

3. Gypsy moth (Asian)—The Asian form of *L. dispar* was introduced to the Pacific Northwest when Russian cargo ships, coming from the Russian Far East, unloaded in Vancouver, B.C., Portland, Oregon, and Tacoma, Washington. The moth also was introduced into North Carolina onboard a cargo ship loaded with military equipment returning from Germany. The Asian gypsy moth is potentially more damaging than the European form because females fly and can disperse over 20 km; they lay more eggs, and have additional hosts—including larch, a conifer. Pheromone-trapping grids are used to locate centers of introduction. These pest foci are then accurately delineated, mapped, and sprayed with *Bacillus thuringiensis* in an effort to eradicate small populations as they occur.

4. Common European pine shoot beetle (Scolytidae)—*Tomicus pineperda* is a major threat to young pines in Christmas-trees plantations. After emerging

from blowdowns or logs, adult pine shoot beetles feed within shoots and terminals of standing trees during a period of sexual maturation, and often cause 20% or more of the shoots to die.

5. Hemlock woolly adelgid (Phylloxeridae)—When this phylloxerid was introduced into Virginia in 1950, it quickly spread north and south throughout the range of eastern hemlock (*Tsuga canadensis*). As eastern hemlock did not coevolve with *A. tsugae,* there is a real danger that this tree species may disappear from North America.

6. Asian longhorned beetle—Unlike most wood borers, *Anolophora glabripennis* infests living trees (Haack et al. 1997) (Figure 25.2). The insect is native to China and other Asian countries and attacks many hardwood trees including maple, poplar, and willow. When discovered in New York in 1996, it already had infested Norway maple, sugar maple, silver maple, sycamore maple, and box elder. This cerambycid spreads quickly as females live for more than 40 days and can fly long distances to find new hosts. If the Asian longhorned beetle were to become established in northeastern North America, severe consequences would follow. The structure of the northern hardwood forests would be altered, there would be a decline in maple syrup production, and fall tourism would be affected. Because of these threats, a major eradication program to remove all infested trees was done in Brooklyn and Amityville, New York (Figure 25.3).

7. Beech scale—Beech bark disease results when bark, infested by the beech scale is altered and allows entrance of the fungus *Nectria coccinea* var *faginata*. Even though the scale came to North America in 1890, the problem associated with the vectored disease was not noticed until the 1930s when stands of beech were being killed in the maritime provinces, Maine, and Massachusetts. The scale and the fungus it vectors are now found in West Virginia and south into North Carolina and Tennessee.

8. Spruce aphid—It's not clear when *E. abietinum* invaded North America, but it must have been early in the 20th century, as there are reports of major Sitka spruce mortality along the tidelines of the Pacific Northwest and the Columbia River drainages during the 1918 and 1930 droughts. Today the spruce aphid is a pest of urban spruces, and is found from Alaska to California and inland into Arizona and New Mexico.

9. Eucalyptus longhorned borer—*Phoracantha semipunctata*, a native of Australia, is a severe pest of eucalyptus in many countries where eucalyptus has been introduced—for example, Israel, Spain, Portugal, Italy, Tunisia, Egypt, South Africa, Uruguay, Chile, and recently the United States (Hanks et al. 1990). Adults of both sexes are attracted to stressed eucalyptus trees and logs of any size as oviposition sites (see Figure 23.3). Outbreaks in California began during the prolonged 1986 to 1992 drought, and thousands of trees were killed during this period. Integrated pest management (IPM) programs show promise of managing this pest.

Figure 25.2
The Asian longhorned beetle (*Anolophora glabripennis*) (**A**), and the damaged trunk of a sugar maple (*Acer saccharum*) growing in Brooklyn, New York (**B**).
(Photo given by Eric Legasa, Washington State Department of Agriculture.)

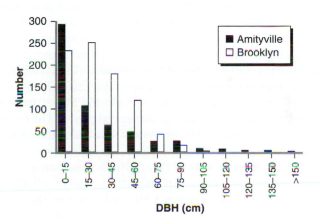

Figure 25.3
Size classes of over 1,400 maple trees cut in Amityville and Brooklyn, New York, during 1997 after they were found to be infested by the Asian longhorned beetle (*Anolophora glabripennis*).
(Source: Data from Haack et al. 1977.)

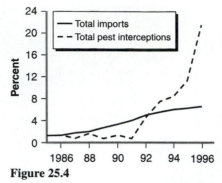

Figure 25.4

Relationship between percent of total U.S. imports of wood products from China and percent of total insect interceptions during 1986–1996.

(Source: Data from Haack et al. 1997.)

Because of restrictions on supply of logs from federal lands, the United States is the world's leading importer of wood and wood products. In 1990, for example, the United States imported over $5 billion of logs, lumber and other unmanufactured wood products, a situation that invites introduction of pests. Between 1985 and 1996 nearly 5,900 insects were intercepted on wood products at ports of entry—for example, the number of insects found on wood products from China closely tracked the growing volume of Chinese imports (Figure 25.4).

The danger of importing timber pests will continue in the coming decades as commercial forest lands available for logging in the United States will decrease about 5% during the next 40 years. Accordingly the timber industry is interested in importing large volumes of logs during this period. The focus will be on conifers from the boreal forests of northeastern Asia and plantation-grown pines and Douglas-firs from the southern hemisphere. Ecologists, however, feel that the greatest threat to the integrity and viability of North American forests is the continued introduction of exotic pests, especially from parts of the world with temperate climates. For example, during a 1991 conference on the threat to North American forests posed by introducing exotic pests with the importation of larch logs from Russia, 175 different insects and pathogens that use larch as a host were identified. Six of these organisms were considered as a documentable risk to trees of western North America.

REGULATIONS ON IMPORTATION OF LOGS AND WOOD PRODUCTS

Recent trade agreements, such as the North American Free Trade Agreement (NAFTA) and the General Agreement on Tariffs and Trade (GATT), strongly encourage free trade, thus the potential of pests entering the United States has drastically increased. Federal regulations dealing with the introduction and spread in the United States of damaging insect pests are the responsibility of the U.S. Department of Agriculture, Animal and Plant Health Inspection Service (APHIS). One of the ways this agency accomplishes its mission is by regulating imports such as agricultural items, wood products, and logs—plant-related materials that could harbor pests.

As the United States has not imported appreciable amounts of logs and lumber (except from contiguous Canadian forests), APHIS had to establish protocols and quarantine regulations before logs and lumber could be imported. Before promulgating new regulations, the agency felt that, with the help of the U.S. Forest Service, detailed risk assessments for movements of logs and unmanufactured wood articles into the United States had to be made. In 1990 risk assessments were made for potential importation of logs and lumber from Siberia. Because of clear indications of serious pest problems, APHIS would not allow importation of timber from Siberia.

During 1993 the U.S. Forest Service completed risk assessments associated with importation of radiata pine, coigüe (*Nothofagus dombeyi*), and tepa (*Laurelia philippiana*) from Chile as well as radiata pine and Douglas-fir from New Zealand. These assessments identified insects and fungi that could pose moderate- to high-risk potentials if they were introduced into forests of North America. Moreover, these analyses recognized that little was known on the biology and ecology of these likely pests. Nevertheless, APHIS, utilizing these two pest risk assessments as well as cost-benefit analyses and other inputs, developed log treatments and handling protocols for importation of these materials (see Chapter 16) (Table 25.2). In 1995 the agency published regulations on how these material could be imported into the United States. Toward the end of 1995, the U.S. District Court for the Northern District of California ruled that the environmental impact statement (an EIS is an analytical process that is a requirement of the 1969 National Environmental Policy Act) upon which the regulations were based, was inadequate. Accordingly APHIS developed a supplemental EIS, which may lead to new regulations (or restrictions) on importation of logs, lumber, and unmanufactured wood articles from the southern hemisphere (USDA/APHIS 1997).

In essence, policies on whether or not logs and their products should be allowed to enter North American ports are based on biological information, risk ratings, cost-benefit analyses and ultimately on political imperatives. Many interests influence quarantine decisions (Figure 25.5). Each interest is not a monolith; as much difference of opinion can occur within an interest group as between them. Not all these interest groups focus on one quarantine issue, nor do all issues carry equivalent weight. For example, a large or politically influential constituency that favors a particular import may far outweigh the positions of a small number of expert scientists who caution against the dangers of introducing pests. Nevertheless, there is an emerging system of quarantine protection that is a blend of scientific information, governmental regulations, public concerns, and an effective use of the legal system by all concerned parties.

T A B L E **25.2** | **Example of APHIS Importation Requirements for Logs, Lumber, and Selected Wood Products**

Article	Requirements
Logs	Prior to importation, they must be debarked and heat treated.
	During the entire interval between treatment and export, logs must be handled and stored to exclude pest access to them.
Lumber (heat treated or treated with moisture reduction)	During shipment, lumber must be segregated from all other regulated articles (except solid wood packing materials) in separate holds or separate sealed containers unless all regulated articles also have been heat treated.
	Lumber on a vessel's deck must be in a sealed container unless it has been heat treated with moisture reduction.
	The treatment used must be stated in the importer document or permanently marked on each piece of lumber.
Lumber (raw)	Any raw lumber (including associated solid wood packing material) from those places in Asia that are east of 60° east longitude and north of the tropic of Cancer is ineligible for import.
	No other regulated article other than packing materials may be on the same means of conveyance with raw lumber unless the raw lumber and other regulated articles are in separate holds or separated sealed containers.
	Raw lumber on a vessel's deck must be in sealed containers.
	All raw lumber imported must be consigned to a facility operating under an APHIS compliance agreement, must be heat treated (either with or without moisture reduction) within 30 days of import, and must be heat treated prior to any process.
Wood chips/bark chips	Wood or bark chips from Asian countries that are wholly or partially east of 60° east longitude and north of the tropic of Cancer are ineligible for import.
	The importer document must certify that the chips are from live, healthy plantation-grown trees grown in the tropics or have been fumigated with methyl bromide or have been heat treated.
	The chips must be free from rot or accompanied by importer document certifying fumigation with methyl bromide.
	During shipment, no other regulated articles other than solid wood packing materials are permitted in holds or sealed containers with chips.
	Chips on a vessel's deck must be in a sealed container (certain specified exceptions are permitted).
	All imported chips must be consigned to a facility operating under an APHIS compliance agreement and must be burned, heat treated, or processed within 30 days of arrival at the facility.
Mulch, humus, compost, and litter	Decomposing, organic material must be fumigated or heat treated prior to importation.
Cork bark	Bark must be free of rot.

Source: From USDA (1997).

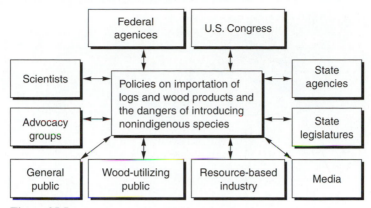

Figure 25.5

Major interests involved in shaping quarantine policies with regard to logs and wood products imports into the United States.

(Source: From U.S. Congress, Office of Technology Assessment, 1993.)

REFERENCES

Barbosa, P., and J. C. Schultz. 1987. *Insect outbreaks.* Academic Press, San Diego, Calif.

Ciesla, W. M. 1993. Recent introductions of forest insects and their effects: a global overview. *FAO Plant Protection Bulletin* 41:3–13.

Dahlsten, D. J., and Richard Garcia. 1989. *Eradication of exotic pests.* Yale University Press, New Haven, Conn.

Haack, R. A., and J. W. Byler. 1993. Insects and pathogens: regulators of forest ecosystems. *Journal of Forestry* 91(9):32–37.

Haack, R. A., K. R. Law, Victor C. Mastro, H. S. Ossenbruggen, and B. J. Raimo. 1997. New York's battle with the Asian long-horned beetle. *Journal of Forestry* 95(12):11–15.

Hanks, L. M., J. G. Millar, and T. D. Paine. 1990. Biology and ecology of the eucalyptus long-horned borer (*Phoracantha semipunctata* F.) in southern California, pp. 12–16. In D. Adams and J. Rios (eds.). *Proceedings, 39th Meeting of the California Forest Pest Council,* 14–15 November 1990. California Department of Forestry and Fire Protection, Sacramento, Calif.

Heiner, H. 1995. Issues in international forestry. *Journal of Forestry* 93(10):6–11.

Morrell, J. J. 1995. Importation of unprocessed logs into North America: a review of pest mitigation procedures and their efficacy. *Forest Products Journal* 45:41–50.

U.S. Congress, Office of Technology Assessment. 1993. *Harmful non-indigenous species in the United States.* U.S. Government Printing Office, OTA-F-565.

USDA Animal and Plant Health Inspection Service. 1997. *Importation of logs, lumber, and other unmanufactured wood articles: draft supplement to the environmental impact statement, December 1997.* USDA Marketing and Regulatory Programs, APHIS, Riverdale, Md.

USDA Forest Service. 1997. *Forest insect and disease conditions in the United States 1996.* USDA Forest Service Forest Health Protection, AB-2S, Washington, D.C.

10 A.M. policy   A fire control policy that specified planning for control by 10 A.M. the next morning, with establishment of 10 A.M. the following day if control was not successful.

abdomen   The most posterior of the three principal body divisions of the insect (the digestive and reproductive center).

abiotic disease   Disease resulting from nonliving agents.

acaracides   Pesticides that kill mites.

Acarina   Order of the class Arachnida, which includes the spiders, mites, and ticks.

acervulus (pl. acervuli)   An open saucerlike fungal fruiting structure bearing conidia.

acetylcholine   A compound that allows pulses to cross nerve synapses.

acetylcholinesterase   (shortened to cholinesterase) An enzyme involved in the removal of acetylcholine.

active ingredient (A.I.)   The actual compound responsible for the toxic effect of a pesticide.

adaptations   Genetic change in a population due to genetic selection in response to changes in the environment of that population.

adiabatic   Of or denoting change in volume or pressure, with no heat gain or loss with the surrounding air.

aeciospore (pl. aecia)   Rust spore borne inside an aecium.

aedeagus   The intromittent organ of male insects (i.e., the penis of higher organisms) that transfers sperm into the reproductive tract of females.

aerobic   Requiring oxygen to grow.

aggregant   A natural or synthetic semiochemical that attracts aggregations of males and females (e.g., the aggregating pheromones of bark beetles).

aggregation pheromone   See aggregant.

allelochemic   Defensive chemicals.

allogenic succession   Change in plant communities produced by alterations of the environment or community members by factors outside of the system (such as sediment filling a pond, disturbance, etc.)

alternate host   Used in reference to one of the hosts of a heteroecious rust fungus; usually the noneconomic host. Two or more hosts are needed to complete the life cycle of an organism.

anaerobic   Requiring no oxygen to grow.

anaerobic metabolism   Metabolism in the absence of oxygen.

anamorph   The asexual stage in the life cycle of fungus that produces sexual spores.

antennae   Sensory appendages found on the insect head and generally bearing olfactory and sensory receptors.

anterior   Pertaining to in front or ahead of.

antheridium (pl. antheridia)   A male gametangium.

anthracnose   Fungal disease of broad-leaved trees with symptoms ranging from distorted, ulcerlike lesions on leaves to cankers and diebacks of twigs and branches.

antiaggregant   A natural or synthetic semiochemical that repels aggregations of males and females of the same species.

antibiotic   A substance produced by a living organism that can inhibit another organism.

anus   Posterior opening of the digestive tract (alimentary canal).

apocrita   Suborder of the Hymenoptera containing the ants, parasitoids, wasps, hornets, velvet ants, and other mostly beneficial insects.

apothecium (pl. apothecia)   A open, saucerlike ascocarp in which the asci are borne.

appendage   A part or an organism, such as a leg, that is joined to the axis or trunk of the body.

appressorium (pl. appressoria)   A flattened, hyphal, pressing organ from which an infection peg usually grows and enters the host.

Arachnida   Class of arthropods within the subphylum Chelicerata including spiders, mites, and ticks.

arbuscules   Minute, treelike, hyphal branches in host cells produced by endomycorrhizal fungi.

arrestant   A chemical that stimulates a flying insect to land or a crawling insect to stop.

arthropod   An invertebrate animal with jointed appendages and other features; organisms belonging to the phylum Arthropoda.

Arthropoda   An immense phylum of invertebrates containing the subphyla Chelicerata and Mandibulata.

arthrospore   A spore resulting from the fragmentation of hyphae.

ascocarp   A structure in or on which asci are produced in the Ascomycota.

ascospore   A sexual spore borne in an ascus.

ascus (pl. asci)   A saclike structure, usually containing eight ascospores or the sexual spores of the Ascomycota, formed by karyogamy and meiosis.

aseptate   Without septa or cross-walls.

asexual   Reproduction not involving the fusion of two nuclei.

aspect   The direction a slope faces.

atmospheric stability   See stability.

atrophy   Lack of growth.

auditory   Pertaining to the sense of hearing.

autoecious   A characteristic of rust fungi that complete their life cycle on one host.

autogenic succession   Change in plant communities produced by alterations of the environment caused by the resident organisms.

avoider   A life-history strategy of plants that have little adaptation to fire.

axillary sclerites   Flexible plates (sclerites) found in the axillary region of the wings associated with wing flexing or folding.

Bacillus thuringiensis **(Bt)**   A spore-forming bacterium that, when ingested, is toxic to insects, in particular, caterpillars.

backfiring   A fire control tactic involving burning out of fuel within a control line.

backing fire   See Table 6.5.

bacteriophage   A virus associated with bacteria.

bacterium   A unicellular, prokaryotic microorganism that reproduces by fission.

basal area   Area of the cross section of a tree stem, inclusive of bark, measured at breast height (1.3 m or 4.5 ft). Usually expressed as square feet per acre or square meters per hectare.

basidiocarp   Fruiting body of fungi in the Basidiomycota (e.g., mushrooms and conks).

basidiospore   A sexual spore produced on a basidium.

basidium (pl. basidia)   A structure bearing on its surface usually four basidiospores, the sexual spores of the Basidiomycota, following karyogamy and meiosis.

Beaufort Scale   A scale of windspeed categories that ranges from calm to hurricane (see Table 7.1).

BEHAVE   A fire behavior prediction model (see Chapter 3).

bilateral symmetry   A type of symmetry where body parts are aligned symmetrically on either side of a plane.

biological control   Control of pests by use of predators, parasitoids, and disease-producing microorganisms. Control of diseases by fungi and bacteria.

biome   A set of plant communities that have similar life-forms, generally expressed at a continental scale (e.g., the coniferous biome).

biotic disease   Disease caused by living organisms.

biotic potential   A measure of the maximum rate of increase of a population when isolated from factors of natural mortality.

blight   A disease symptom characterized by rapid discoloration and death of all parts of a plant.

broadcast burning   Burning forest fuels as they are; no piling or windrowing.

brood   A cohort of individuals that hatch at about the same time from eggs laid by males and females that matured together.

brown rot   A condition where the cellulose and hemicellulose of wood are decomposed, leaving the brown-colored lignin (cf., white rot). Also a flower and twig disease of cherries.

cambial heating   Rise of temperature at the cambium due to flaming or smoldering combustion around the tree stem.

canker   A visible dead area in the cortex or bark of a plant, usually of limited extent.

canopy   The part of any stand of trees represented by the tree crowns; canopies may occur in layers.

carapace   A hard dorsal covering consisting of fused dorsal segments characteristic of crustaceans.

carnivores   Secondary consumers that eat other animals.

castes   The various forms that develop in colonial insects such as termites and bees.

caudal   Belonging to the tail or the posterior end of the body.

cellulose   A linear polymer of glucose molecules linked by 1,4-beta-glycosidic bonds, making up the cell walls of plants.

center fire   See Table 6.5.

cephalothorax   The united head and thorax found in the classes Arachnida and Crustacea.

cercus (pl. cerci)   Sensory appendage located at the posterior tip of the abdomen.

chelicera (pl. chelicerae)   Pincerlike first pair of appendages of adult members of the Chelicerata.

Chelicerata   A subphylum of the Arthropoda, with distinct characteristics, an important one being that the mouthparts are not incorporated within the mouth of the organism (e.g., spiders, scorpions, ticks, and mites).

Chilopoda   The centipedes, members of the subphylum Mandibulata, and characterized by having one pair of legs per body segment.

chlamydospore   A thick-walled resting spore usually found in an intercalary position in fungal hyphae.

chlorosis   The yellowing of green leaves due to failure of chlorophyll to form.

chlorotic Yellowish foliage instead of the normal green color.

cholinesterase A shortened word for acetylcholinesterase.

chorion The eggshell of insects.

chrysalis The pupal stage of lepidopterans that do not spin cocoons.

circumesophageal connectives Paired nerves, one on each side of the esophagus, that run between the brain and the subesophageal ganglion.

clamp connection A bridgelike hypal connection involved in maintaining the dikaryotic condition of many Basidiomycota.

clearcut A unit of forest where all the trees are harvested. It is intended to produce a new even-aged forest stand.

cleistothecium (pl. cleistothecia) A closed ascocarp.

climax Species or communities representing the final (or an indefinitely prolonged) stage of a sere.

coalesce Grow together or fuse.

coarse filter Conservation of land areas and representative habitats, and the processes associated with them, with the intent that the needs of all associated species and communities will be met (see fine filter).

coarse woody debris Large deadwood in the forest, that may be either snags or down logs.

cocoon Silken case in which a pupa develops.

codominants See crown class.

coenocytic Nonseptate; nuclei present in cytoplasm without being separated by septa.

coevolved Pertains to coevolution: evolution of a specific relationship between two or more different species.

coexist To exist together at the same time or in the same place.

combustion Rapid oxidation of biomass by burning, resulting in carbon dioxide, water vapor, and other materials, with the release of large amounts of heat.

complete metamorphosis (holometabola) Change in form in which the immatures (larvae, caterpillars in moths and butterflies, or maggots in the flies) are completely different in form and function than the adults; the pupal stage (the pupa [pl. pupae]) bridges the larval stage with the adult stage.

compound eye An eye composed of many individual elements (ommitidia), each of which is seen externally by a facet.

compression wood Reaction wood in conifers that develops on the lower side of tilted stems and on the underside of branches.

conduction Transfer of energy from one molecule to another, such as through the soil or through bark.

confine strategy A fire suppression strategy defined by certain geographic or fire prescription limits.

conidiophore Branched or simple hypha arising from somatic hypha and bearing at its tip or side conidia.

conidium (pl. conidia) A nonmotile asexual spore borne on a conidiophore.

conk Fruiting body of a wood-destroying fungus.

consumers Organisms that eat either green plants (primary consumers) or eat primary consumers (secondary consumers).

contain strategy A fire suppression strategy involves construction of a fireline completely around a fire, with unburned fuel inside of the line.

container and bare root nurseries Container nurseries are generally large greenhouses that grow trees in containers with specially prepared potting media; trees grown in outdoor nurseries are termed bare root seedlings.

context   The sterile portion above the hymenium in the Basidiomycota.

control strategy   A fire suppression strategy where a line has encircled the fire and the fire has no chance of flareups that might cross the fireline.

convection   Transfer of energy by movement of liquid or gas, such as the movement of hot air into tree canopies.

cooperage   Pertains to the barrel-making industry.

copulatory   Pertaining to mating.

Coriolis force   An apparent force due to the rotation of the earth that deflects air to the right (in the northern hemisphere) as the earth rotates on its axis.

cornicles   Tubular structures on the posterior part of the abdomen of aphids.

cover type classification   A vegetation classification based on current dominant species of plants.

coxa (pl. coxae)   The ball-like basal segment of an insect leg.

crotchets   Tiny hooks found on the prolegs (pseudopods) of caterpillars.

crop   A dilated portion of the foregut immediately behind the esophagus.

crown class   The social position of trees of different heights within a forest stand (e.g., the tallest trees are called dominants, less tall ones are codominants, those below the codominants are intermediates, and trees that are yet shorter and lack sunlight for growth and maintenance are the suppressed trees).

crown fire   A fire that burns into the crowns of the vegetation, often dependent on intense understory fire.

crown scorch   Leaf mortality created by heating of foliage, with subsequent browning of that portion of the crown.

Crustacea   A member of the subphylum Mandibulata that includes the crabs, lobsters, crayfish, pillbugs, and shrimp.

cull   Wood in a living tree that is not used for lumber because of decay.

cut and scatter   A fuel treatment strategy involving dispersing fuel across a unit to avoid concentrations, with no actual reduction in fuel amount.

damping-off   A disease of seedlings that typically causes them to rot at the soil level and fall over, or not even emerge.

decomposers   Microorganisms found in the forest floor that contribute to the breakdown of organic matter into nutrients.

defensible fuel profile zone   An area of reduced fuel, usually with surface fuel reduction, some pruning of understory fuel to separate surface from crown fuels, and thinning of crown fuels. See fuelbreak.

defoliators   Insect that consume the foliage of plants; they may chew the foliage, feed only on soft tissues and skeletonize the leaves, or live between the leaf or needle cuticle and act as leaf miners.

degrade   Pertains to loss of timber values because of defects (e.g., insect holes or rot).

density-dependent factor   An environmental component that causes a level of mortality that is linked to the level of the population.

density-independent factor   An environmental component that causes a level of mortality independent of the number of individuals in the population.

dermal   Of or pertaining to the skin.

detritus   Dead organic plant and animal matter.

detrivore   An organism that derives its energy from dead organic matter. A decomposer.

diapause   A period of arrested development or suspended animation; an adaptation to survive a period of inclement weather.

diatomaceous earth   An abrasive dust mined from diatom deposits.

dieback   Gradual death of the terminal portions of the crown of a tree.

dikaryotic   A cell with two haploid nuclei that are not fused.

dioecious   Having the male and female reproductive organs on different individuals.

diploid   A nucleus with $2N$ chromosomes.

Diplopoda   The millipedes, members of the subphylum Mandibulata, and characterized by having two pairs of legs per body segment.

direct attack   An aggressive fire suppression strategy involving placement of a control line along the edge of the fire front. See indirect attack.

Discomycete   Common name for an Ascomycete that produces an apothecium.

disease   Sustained physiological or structural damage to plants caused by organisms, such as fungi, bacteria, viruses, phytoplasmas, nematodes, and parasitic plants. Injuries caused by abiotic agents such as air pollution may also be classified as disease.

dispersal   Part of an insectan life cycle where a particular stage (immatures or adults) disseminate the population throughout areas containing their hosts.

disturbance   Any event that alters the structure, composition, or function of terrestrial or aquatic habitats.

diversity   See Table 2.4.

dominants   See crown class.

dorsal   Upper surface or uppermost part of an organism or structure.

dorsal vessel   The tubelike circulatory organ of insects, analogous to the heart.

dwarf mistletoe   A parasitic plant on conifers without true leaves belonging to the genus Arceuthobium.

ecological diversity   An absolute or relative measure of species richness, species abundance, or proportional abundance of species in an ecosystem.

ecological homolog   Organisms that live in geographically separated areas, but because of similar selective pressures have evolved similar characteristics and behaviors.

ecological pyramid   A diagram representing the biomass, numbers, or energy contained in various trophic levels in an ecosystem.

economic damage   The lowest pest population that will cause economic damage to a crop, and would justify the costs of applying pest management tactics (e.g., a pesticide application).

economic damage threshold   Density of pest populations at which control measures must be applied to prevent the increasing population from producing economic damage.

ecosystem   All of the organisms in any system of interest, and the environments that encompass their interactions.

ecosystem classification   A classification system based on multiple characters of an ecosystem (as contrasted to a single factor system such as dominant tree species).

ecosystem management   See Table 2.5.

ectendomycorrhizae   Type of mycorrhiza having characteristics of both ecto- and endomycorrhiza.

ectomycorrhiza (pl. ectomycorrhizae)   Fungus/root association where the fungal hyphae do not penetrate the cell walls of plants, but grow intercellularly.

effused-reflexed   Term used to describe a flattened basidiocarp whose edge rises from the substrate and forms a small shelflike structure.

egg galleries   Tunnels or galleries dug in plant tissues for the purpose of egg laying.

EIS   Environmental impact statement, a document that describes and predicts ecological damage caused by an activity planned on the land in question.

ejaculatory duct   The terminal portion of the male sperm duct.

embryo   An organism before full development or hatching.

emulsion   A suspension of microscopic droplets of one liquid in another; the droplets are so minute that they will not separate out of the mixture.

encapsulation   The covering of foreign bodies in the hemocoel of insects by specialized hemocytes.

endemic disease   Disease constantly infecting a few plants in an area. Low level of infection.

endogenous   Borne within.

endomycorrhiza (pl. endomycorrhizae)   Fungus/root association in which the fungal hyphae penetrate the cell walls of the host. Also called vesicular-arbuscular mycorrhizae or VAM.

endoparasites   Parasites or parasitoids that live inside another organism.

endophyte   Used to describe a fungus that lives inside leaves and/or stems of apparently healthy plants.

endospore   A spore formed within a cell, as in bacteria.

endurer   A life-history strategy of plants to fire where the plant resprouts or endures the effects of fire.

entomophagous   Organisms that kill insects.

enzymes   Proteins that serve as catalysts in the chemical reactions of living beings.

epidemic disease   A disease sporadically infecting a large number of hosts in an area.

epidemic or outbreak   Pertaining to pathogen or insect populations that expand to an extreme level, often disturbing processes and interactions within forested stands and landscapes to the point of causing economic or habitat loss.

epiphytotic   Widespread occurrence of plant disease. Epidemic.

epizootics   Outbreaks of an entomophagous organism (e.g., an insect-killing virus).

equilibrium moisture content   Value that actual fuel moisture content approaches if the fuel is exposed to constant temperature and relative humidity for an infinite amount of time.

eradication   An applied control strategy that aims at complete elimination of an introduced pest.

eruciform   Immature insect resembling a caterpillar (lepidopterous larva) in form and appearance.

esophagus   A narrow part of the digestive tract immediately posterior to the pharynx.

evader   A life-history strategy of plants to fire where long-lived propagules are stored in the soil or canopy and evade elimination for the site after fire.

evaginated   Everted or turned outward.

exogenous   Borne outside.

exoskeleton   External covering of arthropods.

facultative   Optional or unessential.

facultative parasite   An organism that is typically a saprophyte, but under certain conditions may be parasitic.

facultative pathogen   A pathogenic microorganism that can complete its life cycle without invading a host but also can take advantage of a host if one is accessible.

facultative saprophyte   An organism that is typically a parasite, but under certain conditions may be saprophytic.

FARSITE   A fire behavior prediction model that incorporates fire spread through a geographic information system (see Chapter 3 and Figure 6.3).

fasciated   Broadened and flattened.

fecal pellets   Indigestible wastes discharged from the digestive tract of insects.

femur   Third segment of the insectan leg; found between the trochanter and tibia.

fiber saturation point   Water content at which wood cell walls are saturated with water, but there is no free water in the cell lumens.

filter chamber   Bladderlike structure of the midgut, within which the hindgut has looped back into the midgut; the filter chamber removes excess watery sap (honeydew) when hemipterans have their stylets inserted into plant tissues.

fine filter   Specific management for a single species or group of species rather than community or ecosystem-level strategies (see coarse filter).

fire cycle   A fire frequency parameter that either implies a regular return interval, or implies that the distribution of stand ages follows a negative exponential distribution.

fire danger rating   The rating of relative fire danger for a given period of time, usually based on daily and cumulative indexes to fire weather.

fire detection   The process of discovering the existence of a fire.

fire frequency   The return interval of fire.

fire hazard   The amount, conditions, and structure of fuels that will burn if a fire enters an area.

fire intensity   The energy release rate of a forest fire, typically expressed as the energy release per unit length of fire front. Fire intensity can be related to flame length.

fire predictability   A measure of variation in fire frequency.

fire prevention   Techniques that are used to reduce human-caused fire starts.

fire regime   The combination of fire frequency, predictability, intensity, seasonality, and extent characteristic of fire in an ecosystem.

fire retardant   A material that dampens the combustion process, including clay slurries, nitrogen- and phosphorus-based chemicals, or foam.

fire return interval   The average time between fires in a given area (see fire frequency).

fire risk   The chance of a fire starting.

fire scar   A wound near the base of the tree caused by cambial heating and death and subsequent sloughing of the bark. Multiple scars may be present on the same wound face.

fire severity   The effect of fire on plant communities. For trees, usually measured as a proportion of the basal area removed.

fireline intensity   The rate of heat release along a unit length of fireline, measured in kW per m or BTU per sec per ft.

fissures   Longitudinal openings or cracks in the outer bark of trees.

flagellum (pl. flagella)   A structure extending from bacterial cells and some fungal spores used for locomotion.

flagging   A disease symptom where some of the foliage on branches, particularly older foliage, is dead or dying. A tree crown that looks similar to a flag due to persistent mortality of needles on one side of the tree caused by abrasion or desiccation.

flanking fire   See Table 6.5.

fluting   Vertically oriented fissures and ridges spiraling up the stem of a tree.

fluxes   (ecological context) Mass movement of particles or energy.

foehn wind   A dry wind associated with windflow down the lee side of a plateau or mountain range and with adiabatic warming.

foliar biomass   Quantity of foliage of one or more species present on a unit area.

foregut   The most anterior part of the digestive tract (alimentary canal) located between the mouth and midgut.

forest rotations   The period of years required to grow timber crops to produce a desired volume or product.

forest series   A potential vegetation classification level defined on the basis of the most shade-tolerant tree species, and containing numerous plant associations.

formulations   Mixtures of pesticides (A.I.) and inert ingredients that enhance pesticide delivery; proper dilution, suspension or emulsion; and safety.

fossorial   Pertains to digging.

frass   A mixture of fecal matter and chewed plant debris.

frontal winds   Winds associated with the boundary between two air masses of differing temperature and moisture characteristics.

fruiting body   A special fungal structure composed of mycelium which bears spores; sporocarp.

fuel load   The dry weight of combustible materials per unit area.

fuel model   A stylized representation of a fuel complex, represented by fuel amounts, size classes, energy content, and arrangement. See Table 3.4.

fuel moisture   The percentage of water in a fuel particle, measured on a dry weight basis.

fuelbreak   A zone of reduced fuel, generally linear, designed to reduce fire behavior as fires enter the treated area.

fungus (pl. fungi)   Eukaryotic, spore-producing, achlorophyllous organisms with absorptive nutrition that generally reproduce sexually and asexually by spores and have a vegetative body or thallus composed of hyphae.

furcula   Forked abdominal structure used in jumping by the order Collembola.

galea (pl. galeae)   The outer lobe of the maxilla.

gametangium (pl. gametangia)   Structure containing gametes or nuclei that fuse and produce sexual spores.

gamete   A male or female reproductive cell, or the nuclei within a gametangium that fuse and produce sexual spores.

ganglion   An enlargement of a nerve, containing a coordinated mass of nerve cells.

generation   The time it takes for a developing insect to go from egg to adult (e.g., an insect that takes a year to develop from egg to adult has one generation per year).

genital   Pertaining to the sexual organs.

genitalia   Sexual organs and associated structures located outside of the organism.

germ tube   Initial hypha arising from a spore or resting structure.

gills   The lamellae on the underside of the pileus of the mushroom on which spores are borne. Also respiratory organs found in immature stages of aquatic insects.

gonads   The reproductive structures of insects.

gradual metamorphosis (hemimetabola)   Gradual change in form from immatures (called nymphs) to adults; their size increases with successive molts, as does the development of wings and sexual organs.

gradient winds   Winds associated with large-scale atmospheric pressure gradients. Air near the surface flows from high- to low-pressure areas.

grass stage   A juvenile growth form of certain pines, such as longleaf pine, where the tree assumes a perennial grasslike form for several years.

gross primary production   Increases in organic material plus respiration over a period.

ground fire   A fire that smolders in the ground, such as a peat fire.

gustatory   Sense of taste.

habitat   The immediate biotic and abiotic environment occupied by an organism.

habitat type   The land area capable of supporting a single plant association.

halteres   In place of hind wings, small knobbed structures on each side of the metathorax of the flies.

haploid   A nucleus with N chromosomes.

haustorium (pl. haustoria)   Specialized hypha of a fungus that enters a host cell and absorbs food.

hazard   Fuel amounts and distributions.

head   The most anterior of the three principal body divisions of the insects (the sensory center); contains the mouth, single pair of antennae, and the eyes.

heading fire   See Table 6.5.

hemocoel   Insectan body cavity in which the digestive tract and other internal organs are located; the cavity in which the blood (hemolymph) circulates.

hemoglobin   Respiratory pigment found in hemolymph of a few insects, most notably larvae of midges (Diptera: Chironomidae).

hemolymph   Blood of insects containing fluid plasma in which the blood cells (hemocytes) are suspended.

herbivores   Animals that eat living plants. Also known as primary consumers.

heteroecious   A characteristic of rust fungi that complete part of their life cycle on one host and the remainder on another, usually only distantly related.

hibernaculum (pl. hibernacula)   Silken bags spun by the eastern and western spruce budworms (and other *Choristoneura* spp.) in which they spend the winter in a state of diapause.

hindgut   The most posterior part of the digestive tract (alimentary canal).

holarctic species   A species found circumferentially in the northern hemisphere.

homology (pl. homologies)   Comparison of structures (e.g., legs, wings, genitalia) occurring in different organisms that are different in appearance of function due to evolutionary differentiation, but whose origins can be traced to corresponding structures in ancestral organisms.

honeydew   See filter chamber.

host   An organism upon which an organism of a different species grows and from which all or most of its food is derived.

hyaline   Colorless.

hybrid   Offspring of genetically dissimilar parents or stock.

hymenium   Layer in fungal fruiting body where asci or basidia are produced.

hymocytes   Blood cells of insects.

hyoplasia   Underdevelopment of part of a plant.

hyperplasia   Overgrowth of part of a plant due to an abnormally large number of cells.

hypertrophy   Overgrowth of part of a plant due to abnormally large cells.

hypha (pl. hyphae)   A single branch of a fungus thallus; tubular vegetative structure.

immunity   Total nonsusceptibility to a pathogen.

incomplete metamorphosis   Change in form of the primitive aquatic insect orders Ephemeroptera, Odonata, and Plecoptera; the immatures (called naiads) are aquatic, and the adults are terrestrial.

increment   Increase in diameter, basal area, height, volume, quality, or value of a tree or a crop of trees.

incubation   Time between inoculation and appearance of disease symptoms.

indicator species   A species that by its presence indicates a particular set of environmental conditions (wet, dry, cold, warm, etc.).

indirect attack   A less aggressive fire control strategy where fireline is constructed at some distance from the flaming front, often at a defensible location such as a ridge-line, bench, or stream. See direct attack.

infection   Process of pathogen establishing itself on a host and obtaining food.

initial floristics   A process of succession where seeds or plants of later successional stages are present from the outset but are subordinate to other species (see relay floristics).

inoculation   Contact between the disease-producing organism (i.e., inoculum) and the host.

inoculum   Part of a pathogen such as a spore, mycelial fragment, bacterial cell, mistletoe seed, etc.

inoculum potential   The potential of a pathogen to cause infection; related to the number of spores or propagules as well as the ability of the pathogen to invade the host.

instar   The insect itself between molts.

intercellular   Between cells.

intermediate operations   See silviculture.

intermediates   See crown class.

interspecific   Pertains to relationships between different species.

intracellular   Within cells.

intraspecific   Pertains to relationships among the same species.

invader   A life-history strategy of plants to fire where the plant, through highly dispersive propagules, invades the site after fire.

invertebrates   Animals without backbones.

IPM (integrated pest management)   Management of forest pests by developing methods to permanently lower pest populations to a level where fluctuations in their populations will not cause economic damage. This is done through silviculture, use of biorational pesticides when necessary in a manner that conserves natural controls, and through periodic monitoring of pest populations as well as their natural enemies.

kairomone   A natural or synthetic semiochemical that attracts members of another species (e.g., the aggregating pheromone of a bark beetle species that attracts a parasitoid).

karyogamy   The fusion of two nuclei.

kiln drying   Lumber dried in a kiln to a specified moisture content.

krummholz   A shrubby form of tree usually found in environments very marginal for tree growth.

labial palps   Fingerlike appendages of the labium.

labium   The lower lip of insects consisting of fused appendages from an insectan ancestor; labium is just posterior to the maxillae.

labrum   The uppermost mouthpart of insects.

larva (pl. larvae)   The immature stage between egg and pupa in insects with complete metamorphosis.

larval mines   Tunnels dug in plant tissues by insect larvae.

latent population   An insect population that normally remains at a low level and causes no economic damage.

lateral   The side.

LD$_{50}$ That dose of a toxicant expected to kill 50% of a test population of laboratory rats and expressed as mg per kg of body weight.

lenticular   Shaped like the cross section of a lentil

lesion   Localized necrosis or diseased tissue.

lumen   Inner open space or cavity of a tubular organ.

Malpighian tubules   Blind-ended tubes found between the midgut and hindgut of insects used in excretion of nitrogenous wastes.

managed wildland fire (pnf)   The new term for a naturally ignited fire that occurs within an approved management zone and meets prescription criteria (see prescribed natural fire).

mandibles   The first pair of jaws of insects: robust in chewing insects (e.g., woodborers and grasshoppers), hollow needlelike in sap-sucking insects (e.g., aphids and bugs), or hooklike (e.g., certain fly larvae).

Mandibulata   A subphylum of the Arthropoda, with distinct characteristics, an important one being that the mouthparts are incorporated within the mouth of the organism (includes the insects).

mass ignition   Ignition of the entire unit at once.

mass transfer   Movement of energy by physical displacement of burning material. See spotting.

maxilla (pl. maxillae)   Paired mouthpart structures just behind the mandibles.

maxillary palps   Fingerlike appendages of the maxillae.

meristems   Undifferentiated plant tissue from which new cells arise (e.g., apical meristems are cells located at the tips of stems or roots).

mesophyll cells   Cells within plant leaves or needles that contain the photosynthetic pigments in structures called chloroplasts.

metamorphosis   Change in form from immature to adult.

midgut   Middle portion of the digestive tract (alimentary canal).

mites   Members of the Chelicerata found in the class Arachnida and belonging to the order Acarina.

molting   The process of getting rid of the old exoskeleton (cuticle), a process that also is termed ecdysis.

monophagous   An animal species that feeds on one plant species.

morphology   The study of form or structure.

mosaic   A virus disease that causes affected plant parts to become mottled.

multiple pathway models   Alternate successional pathways that an ecosystem may take after disturbance, depending on attributes of the species likely to exist in the ecosystem.

mushroom   Strictly the gill fungi, but also refers to all fleshy fungi.

mutualism   A symbiotic relationship beneficial to both organisms.

mutualistic   A symbiotic relationship that benefits both members of the participating species.

mycangium (pl. mycangia)   A cuticular pouch in insects in which fungal spores are carried.

mycelia   A mass of fungal hyphae.

mycelium (pl. mycelia)   A mass of hyphae composing the vegetative body of a fungus.

mycorrhiza (pl. mycorrhizae)   Fungus/root mutualistic association.

naiad   The immature stages of insects with incomplete metamorphosis.

National Interagency Incident Management System (NIIMS)   A federal approach to emergency situation management (see Chapter 5).

natural fire rotation   A fire return interval calculated as the quotient of a time period and the proportion of a study area burned in that time period.

necrosis Death of host tissue.

nematode Generally microscopic wormlike animals that are parasites of plants and animals or saprophytes in water in soil.

net primary production Gross primary production less respiration, equivalent to net ecosystem biomass gain in a given period.

niche (pl. niches) The functional role of an organism in a community.

no metamorphosis (ametabola) Insects with no change in form from immatures to adults; the most primitive of insects belong to the orders Collembola and Thysanura.

nucleopolyhedrosis virus A virus that attacks the nuclei of insects; the virus particles (virions) are within a proteinaceous, polyhedron-shaped structure.

nucleus The part of the cell made up of chromosomes contained by a nuclear membrane.

nymph The immature stages of insects with gradual metamorphosis.

obligate parasite A parasite that is incapable of existing independently of living tissues.

obligate saprophyte An organism that can only grow saprophytically on dead organic matter.

ocellus (pl. ocelli) A simple eye of an arthropod.

oidium (pl. oidia) Thin-walled, free, hyphal cell derived from fragmentation of somatic hyphae into its component cells or from an oidiophore; used with respect to certain Basidiomycota.

old-growth stage A final stand development stage represented by forest structure of large, old trees, snags, and multiple canopy layers.

oleoresin exudation pressure The pressure that resin is under within resin canals of certain conifers.

olfactory Sense of smell.

oligophagous An animal species that feeds on several plant species.

omnivore An animal that consumes both plants and animals, so is both a carnivore and herbivore. Insect that feeds on a variety of food items.

oogonium Female gametangium in the Oomycota.

ostium (pl. ostia) A slitlike opening of the insect heart (dorsal organ).

ovary (pl. ovaries) Egg-producing organs of female insects.

overstory Those plants that are dominant on a site, usually a tall tree layer.

oviparous Egg laying.

oviposit The act of laying eggs.

ovipositor Egg-laying apparatus and part of the female insect's external genitalia.

ovoviviparous Production of young just after eggs hatch within the female.

paedogenesis Production of young by immature insects.

palatability Pertaining to the tastiness of a food item.

parameter One of a set of measurable factors, such as temperature and light, that define a niche and describe a component of the niche.

parasite An organism that grows part or all of the time on or within another organism of a different species (known as its host), and from which it derives all or part of its food.

parasitic Living as a parasite or parasitoid.

parasitic plant A vascular plant growing on another vascular plant from which it obtains nutrients and water and in some cases photosynthate.

parasitoids Parasites that kill their host (e.g., the insect parasites).

parthenogenesis Reproduction by eggs that develop without being fertilized.

partition resources   Organisms that coexist do so by dividing their habitat in time or space.

pathogenicity   Ability of an organism to cause disease.

pathological rotation   Age of a forest stand at which time heart rot decay fungi decay more wood than is produced by the tree.

pedipalps   Paired bucal structures found in the Chelicerata; second pair of appendages next to the mouth.

peristalsis   Wavelike muscular contraction of the digestive tract by which contents are moved forward toward the anus.

perithecium (pl. perithecia)   A closed ascocarp with a pore at the top and a wall of its own.

pharynx   Anterior part of the foregut, between the mouth and esophagus.

pheromone   A natural or synthetic semiochemical used by insects to communicate with members of the same species (e.g., sex attractant of the gypsy moth).

phloem   Vascular plant tissues that transport photosynthates to metabolic sites, storage sites in the stem or roots.

phylogeny   The evolutionary history of a group of organisms; study of natural groups and their shared characteristics.

phytophagous   Plant feeding organisms.

phytoplasmas   Prokaryotic organisms having no nucleus or true cell wall; usually spread by insects.

pile burn   See Table 6.5.

pileus   Upper portion or cap of certain types of ascocarps and basidiocarps.

pit-and-mound   A hummocky topography caused by repeated upturning of soil by windthrow of trees and associated uprooting.

plant association   The distinctive combination of trees, shrubs, and herbs occurring in a theoretical terminal or climax community, identified by indicator overstory and understory species.

plant community   An assemblage of plant species that occur widely enough across the landscape to be recognized as a unit. This assemblage can be a pioneer group of species, a late-successional group, or a combination of both.

plant series   Aggregations of plant associations having the same overstory dominant.

plasmodium   Naked, multinucleate mass of protoplasm as in the slime molds.

pleomorphism   The ability of a fungus to produce more than one type of spore.

polyphagous   An animal species that feeds on many plant species.

postoral segment   A segment that originates behind the mouth (e.g., the mandibles of insects arise from embryonic appendages located behind the mouth of the embryo). During embryonic development these appendages travel over the head and fuse into the embryonic mouth and develop in paired mandibles.

posterior   Behind, hindmost, or rear.

potential vegetation   The vegetation likely to exist on a site were it to be protected from disturbance for a very long time.

predators   (insect predators) Insects that hunt down and devour their prey.

prehensile   Structure adapted for grasping or seizing.

prescribed fire   A fire ignited under predetermined conditions of fuels, weather, and topography to meet specific management objectives.

prescribed natural fire   A fire ignited by natural processes (usually lightning) and allowed to burn within specified parameters of fuels, weather, and topography to achieve specified objectives (now called managed wildland fire).

primary consumers   Plant-eating organisms (herbivores).

primary insects   Insects that prefer to feed on normally healthy hosts (e.g., various defoliators feed on foliage of healthy trees).

primary succession   A successional sequence starting with no legacy of the previous community, such as after glacial retreat. See secondary succession.

pro-, meso-, meta-   Most forward, middle, and after, respectively.

proboscis   Snoutlike structure associated with the mouthparts of certain insects.

producers   Plants that capture energy from the sun, a process known as primary production.

protozoa   Individual organisms of the kingdom Protozoa.

proventricular teeth   Structures posterior to the crop that bear hardened teeth or spines that further grind the ingested food items.

pruning   Removal of the lower branches from a tree stem in order to improve wood quality (lumber is free of knots) or to reduce the chance of fire moving into the tree crown.

pseudoplacenta   Placentalike filaments that nourish immature insects while still inside the female.

pseudopods   Leglike evaginations of the body wall of caterpillars and sawfly larvae.

pupa (pl. pupae)   A resting stage between the larva and adult in insects with complete metamorphosis.

puparium   The thickened, hardened last larval (maggot) skin within which some flies go through the pupal stage.

pustule   A small swelling on a host, usually with a mass of spores.

pycnidiospore   A type of conidia borne in a pycnidium.

pycnidium   A spherical fruiting body in the Ascomycota or Deuteromycetes that opens by a pore and bears conidia.

pycniospore   The spore that functions as a gamete in the rust fungi; borne in a pycnium.

pycnium (pl. pycnia)   Another designation for spermogonium in the rust fungi.

quarantine   Enforced isolation or restriction of free movement imposed by governments to prevent the spread of pest organisms.

race   A subdivision of a variety based on physiological rather than morphological characteristics.

racemic   A stereo isomer that does not rotate polarized light (light vibrating in a single plane).

radiation   Transfer of energy through electromagnetic wave motion, such as being warmed by a campfire.

rasping-sucking mouthparts   Mouthparts of the order Thrips, where the top surface of fused mandibles are etched like a file, which they use to break down tender plant tissues.

rate of spread   The movement of fire across a landscape, often measured as m per sec or ft per sec.

rectal papillae   Padlike structures projecting into the lumen of the rectum.

relay floristics   A process of succession where one set of species prepares the site and is replaced by a new set of species (see initial floristics).

reproductive potential   See biotic potential.

resister   A life-history strategy of plants in which the plants, through adaptation like thick bark, survive low-intensity fire relatively unscathed.

restoration   The maintenance or recovery of original elements, structures, processes, and interactions of an ecosystem.

resupinate   Lying flat on the substrate.

rhizomorph   Thick strand of mycelium consisting of strands of hyphae twisted together and resembling a root.

Rhynchophora   A group of Coleoptera that contain the Curculionidae and the Scolytidae.

riparian   Pertaining to land that is next to water, where plant communities dependent on a permanent water source are found.

risk   The chance of fire starting.

roguing   Removal of seed-orchard trees whose seeds do not produce trees of selected criteria, (e.g., fast growth, straightness, little tapering, and small crown to stem ratios).

rosette   Crowded or clustered foliage resulting from less than normal elongation of the internodes.

rotation   Number of years required to establish and grow timber crops to a size and quality that is ready to harvest.

sap   Water and dissolved nutrients moving through conducting tissues of plants.

saprophyte   An organism that lives on dead organic matter.

scavengers   Insects feeding on dead plants or animals, on decaying organic matter, or on dung.

secondary insects   Insects that prefer to feed on weakened hosts (e.g., many bark beetles attack moribund trees, dying or dead trees, or large branches).

secondary succession   Successional development of an ecosystem where substantial legacy (organic matter, nutrients, boles, logs, plants, etc.) remains after a disturbance.

seed orchard   Tree plantation of superior trees (often grafted seed-bearing branches on root stock) whose seeds are used to establish plantations of superior growth and wood qualities.

seminal vesicle   Tubes of the male reproductive system that transport the sperm.

semiochemical   Any chemical that elicits communication between organisms.

senescent   Growing old or aging.

septate   With septa or cross walls.

seral or seral stage   A plant species or community that will be replaced by another plant community if protected from disturbance.

sere   The product of succession: the entire sequence of plant communities that successively occupy and replace one another in a particular environment over time.

serotinous cones   Conifer cones whose scales are sealed with a thin coat of resin. Intense heat melts the resin and allows the cone to open.

shade tolerance   Ability of trees to reproduce and grow in the shade of and in competition with other trees.

shelterbelt   A row of trees that is planted specifically to shelter a downwind area from deleterious effects of wind.

sign   The presence of biological causal agents of disease or insects.

silviculture   Manipulation of forest stands to accomplish specific objectives; includes procedures to establish reproduction; intermediate operations such as thinning, pruning, fertilizing, and vegetation management; and protection from insect outbreaks, diseases, wildfires, and weeds (competing vegetation).

size-up   The first part of initial attack where firefighters assess the best way to control the fire.

slime mold   An organism originally classified as a fungus (now classified in the kingdom Protozoa) which consists of plasmodium rather than hyphae and that reproduces by spores.

slope   The amount of pitch on a hill or mountain, usually measured as a percentage: the vertical gain divided by the unit horizontal length times 100. A 100% slope is equal to a 45° slope.

snag   A standing dead tree.

solution   A preparation made by dissolving a solid, liquid, or gas into a liquid without a chemical change taking place (e.g., sugar in water).

somatic   The vegetative or mycelial stage of a fungus.

spermagonium (pl. spermagonia)   A fruiting structure in which spermatia are produced, sometimes referred to as pycnium in some rust fungus literature.

spermatheca   A structure or receptacle in the reproductive tract of females that receives the sperm during mating.

spermatophore   A capsule produced by males that contains the sperm.

spiracles   External openings of the insectan breathing system.

sporangiophore   The structure bearing a sporangium.

sporangiospore   Asexual spore of the Zygomycota borne in a sporangium.

sporangium   Structure enclosing sporangiospores.

spore   The reproductive structure of fungi corresponding to a seed in flowering plants.

sporocarp   See fruiting body.

sporodochium   A cushion-shaped conidial fruiting structure in which the spore mass is supported by a stroma covered by short conidiophores.

sporophore   A spore-producing or spore-supporting structure.

spot headfire   See Table 6.5.

spotting   The convective movement of firebrands carried over the main perimeter of a fire.

stability   The degree to which an ecosystem can resist change or rebound from change. Condition of the atmosphere related to the temperature profile above the ground; increasing temperatures with height indicate that the atmosphere is stable, while decreasing temperatures indicate it is unstable.

stadium   The period between molts in a developing insect.

staminate flowers   Male flowers that will be developing the pollen-bearing structures.

stand   A vegetation group occupying a specific area and sufficiently uniform in composition, size, arrangement, structure, and condition to be distinguished from adjacent vegetation units.

stand initiation stage   The initial stage of secondary forest succession after major disturbance, represented by continued establishment of tree species.

standing crop   Used in production ecology to denote all of the current above- and belowground biomass at a site.

stem buttressing   Basal swelling by wood production at the tree base, representing a tree response to motion such as by wind.

stem exclusion stage   The second stage of secondary forest succession after major disturbance, represented by cessation of tree establishment and thinning of the existing stems.

sterigma (pl. sterigmata)   A tapering projection on a basidium on which basidiospores develop

sternum (pl. sterna)   Ventral (bottom) surface of any body segment.

stipe   A stalklike or stemlike structure that supports the pileus of a Basidiomycete fruiting body.

strain   A genetic race of a pathogen usually differing in pathogenicity from other races.

stream buffer   Strips of trees either left uncut, or selectively cut, in the vicinity of streams.

strip headfire   See Table 6.5.

stylets   The tubelike mandibles and maxillae of aphids, bugs, scales, and other members of the Hemiptera; main elements of piercing-sucking mouthparts that probe into plant or animal tissues and suck up sap or blood. Also possessed by plant-parasitic nematodes.

subcortical   Beneath the bark (e.g., bark beetles live beneath the bark and feed on phloem tissues).

subesophageal ganglion   A nerve swelling at the anterior end of the ventral nerve cord located immediately below the esophagus.

subimago   An immature, winged stage of the mayflies that emerges from the last aquatic instar after it crawls out of the aquatic environment; the subimago molts once more and the definitive, sexually mature, winged adult (adult = imago) emerges.

substrate   A surface on which an organism feeds, grows, or is attached.

succession   Change in forest community composition and structure over time including the action of disturbances caused by storms, wildfires, insects, and diseases.

suppressed trees   See crown class.

surface area to volume ratio   The ratio of the surface area of a fuel particle to its volume. High values mean that the particle can exchange energy with the atmosphere easily, so it can wet and dry quickly.

surface fire   A fire that moves above the ground in fuels that have contact with the surface, without moving into the shrub or tree layer. Usually a flame length less than 1 m.

sustainability   A conceptual approach to forest management that incorporates ecological, social, and economic states and outputs that are sustained over time.

switching mechanism   A chainlike series of tree infestations common to certain bark beetles. For example, a pioneering *Dendroctonus frontalis* female finds a susceptible southern pine tree. She emits an aggregating pheromone and the tree is mass attacked. While this focus tree is being attacked, some incoming females land on a neighboring tree. When the original focus becomes less attractive, the pheromones of females on the neighboring tree begin to induce a mass attack on that tree, a recipient tree. In time a large part of the stand comes under attack via this tree-to-tree spread.

symbiosis   The process involving relationships between two genetically different organisms.

symbiotic   Referring to two species living together in an close relationship.

symphyta   Suborder of the Hymenoptera containing the sawflies and horntailed wood wasps.

symptom   The noticeable evidence of change in the physiology or morphology of a host as a result of disease or insect attack.

synnema (pl. synnemata)   Conidiophores cemented together to form an elongated spore-bearing structure.

systematics   Biological classification of organisms based on homologous structures, the fossil record, and genetics.

systemic   An insecticide, fungicide, insecticide, or herbicide that is absorbed by the roots and is translocated to other parts of the plant. Also a parasite that spreads through the host.

tagmosis   The fusing of body segments in response to forces of evolution.

tarsus   The part of the insect leg beyond the tibia, consisting of one or more segments.

taxon (pl. taxa)   Any taxonomic unit (e.g., a species, a genus, a family, or simply a group of similar organisms).

taxonomists   Scientists dedicated to classification of organisms (science of taxonomy) into various ranks or categories based on similarities and differences: they describe and name these categories (e.g., a new species).

teliospore   The sexual, thick-walled resting spore of the rust and smut fungi.

tellum (pl. telia)   A structure producing teliospores; refers to rust fungi.

tenaculum   Clasplike structure that keeps the furcula under tension until the furcula is unclasped, which propels the Collembola into the air.

tension wood   Reaction wood in hardwoods that develops on the upper side of tilted stems and on the upper side of branches.

tergum (pl. terga)   Dorsal (top) surface of any body segment.

testis (pl. testes)   Sexual structure of the male that produces the sperm.

thallus   A vegetative body showing no characterization into stems, roots, or leaves as in fungi.

thigmotactic   A sense of touch or contact.

thinning   Silvicultural treatment in which stand density is reduced to accelerate diameter growth on the remaining trees, the leave trees; the planned removal of trees during the development of a forest, used to regulate characteristics of tree growth through adjustments in tree spacing and density without creating a new age class.

thorax   The body region of insects (the locomotor center) between the head and abdomen; contains the wings and legs.

tibia   Fourth segment of the insectan leg; found between the femur and tarsus.

timelag class   A method of categorizing fuels by the rate at which they are capable of moisture gain or loss, indexed by size class of fuel.

tolerance   Ability of a host to withstand infection and/or development of a pathogen.

tornado   A localized, violently destructive windstorm.

toxicants   Poisons.

toxicity   The inherent ability of a toxicant to damage plants and animals.

trachea (pl. tracheae)   Tube of the insectan respiratory system. Tracheae branch into ever-smaller tubes, the tracheoles.

tracheoles   Fine terminal branches of the respiratory tubes or tracheae.

tree improvement   A part of silviculture that maximizes tree growth and tree form (minimizing taper) through applied genetics.

trochanter   Second segment of the insectan leg; found between the coxa and femur.

trophic level   The organisms that consume the same general type of food in a food chain or web.

underburn   A low-intensity surface fire burning under a forest canopy, with a flame length less than 1 m.

understory   Those plants beneath the overstory.

understory fire   A fire that burns surface fuels and torches into the understory, often with flame lengths of 1 to 3 m.

understory reinitiation stage   The third stage of stand development represented by sufficient overstory mortality that a new cohort of reproduction occurs.

urediniospore   A binucleate spore borne in a uredinium and capable of infecting the same host on which it originated; refers to rust fungi.

uredinium (pl. uredinia)   A structure that produces urediniospores; refers to rust fungi.

vagina   A tubular and terminal part of the female reproductive system that opens to the outside.

values   The net change in resource condition when a fire occurs.

variety   A genetically different plant of the same species.

vascular wilt   See wilt.

vector   The agent that transports viruses, phytoplasmas or fungal spores; usually insects or nematodes. Wind also may act as vector. Birds may vector mistletoe seeds.

vegetation management   Removal of vegetation that competes with commercially valuable trees—generally by use of herbicides—so that they will have maximum room to grow.

vegetation zone   A land area with a single overstory dominant as the primary climax dominant. Occasionally zones are named after major seral species. Other climax types may exist in the zone.

vein   A thickened line in the wing of an insect.

veneer   Thin sheets of wood peeled from a log that is being rotated on a lathe.

ventral   Underside, lower surface or lowermost part of an organism or structure.

vestigial   Through the process of evolution, an organ or structure no longer used.

viroid   Small, low-molecular-weight RNA that can infect plants, replicate themselves, and cause disease.

virulence   Degree of pathogenicity of a pathogen; the relative capacity of a pathogen to cause disease.

virus   Naked RNA or DNA in a protein coat that causes disease.

viviparous   Insects that give birth to living young, rather than egg-laying.

wildfire   An unwanted wildland fire.

wildland fire situation analysis (WFSA)   An formalized assessment procedure on a wildland fire that helps to evaluate the best control strategy.

wildland fire   Any nonstructure fire, other than a prescribed fire, that occurs in a wildland.

wilt   Symptom of lack of water in the plant vascular system whereby leaves loss their turgidity and droop; also termed vascular wilt.

windsnap   Trees that have broken boles due to wind.

windthrow   Trees blown over by the wind.

witches'-broom   Malformed vegetation on a branch in the shape of a "witches'-broom"; typically refers to the effect of dwarf mistletoe infections on conifers.

wood pores   The exposed water-conducting tissues (xylem or vessel elements) of shardwood tree species.

xylem   Vascular plant tissues consisting of tracheids, vessels, parenchyma cells, and fibers that transport water and minerals from the soil to the photosynthesizing tissues in the crown (foliage of plants or the leafy parts of trees); wood.

yolk   Nutritive matter within an egg consisting mainly of proteins and fats.

zone lines   Narrow, dark-brown or black lines in decayed wood, generally resulting from the interaction of different strains or species of fungi or the host reaction.

zoospore   Asexual swimming spores of Oomycota that possess flagella.

zygospore   Sexual spore of a fungi in the phylum Zygomycota.

zygote   A fertilized egg, just before cleavage and embryo formation.

CREDITS

CHAPTER 2

Figure 2.1   From Barbour and Billings, 1988. *North American Terrestrial Vegetation,* 1st edition. Cambridge University Press. Reprinted with permission.

Figure 2.2   From Greller, A.M., 1988, in Barbour and Billings (eds.). *North American Terrestrial Vegetation,* 1st edition. Cambridge University Press. Reprinted with permission.

Figure 2.3, 2.12   Courtesy of the National Geographic Society

CHAPTER 4

Pages 92–98 (excluding box 4.3)   From Agee, J.K., "Fire and Pine Ecosystems," in Richardson, D.M., *Ecology and Biogeography of Pinus,* ©1998 by Cambridge University Press. Reprinted with permission of Cambridge University Press.

CHAPTER 7

Figure 7.1   Adapted from Foster, 1988, *Journal of Ecology,* with permission of Blackwell Science, Ltd., Oxford, England.

CHAPTER 8

Figure 8.1   Natural Resources Canada, Canadian Forest Service. Reproduced by permission.

Figure 8.2   From Agrios, G.N., *Plant Pathology,* 4th ed., ©1997 by Academic Press.

CHAPTER 9

Figure 9.2   Natural Resources Canada, Canadian Forest Service. Reproduced by permission.

Figure 9.4   From Aber, J.D. et al., *BioScience,* 48(11), p. 922. ©1998 American Institute of Biological Sciences.

Figure 9.5   From Lowry, W.P., *Weather and Life: An Introduction to Biometeorology,* ©1969, by Academic Press.

CHAPTER 10

Figures 10.1, 10.6, 10.8, 10.10, 10.12   From Agrios, G.N., *Plant Pathology,* 4th ed., ©1997 by Academic Press.

Figure 10.2   From Killham, K., *Soil Ecology,* Cambridge University Press, 1994.

Figure 10.7   *Tree Disease Concepts,* 2/e by Manion, Paul D., ©1991, Reprinted by permission of Prentice-Hall, Inc., Upper Saddle River, NJ.

Figure 10.11   From, Alexopoulos, C.J., C.W. Mims and M. Blackwell, *Introductory Mycology,* ©1996 by John Wiley & Sons, Inc. Reprinted by permission of John Wiley & Sons, Inc.

CHAPTER 11

Box 11.1, 11.3, Figures 11.1, 11.2, 11.3, 11.4   Natural Resources Canada, Canadian Forest Service. Reproduced by permission.

Figure 11.5   From Richards, B.N., *The Microbiology of Terrestrial Ecosystems,* ©1987 by Longman Scientific and Technical.

Figure 11.6   From Paul, E. A. and F.E. Clark, *Soil Microbiology & Biochemistry,* 2nd ed., © 1996 by Academic Press.

CHAPTER 12

Figures 12.3, 12.12, 12.15   Natural Resources Canada, Canadian Forest Service. Reproduced by permission.

CHAPTER 13

Figure 13.2, 13.9, 13.14   Natural Resources Canada, Canadian Forest Service. Reproduced by permission.
Figure 13.3   From Agrios, G.N., *Plant Pathology,* 4th ed., ©1997 by Academic Press.
Figures 13.4, 13.7, 13.8   Department of Plant Pathology, Cornell University.

CHAPTER 14

Figures 14.5, 14.6   Department of Plant Pathology, Cornell University.
Figure 14.9, 14.10, 14.11, 14.12   Natural Resources Canada, Canadian Forest Service. Reproduced by permission.
Figure 14.14   From Agrios, G.N., *Plant Pathology,* 4th ed., ©1997 by Academic Press.

CHAPTER 15

Figure 15.7   From Franklin et al., *BioScience,* 37(8), p. 556. ©1987 American Institute of Biological Sciences.

CHAPTER 16

Figure 16.3   Natural Resources Canada, Canadian Forest Service. Reproduced by permission.

CHAPTER 17

Figure 17.4   From Robinson, W.H., *Urban Entomology,* Stanley Thornes (Publishers) Ltd., ©1996 by W.H. Robinson. Reprinted with permission.
Figure 17.6   With permission from "History of Entomology," ©1973 by Annual Reviews.

CHAPTER 19

Figure 19.5   Drawings of nerves modified from *Principles of Insect Morphology,* R.E. Snodgrass. Copyright 1935 by McGraw-Hill Book Co., Inc. Used by permission of Cornell University.
Table 19.3   Courtesy of Michael Hulme and the Commonwealth Agricultural Bureaux.

CHAPTER 20

Figure 20.3   Reprinted with permission from H.H. Shugart, Jr., Editor, *Time Series and Ecological Processes,* p. 280. Copyright 1978 by the Society for Industrial and Applied Mathematics.
Figure 20.5   From Dorais, L. and E.G. Kettela, 1982, A review of entomological survey and assessment techniques used in the assessment of operational spray programs, Figure 7, page 37.

CHAPTER 21

Figure 21.2   Boyce Thompson Institute for Plant Research.

CHAPTER 23

Figure 23.2   Photo by H. Lyon, Plant Pathology Department, Cornell University.
Figure 23.6   Diagrams courtesy of M.A. Deyrup.
Figure 23.13   From Robinson, W.H., *Urban Entomology,* Stanley Thornes (Publishers) Ltd., ©1996 by W.H. Robinson. Reprinted with permission.

CHAPTER 24

Figure 24.13   From Alfaro et al., 1995, The white pine weevil in British Columbia: basis for an integrated pest management system. Forest Chronicle 71:66-173.

AUTHOR INDEX

Page numbers followed by *b, f,* and *t* refer to boxes, figures, and tables, respectively.

Abbott, L. K., 255, 271, 273
Aber, J., 206, 207*f,* 375
Aber, J. D., 206, 212, 220
Aber, J. W., 14
Adams, D. L., 2, 9
Addy, N. D., 486
Agee, J. K., 10, 27, 36, 40, 50, 57*f,* 58, 71, 75, 76, 77, 78, 80*f,* 82, 83, 84, 87, 89, 104, 105*f,* 124, 131, 132, 144, 147, 154, 169, 170, 221, 356, 408*f*
Agrios, G. N., 187, 191, 192, 192*f,* 228, 229*f,* 230, 233, 234, 235, 240, 241*f,* 242, 244*f,* 248, 250, 250*f,* 251, 252, 253, 309, 310, 313, 316, 323, 327, 333, 358, 360*f*
Albini, F. A., 56, 59, 62*f,* 67
Aldous, J. R., 254, 255, 256, 264, 265
Alexander, R. R., 178, 179*f*
Alexopolous, C. J., 228, 230, 231, 237, 240, 241, 242, 246, 249*f,* 251, 252, 313
Alfaro, R. I., 556, 558*f*
Allan, G. G., 446
Allen, E., 196, 283*t,* 293*f,* 298*f,* 309, 312*f,* 325*f,* 326*f,* 333, 350, 350*f,* 351*f,* 354*f*
Allen, M. F., 264, 266, 270
Amaranthus, M. P., 254, 270, 272, 273
Amerson, H. V., 330
Anderson, F. K., 204, 205, 209
Anderson, J. F., 452, 455
Anderson, N. H., 169
Andrews, P. L., 49, 59, 66, 68, 69*f,* 102, 104, 105, 108, 112, 114, 115, 117*f,* 120, 121, 123, 126, 139, 142, 144
Angus, T. A., 452
Aplet, G. H., 43
Arno, S. F., 7, 9, 24, 46, 88, 126, 131, 139
Asbjornsen, H., 2
Ashraf, M., 491
Atkins, M. D., 414
Auclair, A. N., 369, 370, 372*f,* 373
Aumen, N. G., 169
Axelrod, M. C., 76
Ayers, H. J., 7, 14

Bailey, A. W., 82
Bailey, M. J., 452
Bailey, R. G., 28
Baker, B., 486, 487
Baker, F. A., 185, 187, 188, 191, 191*f,* 196, 199, 200, 202, 203, 204, 206, 209, 210*f,* 215, 216, 217, 217*t,* 218*f,* 219*f,* 220, 221, 222, 228, 233, 235, 240, 262, 338, 341, 344, 346, 347, 357, 358, 361, 362
Baker, F. E., 383, 388, 390, 392*f,* 393*f*
Baker, W. L., 98, 406, 520, 526
Barbosa, P., 512, 561
Barbour, M., 24, 25*f,* 26*f*
Barnett, J. P., 546
Barras, S. J., 491
Barrett, J. W., 81
Bartlette, R. A., 123
Bartuska, A. M., 2
Basabe, F. A., 209, 214
Baxter, D. V., 196
Beaufait, W. R., 96
Beeche, T. J., 3, 6
Belyea, R. M., 469
Benda, L. E., 3, 6
Benjamin, D. M., 477
Bennett, G. W., 539
Bennett, W. H., 530*f,* 555*f*
Berg, D. R., 3, 6
Berisford, C. W., 442, 447
Bernston, G., 14, 375
Beroza, M., 557
Berryman, A. A., 491, 495
Bilby, R. E., 169
Billings, D., 24, 25*f,* 26*f*
Billings, R. F., 500, 500*t*
Billings, W. D., 89
Binkley, D., 206, 208, 210, 212
Bisson, P. A., 3, 6
Biswell, H. H., 143
Black, H. C., 199, 222, 222*t,* 224, 225, 226
Blackwell, M., 228, 230, 231, 237, 240, 241, 242, 246, 249, 251, 252

Blanchard, R. O., 196
Blanche, C. A., 494
Blank, R. W., 102
Bloomberg, W. J., 392, 394*f*
Bohm, M., 206, 208, 210, 212
Bohmont, B. L., 442
Bonello, P., 347
Boone, R. D., 206, 207*f*
Borchers, J. G., 254, 270, 272, 273
Borchers, S. L., 254, 270, 272, 273
Borden, J. H., 517, 518, 520, 521*t*, 556, 558*f*
Bormann, F. H., 33*f*, 378, 380*t*
Borror, D. J., 414, 419*f*
Boyce, J. S., 170, 171*f*, 181, 196, 348, 362
Bradley, G. A., 177
Bradshaw, L. S., 123, 144
Brainerd, R. E., 254, 270, 272, 273
Briggs, G., 105
British Columbia Ministry of Forestry and
    Lands, 224
Britton, K. O., 315
Brown, A. A., 103, 110
Brown, H. D., 196, 199, 200*f*, 284*f*, 295, 309,
    312*f*, 318*f*, 319*f*, 325*f*, 326*f*
Brown, J. K., 7, 9, 131, 148
Browning, J. E., 290, 302
Brussard, P. F., 154
Buck, C. C., 50, 51, 52, 52*f*, 54, 55*f*, 56*f*, 57
Buelow, G. P., 378
Bull, E. L., 190, 333, 356, 384
Bunting, S. C., 91
Burgan, R. E., 54, 59, 60, 67, 131
Burkman, W., 15
Burkman, W. G., 16
Burns, M. F., 144
Bush, R. R., 12
Butler, B. W., 123
Butts, W. L., 539
Byler, J., 7, 9
Byler, J. W., 561

California Division of Forestry, 118*f*, 119*f*
Callan, B., 189*f*
Callicot, J. B., 2
Camann, M. A., 447
Cameron, A. E., 196, 199, 200*f*, 284*f*, 295, 309,
    312*f*, 318*f*, 319*f*, 325*f*, 326*f*
Cameron, R. S., 544, 545*f*, 549*t*
Campana, R. J., 196, 358
Campbell, R. W., 457, 473
Campbell, S., 7, 9, 11, 15, 16, 195, 390
Campbell, S. J., 254, 255, 256, 264, 265
Canfield, J. E., 378
Carlson, C. E., 7, 9

Carolin, V. M., 81, 406, 465, 469, 472*t*, 480,
    485, 486, 506, 520, 526, 535*t*
Carpenter, S., 2
Carter, R., 218, 220*f*
Castello, J. D., 1
Chamberlain, P., 122
Chandler, C. C., 59, 116, 117
Chapman, H. H., 93
Chapman, J. A., 517, 519*f*
Chapman, R. F., 414
Chappell, H. N., 220
Chastagner, G. A., 327
Cheney, P., 59, 116, 117
Chong, L., 520
Christensen, N. L., 2, 34, 92, 93, 154
Cibrían, D. C., 406
Ciesla, W. M., 8, 364, 367*t*, 368, 561
CINTRAFOR, 516
Clark, F. E., 240, 246, 252, 269*f*
Clark, H. J., 2
Clements, F. E., 34
Clendenen, G., 439, 465*f*
Cline, S. P., 169
Cobb, F. W., Jr., 277, 295
Cohen, J. D., 67, 123
Commission of the European Communities, 17
Conner, R. N., 172
Conrad, C. E., 57*f*
Constanza, R., 2, 7
Cook, S., 88
Cooper, C. F., 81
Cooper, J. I., 234
Cordell, C. E., 271
Cory, J. S., 452
Coster, J. E., 499
Coulter, W. K., 465
Coutts, B., 174
Coutts, M. P., 161, 165, 166
Covington, W. W., 2, 82
Cowling, E. B., 375
Craighead, F. C., 406, 469, 480
Cromack, K., Jr., 169
Cumbertirch, P. M., 392
Cummins, K. W., 169
Cunningham, J. C., 455, 457*t*
Currie, W., 14, 375

Dahlsten, D. J., 563
Dahlsten, D. L., 442
Dana, S. T., 103
D'Antonio, C., 2
Daterman, G. E., 448, 557
Daubenmire, R., 27
Daughtrey, M. L., 315

Davis, D. D., 196, 199, 200*f,* 284*f,* 295, 309, 312*f,* 318*f,* 319*f,* 325*f,* 326*f*
Davis, K. P., 103, 110
Davis, M. B., 365, 368
DeAngelis, J. D., 494
DeBano, L. F., 57*f*
De Barr, G. L., 544
deCalesta, D., 214
Deeming, J. E., 67
De Groot, P., 545
De Gryse, J. J., 409
Dell, J. D., 139, 141*f,* 142*f,* 143
DeLong, D. M., 414, 419*f*
Denke, I. J., 401
Department of the Environment Canada, 407, 469
Despain, D., 51, 153
Dickman, A., 88
Dieterich, J. H., 81
Dixon, G. E., 391
Dochinger, L. S., 196, 199, 200*f,* 284*f,* 295, 309, 312*f,* 318*f,* 319*f,* 325*f,* 326*f*
Donaubauer, E., 364, 367*t,* 368
Dorias, F., 465, 466*t,* 469*f*
Dost, F. N., 155, 156
Downs, M., 206, 207*f*
Drake, L. L., 544
Driver, C. H., 356, 408*f*
Drooz, A. T., 477
Drummond, D. B., 196, 199, 200*f,* 284*f,* 295, 309, 312*f,* 318*f,* 319*f,* 325*f,* 326*f*
Dubois, N. R., 452
Dyer, E. D. A., 517
Dyrness, C. T., 24, 26, 83, 87, 169

Ebel, B. H., 544
Ebel, F. W., 450
Ebling, W., 532
Ecological Society of America, 44
Edminster, C. B., 504
Edmonds, R. L., 209, 212, 214, 220, 290, 291, 302, 393
Egler, F., 34
Embree, D. G., 449, 451
Emmett, B. A., 206
Entwisle, P. F., 452
Epstein, P. R., 2, 7
Evans, G. W., 94
Evans, H. E., 414, 427*f,* 428*f,* 429*f*
Evans, H. J., 12, 237

Fahnestock, G. R., 57*f,* 84, 90
Fairfax, S., 103
Farr, W. A., 162

Fellner, R., 273
Felt, E. P., 399*t,* 400
Ferguson, D. E., 2
Fernandez, I., 14, 375
Fettes, J. J., 469
Fiddler, G. O., 507
Fiddler, T. O., 507
Fielding, N. J., 12, 237
Filip, G. M., 187, 195, 276, 280, 281*t,* 285, 290, 293, 295, 296, 297*f,* 299*f,* 303, 389, 390, 392, 512
Finney, M. A., 67, 68*f*
Fischer, W. C., 126, 148
Flavell, T. H., 544
Flinn, M., 97
Foster, D. R., 41, 162, 162*f,* 171
Foster, R. E., 196, 200, 201*f*
Francis, R., 2
Frankel, S. J., 304, 306*f,* 393
Franklin, J. F., 2, 24, 26, 42, 43, 83, 84, 87, 169, 185, 364, 376, 378, 379*t,* 380*t*
Fraser, R. G., 556, 558*f*
Friskie, L. M., 520
Frost, J. S., 102
Fuhrer, B. A., 187
Funk, A., 189*f,* 309, 333
Furniss, M. M., 406
Furniss, R. L., 81, 406, 469, 472*t,* 480, 485, 486, 506, 520, 526, 535*t*
Fuxa, J. R., 454, 455

Gallucci, V. F., 439, 465*f*
Gara, R. I., 356, 408*f,* 418, 439, 446, 465*f,* 495, 499, 500, 504, 534, 556, 559
Garcia, R., 563
Gardiner, B., 174
Gaudet, C., 2, 7
Gaugler, R., 456
Geiszler, D. R., 356, 408*f,* 495, 504
Gessel, S. P., 219
Gibbs, J. N., 185
Gibson, K. E., 196
Gilbert, G. S., 185, 189
Gilligan, C. J., 196
Gleason, H. A., 34
Glitzenstein, J. S., 93
Goheen, D., 493
Goheen, D. J., 276, 280, 281*t,* 291, 293, 296, 297*f,* 299*f,* 302, 303, 392
Goheen, D. M., 290
Gordon, D. T., 180
Gordon, J. C., 2
Gordon, T. R., 347
Gosz, J. R., 33*f*

Gottschalk, K. W., 12
Goyer, R. A., 447, 449, 498, 500, 508
Grace, J., 161, 165, 166
Gratkowski, H. J., 162, 178, 178f
Green, G. W., 451, 453t
Green, L. R., 148
Gregory, S. V., 42, 169
Greller, A. M., 24, 26f
Gresham, C. A., 162, 172, 173, 173f
Greulich, F. E., 418
Grey, G. W., 401
Grier, C. C., 31
Griffin, D. H., 345
Griffin, J. R., 83
Grigel, J. E., 96
Grumbine, E., 44, 45f
Gundersen, P., 206
Guries, R. P., 361

Haack, R. A., 561, 564, 565f, 566f
Hadfield, F. S., 276, 280, 281t, 293, 296, 297f, 299f, 303, 385
Hadley, J. L., 164
Hager, J., 7, 14
Hagle, S. S., 196
Haines, D. A., 102
Hajek, A. E., 454
Hall, J. N., 144
Hall, J. P., 15, 18, 19
Hallett, R., 206, 207f
Halpern, C. B., 86
Hamm, P. B., 254, 255, 256, 264, 265
Hanks, L. M., 527, 528, 564
Hansen, E., 214, 275, 296
Hansen, E. M., 8, 196, 254, 255, 256, 264, 265, 277, 280, 300, 309, 318, 319, 320f, 321, 322, 333, 341, 346, 349, 350, 362, 383, 386, 387, 388, 492
Hanson, A. D., 144
Harcombe, P. A., 169
Hard, J. S., 487, 488t, 505
Hardy, C. C., 139
Harmon, M. E., 42, 169, 364, 376, 378, 379t, 380t
Harrington, T. C., 277, 295
Harris, A. S., 161, 165, 166f, 167f, 174, 180f
Hart, D. R., 507
Harvey, A. E., 2, 285, 302
Harvey, R. D., 196, 276, 280, 281t, 293, 296, 297f, 299, 303, 392
Harvey, R. D., Jr., 277, 295
Hawksworth, F. G., 231, 232f, 334, 335f, 337t, 338, 339, 339f, 385, 391
Hedden, R. J., 500

Hedden, R. L., 495, 501
Hedlin, A. F., 544, 550f
Heikkenen, H. J., 405, 495
Heiner, H., 561
Heinrichs, J., 84
Heinselman, M. L., 98
Helgerson, O. T., 214
Helms, J. A., 132
Hemstrom, M. A., 84
Hennon, P. E., 356, 368, 370
Hennon, P. H., 487
Hepting, G. H., 196
Hertel, G. D., 15
Hessberg, P. F., 196
Hibben, C. R., 315
Hickin, N. E., 526, 536
Hiekennen, H. J., 501t
Higgs, S., 452
Hinckley, T. M., 162, 556
Hiruki, C., 234
Hodges, J. D., 494, 501
Hoff, R., 46
Holmes, R. T., 33f
Holroyd, E. W., 164
Holsten, E. H., 487, 505
Holt, R. A., 378
Hom, J. L., 528
Hopkins, A. D., 398
Hosman, K. P., 457
Houghton, J. T., 203, 214
Houston, D. R., 346, 368, 370
Howarth, R. W., 206
Howse, G. M., 455, 457t
Hubbell, S. P., 185, 189
Hubert, E. E., 196
Hudler, G. W., 369
Huettl, R. F., 14, 214
Hughes, J., 154
Hughes, S. J., 317
Hulme, M. A., 451, 453t
Hunter, M., 96
Hunter, M. L., Jr., 45
Huttermann, A., 275, 276, 288, 298, 300, 301, 302
Hynum, B. G., 495

Ignoffo, C. M., 455
Ikeda, T., 477
Innes, J. L., 7, 367t, 368

Jenkins, W. R., 235
Johannson, M., 275
Johnsey, R., 393
Johnson, D. W., 206, 212

Johnson, E. A., 88, 95
Johnson, J., 168
Johnson, J. M., 559
Johnson, N., 43
Johnson, W. T., 187, 196, 199, 200, 201, 203, 212, 213*t*, 216, 275, 309, 310, 315*f*, 317, 318, 320*f*, 321*f*, 333, 341, 343*f*, 344*f*, 347, 350, 526, 528
Julin, K. R., 165

Kamakea, M., 14, 375
Kandler, O., 375
Karjalainen, R., 275, 276, 288, 298, 300, 301, 302
Karr, B. L., 494, 501
Karr, J. R., 3
Kastner, W. W., 291, 302
Kauffman, J. B., 74*f*
Kaya, H. K., 452, 455, 456
Keane, R. E., 76
Keaton, W. S., 2
Keen, F. P., 81, 506, 508, 509*f*
Kercher, J. R., 76
Kerouac, J., 110
Kettela, E. G., 465, 466*t*, 469*f*
Keyes, M. R., 31
Kiil, A. D., 96
Kile, G. A., 276, 280, 281, 282, 284, 285*f*, 286*f*, 287, 288
Kilgore, B. M., 105, 148
Kilham, K., 230*f*
Kimmins, J. P., 23, 24, 27, 29, 30, 33, 163, 165
Kinghorn, J., 487
Kinghorn, J. M., 517, 519*f*
Kjonaas, O. J., 206
Kliejunas, J. T., 277, 295
Knight, 501*t*
Knight, D., 91
Knight, D. H., 154
Knight, D. L., 162
Knight, F. B., 405, 442
Knopf, F. L., 2
Koch, P., 530
Koerber, T. W., 448, 544, 557
Kohm, K. A., 2, 185
Kolb, T. E., 2
Komarek, E. V., 92
Koopmans, C. J., 206
Krebs, C. J., 88
Krupa, S. V., 204, 205, 205*f*, 206, 212
Kuchler, A. W., 28
Kuhlman, G. E., 330
Kuijt, J., 333
Kulhavy, D. L., 170

Kulman, H. M., 462, 464*f*
Kydonieus, A. F., 557

Lachance, D., 364, 365, 368, 370, 372*f*, 373, 373*f*, 374*f*
Landis, T. D., 548
Landres, J., 151*f*
Lange, O. L., 206, 212
Larcher, W., 201, 202, 203, 209, 210, 222
Larson, B. C., 90, 168
Latjtha, K., 212, 220
Lattin, J. D., 42, 169
Laven, R. D., 49, 59, 66, 68, 69*f*, 102, 104, 105, 108, 112, 114, 117*f*, 120, 121, 126, 139, 142
Law, K. R., 564, 565*f*
LeBarron, R. K., 97
Legge, A. H., 204, 205, 205*f*, 206, 212
Leopold, D. J., 1
Lertzman, K. P., 88
Leuschner, W. A., 497
Levins, R., 2, 7
Lewis, F. B., 452
Lewis, H. T., 83
Lewis, K. J., 8, 196, 277, 280, 300, 309, 318, 319, 320*f*, 321, 322, 333, 341, 346, 349, 350, 362, 383, 386, 387, 388
Liebold, A. M., 12
Liegel, L., 7, 9, 11, 15, 16, 195, 390
Lienkamper, G. N., 169
Likens, G. E., 33*f*, 206
Lindahl, D. L., 495
Lindberg, S. E., 206, 212
Lindgren, B. S., 520
Linsely, G., 530
Lipscomb, D. J., 162, 170, 173, 173*f*
Littke, W. R., 290, 356, 408*f*, 495
Little, C. E., 7, 14
Liu, B. H., 330
Lotan, J. E., 148
Lowry, W. P., 202, 207, 208, 208*f*
Lyon, H. H., 187, 196, 199, 200, 201, 203, 212, 213*t*, 216, 309, 310, 315*f*, 317, 318, 320*f*, 321*f*, 333, 341, 343*f*, 344*f*, 347, 350, 526, 528
Lyon, H. T., 275

MacArthur, R. H., 46
MacCleery, D., 105*f*
MacDonald, L. H., 3, 6
Maclean, N., 123
MacLeod, D., 486
MacMahon, J. A., 2
Magill, A., 14, 375

Magill, D. A., 206, 207*f*

Magurran, A. E., 42

Malajczuk, N., 255, 271, 273

Malany, H. S., 2

Maloy, O. C., 330

Mandzak, J. M., 2, 302, 387

Mangan, R. J., 123

Manion, P. D., 187, 196, 228, 242, 243*f*, 247, 275, 309, 313, 328*f*, 333, 345, 348, 358, 364, 365, 368, 369, 371*f*, 373*f*, 374*f*, 378

Mann, D. H., 168

Marks, G. C., 189

Martin, H. C., 370, 372*f*, 373

Martin, N. E., 285, 302

Martin, R. E., 139, 141*f*, 142*f*, 143

Marx, H. G., 354, 355*f*

Mason, D. A., 12

Mason, R. R., 465, 466, 467*f*, 468*t*, 471*f*, 483

Mason, W. L., 254, 255, 256, 264, 265

Mastro, V. C., 564, 565*f*

Mata, S. A., 504

Matson, P. A., 206

Matsumura, F., 477

Mattoon, W. R., 172

Mattson, W. J., 486

McBride, J. R., 7, 13, 16, 206, 210, 212

McCarter, J. B., 393

McClean, J. A., 516, 520

McDonald, G. I., 285, 302

McDonald, K. J., 2

McDonald, P. M., 507

McDonald, S. E., 546

McDowell, W., 14, 375

McElfresh, J. S., 527

McFadden, M. W., 442

McNamee, P., 275, 305*f*

McNulty, S., 14, 375

McNulty, S. G., 206, 207*f*

Merkel, E., 544

Merler, H., 282, 298, 301, 386

Merrill, W., 196, 199, 200*f*, 284*f*, 295, 309, 312*f*, 318*f*, 319*f*, 325*f*, 326*f*

Metcalf, R. A., 526

Metcalf, R. L., 526

Metterhouse, W. M., 442

Millar, J. G., 527, 528, 564

Miller, H. A., 98

Miller, J. M., 81, 506, 508, 509*f*

Miller, K. F., 174, 175*f*

Miller, P., 7, 13, 16, 206, 210, 212

Mims, C. W., 228, 230, 231, 237, 240, 241, 242, 246, 249, 251, 252

Minshall, G. W., 154

Mitchell, R. G., 504

Moeck, H. A., 495

Moehring, D. M., 501

Moeller, R. F., 368

Moeur, M., 133

Molina, R., 275, 296

Monnig, E., 7, 9

Moore, J. A., 302, 387

Moore, M. M., 82

Morell, J. J., 389

Morgan, P., 76, 91

Morrell, J. J., 561

Morris, R. F., 469

Morrison, D., 196, 283*t*, 293*f*, 298*f*, 309, 312*f*, 325*f*, 326*f*, 333, 350, 350*f*, 351*f*, 354*f*

Morrison, D. J., 196, 275, 276, 277, 282, 287, 298, 301, 386

Morrison, O. J., 357

Morrison, P., 84, 85*f*

Mueller-Dombois, D., 378

Munger, T. T., 86

Mutch, R. W., 2, 7, 9, 148

Myers, R. L., 94

Nadelhoffer, K., 14, 375

Nadelhoffer, K. J., 206, 207*f*

Naiman, R. J., 3, 6, 169

Nair, V. M. G., 362

National Acid Precipitation Assessment Program, 13, 14

National Advanced Resource Training Center, 113

National Atmospheric Precipitation Assessment Program, 212

National Park Service, 146

National Research Council, 108

Natural Resources Canada, 401*f*

Nebeker, T. E., 494, 501

Newcombe, G., 327

Newton, M., 170, 214

Noble, I. R., 37

Nord, J. C., 522

Norlund, D. A., 447

Norris, D., 282, 298, 301, 386

Noss, R. F., 2

O'Connor, M. D., 3, 6

O'Hara, J. L., 2

Old, K. M., 11, 277, 294

Oliver, C. D., 2, 90, 168, 393

Olson, J. T., 43

Olson, P. A., 3, 6

Olson, R. K., 206, 208, 210, 212

O'Malley, D. M., 330

Oren, R., 206, 212

Orr, D. B., 520
Orr, P. W., 170
Orr, R., 436
Ossenbruggen, H. S., 564, 565*f*
Ostmark, H. E., 530*f,* 555*f*
Ottmar, R. D., 7, 9, 144
Otvos, I., 486
Overhulser, D. L., 448, 556, 557

Packard, A. S., 398, 406
Paine, T. D., 528, 564
Paine, T. M., 527
Palmiotto, P. A., 2
Parks, C. G., 190, 333, 356, 392
Parmeter, J. R., Jr., 168
Parsons, D. J., 2, 151*f*
Patel-Weynard, T., 2
Paul, E. A., 240, 246, 252, 269*f*
Paul, H. G., 468*t*
Peace, T. R., 196
Pedigo, L. P., 414, 442, 449, 455
Peek, J. M., 154
Perry, D. A., 254, 270, 272, 273
Perry, D. H., 24, 29, 30, 31, 32, 34, 43
Perry, T. J., 491
Peskova, V., 273
Peterson, C. H., 2
Peterson, D. L., 74*f,* 75, 76, 78*f*
Peterson, J. L., 7, 9
Petruncio, M. D., 559
Pfister, R. D., 24
Pickett, S. T. A., 35, 36*f*
Pirone, P. P., 196
Pitman, G. B., 502, 504
Platt, W. J., 93, 94
Powell, D. C., 457, 481, 483
Putnam, T., 123
Pyatt, G., 174
Pyne, S. J., 49, 59, 66, 68, 69*f,* 71, 102, 103,
    104, 105, 108, 112, 114, 117*f,* 120, 121,
    126, 139, 142, 154

Quine, C. M., 174
Quirk, W. A., 98

Raffa, K. F., 476
Raimo, B. J., 564, 565*f*
Rapport, D. J., 2, 7
Rasmussen, L., 14
Rathbun, S. L., 94
Reardon, R. C., 447
Redin, S. C., 315
Reed, D. J., 266, 270, 271
Reinhardt, E. D., 74, 76

Reynolds, K. M., 505
Richards, B. N., 266, 268*f,* 269
Robinson, W. H., 403*f,* 539, 540*f*
Robson, A. D., 255, 271, 273
Roeloff, W. L., 447
Roland, J., 451
Romme, W. H., 51, 91
Romme, W. L., 153
Romoser, W. S., 442
Ross, D. W., 196, 357
Roth, L. F., 277, 295
Rothermel, R. C., 54, 59, 60, 60*f,* 62, 64*f,* 66*f,*
    115, 131, 133
Rowe, J. S., 73, 97
Rowoser, W. S., 414
Rudinsky, J. A., 491, 493*t,* 507*t*
Rudolph, D. C., 170
Running, S. W., 76
Russell, K., 393
Russell, K. W., 290
Rustad, L., 14, 375
Ruth, D. S., 544
Ryan, K. C., 74, 75, 76, 78*f*
Ryan, P., 170
Ryan, R. B., 449, 450, 480

Sackett, S., 94
St. Leger, R. J., 454
Sample, V. A., 43
Sampson, F. B., 3, 4
Sampson, R. N., 2, 9
Sandberg, D. V., 155, 156
Sartwell, C., 448, 504, 506*t,* 557
Sawyer, J. O., 83
Scharpf, P. F., 231, 232*f*
Scharpf, R. F., 199, 201, 211*f,* 282, 348*f*
Scharpf, R. T., 196, 309, 320
Schindler, D. W., 206
Schlesinger, W. H., 206
Schmid, J. M., 504
Schmidt, C. L., 390
Schmidt, F. H., 457
Schmidt, W. C., 2
Schmitt, C. L., 276, 280, 281*t,* 293, 296, 297*f,*
    299*f,* 303, 392
Schowalter, T., 214, 275, 295, 296
Schowalter, T. D., 187, 195, 285, 447, 449,
    498, 512
Schroeder, M., 50, 51, 52, 52*f,* 54, 55*f,* 56*f,* 57
Schultz, J. C., 561
Schulze, E. D., 206, 212
Schutt, P., 375
Schwandt, J. W., 321
Schwerdtfeger, F., 401, 402

Scott, D. W., 468*t*, 534
Scotter, G. W., 97
Sedell, J. R., 169
Sederoff, R. R., 330
Shaw, C. G., III, 162, 275, 276, 280, 281, 282, 284, 285*f*, 286*f*, 287, 288, 296, 305*f*, 368, 370
Shaw, D. C., 290
Sheehan, K. A., 446, 467
Shepard, R. F., 481
Shigo, A. L., 196, 354, 355*f*, 368
Shrimpton, G. M., 254, 255, 257, 258, 259, 259*f*, 260*f*, 261, 262, 262*f*, 264, 327, 330*f*
Shugart, H. H., 364, 376, 378, 379*t*, 380*t*
Simard, A. J., 102
Sinclair, W. A., 187, 196, 199, 200, 201, 203, 212, 213*t*, 216, 275, 309, 310, 315*f*, 317, 318, 320*f*, 321*f*, 333, 341, 343*f*, 344*f*, 347, 350, 369
Skelly, J. M., 196, 199, 200*f*, 284*f*, 295, 309, 312*f*, 318*f*, 319*f*, 325*f*, 326*f*, 365
Skillen, E. L., 447
Skinner, C. N., 84, 148
Slayter, R. O., 37
Smalley, E. B., 361
Smallidge, P. J., 1
Smith, D., 167, 176, 177*f*
Smith, D. M., 402*f*
Smith, L., 89
Smith, R. F., 449
Smith, S. E., 266, 270, 271
Smith, S. M., 457, 457*t*
Smith, W. H., 12, 196
Smith, W. K., 164
Snodgrass, R. E., 445*f*
Snow, A. E., 170
Soares, R. V., 195
Society of American Foresters, 27, 28
Sollins, P., 42, 169
Solomon, J. D., 406, 522, 523, 526, 527
Somerville, A., 176
Sower, L. L., 448, 557
Speight, M. R., 457, 508
Spoon, C. W., 148
Sprugel, D. G., 1, 3, 5*f*, 378, 380*t*
Srago, M., 168
Stage, A. R., 275, 305*f*
Stahelin, R., 89
Steel, E. A., 3, 6
Stenlid, J., 275, 276, 288, 298, 300, 301, 302
Stern, V. M., 496
Stevens, R. E., 504, 506*t*
Stipes, R. J., 196, 358
Stocks, B. J., 95

Stoffolano, J. G., 414
Stoffolano, J. G., Jr., 442
Stolte, K. W., 15, 16
Storer, A. J., 347
Street, R. B., 95
Streng, D. R., 93
Stuart, D. R., 408*f*
Stuart, J. D., 356
Sturrock, R. N., 254, 255, 257, 258, 259, 259*f*, 260*f*, 261, 262, 262*f*, 264, 277, 291, 292, 293, 298, 301, 302, 304, 327, 330*f*, 386
Sutherland, J. R., 254, 255, 257, 258, 259, 259*f*, 260*f*, 261, 262, 262*f*, 264, 327, 330*f*
Swanson, F. J., 42, 84, 85*f*, 154, 169
Swezy, D. M., 147
Sykes, D. J., 98
Sze-ling, Y., 455

Taber, D. A., 368
Tainter, F. H., 185, 187, 188, 191, 191*f*, 196, 199, 200, 202, 203, 204, 206, 209, 210*f*, 215, 216, 217, 217*t*, 218*f*, 219*f*, 220, 221, 222, 228, 233, 235, 262, 338, 341, 344, 346, 347, 357, 358, 361, 362, 383, 388, 390, 392*f*, 393*f*
Talbert, J., 543
Tappeiner, J., 170
Tattar, T. A., 196
Taylor, A. R., 84
Taylor, D. P., 235
Taylor, J. E., 321
Thies, W. G., 277, 291, 292, 293, 298, 301, 302, 304, 386
Thomas, J. W., 154
Thomas, P., 59, 116, 117
Thomas, T. L., 132
Thompson, C. G., 483
Thompson, W. T., 442
Thornburgh, D. A., 83
Tietema, A., 206
Tilman, G. D., 206
Tinus, R. W., 546
Tomback, D. R., 91
Torgensen, T. R., 190, 333, 356, 392
Torgerson, R. F., 457
Torgerson, T., 486, 487
Torgerson, T. R., 457
Tovar, D. C., 544
Trabaud, L., 59, 116, 117
Treshow, M., 204, 205, 209
Trimble, R. M., 472
Triplehorn, C. A., 414
Truman, L. C., 539
Tunnock, S., 196, 450

Turgeon, J. J., 545
Turner, K. B., 469
Turner, M. G., 2
Tyrell, D., 486

Ugolini, F. C., 168
U.S. Congress, Office of Technology
    Assessment, 569f
USDA, 406, 466, 467, 470, 473, 481, 482, 483,
    484f, 562t, 568t
USDA Forest Service, 109f, 111f, 115, 118f,
    119f, 147, 354, 400, 561
USDA-USDI, 124f

Van Arsdel, E. P., 182, 331
van den Bosch, R., 449
Van Wagner, C. E., 43, 63, 65, 74, 95, 97,
    131, 133
van Wagtendonk, J. W., 131, 135, 136f,
    137f, 154f
Viereck, L. A., 95
Vité, J. P., 495, 502, 507
Vitousek, P. M., 206
Vogt, D. J., 2
Vogt, K. A., 2

Wagner, M., 476, 477
Wagner, M. R., 2, 447, 449, 498, 512
Wainhouse, D., 185, 457, 508
Wakimoto, R. H., 106, 148
Wakimoto, R. W., 149
Walker, R. B., 219
Wallis, G. W., 196, 200, 201f, 283t, 293f, 298f,
    309, 312f, 325f, 326f, 333, 350, 350f,
    351f, 354f, 357, 392
Walters, C., 146
Walters, N. E. M., 187
Ware, G. W., 442
Wargo, J. P., 2
Wargo, P. M., 280, 296
Watling, R. H., 504
Warkentin, D. L., 556
Weatherspoon, C. P., 148

Weaver, H., 81
Wein, R., 97
Wellington, W. G., 469, 472
Wells, S., 154
Werner, R. A., 487, 505, 511
West, L., 168
Whelan, R., 73
White, A. S., 81, 82
White, P. S., 35, 36f
Whittaker, R. H., 31, 32f
Whytemare, A. B., 212, 220
Wickman, B. E., 406, 407f, 439, 483
Wiens, D., 231, 232f, 334, 335f, 337t, 338,
    339, 385
Wilcox, P. L., 330
Wilkins, R. M., 446
Williams, D., 59, 116, 117
Williams, J. T., 62, 64f
Williams, S. E., 154
Williams, T. M., 162, 170, 173, 173f
Wilson, C. C., 122
Wilson, E. O., 46
Windham, M. T., 315
Witten, E., 2
Wood, D. L., 347, 493, 495
Woodmansee, R. G., 2
Woodward, S., 275, 276, 288, 298, 300,
    301, 302
Worrest, R. C., 370, 372f, 373
Wright, H. A., 82, 154
Wright, R. F., 14

Yamaguchi, D. K., 39, 41f
Yanchuk, A., 556, 558f
Yates, H. O., III, 544

Zak, B., 264, 270f
Zasada, J., 170
Zhang, Y., 275, 296
Ziller, W. G., 323
Zobel, B., 543
Zoettl, H. W., 14, 214
Zweig, G., 557

SUBJECT INDEX

Page numbers followed by *b, f,* and *t* refer to boxes, figures, and tables, respectively.

Abiotic diseases/injuries, 187, 199–222. *See also*
   Air pollution
  declines caused by, 186*t*
  fire-related, 221
  global warming-related, 203–204
  herbicide-related, 214–215
  mechanical, 215–217
  moisture-related, 200, 202–203
  of nursery seedlings, 255–256, 264
  nutrient deficiencies, 217–220
  nutrient imbalances, 220
  nutrient toxicity, 221
  patterns of, 199–200
  salt-related, 221–222
  temperature-related, 200–202, 256
Acaracides, 443, 444
Acarina, 416–417
Acephate (Orthene), 444
Acervuli, 242, 243*f*
Acid Deposition Control Program, 14
Acid precipitation, 13–14, 204, 205*t*, 206, 212,
   221, 375
Acid Rain National Early Warning System
   (ARNEWS), 18–19
Acid soils, 221, 225
Acrididae, 431
Active crown fires, 66
Active ingredient (A.I.), 446
Adaptations, of plants to fire, 71–73, 72*f*, 73*f*, 74*t*
Adaptive management, 146, 147*f*
Adiabatic lapse rates, 53
Adiabatic warming/cooling, 51
Aedeagus, 424
Africa, 7, 287, 367*t*
Agathis pumila, 450, 451*f*
Aggregants, 449. *See also* Pheromones
Agrobacterium tumefaciens, 347
Agrypon flaveolatum, 450–451
A.I. *See* Active ingredient
Air moisture, 51
  and precipitation, 53
  and temperature, 51, 53*f*
Air pollution, 204–214
  acid rain. *See* Acid precipitation
  acute, 209–210
  and atmospheric stability, 207, 208*f*
  chronic, 209

Air pollution—*Cont.*
  damage caused by, 209–212
  dispersion of pollutants, 207–209
  and forest decline, 7, 206, 375–376
  and forest health, 12–14
  management, 212–214
  pollutant types and sources, 204–206, 205*t*
  tree damage symptoms, 210*b*
Al toxicity, 221
Alderflies, 435
Allelochemics, 448*f*
Allgemeine Oeconomische Forst-Magazin, 403
Allomones, 448*f*
Alpha diversity, 42
Ambrosia beetles, 491, 492*f*, 516–524, 518*f*
  damage by, 516–517, 518*f*
  in hardwoods, 522–524
  in softwoods, 517–521
Ambrosia fungi, 518, 520, 522
American dog tick (*Dermacentor variabilis*), 416
Ammonia (NH$_3$), 204
  sources, 204, 205*t*
  tree damage by, 212
Anamorphs, 242
Angiosperms. *See also individual species*
  ambrosia beetle infestations, 517–520
  wind effects, 165
Animal and Plant Health Inspection Service
   (APHIS), 389, 567
Animals
  building post-fire cavities for, 125
  tree damage by, 222–226
   management, 222–223, 223*t*
Annosus root and butt rot. *See Heterobasidion*
   annosum
Annual grass (Fuel Model 1), fire behavior
   modeling, 61–62, 62*t*, 63*t*
Anobiidae, 536–538
Ant(s), 428*f*, 434
  carpenter, 537*f*, 538–539
  as insect pest control, 457
  leafcutting, 195–196, 559
Ant lions, 428*f*, 435
Antheridia, 247
Anthracnose, 310
  of dogwood, 315–316
  fungi causing, 245*t*, 311*t*, 313, 316

Anthracnose—*Cont.*
  of oak, 314
  of sycamore, 313–315, 315*f*
Antiaggregating pheromones, 449
Aphidlions, 435. *See also* Lacewings
Aphids, 427*f,* 429*f,* 432
  classifications, 427*f,* 429*f*
  in plantations, 553
  root, 553*f*
  seedling-damaging, 549, 552, 553*f*
  spruce, 564
APHIS. *See* Animal and Plant Health Inspection
  Service
Apiogynomonia veneta, 314–315
Apocrita, 434
Apothecia, 242, 243*f*
Appressoria, 252
Appropriate response policy, 105–106
Arachnida, 414–415, 416*f*
Araneida, 414
Arbuscular mycorrhizae, 266–269, 269*f*
Arceuthobium, 333, 336*t*–337*t. See also* Dwarf
  mistletoes
  life cycle, 334–335, 335*f*
Archae, 228
Archiascomycetes, 241–242
Argasidae, 416
Armillaria, 276, 278*t,* 280
  high-hazard sites, 302
  insect associations, 297*f,* 304–305, 305*f*
  role in natural forests, 282–287, 285*b*
  slash burning and, 303
Armillaria root disease, 276, 278*t,* 280–288
  conifer susceptibilities, 299*t*
  distribution, 278*t,* 286*f,* 287–288
  fertilization and, 302
  management, 298, 300–303, 385, 386
  modeling of effects, 304–305, 305*f*
  spread, 280–281, 282*f*
  symptoms and signs, 281*t,* 282, 282*f,* 283*f*
ARNEWS. *See* Acid Rain National Early Warning
  System (ARNEWS)
Arthropoda, 414–418, 415*f*
Arthrospores, 231
Ascocarps, 242, 243*f*
Ascomycota, 237, 238*t*–239*t,* 241
  classification, 241–242
  in cones, 257
  diseases caused by, 238*t*–239*t,* 242–244, 245*t*
    cankers, 341
    decay/rots, 349
    Dutch elm disease, 358
    foliage, 309, 313, 316, 318
    root, 277–280
  filamentous, 241, 242, 243*f,* 244*f*
  fruiting bodies (ascocarps), 242, 243*f*

Ascomycota—*Cont.*
  life cycle, 242, 244*f*
    foliage-pathogenic, 313, 314*f*
  and mycorrhizae, 267*t*
Ascospores, 241
Ascostroma, 242
Ascus, 241
Ash
  dieback, 370–373, 372*f,* 374*f*
  yellows, 234
Asia
  Armillaria root disease, 288
  forest diebacks and declines, 367*t*
  forest stress, 7
  insect pest control, 442, 455
Asian gypsy moth, 12, 563
Asian longhorned beetle (*Anolophora
  glabripennis*), 564, 565*f*
Aspect, and forest zonation, 26, 28*f*
Aspen rust, 264
Aspergillus, 257, 264
Atrophy, 194, 194*t*
Auclair model of forest decline, 369–374
Australia
  animal damage to trees, 223
  forest conservation policy, 7–8
  forest diebacks and declines, 367*t,* 374
  forest health problems in, 7–8
  introduced pathogens and insects, 11, 12
  nitrogen toxicity, 221
  plantations, 8, 9*f*
  prescribed fire, 303
  root diseases, 278*t*–279*t,* 287
  management, 303–304
Autosuccession, 378
Autotrophic organisms, 192
Autumn frost, 256
Avoiders, 73

Bacillus thuringiensis (Bt), 452–453, 454*f*
  as defoliator control, 453, 468–469, 476, 480
Backfiring, 114
  pros and cons, 116, 119*f*
Backing fire, 139
Bacteria, 187, 228, 233
  disease-causing, 233
  in forest energy flow, 32, 33*f*
  galls caused by, 347
  as insect pest control, 452–453, 454*f*
    defoliators, 453, 468–469, 476, 480
    gypsy moth, 475
  morphology and reproductive features, 229*f,* 233
  symptoms and signs, 193, 194*t,* 313
BAER teams. *See* Burned Area Emergency
  Rehabilitation (BAER) teams

Ballooning, 472
Balsam fir (*Abies balsamae*), 5*f*
  dieback
    climate change and, 373, 374*f*
    mortality waves, 378, 380*f*
  eastern spruce budworm attacks, 471–473,
    472–473, 473*f*
  fire adaptations, 97
Balsam woolly adelgid (*Adelges piceae*), 14,
  561–562, 563*f*
Barbara colfaxiana, 547, 550*f*
Bark
  animal damage, 224, 225, 225*f*
  cankers, 341–347
  fire adaptations, 74*t*
  fire damage, and tree mortality, 74, 75*f*, 75–76
Bark beetles, 491–514, 492*f*. *See also individual
    species*
  aggregating pheromones, 449
  biology, 493–495
  Dendroctonus species, 498–508
  egg gallery and larval mine patterns,
    493, 494*f*
  engraver species, 508–514
  fungi interactions, 491–493
    and blue-stain fungi, 395
    in Dutch elm disease, 358–361
    in root diseases, 296, 297*t*
  host selection behavior, 495, 496*f*
  management, 495–498
    control, 449, 497–498
    damage thresholds, 496–497
    detection, 497, 498*f*, 499*f*, 500*t*
  wind interactions, 181
Basamid, 265
Basidiomycota, 237, 239*t*, 240, 242
  basidiospores, 246, 247*f*, 251
  as biological control, 301
  classification, 244–246
  diseases caused by, 239*t*, 246, 249*t*
    cankers, 341
    decay/rots, 349, 350–354, 352*t*–353*t*
    foliage, 309
    root, 276–277
    rusts, 323
    seedling, 257
  life cycles, 246, 248*f*, 249*f*
  and mycorrhizae, 267*t*
Basidiospores, 246, 247*f*
  hyaline, release of, 251
  structures bearing, 246, 247*f*
Bears, tree damage by, 225
Beaufort Scale (of wind velocity), 160–161, 161*t*
Beauvaria bassiana, 454–455, 475
Beavers, tree damage by, 223, 225–226
Beech bark disease, 341–344, 342*t*, 345*f*, 346
Beech scale, 564

Bees, 428*f*, 434
  carpenter, 541
Beetles, 428*f*, 433. *See also* Ambrosia beetles;
    Bark beetles; *individual species;* Insect(s);
    Wood borers
BEHAVE model (of fire spread), 60–63, 61*f*,
    63*t*, 136*t*
Bessey, Ernst, 188
Beta diversity, 42, 43*f*
Biodiversity. *See* Ecological diversity
Biogeoclimatic classification, 29
Biological control, 384, 388. *See also specific
    methods and organisms*
  of dwarf mistletoe, 340
  of gypsy moth, 475–476
  of hemlock looper, 486–487
  of insect pests, 449–457, 453*t*
  of nursery diseases, 266
  permanent, 452
  of root disease, 301
Biology, 413
Biomass fuel treatments, 136, 136*t*, 137*f*. *See also*
    Thinning, for fire spread prevention
Biome, 24
Birch
  dieback, 370–373, 372*f*, 374*f*
  fire adaptations, 97
Birds, gypsy moth predators, 475
Black bears, tree damage by, 225
Black root disease, 260–261
Black spruce (*Picea mariana*), 94, 96
  fire adaptations, 97
Black stain root disease. *See Leptographium
    wageneri*
Black walnut canker, 346
Black widow (*Lactrodectus mactans*), 415
Blackheaded budworm (*Acleris gloverana*),
    487–488
  predicted defoliation, 488*t*
Blackheaded sawfly (*Neodiprion excitans*), 471
Blackman, M. W., 406
Blatteria, 429*f*
Blattidae, 432
Blight(s), 310, 311*t*, 364. *See also* Declines
  canker, 341
  chestnut, 2, 341, 342*t*, 343*f*, 344–345
  fungi causing, 311*t*
  needle. *See* Needle casts
  potato, 250–251
  seedling, 262, 263*b*
Blue stain, 245*t*, 395, 395*f*
"Bootstrapping" hypothesis, 272
Borax, 300
Boreal forest region, 94
  fire regimes, 94–95
  stand development patterns, 96–98
Borrelinavirus reprimes virus, 475

Bostrichidae, 538
Botanical insecticides, 443
Botrytis cinerea, 261, 264
Branch diseases. *See* Stem and branch diseases
Brazil, 8, 195–196
Bread mold fungus (*Rhizopus*), 241, 264
Brown felt blights, 323
Brown recluse (*Loxosceles reclusua*), 415
Brown root and butt rot, 279*t*
Brown rots, 348*b*, 349, 356
Brown spot needle blight, 322
Bt. *See Bacillus thuringiensis*
Bugs. *See* Insect(s); True bugs
Buprestids, 528–530, 531*f*
Bureau of Land Management, lands managed, 151*t*
Burls, 347
Burned Area Emergency Rehabilitation (BAER)
    teams, 123
Butt log pruning, 176–177
Butt rots, 349, 352*t*–353*t*. *See also Heterobasidion
    annosum*
    ecological roles, 356–357
    fungi causing, 249*t*, 350, 352*t*–353*t*
Butterflies, 420, 428*f*, 433–434
Butternut canker, 16, 342*t*, 344, 346

Caddis flies, 428*f*, 433
Caelifera, 429*f*
Caloscyphe fulgens, 257
Cambial heating/kill, 74, 75*f*, 75–76
Cambium, and cankers, 341
Camponotus ants, 537*f*, 538–539
Canada
    Acid Rain National Early Warning System
        (ARNEWS), 18–19
    Division of Forest Insect Investigations, 407–408
    eastern, diebacks in, 370–373, 372*f*, 374*f*
    ecosystem classifications of vegetation, 29
    Fire Weather Index System (FWI), 68
    forest entomology, 407–409
    forest health, 7, 19
    forest pathology research, 189
    spruce budworm damage, 401*f*
Canker(s)
    annual, 341
    classifications, 341, 342*t*–343*t*
    conifer, 342*t*–343*t*, 346–347
    fungi causing, 245*t*, 341–347
    hardwood, 341–346, 342*t*
    perennial, 341
Canker blights, 341
Canker rots, 341, 349
Carapace, 418
Carbamate insecticides, 445*f*, 445–446
Carbaryl (Sevin), 446
Carbofuran, 446

Carnivores, 29, 30*f*
    and forest productivity, 33
Carpenter ants, 537*f*, 538–539
Carpenter bees, 541
Carpenter worm moths, 526–527, 527*f*
Cascades, 7
    dwarf mistletoe management, 339–340
    fire mortality variation, 85*f*
    fire return intervals, 84
    integrated disease management
        recommendations, 391*b*
Castes, termite, 431
Caterpillars, 434
    vs. sawflies, 476, 477*f*
Caudal filaments, 430
CCA. *See* Copper chrome arsenate
Center firing, 141, 141*t*, 142*f*
Centipedes, 418, 419*f*
Central America, Armillaria root disease in, 287
Cerambycids, 527–528, 530–533, 531*f*, 536
Ceratocystis fagacearum, 359*t*, 362
Cerci, 430
Chaparral
    fire behavior modeling, 62, 62*t*, 63*t*
    fuel consumption by fire, 57*f*
Check dams, post-fire, 125
Chelicerae, 414
Chelicerata, 414–417, 415*f*
Chemical control, 388. *See also specific
    organisms*
    of decay, 395
    of diseases, 384, 388
        nursery, 264–265
        root, 300–301
        rusts, 331
    of dwarf mistletoe, 340
    herbicides, 214–215, 340, 388
    of insect pests, 442–447
        bark beetles, 497
        in container seedlings, 552*t*
        gypsy moth, 476
        problems with, 409–410
        western hemlock looper, 487
Chestnut blight, 2, 341, 342*t*, 343*f*, 344–345
Chiggers, 417
Chile, 7, 8
China. *See also* Asia
    wood imports and pest introductions from,
        566, 566*f*
Chinooks. *See* Foehn winds
Chlamydospores, 230
Chlorinated hydrocarbons, 443–444
Chlorine (Cl_2), 204
    sources, 204, 205*t*
    tree damage symptoms, 210*b*
Choanephora, 240
Chromista, 239*t*, 240

Chronic pests, 439, 440*f*
Chronosequential gap rotation, 378
Chrysocharis laricinellae, 450, 451*f*
Chrysopidae, 435
Chytriodomycota, 237
Cicadas, 427*f,* 429*f,* 432
Circumesophageal connectives, 424
Clarke-McNary Act (1924), 104
Clean Air Act, 1990 Amendment of, 14, 265
Clearcutting, 7*f,* 177–179, 385. *See also* Thinning
    for decay management, 357
    for dwarf mistletoe management, 339–340
    examples of use, 385*t*
    wind and, 177–180, 178*f,* 179*t*
Cleistothecia, 242, 243*f*
Cleveland, Grover, 103–104
Climate, 50–51
    and biome, 24
    fire, 54
        North American, 54, 55*t,* 56*f*
    in root disease management, 302–303
Climate change, 203, 214
    and tree damage/forest decline, 203–204,
        369–375, 373*f,* 374*f*
Climax, 34
Coarse filter management approaches, 45–46
Coast redwoods, silt deposition and, 39–40
Cockroaches, 427*f,* 429*f,* 432
Cocoon sampling, 466, 468*t*
CODIT. *See* Compartmentalization of decay in trees
Cold front, 51
Coleoptera, 428*f,* 433, 463, 547
Collembola, 426
Colletotrichum blight, 263*b*
Columbia River, 54
Columbian timber beetle (*Corthylus
    columbianus*), 522–523, 524*f*
Combustion, 49–50, 50*f*
Commensalism, 192, 193*b*
Commercial forests. *See* Private forestry
Compartmentalization of decay in trees (CODIT),
    354–356, 355*f*
Compression wood, 165, 215–216
Conduction, 50
Cone(s)
    fungi, 257
    serotinous, 73, 73*f,* 74*t,* 90, 91*f,* 96–97
Cone and seed insects, 544–548
    economic damage by, calculation of,
        439–440, 441*b*
    management, 544–548, 545*f*
    species, 547–548
Coneworms, 545, 546*f,* 547–548, 548*f,* 549*t*
Confine strategy, 114
Conidiospores, 230, 230*f*
Conifers. *See also individual species*
    ambrosia beetle outbreaks, 517–520

Conifers—*Cont.*
    bark beetle outbreaks, 491
    cankers, 342*t*–343*t,* 346–347
    cone and seed insects, 544–548
    defoliators, 471, 472*t*
        control, example, 458*t*
    diebacks, declines, and blights, 365
    foliage diseases, 309–310, 311*t,* 318–323
        symptoms, 310, 312*f*
    herbicide sensitivity, 214–215
    log exports, 516, 517*f*
    mixed forests, health of, 10–11
    nursery diseases, 261–262, 264
    nursery soil for, ideal, 255
    pitch moths, 559
    root diseases
        Armillaria root disease, 284
        laminated root rot, 292
        management with resistant species, 298–300
        susceptibilities to, 299*t*
    rusts, 264
    stains, 362
    thinning on plantations, 386
    vascular wilts, 359*t*
    wind effects, 165, 215–216
    winter damage, 201
    wood borers, 533, 535*t*
Consumers, 29, 30*f*
Contain strategy, 114
Contour felling, post-fire, 125
Contributing factors, 369, 371*f,* 381
Control strategy, 114
Convection, 50
Cooley gall adelgid (*Adelges cooleyi*), 553, 554*f*
Copper chrome arsenate (CCA), 395
Coremia, 242, 243*f*
Coriolis force, 51
Corky root disease, 261, 262*f*
Corydalidae, 435
Cosmopolitan ambrosia beetle (*Xyleborus
    ferrugineus*), 524
Cossidae, 526
Cottonwood rust, 264
Cover, classifications of vegetation by, 27, 28
Coxa, 421
CPVs. *See* Cytoplasmic polyhedroses
Crab spiders, 414–415
Craighead, F. C., 398
Crater Lake National Park, fire management
    monitoring, 146–147
Creosote, 300, 395
Crickets, 427*f,* 429*f,* 431
Cronartium quercum f. sp. *fusiforme,* 264,
    324*t,* 325
    life cycle, 329
Cronartium ribicola, 323, 324*t. See also* White
    pine blister rust

Cronartium ribicola—Cont.
  life cycle, 328*f*
  management, 330–331
Crop, insect, 421
Crown(s)
  fuel treatments, 131–139
    effectiveness, 135–139
  thinning, 133–134, 135*f*, 135*t*, 135–139
  wind effects on, 163–164, 166
Crown bulk density
  calculation, 133
  and crown fire behavior, 65, 133–134, 134*f*, 135*t*
Crown fires
  behavior, 58*t*, 63–66, 131–134
  classes of, 66
  initiation, 58*t*, 63–65, 66*t*, 131–132, 132*t*
  prevention strategies, 133–139
  spread, 65, 133–134, 134*f*
  tree damage and mortality from, 74–75
Crown gall, 347
Crown mass, 133
Crown scorch, 74, 75*f*
Crown volume, 133
Crustacea, 418
Cryphonectria parasitica, 342*t*, 344–345
Cryptococcus fagisugal, 346
Cultural management. *See also* Silvicultural
    management
  of nursery diseases, 266
Cut and scatter, 136, 136*t*, 137*f*, 138
Cutworms, 549
Cylindrocarpon
  and beech bark canker, 346
  and root rot, 258, 258*b*, 259
  in stored seedlings, 264
Cylindrocladium, 257
Cytoplasmic polyhedroses (CPVs), 455, 456*f*

Damping-off diseases, 250, 257–258, 258*b*
Damping-off fungi, 257
  nitrogen toxicity and, 220
Damselflies, 427*f*, 430
Dead trees. *See also* Decay; Slash; Woody debris
  as leave trees, 390
  roles, 42, 190
deBary, Anton, 188
Decay, 348–358, 393–395
  advanced, 350
  after winter damage, 202
  compartmentalization of decay in trees
    (CODIT), 354–356
  ecological roles, 356–357
  incipient, 350
  increasing, as forest management, 384, 390
  management, 357–358, 387, 395
  types, 348–349, 349*b*

Decay—*Cont.*
  wind and, 168
Decay fungi, 192, 349, 350, 352*t*–353*t*, 393
  ecological roles, 356–357
  fire and insect interactions, 356–357, 407, 408*f*
  fruiting bodies, 350, 350*f*, 351*f*, 354*f*
  hosts, 352*t*–353*t*
  inoculation with, 390
  life cycle, 248*f*, 350
  spread, 350–354
Declines, forest, 185, 186*t*, 364–381, 366*t*–367*t*
  air pollution and, 375–376
  autosuccession and, 378
  causes, 185, 186*t*, 368–381
  characteristics, 364–365
  climate change/climate perturbation theory of,
    369–375
  decline diseases, 368–381
  definition of, Manion's, 369
  disease decline spiral model, 369, 371*f*, 378,
    379*t*, 380*f*
  eastern hemlock, 365–368
  ecological theory of, 376–381
  environmental stress and secondary organisms
    theory of, 368–369, 369*b*, 370*f*
  natural dieback theory of, 378–381
  predisposing, inciting, and contributing factors
    theory of, 369, 371*f*, 379–381
  as species decline, 365
  symptoms, 364
Deer, tree damage by, 224
Defensible fuel profile zone (DFPZ), 148
Defoliation. *See also* Defoliators
  detection and evaluation, 463–467
  in European forests, 17, 17*b*, 18*f*
  intensity classification, 465
Defoliators, 462–488
  control, 453, 455, 458*t*, 467–469. *See also*
    individual insects
  detection and evaluation, 463–467
  in eastern and southern forests, 470–481
  effects, 462–463, 464*f*
  feeding and host preferences, 462
  in pine plantations, 559
  of seedlings, 550, 552, 553*f*
  timing of outbreaks, 463
  types, 463, 469–471
  in western forests, 471, 472*t*, 481–488
DeGryse, J. J., 408–409
Dendroctonus beetles, 498–508. *See also*
    individual species
Density dependency, and insect pest control,
    449, 450*t*
Dermaptera, 427*f*
Detrivores, and forest energy flow, 30*f*, 32
Deuteromycota. *See* Fungi imperfecti
DFPZ. *See* Defensible fuel profile zone

Diapause, 544

Dicofol (Kelthane), 444

Dictyoptera, 427*f*, 429*f*, 431–432

Dieback(s), 364, 365. *See also* Declines
displacement, 378
replacement, 378
stand-reduction, 378

"Dieback threshold," 373, 374*f*

Differential equations, modeling rate of disease
spread with, 390–391

Diffuse cankers, 341

Dikaryomycetes, 240

Dioryctria coneworms, 545, 546*f*, 547, 548*f*, 549*t*

Diplodia canker, 342*t*

Diplodia pinea, 322

Diplodia tip blight, 321–322

Diptera, 428*f*, 435
morphology, 419, 420, 435
seed-damaging, 548

Direct attack (of fire), 116, 117*f*
pros and cons, 118*f*

Discoloration, in European forests, 17, 18*b*

Discomycetes, 238*t*, 242

Discula, 314–315, 316

Disease avoidance, 387–388

Disease decline spiral model, 369, 371*f*
adaptations, 378, 379*t*, 380*f*

Disease(s), forest, 185–197. *See also specific
diseases and causal agents*
appraisal, 383
bacteria as cause of, 233
classification by tree part, 188, 189*f*
definitions of, 187–188
detection, 383, 384*b*
factors in development of, 191
fire and, 195–196
foliage. *See* Foliage diseases
fungi as cause of, 228, 229–230, 237, 238*t*–239*t*,
240. *See also* Fungal diseases; *specific
diseases, parts of tree, phyla, and
organisms*
impacts of, 190, 190*b*
insects and, 187, 195–196
management. *See* Disease management
modeling progression of, 390–393
nematodes as cause of, 235–237
outbreaks since 1875, 185, 186*t*
parasitic plants as cause of, 231, 232*f*, 233
pathogens and, 188
phytoplasmas as cause of, 234–235
recognition, 383, 384*b*
resources on, 196–197
root. *See* Root disease(s)
rusts. *See* Rust(s)
seedling. *See* Nursery diseases
signs and symptoms, 193, 194*t*,
194–195, 195*b*

Disease(s), forest—*Cont.*
stem and branch. *See* Stem and branch diseases
in urban forests and street trees, 196
viruses as cause of, 233
wind and, 182, 196

Disease management, 383–390. *See also specific
diseases, parts of tree, and control
techniques*
biological control as, 388
chemical control as, 388
disease detection and recognition, 383, 384*b*
genetic resistance as, 388
history of, 188–189
increasing decay as, 390
integrated, 390
example recommendations, 391*b*
integrated pest management in, 384
modeling progression, 390–393
passive, 389
quarantine as, 388–389
silvicultural techniques, 384–388, 385*t*
strategies, 384–390

Disease square, 191, 191*f*

Disease tetrahedron, 191, 192*f*

Disease triangle, 191, 191*f*

Disease-causing organisms. *See* Pathogen(s)

Displacement dieback, 378

Distress syndrome, 3*b*

Disturbance(s)
classification of effects, 35–36, 36*t*
definition of, 34–35
effects of, 1
productivity effects, 31
and stability and equilibrium, 43–44
successional effects, 34–42
multiple-pathway modeling of, 36–40,
38*f*–40*f*

Disturbance agents, natural, 1, 24
interaction of, 1

Diversity
ecological, 2, 45–46
and stability, 43
types of, 42, 43*f*

Division of Forest Disease Research, 188

Division of Forest Insect Investigations (Canada),
407–408

Dobsonflies, 428*f*, 435

Dodders, 231, 231*t*

Dogwood anthracnose, 315–316

Dothistroma needle blight, 320–321, 321*f*

Douglas-fir. *See also* Douglas-fir forests/stands
ambrosia beetle damage, 518*f*, 520
Armillaria root disease, 284, 285
bacterial galls, 347, 348*f*
bark beetle/root rot pathogen complexes, 296
biosolids-treated, damage to, 203, 221, 224
corky root disease, 261

Douglas-fir—*Cont.*
    fungal decay in, cull in relation to tree age, 357*b*
    herbicide damage, 215*f*
    laminated root rot and, 292, 294
    rhabdocline needle cast, 318, 318*f*
    shoot diseases, 262–264
    Swiss needle cast, 318–319, 319*f*
    tree mortality by successional stage, 379*t*
    western spruce budworm attacks, 481–482
    windthrow effects, 181
        successional, 169–170
    winter damage to, 201, 202
Douglas-fir beetle (*Dendroctonus pseudotsugae*),
        491, 501*t,* 504–505
    conditions for outbreak, 504–505, 507*t*
    and root disease, 296, 297*t*
    and windthrow events, 181
Douglas-fir chalcid (*Megastigmus
        spermotrophus*), 548, 551*f*
Douglas-fir cone moth (*Barbara colfaxiana*),
        547, 550*f*
Douglas-fir forests/stands
    coastal (*Pseudotsuga menziesii*)
        fire regime, 78, 83–84
        stand development patterns, 84–87
    fire behavior modeling, 62, 62*t,* 63, 63*t,* 135*t*
    fire mortality probabilities, 78*t*
    fuel consumption by fire, 57*f*
    inland western, health problems, 10, 11
Douglas-fir pitch moth (*Synanthedon
        novaroensis*), 559
Douglas-fir tussock moth (*Orgyia pseudotsugata*),
        439, 483–485, 485*f*
    control, 440, 484–485
    feeding habits, 462
    sampling, 466, 467, 467*f,* 468*t,* 471*f*
    timing of outbreaks, 463, 465*f*
Downy mildews, 251, 310
Dragonflies, 425, 427*f,* 430
Drip torch, 121, 142–143
Drought stress, 202–203
Dry wood, borers of, 536–541
Dryness, tree damage by, 200, 202–203
Dusts (pesticidal), 446–447
Dutch elm disease, 338–362, 359*t*
    elm bark beetle and, 513
    life cycle, 360*f*
    management, 361–362
Dwarf mistletoes, 16, 231, 231*t,* 232*f,* 334–340,
        336*t*–337*t,* 338*f*
    damage by, 338
    distribution, 334, 336*t*–337*t*
    ecological relationships, 338–339
    hosts, 336*t*–337*t*
    infection rating scheme, 338, 339*f*
    life cycle, 334–335, 335*f*
    management, 339–340, 385, 386

Earwigs, 427*f*
Eastern Europe, 7, 13
Eastern forests. *See also* Northeastern forests;
        Southern/southeastern forests
    deciduous, vegetation zones, 24, 26*f*
    declines and diebacks, 375–376, 378
    defoliators, 471–481
    health of, 6–7
Eastern hemlock (*Tsugae canadensis*)
    decline, 365–368
    hurricane effects, case study, 170–172
    woolly adelgid attacks, 564
Eastern hemlock looper (*Lambdina fiscellaria
        fiscellaria*), 462–463
Eastern spruce budworm (*Choristoneura
        fumiferana*), 469, 471–473
    damage by, 398, 400*f,* 410*f,* 467, 469*f,* 473*f*
    egg surveys, example, 466*t*
    management, 455, 457*t,* 458*t,* 472–473
Ecological diversity (biodiversity). *See also*
        Diversity
    approaches to maintaining, 2, 45–46
Ecological homologs, 561
Ecological pyramid, 30
Ecology. *See also* Forest ecosystems
    fire. *See* Fire ecology
    production, 29–33
Economic damage by insects, calculation of,
        439–440, 441*b*
Economic models, in forest disease management,
        392–393
Ecosystem classification, of vegetation, 27, 29
Ecosystem health. *See also* Forest health
    and distress syndrome, 3*b*
    and ecological diversity, 2
    and fire, 6
    and watershed health, 3
Ecosystem management, 2, 44–45, 436–439
    coarse filter approaches, 45–46
    definition of, 44
    fine filter approaches, 46
    and island biogeography, 46
Ecosystem perspective of forest health, 2
Ecosystem(s), sustainability of, 43
ECs. *See* Emulsifiable concentrates
Ectomycorrhizae, 246, 266, 269–270, 270*f*
    management, 271–273
    roles, 270–271
    seedling inoculation with, 271–273
    vs. uninfected root, 268*f*
EFSA. *See* Escaped Fire Situation
        Analysis (EFSA)
Egg surveys, 465–466, 466*t*
Elevation
    and forest zonation, 25–26, 27*f,* 28*f*
    and windthrow hazard, 174*t*
Elk, tree damage by, 224

Elm
  Dutch elm disease, 358–362
  phloem necrosis (yellows) in, 234
Elm bark beetle (*Scolytus multistriatus*), 513–514
Elm leaf beetle (*Pyrrhalta luteola*), 462
Elytra, 433
Elytroderma needle cast, 320, 320*f*
EMAP. *See* Environmental Monitoring and
    Assessment Program (EMAP)
Embioptera, 427*f*
Emulsifiable concentrates (ECs), 447
Endocronartium harknessii, 264, 326*f*, 327, 329
Endomycorrhizae, 266
  types, 266–269, 267*t*
Endotoxin-Δ, 453
Endurers, 73
Energy flow (energetics) of forest ecosystems,
    29–33, 30*f*
  northern hardwood example, 33*f*
Engelmann spruce, stand development patterns,
    89, 90–91
Engraver beetles
  Ips, 491, 508–512
  Scolytus, 512–514
Ensifera, 429*f*
Entomology, 413–414. *See also* Forest
    entomology; Insect *entries*
Entomopoxviruses (EPVs), 455, 456*f*
Environmental Monitoring and Assessment
    Program (EMAP), 15
Environmental Protection Agency (EPA), 15
Environmental stress, and forest decline, 368–369,
    369*b*, 370*f*
EPA. *See* Environmental Protection Agency
Ephemeroptera, 425, 427*f*, 430
Epicoccum, in stored seedlings, 264
Epidemics, since 1890, 186*t*
Epiphytotics, since 1890, 185, 186*t*
EPVs. *See* Entomopoxviruses
Equilibrium, 43–44
  and disturbance, 43–44
Equilibrium moisture content, of fuel particles, 57
Equilibrium position, 458
Eriophyidae, 417
Escaped fire, 157
Escaped Fire Situation Analysis (EFSA), 117
Escherich, Karl, 405, 406
Esophagus, insect, 421
Essential elements, in plants, 217, 217*t*, 218*b*. *See
    also* Nutrient(s)
Eucalyptus, ectomycorrhizal vs. uninfected
    root, 268*f*
Eucalyptus longhorned borer (*Phoracantha
    semipunctata*), 527–528, 529*f*, 564
Eurkarya, 228
Europe
  acid precipitation, 206

Europe—*Cont.*
  air pollution damage, 13
  annosus root and butt rot, 288
  Armillaria root disease, 287
  diebacks and declines, 365, 366*t*, 375, 377*b*
  Dutch elm disease, 358
  forest damage control, 14, 214
  forest entomology history, 403–405
  forest health, 7, 17–18, 18*f*
  forest health monitoring, 17–18
  nitrogen pollution, 14, 206, 375
  root diseases, 276, 278*t*–279*t*
European Community, 17–18
European crane fly, 553*f*
European pine sawfly (*Diprion similis*), 479*f*
European pine shoot beetle (*Tomicus pineperda*),
    563–564
European pine shoot moth (*Rhyacionia buoliana*),
    555*f*, 556, 557
European Union, Nitrogen Saturation Experiments
    on Forest Ecosystems (NITREX), 206
Evaders, 73
Exotic organisms. *See* Introduced *entries*
Explosives, in fire control, 120
Exponential curve of disease progression, 390, 392*f*

Facultative parasites, 192, 193, 193*b*
Facultative pathogen, 453
Facultative saprophytes, 192, 193, 193*b*
Fallowing, 266
False powderpost beetles, 538
False spring, and xylem cavitation injury, 373
FARSITE model (of fire behavior prediction), 67,
    68*f*, 136*t*
Fasciation, 347
Fatalities, fire, causes of, 121, 122, 123*b*
Federal agencies
  fire prevention campaigns, 108–110
  and fire protection, 106, 108
Feeder root diseases, 277, 279*t*, 294–295. *See also
    specific diseases*
  management, 302, 303
Felt, Ephraim P., 398, 400–401, 406
Fertilizers, 220, 302, 387
  examples of use, 385*t*
  overfertilization, 220
FHM. *See* Forest Health Monitoring (FHM)
    program
Fiber saturation point, 394
FIDS. *See* Forest Insect and Disease Survey
Filamentous ascomycetes, 241, 242
  fruiting structures, 242, 243*f*
  life cycle, 242, 244*f*
Filter chamber, insect, 422
Fine filter management approaches, 46
Fir(s). *See also individual species*
  defoliator outbreaks, 469–470

Fir(s)—*Cont.*
   insect pests, 561–562
   western spruce budworm attacks, 481–483,
      483*t*, 484*f*
Fir engraver beetle (*Scolytus ventralis*), 512–513
Fire(s)
   behavior. *See* Fire behavior
   causes, 108, 109*f*
   crown. *See* Crown fires
   decay fungi interactions, 356–357, 407, 408*f*
   disease and, 195–196
   dwarf mistletoe and, 340
   ecological effects. *See* Fire ecology
   fatalities in, 121, 122, 123*b*
   and forest dieback/decline, 376–378
   and forest health, 6, 10, 49, 69
   fuels. *See* Fuel(s), fire
   ground, 58*t*, 59
   insect interactions, 10, 81, 296–297, 356–357,
      376–377, 407, 408*f*
   and landscape ecology indices, 78, 80*t*
   and landscape stability and equilibrium, 43–44
   plant adaptations to, 71–73, 72*f*, 73*f*, 74*t*
   prescribed. *See* Prescribed fire(s)
   and root diseases/pathogens, 296–297, 303
   statistics, 105*f*
   and succession, 42. *See also* Stand development
      patterns
   terms for, 107*b*
   tree damage caused by, 221
   tree mortality prediction, 74–76, 75*f*
      models, 76, 78*t*, 79*t*
   wind and, 182
Fire behavior, 49–69
   combustion process, 49–50, 50*f*
   crown, 58*t*, 63–66, 131–134, 132*t*, 134*f*
   definition of, 50
   elements of, 49–59
   in fatal fires, 122
   and fire danger ratings, 67–68, 69*f*
   and fireline intensity. *See* Fireline intensity
   and fuel treatments, 130, 131–139
   and fuels, 54–59
   and geographic information systems, 67, 68*f*
   modeling of, 60–63, 61*f*, 63*t*, 67, 68*f*, 135*t*,
      136*t*, 137*t*
   prediction of, 59–68
      at ecosystem level, 71, 72*f*
      fire characteristics chart, 115*f*, 115–116
      in subalpine forests, 88
   spotting, 50, 66–67
   spread. *See* Fire spread
   surface fires, 59–63
   and topography, 54
   and weather, 50–54
Fire blight, 342*t*, 344
Fire characteristics chart, 115*f*, 115–116

Fire climate(s), 54
   North American, 54, 55*t*, 56*f*
Fire control. *See also* Fire management; Fire
      suppression
   equipment, 117–121, 122
   federal responsibilities, 106, 108
   forest health problems from, 10
   functional organization, 112–114, 113*f*
   policy evolution, 103–106, 149
   safety issues, 121–123
   state-level, 106–108
   strategies, 114–117
   in urban-wildland interface, 126
Fire cycle, 84
Fire danger rating systems, 67–68, 69*f*
Fire ecology, 71–98. *See also* Fire regime(s);
      Fire(s); Stand development patterns
   of coastal Douglas-fir forests, 83–87
   ecosystem productivity changes, 31
   of jack pine forests, 94–98
   and landscape, 78, 80*t*
   of longleaf pine forests, 92–94
   plant adaptations, 71–73, 74*t*
   of ponderosa pine forests, 80–83
   tree mortality prediction, 74–76
   of western subalpine forests, 87–92
Fire equipment, 117–121
   safety-related, 122
Fire exclusion policy, 104–105, 105*f*
Fire frequency. *See also* Fire regime(s)
   and forest dieback/decline, 376–378
Fire ignition devices, 121, 142–143
Fire intensity, 59. *See also* Fireline intensity
   and fire adaptations of plants, 73
   and fire severity, 77–78
Fire lookouts, 110, 112*f*
Fire management, 102–127, 130–158
   constraints to fire use, 155–158
   control. *See* Fire control
   detection, 110–112
   escaped fire and, 157
   federal responsibilities, 106, 108
   by fire type and fireline intensity, 58*t*
   hazard reduction, 103, 131. *See also* Fuel
      treatments
   individual species needs and, 157–158
   integrated, 106, 130
   "let it burn" policies, 148
   managed wildland fire (pnf), 107, 148–154
   objectives, 148–149
   policy, evolution of, 103–106
   prescribed fire. *See* Prescribed fire(s)
   prevention, 104–105, 108–110
   rehabilitation, 123–125
   risk evaluation, 102–103
   safety issues, 121–123
   smoke management, 155–157

Fire management—*Cont.*
  state responsibilities, 106–108
  strategy mix, 102
  suppression. *See* Fire suppression
  terminology, 107*b*
  in urban-wildland interface, 125–127
  values and, 103
Fire management triangle, 102–103, 103*f*
Fire prevention, 108–110
  conceptual approaches, 108, 110*t*
  history of, 104–105
Fire regime(s), 71, 77–78
  of boreal forest region, 94–95
  of coastal Douglas-fir forests, 83–84
  and fire severity, 77–78
  and forest dieback/decline, 376–378
  of jack pine forests, 94–95
  and landscape ecology indices, 80*t*
  of longleaf pine forests, 93
  North American, 77–80
  of ponderosa pine communities, 81
  of subalpine fir forests, 88–89
Fire retardants, 120
Fire return intervals, 84
Fire scars, 81, 195, 221
Fire severity, 77–78, 80*t*
Fire shelters, 122
Fire spread, 59
  BEHAVE and fuel models of, 60–63, 61*f*,
    63*t*, 136*t*
  in crown, 65, 133–134, 134*f*
    fuel treatments for, 133–134, 135–139
  to crown. *See* Crown fire(s), initiation
  FARSITE model of, 67, 68*f*
  patterns, 59, 60*f*
  rate of, 59, 61–62, 65, 133
  by spotting, 50, 66–67
Fire suppression, 102, 112–123. *See also*
    Fire control
  effects, 83, 123, 296–297, 376–378
  equipment and materials, 117–121, 122
  and forest dieback/decline, 376–378
  functional organization, 112–114, 113*f*
  and insect/root rot pathogen complexes,
    296–297
  policy history, 104–106
  safety issues, 121–123
  strategies, 114–117
  tactics, 114–116, 117*f*–119*f*
  and urban-wildland interface, 126
Fire triangle, 49, 50*f*
Fire weather, 50–54
Fire Weather Index System (FWI) (Canada), 68
FIRE-BGC model, 76
Firefighters
  physical conditioning, 122–123
  safety equipment and clothing, 122

Fireline(s)
  construction, 114, 116, 120
  natural barriers as, 116
Fireline intensity, 58*t*, 59–60. *See also* Flame
    length
  and crown fire initiation, 58*t*, 65, 66*f*, 66*t*,
    131, 132*t*
  management, 58*t*, 61*f*
  modeling, 61*f*, 61–62, 63*f*
  and scorch height, 74–75
  and tree damage and mortality, 74–75
Fish and Wildlife Service, wilderness and land
    areas managed, 151*t*
Fitch, Asa, 406
Flagging
  heat- and dryness-related, 200
  wind-related, 163*f*, 163–164, 164*f*
Flame length. *See also* Fireline intensity
  after thinning, 140*f*
  and crown fire initiation, 58*t*, 65, 66*t*, 131, 132*t*
  in fuel treatment strategies, compared, 137*f*
  in prescribed fire planning, 144
Flamethrowers, 121, 143
Flanking attack (of fire), 116, 117*f*
Flanking fire, 139, 141*t*, 142, 142*f*
Flatheaded wood borers, 533
Fleas, 428*f*
Flies, 428*f*, 435
  morphology, 419, 420, 435
Flooding, 39–40, 203, 223
Florel, 340
Floristics models, 34, 35*f*
Fluting, 165
Foams, fire-retardant, 120
Foehn winds, 52–53, 162
FOFEM (First Order Fire Effects Model), 76, 79*t*
Foliage. *See also* Defoliation; Defoliators
  moisture, and crown fire initiation, 63–65, 66*t*,
    131, 132*t*
Foliage diseases, 309–323. *See also* Rust(s)
  conifers, 309–310, 311*t*, 312*f*, 318–323
  fungi causing, 245*t*, 309–310, 311*t*
    life cycle, 313, 314*f*
  hardwood, 309, 310, 311*t*, 312*f*, 313–317
  hosts, 311*t*
  management, 386–388. *See also* specific diseases
  symptoms, 310–313
Foliar mass, 133
Fomes annosus, 277
Fomitopsis annosa, 277
Forest biomes, 24–29
Forest classification, 24–29
  biomes, 24
  by current and potential vegetation, 27–29
  zonation variables, 25–27
Forest declines. *See* Declines, forest
Forest disease(s). *See* Disease(s), forest

Forest ecosystems, 24–46, 436–439, 438*f*
  classifications, 24–29
  diseases and, 185
  disturbance agents, natural, 1, 24
  disturbance effects, 34–42
  ecological diversity in, 42, 43*f*
  fires effects. *See also* Fire ecology
    prediction of, 71, 72*f*
  management. *See* Ecosystem management
  production ecology/energetics, 29–33
  root diseases and, 275, 307
  stability, 43–44
  succession in, 23, 33–44
  and sustainability concept, 43
Forest entomology, 400–410, 411*f. See also*
  Entomology; Insect *entries*
  future of, 409–410
  history of, 401–409
    issues in, 404*f*
  research, 406–407
Forest Experiment Stations, 188
Forest fragments, management as "islands," 46
Forest health, xv, 1–20
  air pollution and, 12–14
  causes of problems, 6
  definitions of, 2, 3*b*
  fire and, 6, 10, 49, 69, 149
  human activity influences on, case studies, 9–14
  ignoring in planning efforts, 46
  indicators of, 3–4
  in inland western U.S., 9–11, 16
  introduced pathogens and insects and, 11–12
  symptoms of problems, 4–6
  tree death and, 3, 5*f*
  and urban-wildland interface, 126
  in U.S., 7, 16
  worldwide, 6–8
Forest Health (organization), 189
Forest Health Assessment Program, 44
Forest health monitoring networks, 15–19, 44
  in Canada, 18–19
  in U.S., 15–16
  in western Europe, 17–18
Forest Health Monitoring (FHM) program, 15–16
Forest Insect and Disease Survey (FIDS), 18–19
Forest management
  adverse effects of practices
    in inland western U.S., 9–10
    in pathogen resistance, 252–253
    in root disease, 275, 285, 288, 295
    in rusts, 325–326
  for air pollution, 212–214
  coarse filter approaches to, 45–46
  of commercial forests, 399–400, 402*f*
  fine filter approaches to, 46
  for fire. *See* Fire management
  intensive, 8, 436

Forest management—*Cont.*
  island biogeography theory in, 46
  objectives, for public lands, 148–149
  for pathogens, 187
  planning approaches to, 46
  wind and. *See* Wind management strategies
Forest pathology
  history, 188–189
  organization for research, 188–189
  resources, 196–197
Forest Pest Control Act of 1947, 188
Forest production, 29–33
  measurements of, 31, 32*t*
  organisms in, 29–30, 30*f*
  primary, 30–31
Forest series, 28
Forest Service (USDA)
  Division of Forest Disease Research, 188
  entomological research, 406–407
  fire management options, 106
  state fire programs, 108
  wilderness and land areas managed, 151*t*
Forest structure, 130. *See also* Fuel(s), fire
  manipulations for fire management, 130,
    131–139
Forest Vegetation Simulator (FVS), 304, 392
Forest zonation, 25–27, 27*f*, 28*f*
Forestry Canada, 189
Formica ants, 457
Formulations, insecticide, 446–447
Die Forstinsekten (Ratzeburg), 403, 405
Die Forstinsekten Mitteleuropas (Escherich), 405
Fossorial legs, 421
France, 285–287
Frass, 462
Freon, 204, 205*t*
Frontal attack (of fire), 116, 117*f*
Frontal winds, 51
Frost damage, 200–201, 201*f*, 256, 373
Frost heave, 256
Fruiting bodies, 242, 243*f*
  decay fungi, 350, 350*f*, 351*f*, 354*f*
Fuel(s) (fire), 54–59
  arrangement, 59
  consumption, 55–59, 57*f*
  and fire control, 68
  and fire spread, 60–63, 62*t*, 63*t*, 65
  hazard reduction strategies using. *See* Fuel
    treatments
  heat energy, 54
  manipulation. *See* Fuel treatments
  moisture, 56, 57–59
Fuel load, 56
Fuel models (of fire spread), 60–63, 62*t*, 63*t*,
  135, 136*t*
Fuel treatments, 68, 102, 130, 131–139
  in boundary areas, 152

Fuel treatments—*Cont.*
  cut and scatter, 136, 136*t*, 137*f*, 138
  effectiveness, compared, 135–139, 136*f*, 137*f*
  fire behavior effects, 130, 132–139
    modeling of, 62, 135, 135*t*, 136*t*, 137*t*, 140*f*
  increase of height to crown base, 131–132
  pile burning, 135–136, 136*t*, 137*f*
  prescribed fires. *See* Prescribed fire(s)
  salvage logging, 138*b*
  surface fuel reduction, 132
  thinning, 133–134, 135*f*, 135*t*, 136*t*, 136–139,
    137*f*, 140*f*
Fuelbreaks, 147–148, 149*f*
  natural, 152, 154*f*
Fumigant insecticides, 446
Fumigation, 207, 265, 297, 388
Fungal diseases, 238*t*–239*t*. *See also specific
    diseases, parts of tree, phyla, and
    organisms*
  ascomycete, 242–244, 245*t*
  basidiomycete, 246, 249*t*
  imperfect fungi, 242–244, 245*t*
  management, 264–266. *See also specific
    diseases*
  mechanism of action, 229–230
  oomycete, 239*t*, 248–251
  in roots, 258–261, 276–307, 278*t*–279*t*
  in seedlings, 257–261
  in seeds, 257
  in shoots, 261–264
  signs of, 193, 194*t*
  zygomycete, 240–241
Fungi, 187, 228–231
  beetle associations, 491–493, 518, 520, 522
  classifications, 192, 193, 193*b*, 237–241, 242
  conditions for penetration and growth, 252
  decay. *See* Decay fungi
  definition, 230
  disease-causing, 228, 237, 238*t*–239*t*, 240. *See
    also specific diseases, parts of tree, phyla,
    and organisms*
  in forest energy flow, 32, 33*f*
  insect interactions, 251–252, 296, 297*t*,
    356–357, 407, 408*f*
  as insect pest control, 453–455
    gypsy moth, 475
  introductions, 388–389
  life cycles, 240. *See also specific phyla and
    organisms*
  morphology and reproductive features, 229*f*,
    230–231, 240
  mycorrhizal. *See* Mycorrhizal fungi
  and nematodes, 236–237
  smothering, 263*b*, 264
  spread of, 251–252
  storage (seed-inhabiting), 257
Fungi (kingdom), 237, 238*t*–239*t*

Fungi imperfecti (Deuteromycota), 239*t*, 240, 246
  classification, 239*t*, 242
  diseases caused by, 239*t*, 242–244, 245*t*
    foliage, 309, 313
    fruiting bodies, 242, 243*f*
  as insect pest control, 454–455
  in seedlings, 257
Fungicides, 265, 388, 395
  natural, 266
  for rusts, 331
Fungus gnats, 553*f*
Furcula, 426
Fusarium
  cankers, 342*t*
  in seedlings, 257, 258, 261, 266
    and black root disease, 260
    and hypocotyl rot, 262–264
    life cycle, 259*f*
    and root rot, 258*b*, 258–259
    stored, 264
    and top blight, 262
Fusiform rusts, 323, 325, 325*f*
  life cycle, 329
  management, 331
FVS. *See* Forest Vegetation Simulator (FVS)
FWI. *See* Fire Weather Index System (FWI)

Galea, 420
Gall(s), 347–348, 553, 554*f*
Gall mites, 417
Gamma diversity, 42, 43*f*
Ganoderma applanatum, 351*f*
Ganoderma lucidum root disease, 277, 279*t*
Gap rotation, 378
Generation, insect, 424
Genetic diversity, 43*f*
Genetic resistance, 384, 388. *See also* Resistant
    strains/species
  breeding trees for, 252, 300, 388
  selection for growth and, 252–253
Geographic information systems (GIS), and fire
    prediction, 67, 68*f*
Germany
  defoliation, 17, 18*f*
  early entomologists, 403–405
  early foresters, 401–402
  forest diebacks and declines, 365, 366*t*
    Waldsterben, 375, 377*b*
  forest health, 7
  plantation forests, 8
Girdling, by animals, 224, 225, 225*f*
GIS. *See* Geographic information systems
Glaze injury, 216
Global warming, 203, 214
  and forest dieback, 369–375, 373*f*, 374*f*
  and tree damage, 203–204

Gnathothrichus beetles, 518*f,* 520
  damage by, 522*f*
  management, 520
Gnats. *See also* Diptera
  fungus, 553*f*
Golden buprestid (*Buprestis aurulenta*), 528–530
Gophers, tree damage by, 225
Gouty pitch midge (*Cecidoymia piniinopsis*), 559
GPP. *See* Gross primary production (GPP)
Gradient winds, 52
Granules, insecticide, 447
Grass seeding, post-fire, 124–125, 125*f*
Grass stage, of longleaf pine, 93
Grasshoppers, 420*f,* 427*f,* 429*f,* 431
Gray mold, 261–262, 263*b*
  control, 265
Great Smoky Mountains National Park, forest
  damage, 14, 14*f*
"Greenup," 139
Gross primary production (GPP), 31
Ground fires, 58*t,* 59
Grylidae, 431
Gypsy moth (*Lymantria dispar*), 12, 13*f,* 470,
  473–476, 475*f,* 563
  damage by, 473–474
  host preferences, 462, 463*t*
  management, 475–476
  susceptibility to, 475

Habitat typing, 28
Halteres, 435
Hammocks, longleaf pine, 94
Hardwoods. *See also individual species*
  ambrosia beetles, 522–524
  cankers, 341–346, 342*t*
  decay, 354
  diebacks/declines/blights, 365, 372*f*
  foliage diseases, 309, 310, 311*t,* 313–317
    symptoms, 310, 312*f*
  hurricane effects, case study, 170–172
  nursery soil for, ideal, 255
  vascular wilts, 359*t*
Hartig, Robert, 188
Haustoria, 252, 310
Haustorial strands, 231
Hawaiian Ohia forests, 367*t,* 378
Hawkins Act (Florida) (1977), 106
Hawksworth scheme (dwarf mistletoe rating),
  338, 339*f*
Hazard reduction, in fire management, 103, 103*f*
Heading fire, 139
Heart rot decay, 352*t,* 356
Heat transfer, 50
Heat, tree damage by, 200, 256
Heavy metal toxicity, 221
Hedgcock, G. G., 188

Height to crown base
  and crown fire initiation, 63–65, 66*t,* 131, 132*t*
  increasing, for fire prevention, 131–132
Hemiptera, 427*f,* 429*f,* 432, 548, 553, 554
  mouthparts, 419–420
Hemlock(s). *See also* Eastern hemlock; Mountain
  hemlock; Western hemlock
  annosus root and butt rot, 289*f,* 290
Hemlock borer (*Melanophila fulvoguttata*), 463
Hemlock looper, 368, 486–487. *See also* Eastern
  hemlock looper; Western hemlock looper
  control, 486–487
Hemlock woolly adelgid, 564
Hemocoel, insect, 423
Hemolymph, 423
Herbicides, 214–215, 388
  for dwarf mistletoe, 340
Herbivores, 29, 30*f*
  and forest productivity, 31–32
Herpotrichia, 323
Heterobasidion annosum (annosus root and butt
  rot), 276–277, 279*t,* 288–291
  distribution, 288
  effects, 289–291
  insect associations, 297*f*
  management, 300, 303, 385, 386, 388
  modeling of effects, 304–305, 305*f*
  relative conifer susceptibility, 299*t*
  spread mechanisms, 288, 289*f*
  symptoms and signs, 281*t,* 288, 289*f,* 290*f*
  thinning and, 301
Heterobasidion root disease, 276, 288
Heteroptera, 429*f,* 432
Heterotrophic organisms, 192
High-hazard site avoidance, 385*t*
  for root disease management, 302
High-hazard time avoidance, 385*t*
High-pressure systems, 51
High-severity fire regime, 78
Hitting the head (of fire), 116, 117*f*
Hobcaw Forest, South Carolina, 173*t*
Homoptera, 429*f,* 432
Honeydew, 317
Hopkins, Andrew D., 406, 407*f*
  insect damage reports, 398, 399*t*
Hornets, 434
Horntailed wasps, 533–534, 535*f,* 535*t*
Hotspotting, 116, 117*f*
Houston model of forest decline, 368–369,
  369*b,* 370*f*
Human activities, case studies of forest effects, 9–14
Hunting spiders, 414
Hurricane(s)
  of 1938 (New England), 162, 162*f,* 171–172
  frequency, 162
  Hugo (1989), 172, 173, 173*t*
  salt water transport by, 221–222

Hurricane(s)—*Cont.*
  successional effects, 41–42
    case studies, 170–172, 172–173
  windspeed, 161*t*
Hydrogen fluoride (HF), 204, 205*t*
  tree damage symptoms, 210*b*
Hylobius weevils, 554
Hylurgopinus rufipes, 358–361
Hymenomycetes, 246, 247*f,* 248*f*
Hymenoptera, 428*f,* 434, 533, 548. *See also*
    individual species
Hypertrophy, 194, 194*t*
Hyphae, 230*f,* 230–231
  aseptate, 237
  septate, 230, 230*f*
Hypocotyl rot, 262–264, 263*b*
Hypodermella laricis, 322
Hypoxylon canker, 341, 342*t,* 344*f,* 345
Hysterothecia, 242, 318

Ice, tree damage by, 216
ICS. *See* Incident Command System
Immunity, to pathogens, 193
Imperfect fungi. *See* Fungi imperfecti
Incense-cedar (*Calocedrus decurrens*), windthrow
    hazard rating system, 168
Incident Command System (ICS), 112–114,
    113*f,* 123
  for defoliator control, 468, 470*f*
Inciting factors, 369, 371*f,* 379
Independent crown fires, 66
Indexes of fire potential
  in FWI (Canada), 67, 68, 69*f*
  in NFDRS (U.S.), 67, 68, 69*f*
Indicator species, 28
Indirect attack (of fire), 116, 118*f*
Individualistic theory of plant succession, 34
Initial floristics model, 34, 35*f*
Inland western forests
  Armillaria root disease, 285
  health/health problems, 7, 9–11, 16
Inoculation
  ectomycorrhizal, 271–273
  increasing decay for, 390
Inoculum, ectomycorrhizal, 271–273
Inoculum reduction/removal, 385*t,* 386
  for root disease management, 297–298
Inonotus tomentosus root disease, 277, 279*t*
Insect(s), 413–435. *See also individual species*
  biology. *See* Insect biology
  classification. *See* Insect classification
  climate change and, 204
  control. *See* Insect pest management
  defoliating. *See* Defoliators
  disease interactions, 187, 195–196
  disease outbreaks caused by, 185, 186*t*

Insect(s)—*Cont.*
  diversity, 413
  economic damage by, calculation of,
      439–440, 441*b*
  fire interactions, 10, 81, 356–357, 376–377,
      407, 408*f*
  and forest productivity, 31–32
  fungi interactions, 251–252, 296, 297*f,*
      356–357, 407, 408*f*
  introduced. *See* Introduced insects
  losses due to, historical, 398, 399*t*
  management. *See* Insect pest management
  mortality, and density dependence, 449, 450*t*
  nematode spread by, 237
  nursery. *See* Nursery insect pests
  pest types, 439, 440*f*
  phylogeny, 413
  plantation, 553–559, 563–564
  population dynamics, 404*f,* 411*f*
  reproductive potential equation, 439–440
  roles in forest, 439
  root disease interactions, 296–297, 297*t*
      modeling of, 304–305, 305*f*
  seed orchard, 543, 544–547, 545*f,* 559
  study of, 413–414. *See also* Forest entomology
  successional effects, 42
  symptoms and signs, 313
  systematics, 413. *See also* Insect biology; Insect
      classification
  and wind, 181, 182
  wind as carrier of, 182
  wood-boring. *See* Wood borers
Insect biology, 418–426. *See also specific order*
  Arthropoda (phylum-level), 414
  circulatory system, 423
  coordination system, 424
  development, 424–426
  digestive system, 421–422
  excretory system, 423
  external morphology, 418–421
  internal morphology, 421–424, 422*f*
  legs, 421
  Mandibulata (group-level), 414, 417–418
  metamorphosis, 424–426, 425*f*
  mouthparts, 419–420, 420*f*
  reproductive system, 423–424
  respiratory system, 423
  thorax, 420–421
Insect classification, 413, 415*f*
  Mandibulata (group), 414, 415*f,* 417–418
  orders, 426–435, 427*f*–429*f*
Insect pest management, 440–458, 449–457. *See
    also specific insects and control methods*
  biological control, 449–457
  chemical control, 409–410, 442–447
  future of, 409–410, 411*f*
  integrated pest management, 457–459

Insect pest management—*Cont.*
  mechanical/physical control, 442
  on plantations, 543
  quarantine, 567, 568*t*, 569*f*
  semiochemical control, 447–449, 448*f*
Insecta (class), 415*f*, 418
Insecticides, 442–447
  botanicals, 443
  carbamates, 445*f*, 445–446
  classifications, 442, 443, 444*t*
  for container tree seedlings, 552*t*
  formulations, 446–447
  fumigants, 446
  organochlorines, 443–444
  organophosphates, 444, 445*f*
  problems with use, 409–410
  synthetic pyrethroids, 446
  toxicity, 442–443, 444*t*
Insects Affecting Park and Woodland Trees
  (Felt), 406
Insects Injurious to Forest and Shade Trees
  (Packard), 406
Instar, 424
Integrated fire management, 106, 130
Integrated pest management (IPM), 384, 440,
  457–459, 459*f*
  of bark beetles, 495–498
  guidelines, 458
  for nurseries, 548–552
  for plantations, 556, 557, 558*f*
  for seed orchards, 544–547, 545*f*
Intensive forestry, 436
  problems of, 8
Intermediate operations, 543
Intracellular freezing, 201
Introduced insects, 11–12, 388–389, 561–564, 562*t*
  control, 185, 449–452
  disease outbreaks, 185, 186*t*
  as insect pest control, 449–452
  quarantine to prevent, 388–389, 567, 568*t*, 569*f*
  wood imports and, 566–567
Introduced pathogens, 11–12, 388–389
  control, 185
  disease outbreaks, 185, 186*t*
Introduced plants, 8
Invaders, 73
Ips beetles, 491, 508–512
  management, 504, 506*t*, 509–512
Ips pini, 508–511
  activity patterns, 511*f*
  flight periods, 512*f*
  management, 509–511
  population fluctuations, 510, 513*f*
Irish potato blight, 250–251
Irrigation systems
  salt damage caused by, 222
  in wildland fire control, 121

Island biogeography theory, 46
Islands, longleaf pine, 94
Isoptera, 429*f*, 431
Ixodidae, 416

Jack pine (*Pinus banksiana*), 94–98
  fire regime, 78, 94–95
  stand development patterns, 96–98
Japan, 237
Jarrah, 354, 374

Kairomones, 448*f*
Katydids, 427*f*
Keep America Green campaign, 110
Kelthane (Dicofol), 444
Kirtland's warbler (*Dendroica kirtlandii*), 98
Koch's postulates, 194–195, 195*b*
Krummholz, 164

Labial palps, 419, 420*f*
Labium, 419, 420*f*
Labrum, 419, 420*f*
Lacewings, 428*f*, 435
Laetiporus sulphureus fruiting body, 354*f*
Laminated root rot. *See Phellinus weirii*
Land management systems, 393
Land-grant institutions, and forest
  entomology, 407
Landscape ecology indices, and fire, 78, 80*t*
Landscape Visualization System (LVS), 393
Landscaping, fire-safe, 126, 127*f*
Larch casebearer (*Coleophora laricella*), 462, 469,
  479–480
  control, 430, 451*f*, 480
Larch needle diseases, 322
Larch sawfly (*Pristiphora erichsonii*), 477–479
  control, 479
Larvae, 426
Larval sampling, 467, 469*f*
Lasiocampidae, 480
Latent pests, 439, 440*f*
Leadcable borer (*Scobicia declivis*), 338
Leaf blisters, 311*t*, 317
Leaf chewers, 462
Leaf scorch, 203
Leaf spots, 310, 311*t*
Leafcutting ants, 195–196, 559
Leafhoppers, 427*f*, 429*f*, 432
Leafminers, 462
Leafy mistletoes. *See* True mistletoes
Leave trees, 180–181, 390
Lecanosticta acicola, 322
Legs, insect, 421

Leopold, Aldo, 42
Lepidoptera, 428f, 433–434, 463
  galea (mouthpart), 420
  seed-damaging, 544, 547–548
  wood-borers, 526
Leptoglossus corculus, 545, 546f, 549t
Leptographium proceri, 277, 279t, 295
Leptographium root diseases, 276, 277, 279t, 295
Leptographium wageneri (black stain root
    disease), 277, 279t, 285, 295, 296
  insect associations, 296, 297f
  relative conifer susceptibility, 299t
  symptoms and signs, 281t, 295, 296f
  thinning for management, 301
"Let it burn" policies, 148
Leucopaxillus cerealis, 271
Lice, 427f, 429f
Light burning, history of, 104
Lightning damage, 216
Lignin, 348, 349
Liming, 14, 214
Limonene, 443
Livestock, tree damage by, 226
Loblolly pine (*Pinus taeda*), wind and,
    172–173, 173t
Lodgepole pine (*Pinus contorta*)
  dwarf mistletoe, 232f
  fungal shoot diseases, 262–264
  interactions of fire, decay fungi, and insects,
    356–357, 407, 408f
  mountain pine beetle, 502–504
  needle miner, 462
  stand development patterns, 89, 90–91, 91f
  weevil, 556
  in western forests, 10, 12f, 88
  western gall rust, 323, 324t, 326f
Log(s)
  decay, 353t
  export, 516, 517f
    and beetle damage, 516–517
  wood borers, 530–536
Logging
  salvage, 138b
  sanitation, 507–508
Logistical curve of disease progression,
    391, 393f
Longleaf pine (*Pinus palustris*), 92–94
  fire regime, 78, 93
  stand development patterns, 93–94
  wind effects, 172–173, 173t
Lophodermium needle cast, 319–320
Los Angeles basin ozone pollution, 208, 210
Low-pressure systems, 51
Low-severity fire regime, 77
LVS. *See* Landscape Visualization System
Lyctidae, 538
Lygus bug, 553f

Machinery, tree damage by, 216
Mack Lake fire of 1980, 102
Macronutrients, 217, 217t, 218b
  deficiency symptoms, 219b
Macrophomina phaseoli, 260–261
Madrone canker, 342t, 346
Maggots, 435
Magnesium (Mg), 217t, 218b
  deficiency, 219b, 220
Mallophaga, 429f
Malpighian tubules, 422f, 423
Managed wildland fire (pnf), 107, 148–154. *See
    also* Fire management
  area burned, 1968-1994, 151f
  constraints to use, 155–158
  escaped fire and, 157
  evaluation, 152, 155f
  future of, 153–154
  operations, 152–153
  planning, 150–152
  smoke and, 155–157
  term, 107b
  zoning strategy, 152, 153f
Mandibles, 419–420, 420f
Mandibulata, 414, 415f, 417–418
Manion model of forest decline, 369, 371f
  adaptations, 378, 379t, 379–381, 380f
Mantidae, 429f, 431–432
Mantodea, 429f
Maples
  climate change and, 373
  dieback, 370–373, 372f, 374f
  monitoring program, 19
  tar spots, 311t, 314f, 317
Mass transfer, 50
  and spotting, 66
Maxillae, 419–420, 420f
Maxillary palps, 419, 420f
Mayflies, 425, 427f, 430
MCH (3-methyl-2-cyclohexen-1-one), 449
Mechanical control, of insects, 442
Mechanical injuries, 215–217
Mecoptera, 428f
Megaloptera, 428f
Melampsora rusts, 263b, 264, 323, 324t, 326f, 327
  life cycle, 329, 330f
Melanophila wood borers, 533, 534f
Meria laricis, 322
Metamorphosis, 424–426, 425f
Metarrhizium anisopliae, 454–455
Methoxychlor, 444
Methyl bromide, 265, 300–301, 388, 389, 446
3-methyl-2-cyclohexen-1-one (MCH), 449
Micronutrients, 217, 217t, 218b
  deficiency symptoms, 219b
Microsclerotia, 231
Midges, 559. *See also* Diptera

Millipedes, 418
Mistletoes, 231*t*, 231–233, 333–340
    dwarf, 16, 231, 232*f*, 333–340
    true, 231–233, 232*f*, 333–334
Mites, 414, 416, 416*f*, 417
    control (acaracides), 443, 444
    symptoms and signs, 313
Mixed-severity fire regime, 78
MLOs (mycoplasma-like organisms). *See*
    Phytoplasmas
Moderate-severity fire regime, 78
Moisture
    air, 51, 53, 53*f*
    foliar, and crown fire initiation, 63–65, 66*t*,
        131, 132*t*
    in fuels, 56, 57–59
    soil, 200, 203
    tree damage caused by, 200, 202–203
    in wood, 362, 394
Molds
    fungi causing, 245*t*, 257, 395, 396*f*
    in seeds (storage fungi), 257
    in stored seedlings, 264
Mono winds. *See* Foehn winds
Monochamus wood borers, 532
Monoclimax theory, 34
Mortality waves, 5*f*
Mosaics, 234
Mosquitoes. *See* Flies
Moths, 420, 428*f*, 433–434. *See also individual*
    species
Mount St. Helens eruption, measurement of
    severity, 38–39, 41*f*
Mountain beavers, damage by, 225–226
Mountain hemlock (*Tsuga mertensiana*), 87, 88
Mountain pine beetle (*Dendroctonus ponderosae*),
    10, 501*t*, 502–504
    and dwarf mistletoe, 339
    interactions with fire and decay fungi, 356–357,
        407, 408*f*
    management, 504, 505*f*, 506*t*
    and root disease, 296, 297*t*
Mueller-Dombois natural dieback theory, 378,
    379–381
Mulching, post-fire, 125
Multiple Use and Sustainable Yield Act, 399
Multiple-pathway succession models, 36–42
    advantages, 36–37
    simulations using, 37*t*, 38*f*–40*f*
Mutual symbiosis, 192, 193*b*
Mycangia, 252
Mycelium, 230, 240, 252
Mycoplasma-like organisms (MLOs). *See*
    Phytoplasmas
Mycorrhizae, 188, 192–193, 255, 266–273, 267*t*
    fungi. *See* Mycorrhizal fungi
    hosts, 267*t*, 269, 270

Mycorrhizae—*Cont.*
    human impacts on, 273
    management, 271–273
    roles, 270–271
    types, 266–270, 267*t*
Mycorrhizal associations. *See* Mycorrhizae
Mycorrhizal fungi, 229, 246, 255, 267*t*, 270
    ectomycorrhizal, 266, 270, 271
        seedling inoculation with, 271–273
    endomycorrhizal, 266, 270
    human impacts on, 273
Mycosphaerella dearnessii, 322
Mycota, 240

NAMP. *See* North American Sugar Maple Project
    (NAMP)
National Acid Rain Precipitation Assessment
    Program (NAPAP), 13, 14
National Fire-Danger Rating System (NFDRS),
    67–68, 69*f*
National forests, 104. *See also* Public-land forestry
    fire control policy history, 104
    Forest Health Monitoring program, 15–16
    multiple-use management, 399
National Interagency Incident Management
    System (NIMS), 112–114
National Park Service (NPS)
    land areas managed, 151*t*
    Western Region, fire monitoring levels, 146
National Wildfire Coordinating Group
    (NWCG), 107
Native Americans, burnings by, 83
Nattrassia mangiferea, 342*t*, 346
Necrosis, 194, 194*t*
Nectria cankers, 342*t*, 343*t*, 346, 564
Needle casts (blights), 310, 318–323
    brown spot, 322
    diplodia tip blight, 321–322
    dothistroma, 320–321, 321*f*
    elytroderma, 320, 320*f*
    larch, 322
    lophodermium, 319–320
    rhabdocline, 318
    snow molds, 323
    Swiss, 318–319, 319*f*
Nematodes, 187, 228, 235
    as insect pest control, 456
    morphology and reproductive features, 229*f*,
        235–237, 236*f*
    plant-parasitic, 235–237, 236*f*, 261
    root diseases from, 260, 261, 261*b*
    signs, 193, 194*t*
Net primary production (NPP), 31, 32*t*
Neuroptera, 428*f*, 435
Neutral symbiosis, 192, 193*b*
New Zealand, 7–8, 278*t*, 367*t*

Newhouse borer (*Arhopalus productus*), 532
NFDRS. *See* National Fire-Danger Rating
   System
NIMS. *See* National Interagency Incident
   Management System
NITREX (Nitrogen Saturation Experiments on
   Forest Ecosystems), 206
Nitrogen (N), 218*b*
   concentration in plants, 217*t*
   deficiency, 219, 219*b*, 220
   fertilization with, 302, 387
   pollution (excess nitrogen), 14, 206
      effects, 206, 207*f*, 375
      sources, 205*t*
   toxicity in plants, effects, 220
Nitrogen oxides (NO$_x$), 204, 205*t*, 206
Nitrogen Saturation Experiments on Forest
   Ecosystems (NITREX), 206
North America
   annosus root and butt rot, 288
   Armillaria root disease, 287
   defoliators, 469–488
   diebacks/declines, 365, 366*t*, 370–371
   Dutch elm disease, 358
   dwarf mistletoes, 334
   fire climates, 54, 55*t*, 56*f*
   fire regimes, 77–80
   forest entomology history, 405–409
   root diseases in, 276, 278*t*–279*t*
   rusts, 324*t*
   vegetation zones, 24, 25*f*
   windstorm forest damage, 162–163
North American Sugar Maple Project
   (NAMP), 19
Northeastern forests
   defoliators, 469–470
   hurricane effects, 170–172
Norway spruce (*Picea abies*), 18*f*, 374
No-see-ums. *See* Flies
"Noxious, Beneficial and Other Insects of the
   State of New York" (Fitch), 406
NPP. *See* Net primary production (NPP)
Nuclear polyhedral viruses (NPVs), 455, 456*f*
Nurseries
   seedling production, 254–255
   site selection, 255
Nursery diseases, 254–266
   abiotic (temperature-related), 255–256, 264
   of foliage (rusts), 331
   management, 264–266, 331
   of roots, 258*b*, 258–261, 295
   of seeds, 257–258, 258*b*
   of shoots, 261–264, 263*b*
Nursery insect pests, 543, 548–552, 551*f*
   damage key, 553*f*
   insecticides, 552*t*
   management, 548–552

Nutrients, plant, 217*t*, 217–218
   concentrations, 217, 217*t*, 218, 220*f*
   deficiencies, 218–220, 219*b*
   functions, 218*b*
   imbalances, 220
   toxicities, 220, 220*f*, 221
NWCG. *See* National Wildfire Coordinating
   Group (NWCG)

Oaks
   anthracnose, 314
   defoliator control, 458*t*
   gypsy moth damage, 12, 13*f*
   powdery mildews, 316–317
   rust management, 331
   wilt, 359*t*, 362
Obligate parasites, 192, 193, 193*b*, 229
Obligate saprophytes, 192, 193, 193*b*
Odonata, 425, 427*f*, 430
Oidia, 231
Old house borer (*Hylotrupes bajulus*), 536, 537*f*
Old-growth stage, 90
Olethreutidae, 556–559
Olympic Peninsula blowdown of 1921, 170, 170*f*,
   171*f*, 181
Omnivores, 29, 30*f*
Oogonia, 247
Oomycota, 237, 239*t*, 240
   diseases caused by, 239*t*, 248–251
      foliage, 309
      root, 277
      seedling, 257
   life cycle, 250*f*
Ophiostoma, 277, 395
Ophiostoma ulmi, 358, 359*t*, 360*f*
OPs. *See* Organophosphates
Orb weavers, 414, 415
Organism types, in forest ecosystems,
   29–30, 30*f*
Organochlorines, 443–444
Organophosphates (OPs), 444, 445*f*
Organothiophosphates, 444, 445*f*
Ornamental trees, mechanical damage, 216–217
Orthene (Acephate), 444
Orthoptera, 427*f*, 429*f*, 431–432
Ostia, 423
Oviparous reproduction, 424
Ovipositor, 422*f*, 423
Ovoviviparous reproduction, 424
Ozone (O$_3$), 13, 204, 206
   control, 214
   pollution, 204, 206, 208–209
   sources, 204, 205*t*
   tree damage caused by, 210–212, 211*f*
      symptoms, 210*b*
   tree susceptibility, 212, 213*t*

Pacific dampwood termite (*Zootermopsis angusticollis*), 539–540
Pacific Northwest
    forest diebacks and declines, 367*t*
    pole blight, 374
    temperature extremes of 1955, 202
    windthrow in, 169–170
Pacific powderpost beetle *Hemicoelus gibbicollis,* 536–537, 537*f*
Packard, A. S., 398, 406
Pales weevil (*Hylobius pales*), 554, 555*f*
Palm yellowing, 235
PAN. *See* Peroxyacetyl nitrate (PAN)
Pandora moth (*Coloradia pandora*), 486
Paper birch (*Betula papyrifera*), 94, 98
    fire adaptations, 97
Parasite(s), 192–193, 193*b*
Parasitic plants, 187, 231–233, 333. *See also* Mistletoes
    disease-causing, 231*t*
        signs, 193, 194*t*, 333
    morphology and reproductive features, 229*f,* 231–233
Parasitic wasps, 434
Parasitoids, 449–451, 457
Park trees, 196, 400–401
Parthenogenesis, 424
Particulates, 205
Passive crown fires, 66
Passive management, 384, 389
    of root diseases, 304
Pathogen(s) (disease-causing organisms), 228–253. *See also specific organisms*
    bacterial, 233. *See also* Bacteria
    classification of, 192–193, 193*b*
    climate change and, 203–204
    definition of, 192
    and disease, 188
    disturbances since 1875, 185, 186*t*
    facultative, 453
    features of, 228–237
    fungal. *See* Fungi
    immunity to, 193
    as insect pest control, 449, 452–456. *See also* Biological control
    introduced, 11–12, 252, 388–389
        control of, 185
        disease outbreaks caused by, 185, 186*t*
    nematode, 235–237
    parasitic plant, 231–233
    resistance, 193. *See also* Genetic resistance
    tolerance, 193
    viral. *See* Viruses
    vs. parasites, 192
    wind as carrier of, 182
Pathogenicity, Koch's rules of proof of, 194–195, 195*b*

Pathological rotation, 357
Peach leaf curl, 317
Pedipalps, 414
Penicillium, 257, 264
Pentachlorphenol, 395
Perenniporia subacida root disease, 277, 279*t*
Periodic pests, 439, 440*f*
Perithecia, 242, 243*f*
Permanent biological control, 452
Peronospora, 251
Peroxyacetyl nitrate (PAN), 204, 205*t*
    tree damage symptoms, 210*b*
Pest control programs. *See also* Insect pest management; *specific pests*
    responsibility for, 188–189
Pesticides, 265. *See also* Chemical control
    active ingredient, 446
    formulations, 446–447
    problems of, 409–410
    toxicity, 442–443, 444*t*
Peterson/Ryan model, 76, 78*t*
Phaeocryptopus gaeumannii, 319
Phaeolus schweinitzii root disease, 277, 279*t*
Phanerochaete gigantea, 388, 393
Phasmida, 429*f,* 431
Phellinus
    fruiting bodies, 350*f,* 351*f*
    root diseases, 276, 277, 279*t*
Phellinus weirii (laminated root rot), 277, 279*t,* 285, 291, 293*f*
    conifer susceptibilities, 299*t*
    distribution, 277, 279*t,* 291
    insect associations, 296, 297*f,* 304–305, 305*f*
    management, 299–304, 385, 387
    modeling of effects, 304–305, 305*f*
    spread mechanisms, 291*f,* 291–294
    symptoms and signs, 281*t,* 291*f*–293*f,* 291–292, 293
Pheromones, 447, 448*f*
    aggregating, 449
    insect control using, 447–449, 448*f,* 467, 476, 477*f. See also individual species*
Phlebia gigantea, 301
Phoma seedling diseases, 262–264, 263*b*
Phomopsis
    canker and blight, 262, 263*b*
    galls, 348
Phoracantha semipunctata, 527–528, 529*f,* 564
Phoradendron, 333–334
Phosphorus (P), 218*b*
    concentration in plants, 217*t*
    deficiency, 219*b,* 219–220, 220*f*
    fertilizers, 387
Phthiraptera, 427*f,* 429*f*
Phycomycetes, 237
Phylloxerids, 553, 561, 564
Phylogeny, insect, 413

Physical control, of insect pests, 442
Phytophthora
    control, 265, 266, 302, 303–304
    diseases caused by, 248, 250–251
        management, 385, 386
        in roots, 277, 279*t,* 281*t,* 294–295, 302
        in seedlings, 257
    and jarrah dieback, 374
Phytophthora cinnamomi, 294–295
    diseases caused by, 251
        in roots, 277, 279*t,* 294–295
    distribution, 11, 279*t*
    ectomycorrhizae and, 271
    hosts, 251, 279*t*
    management, 302, 303–304
Phytophthora infestans, life cycle, 250*f*
Phytophthora lateralis
    insect associations, 296, 297*f*
    management, 300
    root disease caused by (Port-Orford-cedar),
        251, 277, 279*t,* 295
Phytoplasmas, 187, 228, 234–235
    morphology and reproductive features, 229*f,* 234
    signs, 193
Pile burning, 135–136, 136*t,* 137*t*
Pinchot, Gifford, 104
Pine. *See also individual species*
    annosus root and butt rot, 289*f,* 290
Pine butterfly (*Neophasia menapia*), 485
Pine engraver beetles, 508–512
Pine moth, control, 458*t*
Pine plantations, insect pests, 556–559, 563–564
Pine tip moth, 556
Pine wood nematode (*Bursaphelenchus*
    xylophilus), 12, 237
Ping-pong ball (for fire ignition), 121, 143
Pissodes weevils, 554, 556
Pit-and-mound topography, 168
Pitch canker, 342*t*
Pitch moths, 559
Pitholithus tinctorius (Pt), 271
Plane, 313–314, 315
Plant(s). *See* Vegetation
Plant association, 28–29
    prediction of successional change using, 37
Plant community, theories of, 34
Plantation forests, 8, 9*f*
    Armillaria root disease, 286*f,* 287–288
    insect pests, 553–559, 563–564
    introduced pathogen/insect susceptibility,
        11–12
    monospecific, 8
thinning in, 386
Planthoppers. *See* Leafhoppers
Planting, and wind hazard, 176
Platypodidae, 516
Platypus flavicornis beetle, 520–521, 523*f*

Platypus wilsoni beetle, 518*f,* 520
Plecoptera, 425, 427*f,* 430–431
Pocket gophers, tree damage by, 225
Pole blight, 374
Polyclimax theory of plant succession, 34
Ponderosa pine (*Pinus ponderosa*), 80–83
    dwarf mistletoe, 340
    engraver beetles, 508–511, 511*f*
    fire behavior, 62–63
    fire regime, 78, 81
    in inland western U.S., 9–10, 11*f*
    mountain pine beetles, 502–504
    ozone damage, 210, 211*f,* 212
    plantations, midges on, 559
    shoot diseases, 262–264
    stand development, 77*f,* 81–83
    western pine beetles, 506–508, 509*f*
Ponderosa pine bark borer (*Acanthocinus*
    princeps), 531
Ponderosa Way fuelbreak, 148
Ponderous borer (*Ergates spiculatus*), 531–532
Poplar, borer attacks, 527, 528*f*
Poplar rusts, 323, 324*t,* 326*f,* 327. *See also*
    Melampsora rusts
    life cycle, 329, 330*f*
    management, 330
Poplar-and-willow borer (*Cryptorhynchus*
    lapathi), 527, 528*f*
Porcupines, tree damage by, 224–225
Poria weirii, 277
Port-Orford-cedar root disease, 279*t*
    fungus causing, 251, 277, 279*t,* 295
    and insects, 296
    resistant strain breeding for, 300
Potassium fertilizers, 387
Potato blight, Irish, 250–251
Potential vegetation, 27, 28, 34
    classification based on, 27, 28–29
    and multiple-pathway models, 36
Powderpost beetles, 536–538, 537*f*
Powdery mildews, 311*t,* 316–317
Prebble, Malcolm L., 409
Precipitation, 53
Predators
    in gypsy moth control, 475
    in insect pest control, 449, 457
Predisposing factors, 369, 371*f,* 379
Prescribed fire(s), 139–148
    adaptive management, 146, 147*f*
    burn, 145–146
    computer packages, 144
    constraints to use, 155–158
    definition, 139
    for disease management, 387
        root diseases, 303
    escaped fire and, 157
    examples of use, 385*t*

Prescribed fire(s)—*Cont.*
  for fire management, 132
    effectiveness, 135, 136*t,* 137*f*
  firing techniques, 139–142, 141*t,* 142*f*
  fuelbreaks and, 147–148
  history of, 104, 106
  ignition devices, 121, 142–143
  and increase of crown-base height, 132
  logistical information needed, 144–145
  monitoring, 146–147
  natural, 107. *See also* Managed wildland fire
  near residential areas, 126
  objectives, 139, 143, 144*b*
  planning, 143–147
  prescription criteria/variables, 143–144
  prescription development, 145*b*
  safety issues, 121–123
  smoke and, 155–157
  for surface fuel reduction, 132
  term, 107*b*
Prescribed natural fire, 107*b. See also* Managed
    wildland fire
Pressure cells, semipermanent, 51, 52*f*
Primary pollutants, 204–205, 205*t*
Primary scolytids, 491
Primary succession, 33
Prionoxystus robiniae, 526–527, 527*f*
Private forestry, 399–400, 402*f,* 436, 437*f*
Procera root disease (*Leptographium proceri*),
    277, 279*t,* 295
Producers, 29, 30*f*
Production ecology, 29–33
Prognosis. *See* Forest Vegetation Simulator
Protists, 240
Protozoans, 187, 228, 456
Pruning, 386
  for decay management, 358
  examples of use, 385*t*
  of infected roots, 301
  for rust management, 331
  and windthrow, 176–177, 178*f*
Pseudopods, 434
Pseudothecia, 242, 243*f,* 318
Psocoptera, 427*f*
Pt. *See Pitholithus tinctorius* (Pt)
Public-land forestry. *See also* National forests
  concerns of, 436, 437*f*
  ecosystem management, 436–439. *See also*
      Ecosystem management
  management objectives, 148–149
Puget Sound ozone pollution, 209, 209*f*
Pulaskis, 117
Pupa, 426
Puparium, 435
Pycnidia, 242, 243*f*
Pyralidae, 552
Pyrenomycetes, 316

Pyrethrins, 443
Pyrethroids, synthetic, 446
Pythium, 248, 250
  control, 265, 266
  life cycle, 260*f*
  seedling diseases caused by, 257, 258, 258*b,*
      259, 260*f*

Quaking aspen (*Populus tremuloides*), 94
  fire adaptations, 97
Quarantine
  for disease management, 384, 388–389
    root disease, 303–304
    rusts, 330
  for insect pest introductions, 567
    interested parties, 569*f*
    wood import requirements, 568*t*

Radiata pine (*Pinus radiata*) plantations, 9*f*
  introduced pathogen/insect susceptibility, 11–12
Radiation, 50
Rate of spread, 59
  crown fires, 65, 133
  modeling of, 61*f,* 61–62, 63*t*
Ratzeburg, J.T.C., 403–405, 405*f*
Reaction wood, 165, 215–216
Red alder, rotation with, 299–300
Red band needle blight. *See* Dothistroma needle
      blight
"Red belt," 201–202, 202*f*
Reforestation, with nursery seedlings, 254
Regression equations, for modeling rate of disease
      spread, 390, 391–392
Rehabilitation, after wildland fires, 123–125
Relay floristics model, 34, 35*f*
Replacement dieback, 378
Residences, fire near, 125–127
  protective measures, 126–127, 127*f*
Resistance to pathogens, 193
  and genetic selection for growth, 252–253
Resistant strains/species, 214. *See also* Genetic
      resistance
  breeding/developing, 214, 252, 300
  in Dutch elm disease management, 361–362
  in dwarf mistletoe management, 340
  and pollution, 214
  in root disease management, 298–300
Resisters, 73
Retention strategies, 180–181
Reticulitermes termites, 541
Rhabdocline needle casts, 318
Rhizina undulata root disease, 277–280, 279*t,* 303
Rhizoctonia, 257
Rhizomorphs, 230, 271, 280–281
  in Armillaria root disease, 280–281, 282, 284*f*

Rhizomycelium, 230
Rhizopus, 241, 264
Rhyacionia moths, 555*f,* 556, 577
Rhytisma fungi, 314*f,* 317
Ribbed pine borer (*Rhagium inquisitor*), 531, 532*f*
Ribes, rusts on, 323, 327, 329, 331
Ridges, 26, 27
Rigidiporus lignopus root disease, 277, 279*t*
Ring disease of mountain pine, 285–287
Ring firing. *See* Center firing
Ring spots, 234
Risk, 102
   evaluation, in fire management, 102–103
   management, for windthrow, 182
Rocky Mountains, 7
   fuel types and fire behavior, 62, 64*f*
   vegetation zones, 24, 27*f*
Root(s)
   fire damage, prediction of, 74, 75*f,* 76
   grafting, 165
   insect pests, 549, 550, 553*f,* 554
   oxygen stress, 203
   and tree stability, 167, 167*f*
   wind effects, 165–168
   and windthrow hazard, 175*t*
Root aphids, 553*f*
Root associations
   with fungi. *See* Mycorrhizae
   tripartite, 269
"Root balls," 292
Root decay, 352*t*–353*t*
   fungi causing, 352*t*–353*t*
   and wind, 168
Root disease(s), 275–307
   climate and, 302–303
   combination pathogens, 296
   feeder, 277, 279*t,* 294–295, 302
   fire interactions, 303
   and forest ecosystems, 275, 307
   fungi causing, 245*t,* 249*t,* 251, 276–280,
      278*t*–279*t*
     in seedlings, 258–260
   geographic distribution, 278*t*–279*t,* 280–295
   hosts, 278*t*–279*t,* 280–295
   insect interactions, 296–297, 297*t,* 304–305
   management, 275, 297–304, 385–388
   modeling of, 304–305, 305*f,* 306*f*
   nematodes causing, 261, 261*b*
   in seedlings, 258*b,* 258–261, 295
   spread mechanisms, 280–295
   structural, 276–277, 278*t*–279*t,* 280–294
   symptoms and signs, 280–295, 281*t*
   types, 276–280, 278*t*–279*t*
     in seedlings, 258–261
   vascular, 277–280, 279*t,* 295
   wet soil and, 203
Root weevils, 549, 550, 553*f,* 554

Rosellinia, 262, 263*b,* 264
Rotenone, 443
Rumbold, Caroline, 188
Rust(s), 309, 323–331, 324*t*
   management, 329–331, 386, 388
     and clearcutting, 385
   in shoots, 263*b,* 264
Rust fungi, 249*t,* 323, 324*t,* 327
   basidiospore-bearing structures, 246, 247*f*
   life cycles, 246, 249*f,* 327–329
   spore stages, 327*b*
RXWINDOW (computer programs), 144
Ryania, 443
Ryan/Reinhardt model, 76

Sabadilla, 443
Saccharomycetes, 241, 242
Safety, in fire management, 121–123
Salt injury, 221–222
Salvage logging, 138*b*
Salvage system, for beetle management, 501, 503t
Sanitation. *See* Inoculum reduction/removal
Sanitation logging, 507–508
Santa Anas. *See* Foehn winds
Saperda calcarata wood borer, 527
Saprophytes, 192, 193, 193*b,* 238*t*
Saprot, 349
Sapwood
   decay, 352*t*
   stains, 395, 396*f*
Sarcoptidae, 417
Sawflies, 428*f,* 434, 476–479
   control, example, 479
   defoliation by, 469, 470–471, 476–479
   life cycles, 476, 478*f,* 478–479
   morphology, vs. caterpillar, 476, 477*f*
   seed-damaging, 548
   types and hosts, 476–477
Sawyers, 532
Scales (insects), 432
Scandinavia, 7, 302, 303, 387
Scarab beetles, 550
Schäffer, J. C., 401–402
Schirrhia acicola, 322
Sclerotia, 231, 252
Scolytidae, 491, 492*f,* 512–514. *See also*
     Ambrosia beetles; Bark beetles; *individual*
     species
   classification by host, 493*t*
   fungal associations, 491–493
Scolytus multistriatus, 358–361
Scorpionflies, 428*f*
Sculptured pine borer (*Chalcophora*
     angulicollis), 533
Scutum, 416
Seasoned wood, borers of, 536–541

Secondary organisms, and forest decline, 368–369, 369*b,* 370*f*
Secondary pollutants, 204, 206
   sources of, 205*t*
Secondary scolytids, 491
Secondary succession, 33
Sediidae, 559
Seed diseases, 257–258, 258*b*
Seed insects. *See* Cone and seed insects
Seed orchards. *See also* Nurseries
   insect pest management, 543, 544–547, 545*f,* 559
Seed-eating animals, 223
Seedlings. *See also* Nursery *entries*
   inoculation with ectomycorrhizal fungi, 271–273
   nursery production of, 254–255
   outplanting of, 254
Selective delignification, 349
Semiochemicals, 447–449, 448*f*
Seral stages, 34
Sere, 34
Sericidae, 533
Serotinous cones, 73, 73*f,* 74*t,* 90, 91*f,* 96–97
Sevin (Carbaryl), 446
Sex attractants. *See* Pheromones
Shelterbelts, 173–174, 177
Shieldbacked pine seed bug (*Tetyra bipunctata*), 545, 546*f,* 549*t*
Shigo, Alex, 354
Shipworm (*Bankia setacea*), 418
Shoot moths, 557
Shoots
   diseases of, 261–264
   insect pests, 549, 552, 553*f,* 554–559
Shot-holes, 310
Sialidae (family), 435
Signs (of disease), 193, 194*t*
   linking to symptoms, 194–195, 195*b*
Silicon tetrafluoride (SiF$_x$), 204, 205*t*
Silt deposition, 39–40
SILVA model, 76
Silvicultural management, 384–388, 385*t,* 402*f.*
   See also Cultural management; *specific techniques*
   of defoliators, 472–473, 476, 487
Simazine, 214
Simulation models, of rate of disease spread, 390, 392, 394*f*
Sinclair model of forest decline, 369, 379–381
Siphonaptera, 428*f*
Sirex wasps, 12, 533–536, 535*t*
Sirococcus
   cankers, 342*t,* 343*t,* 346
   shoot blight, 262, 263*b,* 343*t*
Site potential, 23
Site preparation, and wind hazard, 176
Site selection. *See* High-hazard site avoidance

Sitka spruce (*Picea sitkensis*)
   blackheaded budworm outbreaks, 487–488, 488*t*
   gall adelgid damage, 554*f*
   insect pests, 556
      management, 558*f*
   windthrow successional effects, 169–170
Size-up, of fire, 115
Skeletonizers, 462
Slash. *See also* Dead trees; Woody debris
   bark beetles infestation, 491, 504–505, 508–512
   burning, for root disease management, 303
Slash rot, 349, 353*t*
Slime molds, 240
Smoke management, 155–157
Smokey Bear campaign, 108, 111*f*
Smothering fungus, 263*b*
Snow molds, 323
Snow, tree damage by, 216
Snowshoe hares, damage by, 226
   Society of American Foresters
   cover type classification system, 28
   on urban forestry, 401
Sodium chloride damage, 221, 222
Soft rots, 348*b,* 349
Softwoods, ambrosia beetle damage, 517–521
Soil
   acidity, 221, 255
   compaction, 216–217
   dry, 200
   fumigation, 265, 388
   mites in, 417
   nursery, ideal, 255
   oxygen reduction, 216–217
   oxygen stress, 203
   pasteurization, 265
   thaw-freeze events, 373
   and tree stability, 167, 167*f*
   wet
      flooding, 203
      and forest dieback, 374–375
Solutions, insecticide, 447
Sooty molds, 310, 311*t,* 317
South Africa, 7, 8
South America, 287, 367*t*
South Canyon fire of 1994, 123, 124*f*
Southern fusiform rust, 264
Southern pine
   ambrosia beetle attacks, 520–521, 523*f*
   defoliators, 470–471
   fusiform rust, 323, 325*f*
   hurricane-level wind effects, 172–173
   seed orchard pests, 545, 546*f*
Southern pine beetle (*Dendroctonus frontalis*), 495*f,* 498–501. *See also* Bark beetles
   characteristics, 501*t*
   damage by, 398, 399*t,* 400*f,* 401*f*
   detection, 497, 498*f,* 499*f*

Southern pine beetle (*Dendroctonus frontalis*)—*Cont.*
  egg gallery and larval mine patterns, 494*f*
  management, 500–501, 503*f*
  outbreak conditions, 500, 502*f*
  successional effects, 42, 297
Southern pine coneworm (*Dioryctria amatella*), 545, 546*f*
Southern/southeastern forests
  defoliators, 470–481
  health/health problems, 6–7, 16
  history of prescribed fire, 104, 106
  temperature extremes of 1950–51, 202
Spaulding, Percy, 188
Species decline, 365
Species evenness, 43*f*
Species richness, 2, 42, 43*f*
Species selection, 387. *See also* Resistant strains/species
  examples of use, 385*t*
"Speed wobbles," 221
Spermagonia, 329
Spermatheca, 424
Spermatophore, 424
Sphaeropsis sapinea, 322, 342*t*
Spider mites, 417, 553*f*
Spiders, 414–415, 416*f*
Spiracles, 420
Sporangia, 247
Spores, fungal, 230*f,* 230–231
  spread of, 251–252
Sporodochia, 242
Spot head fire, 141, 141*t,* 142*f*
Spotting, 50, 66–67
Spring frost, 256
Springtails, 426
Sprinkler systems, for insect pest control, 442, 443*f*
Spruce. *See also individual species*
  annosus root and butt rot, 289*f,* 290
  insect outbreaks. *See also individual insect species*
    defoliator, 469–470
    engraver beetle, 508
Spruce aphid (*E. abietinum*), 564
Spruce beetle (*Dendroctonus rufipennis*), 398, 501*t,* 505–506
  management, 506, 508*f*
Spruce budworm. *See* Eastern spruce budworm; Western spruce budworm
Spruce gall adelgid (*Adelges cooleyi*), 553, 554*f*
Spruce spider mite (*Oligonychus ununguis*), 417
SPs. *See* Water-soluble powders
Squirrels, tree damage by, 224
Stability
  forest ecosystem, 43–44
  tree, factors in, 167, 167*f*

Stains, 362, 395
  fungi causing, 245*t,* 362
Stand development patterns. *See also* Succession
  in boreal forest region, 96–98
  in coastal Douglas-fir forests, 84–87
  fire frequency and, 376–378
  with fire suppression, 376–378
  and forest decline/dieback patterns, 376–378
  in jack pine forests, 96–98
  in longleaf pine forests, 93–94
  in Ponderosa pine forests, 77*f,* 81–83
  stages of, 84–87
  in western subalpine forests, 89–92
Stand inititation stage, 90
Standing crop, 31
Stand-reduction dieback, 378
States
  fire prevention programs, 110
  fire protection responsibilities, 106–108
Stem and branch diseases, 333–362
  cankers, 341–347
  decay, 348–358
    fungi causing, 249*t*
  galls, 347–348
  mistletoes, 333–340
  stains, 362
  vascular wilts, 358–362
Stem exclusion stage, 90
Stem rot, 349
Stem(s), wind effects on, 165–168
Sterigmata, 246
Stone flies, 425, 427*f,* 430–431
Storage fungi, 257
Stored seedlings, fungal diseases, 263*b,* 264
Stramenopila, 237, 239*t,* 240
Stream buffers, wind and, 181
Street trees, 196, 400–401
Strip cutting, progressive, 180, 180*f*
Strip head fire, 140, 141*t,* 142*f*
Striped ambrosia beetle. *See Trypodendron lineatum*
Structural diversity, 42, 43*f*
Structural root diseases, 276–277, 278*t*–279*t,* 280–294. *See also specific diseases*
  symptoms and signs, 280, 281*t*
Stumps
  chemical treatment, 300–301
  removal, 297–298, 298*f*
Stylets, 419
Subalpine fir (*Abies lasiocarpa*), 87–88
  fire behavior, 63
  stand development patterns, 89, 91
Subalpine forests, western
  fire regime, 78, 88–89
  stand development patterns, 89–92
Subesophageal ganglion, 424

Succession, 23, 33–44, 436–437, 438*f*
  autosuccession, 378
  decay and, 356
  definition of, 33
  diseases and, 185
  disturbance effects on, 34–42
  ecological diversity and, 42–43
  fire and. *See* Stand development patterns
  forest declines/diebacks and, 376–378
  and gap rotation, 378
  laminated root rot and, 292
  multiple-pathway models, 36–42
  mycorrhizal fungi and, 272
  primary, 33
  secondary, 33
  theories of, 34, 35*f*
  and tree mortality, 379*t*
  wind effects, 168–173
Successional species, 34
Sulcatol, 520
Sulfur dioxide (SO₂), 12–13, 204
  control, 14, 213–214
  dispersion, 207
  sources, 204, 205*t*
  tree damage by, 199, 200*f*, 209–210,
    210*b*, 211*f*
Surface fires
  behavior, 58*t*, 59–63
    modeling of, 60–63, 61*f*
  management, and fireline intensity, 58*t*
Surface fuel reduction, 132
Sustainability of ecosystems, 43
Swaine, J. M., 406, 407
Swaziland, 321
Swiss needle cast, 318–319, 319*f*
Switching mechanism, 499
Switzerland, 375, 376*b*
Sycamore anthracnose, 313–315, 315*f*
Symbiosis, 192, 193*b*
Symphyta, 434
Symptoms
  of disease, 193, 194*t*
    linking to signs, 194–195, 195*b*
  of forest health problems, 4–6
Synanthedon clearwing moths, 526
Synnemata, 242
Systematics, insect, 413

Tachinid fly (*Cyzenis albicans*), 450–451
Tagmosis, 418
Tannensterben, 365
Taphrina, 317
Tar spots, 310, 311*t*, 314*f*, 317
Target cankers, 341
Tarsus, 421
"Tea pot fungus," 280

Teleomorphs, 242
Temperature(s), 51–53, 53*f*
  and air moisture, 51, 53*f*
  extreme, tree damage from, 200–202, 256
    protection from, 264
  global rise in, tree damage from, 203–204
  in root disease management, 302–303
  thinning and, 302
Temperature inversion, 207–208
Temperature lapse rate, 51, 53
10 A.M. policy, 104
Ten Standard Firefighting Orders, 121, 122*t*
Tenaculum, 426
Tension wood, 165, 215–216
Tent caterpillars, 470, 480–481
Tentranychidae, 417
Termites, 427*f*, 429*f*, 431, 539–541, 540*f*
Tetyra bipunctata, 545, 546*f*, 549*t*
Texas Forest Service Operational Information
  System, 500*t*
Thaw-freeze events, 373, 374
Thelephora terrestris, 264
Theronia atalantae wasp, 485
Thinning, 385–386. *See also* Clearcutting
  for bark beetle control, 504, 506*t*
  and climate, 302
  for disease avoidance, 387–388
  for dwarf mistletoe management, 339
  examples of use, 385*f*
  for fire spread prevention, 133–134, 135*f*, 135*t*,
    136*t*, 136–139, 137*f*, 140*f*
  "greenup" following, 139
  for root disease management, 301
  for rust management, 331
  tree damage during, 216
  wind damage potential and, 176, 177*f*
Thorax, 420–421
Thrips, 427*f*, 432, 547, 552, 553*f*
Thuricide. *See Bacillus thuringiensis*
Thysanoptera, 427*f*, 432, 547
Ticks, 414, 416*f*, 416–417
Tillamook burns, 84, 86*b*
Timber
  calculation of insect damage to, 439–440
  production, 436
Timber with understory (Fuel Model 10), fire
  behavior, 62, 62*t*, 63*t*
Tip moths, 556
Tipweevils, 556
Tolerance
  to pathogens, 193
  to salt, 222
Tomentosus root rot, 279*t*
  fungus causing (*Inonotus tomentosus*),
    277, 279*t*
Top rot, 349
Topex, 174*t*, 175

Topography
and fire behavior, 54
and forest zonation, 26
pit-and-mound, 168
windthrow effects, 168
and windthrow hazard, 174*t*, 175
Tornadoes, 161, 162
Toxicants, pesticide, 446
Toxicity
insecticide/pesticide, 442–443
nutrient, in plants, 220, 220*f*, 221
Trace elements fertilizers, 387
Tracheal gills, 423
Tree(s)
breeding. *See* Genetic resistance
diseases. *See* Disease(s)
health assessment, in ARNEWS program
(Canada), 19
mycorrhizal, 267*t*, 269, 270
stability factors, 165–168, 166*f*
wind effects, 163–168, 215–216
Tree mortality, 364
causes, 364, 379*t*
from fire, prediction of, 74–76, 75*f*
models, 76, 78*t*, 79*t*
forest declines and, 364
and forest health, 3, 5*f*
impacts on forest, 190
in inland western U.S., 9
waves, 5*f*, 378, 380*f*
Tree pushing, 298, 386
Tree squirrels, damage by, 224
Tremex wasps, 533–536, 535*t*
Trichoderma viride, 388
Trichogramma, 456, 458*t*
Trichoptera, 428*f*, 433
Trochanter, 421
Trombiculidae, 417
Trophic categories, 29–30, 30*f*
Tropical forests, 378
Trouvelot, E. Léopold, 474
True bugs, 427*f*, 429*f*, 432
True firs. *See also* Fir(s); *individual species*
annosus root and butt rot, 289*f*, 290
Douglas-fir tussock moth attacks, 439,
483–485
engraver beetle outbreaks, 512–514
in inland western U.S., 10, 11
western spruce budworm outbreaks, 481–483,
483*t*, 484*f*
True mistletoes, 231*t*, 231–233, 232*f*, 333–334
Trunk rot, 349
Trypodendron lineatum beetle, 516, 517–520,
518*f*, 519*f*
management, 520, 521*t*
Turpentine borer (*Buprestis apricans*), 528,
530, 530*f*

Underburning, history of, 104
Understory fires
behavior, 58*t*
modeling of, 60–63, 61*f*
and fireline intensity, 58*t*
management, 58*t*
Understory reinitiation stage, 90
Union Camp Corporation, 409
United Kingdom, windthrow hazard classification
system, 174*t*–175*t*
United States. *See also specific regions*
federally managed wilderness and land
areas, 151*f*
fire origins, 108, 109*f*
forest declines, 375–376
Forest Health Monitoring program, 15–16
forest health/health problems, 6–7, 7*f*, 8, 16
gypsy moth, 12, 13*f*
Incident Command System, 112–114, 113*f*
insect pest introductions, 561–567
National Fire-Danger Rating System, 67–68, 69*f*
wood imports and quarantine, 389, 566–567
Uprooting. *See* Windthrow
Urban forests
diseases, 196
engraver beetles, 511–512
history of, 400–401
value of, 400–401
Urban-wildland interface, and fire, 125–127
Urediniales, 323
Urocerus wasps, 533–536, 535*f*, 535*t*
Utilitarian perspective of forest health, 2

Valley environments, 26–27
Values, and fire management, 103, 103*f*
VAM. *See* Vesicular-arbuscular mycorrhizae (VAM)
Vascular root diseases, 277–280, 279*t*, 295. *See
also specific diseases*
Vascular wilts, 195, 358–362, 359*t*
fungi causing, 245*t*, 359*t*
spread, 251
Vegetation
classification systems, 27–29
climax, 34
essential elements, 217, 217*t*
fire adaptations, 71–73, 72*f*, 73*f*, 74*t*
potential, 27, 28–29, 34
primary production by, 30–31
Vegetation zones
of eastern deciduous forest, 24, 26*f*
of North America, 24, 25*f*
of Rocky Mountains, 24, 27*f*
Verticicladiella wageneri, 277
Vesicular-arbuscular mycorrhizae (VAM),
266–268, 267*t*, 269*f*, 270, 271
Virions, 455, 456*f*

Viruses, 187, 228, 233–234
  as gypsy moth control, 475
  as insect pest control, 455–456
  morphology and reproductive features, 229*f,*
      233–234
  symptoms and signs, 193, 313
  transmission by nematodes, 237
Viscum, 333
Viviparous reproduction, 424
Volcanic activity, 38–39, 41*f*
Voles, tree damage by, 224
Von Schrenck, Hermann, 188

Waldsterben, 375, 377*b*
Walking sticks, 427*f,* 429*f,* 431
Warm front, 51
Wasps, 428*f,* 434
  parasitic, 434
  wood borers, 12, 434, 533–536, 535*t*
Water molds, 248
Watersheds, 3
Water-soluble powders (SPs), 447
Weather. *See also* Climate; Wind(s)
  fire, 50–54
Webspinners, 427*f*
Webworms, 552, 553*f*
Weeks Act (1911), 104
Weevils, 549, 550, 553*f,* 554, 556
Wellington, William G., 409, 410*f*
Western cedar borer (*Trachykele blondeli*), 528,
      529*f,* 533, 534*f*
Western forests
  coastal, health problems, 16
  defoliators, 471, 472*t,* 481–488
  fire exclusion in, 105*f,* 106
  inland, health/health problems, 7, 9–11, 16
  subalpine
      fire regime, 78, 88–89
      stand development patterns, 89–92
Western gall rust, 264, 323, 324*t,* 326*f,* 327, 329
Western hemlock
  annosum root and butt rot, 290, 290*f,* 291
  blackheaded budworm outbreaks, 487–488, 488*t*
  ectomycorrhizae, 270*f*
  fluting, 103
  windthrow successional effects, 169–170
Western hemlock looper (*Lambdina fiscellaria
      lugubrosa*), 486–487
  control, 486–487
  outbreaks, 368, 463, 464*f*
Western pine beetle (*Dendroctonus brevicomis*),
      501*t,* 506–508
  management, 507–508, 509*f*
Western pine shoot borer (*Eucosma sonomana*),
      557–559
  control, 448

Western poplar clearwing (*Paranthrene
      robiniae*), 526
Western redcedar
  deer damage, 224
  rotation with, 300
  windthrow successional effects, 169–170
Western Root Disease Model (WRDM), 304–305,
      305*f,* 306*f,* 392
Western spruce budworm (*Choristoneura
      occidentalis*), 481–483
  control program, 467–469, 470*f*
  damage by, 398, 400*f,* 401*f*
  feeding habits, 462
  hosts, 481, 483*t*
  outbreaks, 481, 482*f*
  successional effects, 42
  susceptibility to, 481–483, 483*t,* 484*f*
  wind as carrier of, 182
Wettable powders (WP), 474
Wetwood, 362
Weyerhauser Company, 409
WFSA. *See* Wildland Fire Situation Analysis
      (WFSA)
Wheat rust, 253
White pine
  hurricane effects, case study, 170–172
  in inland western U.S., 10, 11
  pole blight, 374
  rotation with, 300
White pine blister rust, 2, 4*f,* 323, 324*t,* 325*f,*
      327–329
  life cycle, 328*f*
  management, 330–331
  wind as carrier of, 182
White pine weevil (*Pissodes strobi*), 556, 557*f*
White rots, 348*b,* 348–349, 356
White spruce (*Picea glauca*), 94
  fire adaptations, 97
Whitebark pine (*Pinus albicaulis*)
  blister rust, 2, 4*f,* 182
  forest fires, 88
  stand development patterns, 91
Whiteflies, 553*f*
Wilderness areas, 150, 151*t*
  management objectives, 148–149
Wildfire, 107*b. See also* Fire(s); Wildland fire(s)
Wildland fire(s). *See also* Fire(s)
  direct effects, 123
  dwarf mistletoes and, 338–339
  future of, 153–154
  "let it burn" policies, 148
  managed (pnf). *See* Managed wildland
      fire (pnf)
  management. *See* Fire management
  near residential areas, 125–127
  terminology, 107*b*
Wildland Fire Situation Analysis (WFSA), 117

Wind(s), 160–182
  direction, 161–162
  foehn, 52–53, 162
  frontal, 51
  gradient, 52
  patterns, 51–53
  as physical process, 160–163
  successional effects, 168–169
    case studies, 169–173
  and topography, 54
  velocities, 160–161, 161*t*
Wind damage
  to crown morphology, 163–164
  to root shape, 165
  to stems, 165–168
  synergistic effects, 181–182, 196
  to trees, 163–168, 215–216
  uprooting. *See* Windthrow
Wind management strategies, 173–181
  clearcutting, 177–180, 178*f*
  pruning, 176–177, 178*f*
  retention, 180–181
  shelterbelts, 173–174, 177
  site preparation and planting, 176
  thinning, 176, 177*f*
  windthrow hazard classification systems, 168,
    174*t*–175*t*, 174–175
  windthrow hazard factors, 179*t*
Windsnap, 165–168, 169
Windstorms. *See also* Hurricane(s); Windthrow
  forest damage by, 162–163
  salt water transport by, 221–222
  windspeeds, 161*t*
Windthrow, 165–168. *See also* Slash
  and bark beetle infestation, 181, 491, 504–505,
    508–512
  clearcutting and, 179–180
  earlier wind damage and, 181
  hazard classification systems, 168, 174*t*–175*t*,
    174–175
  hazard factors, 179*t*
  leave trees and, 181
  pruning and, 176–177
  risk management, 182
  site preparation and planting and, 176
  successional effects, 168–173
    case studies, 169–173
  thinning and, 176
  topographical effects, 168
  and wood borer infestation, 530–536
Winter damage, 200, 201–202, 256
Winter drying, 203
Winter moth (*Operophtera brumata*), parasitoid
  control of, 450–451
Wipfelkrankheit, 475

Witches'-brooms, 334, 339, 347
Wood
  chemistry of, 348
  decay. *See* Decay
  moisture content, 362, 394
  preservatives, 395
  stains in, 245*t*, 362, 395
Wood borers, 526–541
  of dry and seasoned wood, 536–541
  of living trees, 526–530
  of weakened trees, windthrows, and logs,
    530–536
Wood exports, 516, 517*f*
  and ambrosia beetle damage, 516–517
Wood imports
  APHIS requirements, 389, 568*t*
  and insect/pathogen introductions, 389,
    566–567
Wood wasps, 12, 434, 533–536, 535*t*
Woody debris, coarse, 6, 169. *See also* Dead
    trees; Slash
WP. *See* Wettable powders
WRDM. *See* Western Root Disease
    Model (WRDM)

Xeris wasps, 533–536, 535*t*
Xiphinema bakeri, 261, 262*f*
Xyleborus beetles, 524
Xylem cavitation injury, 371–373
  and forest dieback, 374–375

Yakama Indian Nation
  pine butterfly outbreak, 485
  spruce budworm control program,
    467–469, 470*f*
Yeasts, 241, 242
Yellow root rot, 279*t*
Yellow-banded timber beetle (*Monarthum
    fasciatum*), 523
Yellows, 234–235
Yellowstone National Park fires of 1988, 10, 92*b*
  forest recovery from, 12*f,* 90–91
  manageability problems, 152–153, 156*f*
Yosemite National Park, fire-created fuelbreaks,
    152, 154*f*

*Zeitschrift für angewandte Entomologie (Journal
    of Applied Entomology),* 405
Zygomycota, 237, 238*t,* 240
  diseases caused by, 238*t,* 240–241
  life cycle, 240, 241*f*
  mycorrhizal, 267*t,* 270